Food Webs at the Landscape Level

Food Webs at the Landscape Level

Edited by Gary A. Polis, Mary E. Power, and Gary R. Huxel

THE UNIVERSITY OF CHICAGO PRESS

CHICAGO AND LONDON

GARY A. POLIS was chair of the Department of Environmental Sciences and
Policy at the University of California, Davis. He edited *Biology of Scorpions, The
Ecology of Desert Communities,* and *Food Webs: Integration of Patterns and Dynamics.*
MARY E. POWER is professor of integrative biology at the University of California,
Berkeley. GARY R. HUXEL is assistant professor of biology at the University of
South Florida.

The University of Chicago Press, Chicago 60637
The University of Chicago Press, Ltd., London
© 2004 by The University of Chicago
All rights reserved. Published 2004
Printed in the United States of America

13 12 11 10 09 08 07 06 05 04   1 2 3 4 5

ISBN: 0-226-67325-1 (cloth)
ISBN: 0-226-67327-8 (paper)

Library of Congress Cataloging-in-Publication Data

Food webs at the landscape level / edited by Gary A. Polis, Mary E. Power, and
   Gary R. Huxel
p. cm
Includes bibliographical references (p.) and index.
ISBN 0-226-67325-1 (cloth : alk.paper)—ISBN 0-226-67327-8 (paper : alk.paper)
1. Food chains (Ecology) I. Polis, Gary A., 1946– II. Power, Mary E. III. Huxel,
   Gary R.
QH541.F65 2004
577′.16—dc22

                                                                2003018992

Dedicated to the memory of Gary A. Polis

Contents

PART II. FOOD WEB DYNAMICS ACROSS THE LAND-WATER INTERFACE

PART III. SUBSIDIES AT REGIONAL AND GLOBAL SCALES

      The Importance of Allochthonous Carbon Inputs to the
      Metabolism of Lakes and Rivers   301
      Nina Caraco and Jonathan Cole

21   Trophic Flows and Spatial Heterogeneity in Agricultural
      Landscapes   317
      Jacques Baudry and Françoise Burel

PART IV. SYNTHESIS

22   The Influence of Physical Processes, Organisms,
      and Permeability on Cross-Ecosystem Fluxes   335
      Jon D. Witman, Julie C. Ellis, and Wendy B. Anderson

23   Feast and Famine in Food Webs: The Effects of
      Pulsed Productivity   359
      Anna L. W. Sears, Robert D. Holt, and Gary A. Polis

24   Subsidy Effects on Managed Ecosystems: Implications
      for Sustainable Harvest, Conservation, and Control   387
      Mary E. Power, Michael J. Vanni, Paul T. Stapp, and Gary A. Polis

25   Subsidy Dynamics and Global Change   410
      Ralph H. Riley and Robert L. Jefferies

26   At the Frontier of the Integration of Food Web Ecology
      and Landscape Ecology   434
      Gary R. Huxel, Gary A. Polis, and Robert D. Holt

      References   453
      Contributors   527
      Index   531

Preface

On 27 March 2000, Gary Polis, Michael Rose, Shigeru Nakano, Masahiko Higashi, and Tayuka Abe perished when their boat capsized in a violent, unexpected storm offshore from Bahía de Los Angeles in the Sea of Cortez. Gary had been leading an expedition to the islands in the Gulf of California off the central Baja coast, where he had studied desert food webs and their interaction with the marine ecosystem for 11 years. The scientists lost in the Sea of Cortez accident will be terribly missed by many people all over the world. They leave important legacies to the study of food webs. This book, which Gary began as an outgrowth of a session on food webs and landscapes at the 1998 INTECOL conference in Florence, Italy, attempts to capture some of the momentum and excitement in this field, to which Gary contributed so much energy and insight. In this preface, we attempt to set this contribution into the context of Gary's unfolding career in food web ecology.

Gary grew up in southern California, and did his Ph.D. at the University of California, Riverside, where he was supervised by a physiologist, not an ecologist. His self-instruction in ecology began in deserts, where he immersed himself in intense study of natural history, particularly of arachnids. His early papers and a book focus on the interactions of spiders, scorpions, and solpugids, which compete for prey and frequently eat one another. These interactions motivated his development of the intraguild predation concept, and its theoretical extension, in collaboration with Bob Holt, to analysis of the ecological dynamics and evolutionary implications of this widespread interaction.

His deep grounding in natural history prepared Gary to take a leadership role in directing the field of food web ecology away from previously entrenched theory that had been derived from the study of books, not nature (to paraphrase Agassiz). The food web papers that dominated the pages of *Nature* and *Science* through the 1970s and 1980s were largely contributed by theoreticians seeking to test May's (1973) prediction that food webs should persist, or recover quickly from perturbations, only if their richness (S), connectance (C, the number of nonzero interactions) and average interaction strengths (β) were constrained, so that $\beta(SC)^{1/2} < 1$. The theoreticians most interested in this idea emphasized richness and connectance, recognizing that interaction strength was difficult to measure. Empirical evidence, from published data on food webs collected for reasons other than to test May's theory, indicated that connectance and richness were often inversely correlated. Many publications interpreted these trends as support for May's theory. A skeptic pointed out, however, that the literature reported on food webs that were arbitrarily delimited and inconsistently described, and that more effort devoted to a large system encompassing many species meant that less could be allocated to detailed diet analyses (Paine 1988). Food web theoreticians at this time also asserted that omnivory, cannibalism, and loops (A eats B eats C eats A) were rare in nature, as they expected, because these features destabilized this generation of models. These assertions were deeply irritating to field biologists working in diverse systems, who saw such interactions everywhere in nature.

Gary Polis galvanized these grumblings in a talk he gave at the 1990 INTECOL meetings in Yokahama, Japan, where he first met Nakano and other Japanese colleagues. The content of the talk was later published as his 1991 *American Naturalist* paper. Here, he forcefully pointed out that omnivory, cannibalism, and loops were not rare, but rampant in nature, that many predators that compete for food eat one another when possible (intraguild predation), and that literature-based descriptions were not adequate for deriving or testing food web theory. The paper's effectiveness may, in part, derive from its positive, almost exuberant tone—in it, Gary celebrates the complexity of the Coachella Valley desert food web, which he and other field biologists had richly documented. This work by Gary led theoretical ecologists to revisit the relationship between the complexity and stability of communities. For example, McCann and Hastings (1997) developed a model in which omnivory stabilized food webs, in contrast to earlier models (e.g., Pimm and Lawton 1977).

While the 1991 paper steered food web studies toward a more realistic view of nature, it also was somewhat deconstructionist. Gary commented at this time that he didn't want to be just a "nay-sayer" (personal communication to MEP). He was evolving his own view of the larger picture, which fused food web dynamics with spatial or landscape ecology. As early as 1979, he and David Spiller, also a spider ecologist, had shared their common observation that there were extraordinary densities of spiders on beaches (Spiller, personal communication to MEP). This was true on the Baja islands, among other places, and especially true on the smaller islands. Gary entertained three hypotheses to explain this pattern: a lack of predation, including intraguild predation on island spiders; highly suitable habitats for webs, possibly related to island vegetation; and food subsidies from the ocean. The last hypothesis was the one best supported by field evidence. Beach wrack washed up on the Baja shores was teeming with kelp flies and amphipods, which in turn fed the spiders. These spiders attained extraordinarily high densities, and as a consequence, protected desert vegetation from herbivory, although terrestrial primary production was not sustaining them energetically. Gary coined the term "apparent trophic cascades" in analogy to apparent competition (Holt 1977) to describe this interaction.

Gary championed the idea that flows of energy, materials, or organisms from one habitat to another could strongly influence the structure and dynamics of food webs, igniting the interest of many other food web ecologists. Similar insights had occurred long ago to the great field ecologist Charles Elton (Summerhayes and Elton 1923), but in the years that followed, food web studies had taken two different tracks, both veering away from landscape scales. Community ecologists who were experimentalists studied the effects of species interactions or physical factors by manipulating conditions or densities of organisms within small (<1–100 m²) areas within single habitats, for obvious logistical reasons. As river, lake, terrestrial, intertidal, open ocean, or subtidal marine ecologists, we were not focusing on processes affecting communities that operated across landscape boundaries (but see Duggins et al. 1989). The theory that dominated food web ecology in the 1970s and 1980s did not portray much physical or temporal context. Interest in habitat boundaries and fluxes across them was maintained in ecosystem ecology (Likens et al. 1977; Jackson and Fisher 1986) and landscape ecology (e.g., Turner 1989), but among scientists working at large scales without resolving population dynamics or interactions of species in food webs. Polis, struck by the importance of regional oceanic processes for food webs on the small desert islands he studied in the Gulf of Mexico, exhorted community ecologists to "stop looking at our feet." He and his colleagues wrote several highly influential reviews (e.g., Polis, Anderson, and Holt 1997) pointing out that subsidies (fluxes of organisms, energy, or materials from productive to less productive habitats) strongly influenced the structure and dynamics of recipient food webs in a wide range of ecosystems. Theoretical ecologists again responded to his work by examining the potential effect of allochthonous resources or spatial subsidies on food web stability. For example, Huxel and McCann (1998; see also McCann et al. 1998) found that low to moderate levels of allochthonous resources could stabilize communities (see also Schoener 1973; Sommer 1984; Nisbet et al. 1997).

Gary Polis remained deeply grounded in and inspired by natural history throughout his career. He also developed an uncommonly broad vision that over the last several years has been catalyzing useful syntheses of community, ecosystem, and landscape ecology. The Japanese scientists who perished with Gary, Shigeru Nakano, Masahiko Higashi, and Tayuka Abe, shared this vision because of their own deep understanding of the natural history of their systems and of the larger scales over which exchanges and interactions among organisms can occur (Fausch 2000; Power 2001; Fausch et al. 2002). We hope that this book records and communicates some of the energy and excitement that Gary Polis infused into this international endeavor.

The contributed chapters in this book reflect similarly ambitious research efforts of scientists, many of whom are grounded in particular subdisciplines of ecology, but are striving to integrate and expand the spatial and temporal scales of understanding while still resolving process mechanisms and species interactions. These chapters are a sample of a rapidly increasing number of studies of food webs and landscapes in a burgeoning area of ecology. They draw on work in a wide range of ecosystems (in table 1 and in the description that follows, the chapters are ordered by landscape elevation, from marine subtidal to intertidal to terrestrial and freshwater, upstream, and upslope). The influences of marine resources are examined in subtidal ocean food webs where food and propagules are advected by internal waves (Witman et al., chap. 9); in intertidal food webs that respond to resources from regions in the open ocean that vary in productivity (Menge, chap. 5); in the now famous terrestrial food webs of small desert islands in Baja (Anderson and Polis, chap. 6); and in arctic riparian and upland habitats thousands of kilometers upstream from the ocean, to which nutrients and energy are vectored by anadromous salmon (Willson et al., chap. 19). Fluxes flowing downstream are studied by Riley et al. (chap. 16) in their examination of how watersheds affect estuarine food webs; by Caraco and Cole (chap. 20) in their large-scale assessment of the influence of terrestrial carbon on the earth's lakes and rivers, and by DeAngelis and Mulholland (chap. 2), who model how vertical flow separations partition sources and sinks over very small spatial scales, influencing the uptake of nutrients by attached algae. Organisms can sometimes vector resources against the physical flow, as described by Willson et al. (chap. 19) as well as by Winemiller and Jepsen (chap. 8), who examine the trophic effect of prochilodid migrations on fluxes from rich whitewater to poor blackwater habitats in South American rivers; by Vanni and Headworth (chap. 4), who discuss the importance of gizzard shad in resuspending sedimented nutrients into the photic water column of reservoirs; and by Power et al. (chap. 15), who document the importance to terrestrial consumers of river-to-forest fluxes, mediated by emergent aquatic insects. On land, Cadenasso et al. (chap. 10) show that fluxes between forests and meadows vectored by animals such as deer and mice depend on the structure of the forest boundary; Jefferies et al. (chap. 18) document huge, probably irreversible, effects on Arctic marshes of continental-scale agricultural subsidies to snow geese; and Baudry and Burel (chap. 21) describe the largest fluxes of energy and nutrients ever to occur on Earth, mediated by global commerce in industrial agriculture, with future consequences that we can only guess.

These chapters show the generality across scales (microns to thousands of kilometers; seconds to millennia) and ecosystems of the strong effects of fluxes of resources and organisms across traditional habitat boundaries. Holt (chap. 7), pointing out that local systems can be open with respect to some components but closed with respect to others, models the consequences for populations in food webs that receive allochthonous inputs at different trophic positions. Schindler and Lubetkin (chap. 3) review the use of stable isotopes for spatially tracing fluxes and determining the quantitative importance of different sources of elemental constituents to organisms that assimilate them. They preview an exciting modeling breakthrough by Lubetkin that promises to relax the constraint that one cannot determine the relative contributions from more sources than the number of isotopes analyzed.

Modeling and tracer studies like these are together supporting the efforts of ecologists to answer general questions about the community-level and ecosystem-level consequences of cross-habitat fluxes. What are characteristic temporal and spatial scales for these fluxes? How does the contrast in productivity, or the timing of peak productivity in linked habitats (Sears et al., chap. 23; Nakano and Murakami 2001) influence the interaction? Will allochthonous inputs stabilize or destabilize communities? Theory has suggested that this outcome depends on the amount, quality, and edibility of the resource (Huxel and McCann 1998). Chapters in this book include examples of destabilizing subsidies (seabird guano that results in larger-amplitude fluctuations in plant and arthropod populations on Anderson and Polis's bird islands, chap. 6; shad-stirred phosphorus that can initiate a positive feedback toward eutrophication in Vanni and Headworth's reservoirs, chap. 4), as well as examples in which subsidies may stabilize interactions among recipient predators and their

Table 1 Source and recipient habitats, boundaries, vectors, fluxes, and recipient web members described in the chapters of this volume

Author(s)	Source habitat	Recipient habitat	Boundaries /corridors	Vectors	Flux	Recipient organisms
Witman (chap. 9)	Open ocean	Benthic subtidal		Internal waves	Nutrients, plankton, propagules	Attached benthic invertebrates and algae
Menge (chap. 5)	Open ocean	Benthic intertidal		Surface waves, tides	Nutrients, plankton, propagules	Benthic invertebrates and algae
Anderson and Polis (chap. 6)	Open ocean	Desert islands	Seashore	Seabirds, beach wrack	Nutrients, carrion, plant detritus, feces	Marine and terrestrial arthropods and their vertebrate predators
Willson et al. (chap. 19)	Open ocean	Terrestrial watersheds	Rivers	Migrating anadromous fishes	Marine-derived nutrients and energy in fish bodies	Vertebrate predators and scavengers, terrestrial and riverine plants and invertebrates
Riley et al. (chap. 16)	Rivers and watersheds	Estuaries		River discharge	Nutrients, detritus, terrestrial invertebrates	Seaweeds, macroinvertebrates, and vertebrates
Caraco and Cole (chap. 20)	Terrestrial watersheds	Lakes and rivers	Shorelines	Runoff, groundwater, wind, gravity	Organic carbon	Aquatic microbes, detritivores, and grazers
DeAngelis and Mullholland (chap. 2)	Upstream chemostat Free-flowing upper stream	Downstream Sluggish near-bed storage zone	Vertical flow layers	Diffusion, mixing of layers, leaching from particles	Nutrients	Phytoplankton Periphyton

Winemiller and Jepsen (chap. 8)	Whitewater Neotropical rivers	Blackwater Neotropical rivers		Migrating Semaprochilodus	Migrating Semaprochilodus	Resident predatory Cichla in blackwater rivers
Vanni and Headworth (chap. 4)	Watersheds, bottom sediments	Reservoir planktonic and pelagic zones	Streams, groundwater	Gizzard shad	Dissolved and particulate nutrients (P)	Reservoir plankton and their consumers
Power et al. (chap. 15)	Upland river	Forested watershed	River surface, floating algal mats		Emergent aquatic insects	Insectivorous bats, lizards, and spiders
Cadenasso et al. (chap. 10)	Meadow	Interior forest	Forest edge	Wind, foraging deer and voles	Nutrients (N), detritus, seeds	Forest plants and consumers
Jefferies et al. (chap. 18)	U.S. agricultural fields	Arctic salt marshes	Migratory flyways	Lesser snow geese	Nutrients (N), grazing geese	Arctic graminoids, presently severely overgrazed
Baudry and Burel (chap. 21)	Globally derived agro-chemicals and crops	European agro-ecosystems	Hedgerows, grassy strips, earth banks	Industrial agriculture, world trade, wind, water	Nutrients (N), organic matter, crops	Local farmers, consumers in human and natural local food webs
Schindler and Lubetkin (chap. 3)	Watershed	Fresh water		Water movement	Nutrients	Zooplankton, fish
Rasmussen and Vander Zanden (chap. 11)	Watershed	Fresh water		Water movement	Contaminants	Zooplankton, fish

local prey over time scales of weeks to months (e.g., aquatic insects decrease lizard predation on riparian spiders; Sabo 2000, cited in Power et al., chap. 15).

Other general patterns of theoretical interest should emerge as more works reveal the spatial linkage of trophic interactions across scales and ecosystems. These models, analytical tools, and empirical studies will be critical in assessing how human distortion of patterns and fluxes affects food webs and ecosystems, and what players, processes, and scales of protection or restoration are essential if we are to maintain on Earth the intricate and diverse food webs that Gary Polis so appreciated.

Mary E. Power
Gary R. Huxel

PART I.

FLUXES OF NUTRIENTS
AND DETRITUS
ACROSS HABITATS

Chapter 1

Overview: Cross-Habitat Flux of Nutrients and Detritus

Michael J. Vanni, Donald L. DeAngelis,
Daniel E. Schindler, and Gary R. Huxel

Ecologists have long known that all ecosystems receive considerable quantities of materials from outside their boundaries (e.g., Elton 1927), and quantifying the magnitude of such fluxes has long been a central tenet of ecosystem ecology (e.g., Odum 1971). Thus, one might think that the consequences of such fluxes for food webs would be well understood. However, food webs have traditionally been viewed as if they were isolated from surrounding habitats, a habit that has been particularly persistent in the modeling of food webs. When fluxes from the outside have been considered, they have largely been restricted to constant inputs directly affecting the base of the food web (e.g., solar energy or nutrients), and usually only such issues as their effects on equilibrium conditions have been considered (e.g., the well-known relationships between nutrient inputs and average densities of various food web members).

Only recently have ecologists begun to ask explicitly how variable flows across habitat or ecosystem boundaries affect interactions within recipient food webs (Polis, Anderson, and Holt 1997). This recent interest in cross-habitat fluxes and their consequences probably stems in part from two findings. First, ecologists have discovered that subsidies of materials from outside the boundaries of a food web can vary greatly over space and time, can enter a food web at various trophic levels, and can greatly influence food web processes and dynamics (Polis, Anderson, and Holt 1997). In other words, these fluxes can affect the strength of the interactions typically studied by food web ecologists, such as predation and competition. Therefore, cross-habitat fluxes may help explain some of the observed variation in food web regulation and structure. The second important finding is that organisms can be important in moving materials across habitat or ecosystem boundaries, either by direct movement of the organisms themselves or by their physical or metabolic activities (e.g., Bilby et al. 1996; Vanni 1996; Kitchell et al. 1999). The extent to which organisms transport materials from one food web or habitat to another often depends on species interactions-processes within the typical domain of food web ecologists— especially as they affect trophic structure, abundance, size structure, and behavior in the donor ecosystem. Understanding the dynamics and consequences of material and energy flows across habitat boundaries has thus emerged as an important focal point in ecology (Polis, Anderson, and Holt 1997).

Part 1 focuses on the transport of abiotic materials (nutrients and detritus) across habitat boundaries to recipient food webs. While the chapters are quite diverse—representing different approaches, ecosystem types, and organisms—there is a basic set of features that relate to each of the examples considered in the chapters. First, there is the magnitude (how much material) and quality (what kinds of material) of the cross-habitat flux, usually a nutrient or organic matter transported from one system to another. Second, there is always a mediating agent or agents that carry the fluxes or facilitate their effects. Third, there are variations in the effects of these allochthonous fluxes, which may occur at a variety of spatial and temporal scales. Fourth, these subsidies have consequences for species interactions in recipient food webs.

MAGNITUDE AND QUALITY OF CROSS-HABITAT FLUXES, MEDIATING AGENTS, AND EFFECTS OF SCALE

For some materials and in some ecosystems, flux rates are well quantified. For example, considerable research has been conducted to quantify the nutrients exported from watersheds to streams and lakes. However, within the context of how fluxes affect food web dynamics, we know surprisingly little about the quantity and variability of many potentially important fluxes. This is particularly true when the recipients of fluxes are animals; in many ecosystems the focus has been on inputs of nutrients available to primary producers. The ten chapters in part 1 show that the magnitude of fluxes can vary greatly within and among ecosystem types. They also illustrate the diversity of cross-habitat fluxes in terms of what materials are transported and by which processes.

It is not surprising that the land-water interface is an important one for the occurrence of important allochthonous fluxes. It is well known that terrestrial systems often export nutrients and organic matter to coastal oceans and fresh waters, but reverse fluxes can also occur. Anderson and Polis (chap. 6) discuss the importance of ocean-derived material to terrestrial food webs on islands. Fluxes from the ocean to these small islands can be in the form of marine algae and detritus that wash ashore or seabird guano that is derived from predation on marine prey. The fertilization of island plants by guano has several effects on the food web. Thus ocean currents and nesting seabirds are obvious mediating agents that link terrestrial food web processes to marine ecosystem dynamics. In addition, the timing of rainfall determines the temporal variation in nutrient remineralization from guano. Earlier work by these authors and their collaborators showed that total inputs of nutrients are much greater on bird islands (i.e., those with significant seabird colonies) than on non-bird islands (Polis and Hurd 1995, 1996a; Anderson and Polis 1998, 1999). The relative effects of "shore drift," the marine materials that wash ashore on the islands, illustrate the need to consider the spatial scales involved. The relative effect of this marine subsidy on the interaction web of the island is a function of the size of the island or, more specifically, the coast-to-inland ratio of the island. If this ratio is large, then the effect of the subsidy is likely to be proportionally large.

Menge (chap. 5) also discusses the consequences of oceanic fluxes. In this case, ocean currents deliver phytoplankton and larval invertebrates long distances to the intertidal zone. These planktonic organisms are then incorporated into, and can have strong effects on, intertidal food webs. The effects of spatial scale also figure strongly in Menge's analysis of differences in rocky intertidal interaction webs. Allochthonous inputs (of larvae, nutrients, and primary production) are relatively uniform at small scales (hundreds of meters), so that differences in interaction webs at these scales are functions of factors that are also variable at small scales, such as wave action. However, on a larger, regional scale, the magnitude and quality of allochthonous inputs may vary substantially. Thus, communities that are otherwise subjected to the same level of wave action and other physical conditions can show sharp differences in species composition and interaction webs in response to different allochthonous inputs.

Three of the chapters illustrate how mathematical models can be used to predict the magnitude of fluxes across habitat boundaries and to explore their consequences in recipient systems. DeAngelis and Mulholland (chap. 2) develop a two-layer model that explores the consequences of a transient storage zone for stream food webs. A transient storage zone, such as a layer of relatively still water near the stream bottom, is an important stream ecosystem compartment that stores dissolved (available) nutrients before they are used by organisms. The same type of modeling approach used to consider the movement of nutrients into the stream transient storage zone is also used to analyze nutrients moving from a common pool into the rooting zones of plants. The fluxes of nutrients between spatial compartments are through simple advection and diffusion, with diffusion often being the mediating agent for movement of nutrients from the free-flowing water to the periphyton in the storage zone of a stream, or from a common nutrient pool to the rooting zones of plants. The slow time scale of diffusive transport from the overlying water to the storage zone compared with the rapid rates of biological uptake results in nutrient gradients on small spatial scales in streams and around terrestrial plants. Such local-scale variations in nutrient availability can affect a number of food web properties. Thus, a seemingly straightforward parameter such as nutrient input rate can be misleading, depending on how efficiently externally derived nutrients enter the food web as they are moved from different spatial components of ecosystems.

Schindler and Lubetkin (chap. 3) review the use of stable isotopes to elucidate the diets and trophic positions of consumers. In some cases, these analyses can help determine in which habitats animals feed and the extent to which they move among habitats and ecosystems. This information can be used to quantify the movement of materials into habitats via mobile consumers and the extent to which resident consumers rely on material from within and outside of their food webs. An excellent example is that of anadromous fishes such as salmon. Salmon accumulate nearly all of their body mass in the ocean and return to fresh waters to spawn and die. Stable isotope analyses have shown that marine-derived nutrients (e.g., C and N) from decaying salmon carcasses represent an important subsidy to both the recipient freshwater and the adjacent terrestrial (riparian) food webs. Schindler and Lubetkin suggest that food web ecologists could increase the utility of stable isotope data by using them to quantitatively estimate the contributions of various prey to the diets of consumers and the flows of subsidies through recipient food webs. Spatial scale is an important consideration in the examples of waterfowl- and fish-mediated fluxes of nutrients. The effects of these allochthonous nutrient inputs depend on the size of the areas to which the transport is directed. Because these areas can be quite localized in the case of roosting waterfowl or spawning fish, the effects can be strong.

Vanni and Headworth (chap. 4) model linkages between watershed-derived and organism-derived nutrient fluxes in lakes (reservoirs of the midwestern United States). Reservoirs can receive substantial quantities of nutrients from their watersheds (via surface runoff) and from a sediment-feeding fish, the gizzard shad (via its ingestion of sediment-bound nutri-ents and subsequent excretion of dissolved nutrients into the water). As watersheds are degraded from forest to agriculture, both fluxes increase because of increased nutrient runoff and increased gizzard shad abun-dance. Furthermore, the magnitude of the fish-mediated flux is dependent on watershed inputs: increased agriculture in watersheds results in in-creased inputs of particulate (detrital) nutrients, which provide a food source for gizzard shad. Because in this case the gizzard shad plays two roles, that of mediating agent and that of a population within the reservoir interaction web, it can have a positive feedback on itself. Thus the two fluxes are linked, with potentially synergistic effects on food webs. Spatial scale is also an implicit consideration in the reservoir system: the importance of allochthonous fluxes to the reservoir is generally an increasing function of the ratio of the size of the watershed to that of the reservoir.

These chapters illustrate the importance of quantifying cross-ecosystem or cross-habitat fluxes as well as spatial and temporal variation in flux rates. In addition, these chapters show that both physical processes (currents, stream flow) and organisms can be significant sources of detritus and nutrients.

MECHANISMS OF AND LIMITS TO FLUXES

Recent studies, including those in this volume, have strongly demon-strated that the movement of energetic and nutrient resources, as well as that of predators and prey, across habitat boundaries (mostly artificial boundaries) plays an important role in the structure of communities and in population-level responses. The work of Witman et al. (chap. 9) and Menge et al. (chap. 5) demonstrates the interconnections between systems that have been traditionally studied in the disparate fields of oceanography, intertidal ecology, and terrestrial ecology. However, the interconnections between marine upwelling and subtidal, intertidal, supralittoral, and terres-trial habitats require that the flow of energy and nutrients between these connected systems be considered. These interconnections are common fea-tures in nature and occur across a variety of scales. Therefore, the traditional view that ecosystems are closed needs to be replaced by an open-system view. This shift would require the linking of experimental methods of studying local communities and populations with landscape concepts of interconnection between habitats of different types. For example, Witman et al. (chap. 9) demonstrate that internal waves may mix nutrients above and below the pycnocline or cause large vertical displacements of subsurface chlorophyll maximum layers containing concentrated phytoplankton. This mixing creates strong benthic-pelagic linkages, resulting in increased productivity in benthic zones.

The flux of nutrients and organisms between habitats involves either real or artificial boundaries. The characteristics of these boundaries play a major role in regulating the flux rates between habitats. Cadenasso et al. (chap. 10) synthesize the current understanding of the role of boundaries in cross-habitat fluxes. In general, they find that fluxes between two systems or patches must cross the boundary, and that the boundary may inhibit the flux, enhance the flux, or be neutral to the flux. Furthermore, regulation of fluxes by a boundary may be affected by the structure and dimensionality of the boundary, by the type of flux, and by the types of systems or patches in contact at the boundary (Pickett and Cadenasso 1995).

CONSEQUENCES OF CROSS-HABITAT FLUXES FOR FOOD WEBS

All of the chapters in this section explore the consequences of cross-habitat fluxes for recipient food webs. DeAngelis and Mulholland (chap. 2) show, using the two-layer model, that the flux rate of dissolved nutrients through the transient storage zone of streams affects the dynamics of algae and the food web it supports. This model, extended to competing autotrophs in a more general context, shows that a spatial gradient in nutrients, caused by the rapid uptake of nutrients by biota in the storage zone, can increase the number of coexisting species. This study supports other work suggesting that increased spatial or temporal heterogeneity in resource supply can promote species coexistence and increase biodiversity (Tilman 1982; Sommer 1985; Grover 1988). Thus, species diversity may be a function of the flux of nutrients from outside the system, the size and nature of the transient storage zone, and the rate at which nutrients are taken up from the storage zone. The last factor may be affected greatly by food web interactions. For example, in streams, uptake rates may be higher when there is a high algal biomass and severe algal nutrient limitation; these are traits that may be affected by grazing, light intensity, and disturbance in addition to nutrient supply. One interesting feature of the DeAngelis and Mulholland model is that local depletion of nutrients does not necessarily lead to depletion of the regional nutrient pool. As mentioned above, some simple empirical models attempt to relate food web attributes (e.g., biomass, number of trophic levels) to average (= regional) nutrient concentrations, but the DeAngelis and Mulholland model shows that this may be an oversimplification.

Schindler and Lubetkin (chap. 3) review how stable isotopes can be used to trace the fates of allochthonous inputs (e.g., sewage, energy from hydrothermal vents) through recipient food webs. In both cases, the dynamics of recipient webs can be greatly affected by the relative extent to which these sources provide them with nutrients and energy compared with other, in situ sources.

Menge (chap. 5) shows that phytoplankton delivered to the intertidal food web by oceanic currents and upwelling can amplify food web interactions by stimulating the abundance of recipient food web members. Filter feeders (barnacles and mussels) and grazers (limpets and chitons) respond positively to fluxes of plankton because this represents either an increased food supply or a source of larval propagules. Organisms directly preyed upon by these subsidized consumers (e.g., attached algae) may suffer increased mortality, and a number of indirect interactions are possible. Thus, oceanic conditions (currents, upwelling) can affect the entire intertidal interaction web in complex ways.

Holt (chap. 7) considers two general conceptual issues that arise whenever one considers open systems. Holt looks at two cases of local coexistence. First, he examines how the rate of competitive exclusion is influenced by recurrent immigration of an inferior competitor and by an influx of resources. Second, he explores how apparent competition is influenced, again by recurrent immigration of an inferior competitor. Coexistence in both cases is strongly influenced by immigration rates and spatial subsidies.

Three chapters show the various means by which food webs are affected by consumer-mediated cross-boundary fluxes. Schindler and Lubetkin (chap. 3) discuss several examples of how stable isotopes have aided in elucidating the effects of consumers on cross-habitat fluxes of nutrients and energy. The salmon example has already been mentioned. Another case study showed that waterfowl can be important nutrient vectors. Stable isotope analyses showed that geese, which consume nutrients from terrestrial areas such as agricultural fields, can deposit those nutrients (via guano) in wetland areas, and that this flux represents a large proportion of total nitrogen and phosphorus fluxes to the wetlands.

The model of Vanni and Headworth (chap. 4) links watershed-derived and organism-mediated fluxes in reservoirs of the midwestern United States. Their model predicts that the importance of nutrient flux through gizzard shad populations should increase with lake productivity (i.e., with the relative extent of agriculture in watersheds), both in terms of total nutrient flux through the fish and in terms of the proportion of primary production sustained by fish-mediated flux. Because fluxes from the watershed and through gizzard shad both increase with the increasing prevalence of agriculture in the watershed, the two fluxes may act synergistically to regulate primary productivity. In other words, watershed degradation not only makes lakes more productive through direct nutrient inputs, but also stimulates gizzard shad biomass through increased detritus inputs. Increased gizzard shad biomass makes lakes even more productive by further increasing nutrient supply. Increased productivity, in turn, can have a number of effects on limnetic food webs, including increased phytoplankton abundance, shifts in phytoplankton and zooplankton species composition, and increased frequency of nuisance blooms of cyanobacteria (blue-green algae).

Anderson and Polis (chap. 6) show that pulses of marine-derived nutrient inputs to islands, arriving via seabird guano, can magnify interactions between members of the island food web. The fertilization effect of guano leads to increased primary productivity and increased densities of herbivores and other consumers. Increased consumer densities can strengthen top-down effects within the island food webs (Polis and Hurd 1995, 1996a). Nutrient pulses delivered by bird guano can also reduce temporal stability in primary productivity, in herbivore densities (i.e., greater inputs lead to more pronounced temporal fluctuations), and in plant species richness.

Winemiller and Jepsen (chap. 8) examine the influence of nutrient subsidies on fish communities in nutrient-poor blackwater rivers. Whitewater systems, in contrast to blackwater systems, have high primary productivity. Blackwater systems can also support important fisheries, but, as in other systems, Winemiller and Jepsen find that the amount of biomass supported cannot be explained by in situ primary productivity (Allen's paradox; Allen 1951). Winemiller and Jepsen find that blackwater systems are supported largely by young-of-the-year fish migrating from the nutrient-rich whitewater systems during periods of large-scale flooding.

Connectivity among habitats within watersheds can have negative as well as positive effects. Contaminant loading via runoff and stream flow carrying water from adjacent polluted habitats can be highly detrimental to lake ecosystems. Rasmussen and Vander Zanden (chap. 11) demonstrate the serious effects of the flux of pollutants into Canadian lake ecosystems and the consequences of bioaccumulation of those pollutants within lake food webs.

CONCLUSIONS

It seems clear that cross-habitat and cross-ecosystem subsidies of food webs are ubiquitous (Polis, Anderson, and Holt 1997). Ecologists have made excellent progress in quantifying these fluxes and their consequences in a number of habitats (e.g., compare the chapters in this book with those in another book on food webs [Polis and Winemiller 1996] published 7 years ago). The chapters in part 1 illustrate the variety of cross-habitat fluxes to food webs. Agents that range from organism behaviors to various physical processes can be important in mediating these fluxes. Furthermore, the materials transported across ecosystems can vary qualitatively with respect to whether they are available to plants, herbivores, carnivores, or detritivores. The trophic position at which subsidies enter food webs has important implications for food web dynamics (Polis, Anderson, and Holt 1997). This view of food webs can be very different from more traditional models that consider only inputs of nutrients at the bottom of food webs.

The variety of types and magnitudes of allochthonous fluxes, mediating agents, parts of the food webs that can be affected by fluxes, and spatial and temporal scales at which effects might occur lead us to wonder whether any generalizations can be drawn with respect to the effects of such fluxes on interaction webs. For example, can we expect cross-habitat fluxes of detritus and nutrients to be important in particular kinds of habitats? Are they likely to be mediated by, or to affect, particular types of organisms? How much variation in the importance of cross-system fluxes are we likely to observe within a given ecosystem type? How often, and to what extent, do different fluxes (e.g., physically mediated and organism-mediated) interact? The answers to these questions probably lie in effective integration of community and ecosystem ecology with recent advances in landscape ecology that account for the spatial structure of the environment. We hope that the chapters in part 1 inspire ecologists to seek answers to these and other questions related to cross-habitat fluxes.

Chapter 2

Dynamic Consequences of Allochthonous Nutrient Input to Freshwater Systems

Donald L. DeAngelis and Patrick J. Mulholland

Open systems are systems that are sustained by the inflow of energy and, usually, nutrients from external sources. This includes all biological systems, of course. The chemostat or bioreactor, a continuous culture vessel through which water flows, carrying nutrients in and interaction products out, is a simple experimental version of an open system. It has been valuable in studying predator-prey and competitive interactions in microbial communities (e.g., Tsuchiya et al. 1972; Jost et al. 1973; van den Ende 1973; Sommer 1983, 1984, 1988).

The mathematical version of the chemostat model describes a chemostat system by means of coupled differential equations for concentrations of nutrients and biotic populations (e.g., Smith and Waltman 1995). The chemostat model allows consideration of more complex cases than may be easily manageable with an actual chemostat. Numerous chemostat models exist in the literature (e.g., Waltman et al. 1980; Fredrickson and Stephanopolous 1981; Cunningham and Nisbet 1983; Smith and Waltman 1995; Andersen 1997; Grover 1997).

In this chapter, we first use the standard chemostat model to examine theoretically the dynamics of periphyton algae and a limiting nutrient in a stream. We show that this chemostat model is often inadequate to describe the simplest possible stream system, even when longitudinal heterogeneity is considered through the use of a series of coupled chemostat models. However, a two-layer version of the chemostat model, in which each longitudinal segment of the stream is divided into a compartment for free-flowing water and a compartment for still water near the bottom of the stream, appears to capture much of the critical dynamics of stream systems. Realistic models can be achieved without sacrificing analytic tractability.

Use of the chemostat model to describe nutrient competition among autotrophic species has led to pivotal models in theoretical ecology (e.g., Tilman 1982). Simple forms of this model type have been widely used, even though they predict, when there is only one limiting nutrient, the elimination of all but one species of autotroph, which is not normally seen in natural systems. We describe a generalization of the two-layer model that allows one to make different predictions that are more in accord with what is observed in real systems.

THE CHEMOSTAT MODEL AND ITS LIMITATIONS

The chemostat model owes its success to both its simplicity and its ability to capture the essential features of an open biological system that is relatively homogeneous. As an example, consider the chemostat model used to describe a water body with phytoplankton species and an input of a limiting nutrient. Let N be the nutrient concentration and B the concentration of the nutrient tied up in biomass per unit volume of water (g m^{-3}). Then a model describing this system is

$$\frac{d(VN)}{dt} = QN_0 - QN - \frac{r(VB)N}{b + N} + e(VB), \tag{2.1}$$

$$\frac{d(VB)}{dt} = \frac{r(VB)N}{b + N} - e(VB) - QB, \tag{2.2}$$

where r = maximum growth rate of the phytoplankton (d^{-1}); Q = flux of water through the system (m^3 d^{-1}); e = recycling coefficient of nutrient from phytoplankton to water (d^{-1}); b = half-saturation coefficient for growth of phytoplankton (g m^3); and V = volume of the stream segment (m^3).

The term QN_0 represents external input to the system, while the terms QN and QB represent washout losses of nutrient and phytoplankton, respectively, from the system in the outflow of water, and $e(VB)$ is the nutrient recycling. These terms incorporate the openness of the system with respect to nutrients. This model has the following analytic solution for the steady state values, N_{ss} and B_{ss}:

$$B_{ss} = (N_0 - N_{ss})\frac{Q}{V}\Big/\left(\frac{rN_{ss}}{b + N_{ss}} - e\right), \tag{2.3}$$

$$N_{ss} = b\left(e + \frac{Q}{V}\right)\Big/\left(r - e - \frac{Q}{V}\right). \tag{2.4}$$

Note that for N_{ss} and B_{ss} to have positive values, it is necessary that Q/V be smaller than the net rate, $(r - e)$, at which phytoplankton grow. Otherwise the phytoplankton will be lost from the water body faster than they can reproduce, so the population will go extinct.

Analytic models of this type have been verified in many cases by data from actual chemostats or bioreactors. This model has its limits, however. An implicit assumption is that complete mixing occurs in the reactor. In that way, one can define a single value of the nutrient concentration, N, and a single value of the algal biomass concentration, B. If mixing is not complete, such that these concentrations vary through the system, then the problem becomes much more complex.

Now suppose that one is modeling a stream, and that we apply the chemostat model to each segment in a chain of segments representing the whole stream. We assume that the autotroph is not phytoplankton but algal periphyton, which cling to the bottom and do not wash out. Consider applying an elaboration of equations (2.1) and (2.2) to this system. The equation for algal periphyton in a segment of stream must differ from equation (2.2), since no algal washout from the system occurs. With this loss term removed, it can be shown that the denominator of equation (2.4) vanishes and algal biomass is predicted to approach infinity—an obviously unrealistic prediction. A more realistic model would include periphyton mortality within the segment, with the dead periphyton going to a detrital compartment from which there is subsequent loss in outflowing water (see fig. 2.1). Then the set of equations becomes

$$\frac{d(VN)}{dt} = QN_0 - QN - \frac{r(VB)N}{b + N} + e_1(VB) + e_2(VD), \tag{2.5}$$

$$\frac{d(VB)}{dt} = \frac{r(VB)N}{b + N} - e_1(VB) - m(VB), \tag{2.6}$$

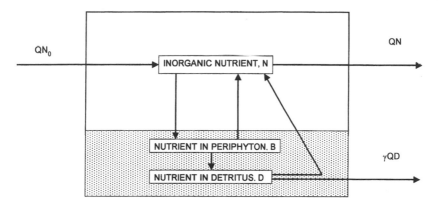

Figure 2.1 Schematic of a three-compartment chemostat model of a stream segment with algal periphyton. The model has the components inorganic nutrient (*N*), nutrient stored in periphyton (*P*), and nutrient stored in detrital biomass (*D*).

$$\frac{d(VD)}{dt} = m(VB) - e_2(VD) - \gamma QD, \qquad (2.7)$$

where D represents dead biomass, or detritus, γQD (where $0 < \gamma < 1$) is the washout of detritus, e_1 and e_2 are nutrient recycling, and $m(VB)$ is mortality. This model differs from equations (2.1) and (2.2) only in having a detrital component. In this version of the model, the steady state biomass, B_{ss}, will remain at a finite value. However, there is still an apparent paradox in that the nutrient concentration in the water, N_{ss}, is held to the constant value

$$N_{ss} = \frac{(e_1 + m)b}{r - e_1 - m}, \qquad (2.8)$$

which depends only on parameters of the algae. If we think about this prediction, it is clearly unreasonable, for it is the same no matter how short the stream segment is or how high Q or N_0 are. That is, no matter how high the input concentration of allochthonous nutrient, or how large the water flow rate, Q, into the system, nutrient concentration in the water is reduced to a level determined purely on the basis of algal growth rates. It is intuitively impossible that the periphyton in one arbitrarily short segment of stream can reduce incoming nutrient this much, a fact easily demonstrated in real streams. The chemostat model, or at least this simple version of it, does not work in this case. Extending the model system to a longitudinal series of joined segments also does not help, because the nutrient would be decreased by this unreasonable amount even in the first segment.

There are at least two possible reasons for the inapplicability of the simple chemostat model to our hypothetical stream system. First, the complete mixing condition is not satisfied. That is, much of the nutrient flows rapidly through the system without even coming into contact with the autotroph, say, in an upper layer above the periphyton, so there is no possibility of uptake. Instead of a uniform concentration of nutrient in the water, N, a gradient between the upper layer and the vicinity of the periphyton exists. Second, other limiting factors start to operate and restrict the increase in autotroph growth. In equations (2.1) and (2.2), only nutrient limitation is modeled. If light, for example, became limiting through self-shading of periphyton, then a term proportional to $-B^2$ would have to be added to equations (2.2) and (2.6), which would change the nature of the solution. This would prevent a reduction of the nutrient to a level determined only by algal parameters, as in equation (2.8).

Stream ecologists would generally agree that the assumption of complete mixing does not hold in real streams. Here we will examine a modification of the chemostat model that accounts for the incomplete mixing. We will begin with the next simplest model, which considers a two-layer system.

TWO-LAYER MODEL OF NUTRIENT DYNAMICS IN A STREAM

It is possible to consider the lack of complete mixing in the model by approximating the vertical gradient in the stream as consisting of two layers, a free-flowing upper, or water column, layer and a layer of still water at the bottom. This bottom still layer is roughly associated with the physical boundary layer of the stream. Bottom friction, especially from irregularities such as rocks, creates these areas of very slowly flowing water. This layer is often called the "transient storage zone" or just "storage zone." Interstitial waters in sediments and waters within algal mats, backwater areas, and pools can also act as transient storage zones (e.g., Bencala 1984; Triska et al. 1990; Bencala and Walters 1983; Runkel and Chapra 1993).

Consider a longitudinal segment of stream with these two compartments (fig. 2.2). Nutrient input enters the segment through free-flowing water from upstream. The nutrients can move back and forth between the free-flowing water and the storage zone through turbulent diffusion. Figure 2.2 shows a schematic of the processes of nutrient transport through a portion of stream short enough to be considered longitudinally uniform. This portion, hereinafter referred to as a "stream segment," is divided into a transient storage

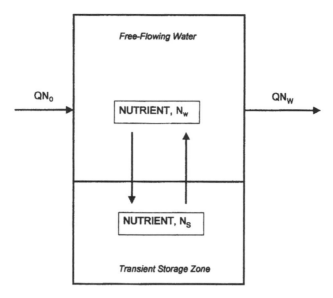

Figure 2.2 Schematic of a two-layer model of a stream segment. The two compartments are nutrient in free-flowing water and nutrient in a transient storage zone. The nutrient is carried through the free-flowing water by flow from upstream, and it diffuses into the transient storage zone.

zone and a free-flowing water zone, with the storage zone along the bottom, where water movement can be approximated as zero flow. The sizes of the two zones and the rate of transfer of nutrients across their interface affect the dynamics of periphyton and the food web it supports in the storage zone.

Good evidence supports such a stream structure of two components, free-flowing water and a transient storage zone. Many studies have used this conceptualization and fit data from experimental injections of tracer solutes to these two-compartment models (e.g., Bencala and Walters 1983; Kim et al. 1992; DeAngelis et al. 1995). In these experiments, a pulse of conservative tracer (i.e., a tracer that is not permanently incorporated into plant biomass or substrate), such as a chloride or tritium solution, was injected at a constant rate into a stream at a point for a known period of time. At a site sufficiently far below the point of injection such that the tracer could be assumed mixed into the stream water, the tracer concentration was measured. The resultant distribution of arrival times of the nutrient to the downstream site could be fit to a model with two spatial compartments (free-flowing water and a storage zone) extremely well.

The model has inorganic nutrient variables for the free-flowing water and the storage zone and living biomass and detritus only in the storage

zone. The unit model used here simulates a short, homogeneous, longitudinal segment, or unit length, of stream. The model for a longer stretch
of stream is made up of coupled segments, each of which has only these
four components:

N_w = nutrient concentration in dissolved, biologically reactive inorganic
form in the free-flowing water

N_s = nutrient concentration in dissolved inorganic form in the water of
the transient storage zone

B_s = concentration of nutrient tied up in the living biomass in the transient storage zone, in units of nutrient per unit volume of water

D_s = concentration of nutrient tied up in the detritus, in units of nutrient per unit volume of water

We assume that there is a constant flux of water of $Q\,\mathrm{m}^3\,\mathrm{d}^{-1}$ into the segment
farthest upstream, and that this flux is conserved along the stream. This
means that we are assuming there are no tributaries or groundwater inputs
along the stretch of stream, no loss of water by evaporation, and so forth.
Thus, Q is the input rate of water into every segment along the stream. The
upstream input of dissolved nutrient to any given segment i along
the stream is then $QN_{w,i-1}$, where $N_{w,i-1}$ is the concentration of nutrient in
the segment immediately upstream. The washout rate of the nutrient is
$-QN_{w,i}$, where $N_{w,i}$ is the concentration of nutrient in the segment i.

A set of equations appropriate for a single segment is as follows (here
we dispense with the i subscript and use "up" to designate the segment immediately upstream):

$$\frac{d(V_w N_w)}{dt} = QN_{up} - QN_w - k_w V_w N_w + k_s V_s N_s, \tag{2.9}$$

$$\frac{d(V_s N_s)}{dt} = k_w V_w N_w - k_s V_s N_s - \frac{r\,V_s B_s N_s}{b + N_s} + e_1(V_s B_s) + e_2(V_s D_s), \tag{2.10}$$

$$\frac{d(V_s B_s)}{dt} = -e_1(V_s B_s) - m V_s B_s + \frac{r\,V_s B_s N_s}{b + N_s}, \tag{2.11}$$

$$\frac{d(V_s D_s)}{dt} = m(V_s B_s) - e_2(V_s D_s) - e_3(V_s D_s), \tag{2.12}$$

where e_3 is the rate of loss of nutrient from the system to recalcitrant forms,
V_s and V_w are the unit segment volumes for the storage zone and free-
flowing water compartments, respectively, $k_w V_w N_w$ and $k_s V_s N_s$ are the nu-

trient fluxes (through turbulent diffusion) from free-flowing water to the storage zone and vice versa, and where $k_w V_w = k_s V_s$ must be assumed in order for the net flux across the boundary between the two zones to be zero when nutrient concentrations are the same. The steady state solution for a given segment in this stream system is given by

$$N_{s,ss} = \frac{b(m + e_1)}{r - m - e_1},$$ (2.13)

$$N_{w,ss} = \frac{1}{Q + k_w V_w}(N_{up}Q + k_s V_s N_{s,ss}),$$ (2.14)

$$B_{s,ss} = \frac{e_2 + e_3}{e_3 m}\left(\frac{Q}{Q + k_w V_w}\right)k_s(N_{up} - N_{s,ss}),$$ (2.15)

$$D_{s,ss} = \frac{m}{e_2 + e_3}B_{s,ss}.$$ (2.16)

The steady state model behavior within a given segment can then be explored for variations in any of the key parameters (see DeAngelis et al. 1995)—for example, the ratio of the volume of the storage zone to the free-flowing water zone, V_s/V_w; the diffusional transfer coefficients, k_w and k_s; and the loss rate of biomass from the system, e_3. To find the values of the four variables N_w, N_s, B_s, and D_s along a longer length of stream, consisting of many segments, we can solve the equations by iteration, starting from a fixed point at the upstream end (see Mulholland and DeAngelis 1999). Steady state values of N_w along the stream are shown, as a function of values of k_w and k_s, in figure 2.3. Note that N_w decreases as a function of downstream distance, eventually approaching N_s. What is interesting is that the higher the transfer coefficients, k_w and k_s, the faster N_w declines. The reason for this behavior is that high transfer rates provide a high input of nutrient into the storage zone. Because the periphyton exerts top-down control on the nutrient, its biomass, B_s, and, therefore, detritus, D_s, build up at a high rate. Thus there is also a high rate of loss of nutrient due to conversion of the detritus into buried or refractory forms, at a rate $e_3 V_s D_s$. This leads to the rapid loss of N_w as a function of distance downstream.

It is useful to look at the stream system from the point of view of temporal and spatial scales. There are two temporal scales in the dynamics: spatial movement across the gradient (approximated by a discontinuity between

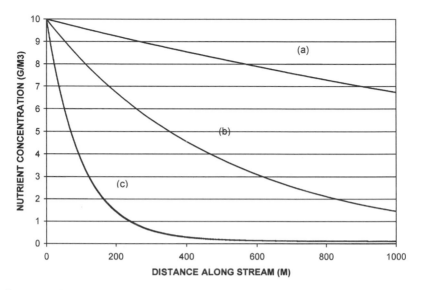

Figure 2.3 The concentration of a nutrient in the free-flowing water zone per 1-meter segment of stream as a function of distance downstream from the point at which nutrient and phytoplankton inputs are fixed. Three different values of k are used: 0.2, 1.0, and 5.0. The other parameter values are $Q = 1,000$, $N_{up} = 10$, $r = 4.0$, $b = 0.5$, $m = 0.2$, $e_1 = 0.5$, $e_2 = 0.2$, $e_3 = 0.01$. The rate of falloff of the nutrient level increases as k_w and k_s are increased. This increase is due to a greater loss of nutrient to recalcitrant forms in the storage layer as the rate of diffusion increases.

compartments here) from the free-flowing water to the storage compartment is slow compared with the local dynamics of nutrient assimilation by the biota. For this reason, the effect of the periphyton in a single small segment is not strong enough to equalize the solute nutrient concentration in the free-flowing water with that in the storage zone, although the effect of many segments is to cause N_w to decrease as a function of distance downstream and approach N_s, as shown in figure 2.3.

ALLOCHTHONOUS NUTRIENT INPUTS AND COMPETITION BETWEEN AUTOTROPHS

Above we showed that the slow rate of nutrient spatial transport relative to biotic processes can lead to spatial gradients of nutrients. One straightforward extension of the simple chemostat model can be applied to the question of species coexistence. The basic chemostat model described in equations (2.1) and (2.2) has long been used to describe two or more autotroph species competing for the same limiting nutrient. Analysis of the model shows that all but one of the species will be eliminated. This conclusion is known as the R^* rule and has been discussed extensively in the ecological literature (e.g., Powell 1958; O'Brien 1974; Tilman 1976, 1977; Hsu et al. 1977; Grover 1997). Experimental verification of this rule has been carried out in some bioreactor experiments (e.g., Hansen and Hubbell 1980).

However, we know that this conclusion violates our everyday observation of high diversity among autotrophs. Many explanations have been proposed to explain this apparent paradox (called the "paradox of plankton" by Hutchinson [1961] with respect to phytoplankton species). One possible reason for this paradox, at least for some situations, is that the mixing condition that is implicit in the model is not fulfilled in nature. Instead of many autotrophs competing in the same pool of a well-mixed nutrient, the slow rate of transport of nutrient through the medium compared with other process rates leads to the formation of spatial gradients of nutrient. This idea probably applies most strongly to terrestrial systems, so we will demonstrate it through a model with such a system of terrestrial plants in mind.

A generalization of the chemostat model of species competition that includes spatial effects was developed by Huston and DeAngelis (1994). Instead of conceiving of autotrophs (terrestrial plants here) in a perfectly mixed medium, we assume that because of the low mobility of nutrients in soil, a plant can cause only local depletions of nutrients. Thus, at low densities of plants, at least, plants do not compete directly for nutrients, but rather indirectly, by lowering the general background pool.

The model for this system has n plants (containing nutrient amounts P_i) and n local pools of nutrients (nutrient concentrations N_i) around these plants. There is also a regional nutrient pool (N_R) connected through diffusional transport to the local pools. The model proposes a hypothesis for how the R^* rule (e.g., Tilman 1976; Holt et al. 1994) might be avoided; in particular, that, although individual plants may be able to reduce their own nutrient pools locally, they cannot reduce the regional pool sufficiently to cause elimination of other plant species through exploitative competition. The equations are as follows:

Regional zone:

$$V_R \frac{dN_R}{dt} = QN_0 - QN_R - nk_{rm}V_RN_R + V_m(k_{mr}N_1 + k_{mr}N_2 + \cdots + k_{mr}N_n) \quad (2.17)$$

Local zones:

$$V_i \frac{dN_i}{dt} = k_{rm} V_R N_R - k_{mr} V_m N_i - \frac{r_i P_i N_i}{b + N_i} \quad (i = 1,2,...,n), \quad (2.18)$$

$$\frac{dP_i}{dt} = \frac{r_i P_i N_i}{b_i + N_i} - d_i P_i \quad (i = 1,2,...,n), \quad (2.19)$$

where $k_{rm} V_R N_R$ and $k_{mr} V_m N_i$ are the transfer rates from the regional pool to the local pools, and vice versa, and V_m is the size, in m^3, of each local pool, where $k_{rm} V_R = k_{mr} V_m$. If the assumption that all local pools are the same physical size were not made, more complex coefficients would be needed.

The above set of equations considers a fixed number of plants, each with a fixed equilibrium size. It is assumed that these characteristics do not change through time. Therefore, this model best describes a relatively short time scale during which the nutrient levels come to equilibrium with the plants present, but the plant population does not change. We can show that over this time scale, at least, it is possible for plants of a variety of species to coexist. The equilibrium levels for this system are

$$N_{i,ss} = \frac{d_i b_i}{r_i - d_i} \quad (i = 1,2,...,n), \quad (2.20)$$

$$N_{R,ss} = \frac{N_0 Q + k_{mr} V_{mr} N_1^* + k_{mr} V_m N_2^* + \cdots + k_{mr} V_m N_n^*}{Q + n k_m V_R}, \quad (2.21)$$

$$P_{i,ss} = k_{rm} V_R (N_{R,ss} - N_{i,ss}) / \left(\frac{r_i N_{i,ss}}{b + N_{i,ss}} \right) \quad (i = 1,2,...,n). \quad (2.22)$$

No species is excluded for which the following inequalities hold:

$$N_{i,ss} > 0 \quad (2.23a)$$

and

$$N_{R,ss} > N_{i,ss} \quad (2.23b)$$

This means that over the short time scale, at least, coexistence of many species is possible when there is allochthonous input of nutrient to the regional pool and subsequent spatial heterogeneity of nutrient concentrations.

Over the long time scale we might expect at least two processes to lead to a gradual reduction in species diversity. First, more plants may colonize the area, so that the population density of plants, n, becomes very high. Under this circumstance, one plant species may be able to lower the general reservoir level of nutrient to the extent that equation (2.23b) is satisfied for only that species (see Holt, chap. 7 in this volume). Second, plants may increase in size clonally, with the same effect. Therefore, other mechanisms are necessary to explain autotroph diversity over longer time scales.

DISCUSSION

This chapter describes two different model systems, periphyton in a stream and competing terrestrial plants, that share important common features. In each case the allochthonous input of a nutrient is the key driving force that structures the system. In each case spatial gradients develop because the rates of nutrient movement and of biotic processes differ. In particular, transport of allochthonous nutrient into the transient storage zone of a stream or into the local rooting zones of terrestrial plants may be much slower than either the external replenishment of the nutrient reservoir or the biotic processing of the nutrient. Thus a gradient, or nonuniform "field," of nutrient availability is formed. Simple models based on the homogeneous chemostat model, such as equations (2.1–2.2), ignore this gradient, effectively assuming instantaneous mixing of the systems—assuming, for example, that the influence of plants on nutrient concentration spreads across space uniformly and instantaneously. The fact that such instantaneous mixing does not happen in most systems reduces the usefulness of the chemostat model. Both the two-compartment model of the stream segment (equations [2.9–2.12]) and the model of terrestrial plants surrounded by local nutrient depletion zones (equations [2.17–2.19]) explicitly model the transfers of nutrient, which changes the nature of the system. Both models explicitly allow spatial gradients (approximated as a discontinuity between compartments) to emerge whose effects can be approximated by dividing the abiotic nutrient into two or more compartments. If, instead of creating general depletions of nutrient, a plant creates only local depletions or gradients, then competition will be affected and the R^* rule may be violated temporarily.

Recent studies have emphasized the importance of allochthonous contributions of resources to food webs, which can affect their structure and stability (e.g., Polis and Hurd 1996a; Polis, Anderson, and Holt 1997). It is also important to note the manner and rates at which these external resources enter a local system. The simple chemostat model may not always be a good model for the input of external resources. Resource movement and uptake by the biota can occur at rates that can differ by orders of magnitude, leading to spatial gradients of the resource and possible effects on the whole system.

ACKNOWLEDGMENTS

This work was supported by the U.S. Geological Survey's Florida Integrated Science Centers and by Oak Ridge National Laboratory.

Chapter 3

Using Stable Isotopes to Quantify Material Transport in Food Webs

Daniel E. Schindler and Susan C. Lubetkin

Food webs are models of the flows of materials and energy through communities that result from consumer-resource interactions. Most food webs have been based on observations of predator-prey interactions or on direct analyses of consumer diets. Many food webs have been based merely on binary data linking known consumers and resources (Cohen 1978; Pimm 1982). More realistic food webs use highly detailed and temporally extensive diet analyses to estimate the relative contributions of different resources to consumer diets, and thus better portray the flows of energy and materials through the biotic component of ecosystems (DeAngelis 1992). Direct diet analyses can be problematic due to the different digestion rates of various prey, due to difficulties in identifying amorphous prey (e.g., algae, detritus), and because diet samples represent "snapshots" of an organism's diet. Even intensive diet analyses can be inadequate for determining energy and material flow through food webs because diet composition rarely accounts for differential assimilation of alternative prey.

In the past decade, ecologists have increasingly used natural stable isotope distributions to evaluate ecological interactions in ecosystems (fig. 3.1). Stable isotopes can track the fates of different sources of energy and materials if those sources have distinct isotopic signatures and if the

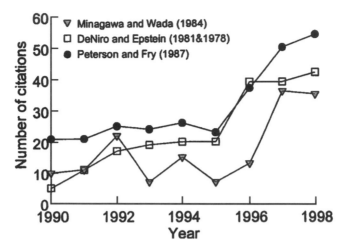

Figure 3.1 Time trend in the number of papers that cite four influential works in stable isotope ecological research in the last decade. The above data are extracted from the *Science Citation Index Expanded, Social Sciences Index,* and *Arts & Humanities Citation Index* databases of the Institute for Scientific Information, Inc. (ISI), Philadelphia, Pennsylvania, USA (copyright 1990).

isotopic signatures change in a predictable fashion as material moves through food webs. Stable isotope distributions are powerful complements to diet analyses because they provide direct estimates of the prey assimilated by consumers and provide a longer-term integration of the feeding history of a consumer than diet analysis does. Stable isotope distributions often vary among different habitats within ecosystems and, therefore, can provide information about the relative contributions of resources from different habitats to the structure and function of food webs. Stable isotope distributions represent a robust source of data to studies of spatial subsidies to food webs because they provide a convenient and informative means to integrate resource flows across the spatial and temporal scales appropriate to food web dynamics (Paine 1988; Polis, Anderson, and Holt 1997).

In this chapter we provide a brief overview of the principles and methods used in stable isotope research in food web ecology. We then present four case studies of systems in which stable isotope distributions have been used to evaluate material flow through food webs and, more specifically, to provide estimates of the magnitude of material flux from spatially segregated habitats to food webs. We finish by suggesting some important avenues for future use of stable isotopes in food web ecology.

OVERVIEW OF ISOTOPE TERMINOLOGY AND PRINCIPLES

Most elements occur naturally in several isotopic forms, some of which are unstable and are diminished through radioactive decay. Other isotopes do not exhibit radioactive decay and are considered stable isotopes. For example, the stable isotope ^{12}C, which dominates the natural pool of carbon in nature, occurs with trace amounts of stable ^{13}C and radioactive ^{14}C. In an ecological and physiological context, rare stable isotopes are virtually identical to the more common stable isotopes except for their increased atomic masses. A stable isotope ratio, denoted as δ, is a measure of the amount of a heavier isotope present in a sample relative to the more common natural isotope of the element, set relative to a standard. For example, the δ-value of ^{13}C is

$$\delta^{13}C = \frac{\dfrac{^{13}C}{^{12}C_{sample}} - \dfrac{^{13}C}{^{12}C_{reference}}}{\dfrac{^{13}C}{^{12}C_{reference}}} \times 1{,}000.$$

Values for $\delta^{15}N$ and $\delta^{34}S$ are calculated similarly. The standard references for C, N, and S are PeeDee limestone, atmospheric nitrogen, and the Canyon Diablo meteorite, respectively (Peterson and Fry 1987). Samples may have positive or negative δ-values, indicative of relative enrichment or depletion of the heavy isotope. For example, a consumer with a $\delta^{13}C$ of $-15‰$ is referred to as enriched relative to a $-20‰$ food source.

Although heavy and light isotopes are chemically similar, they have slightly different thermodynamic properties owing to their different atomic masses. These thermodynamic differences can lead to slightly different biochemical reaction rates. Different kinetic rates of heavy and light isotopes in physiological reactions result in biofractionation, or a change in the ratio of isotopes found in an organism relative to its resources. The first biofractionation of carbon in a food web occurs during photosynthesis, in which ^{12}C is preferentially taken up over ^{13}C. After a fractionation effect of about $-6‰$ relative to atmospheric carbon dioxide ($-7‰$), C_4 tropical plants and sea grasses have $\delta^{13}C$ values averaging about $-13‰$ (Haines and Montague 1979; Peterson and Fry 1987; Sackett 1989). In contrast, terrestrial C_3 plants show a much larger fractionation($\sim -21‰$), leaving them with $\delta^{13}C$ values of about $-28‰$ (Peterson and Fry 1987; Sackett 1989; Forsberg et al. 1993). Water use efficiency, leaf thickness, and temperature can affect the amount of fractionation in C_3 plants by changing the size of the carbon dioxide pool available for uptake or by modifying the difference in diffusion rates between $^{13}CO_2$ and $^{12}CO_2$ (Lajtha and Marshall 1994). Phytoplankton have $\delta^{13}C$ values ranging from −19‰ to −40‰ (Peterson and Fry 1987; Sackett 1989; Yoshioka et al. 1994; Schindler et al. 1997). Benthic algae and especially periphyton often exhibit relatively low biofractionation of C due to boundary layer effects that can reduce C availability near solid substrates (Hecky and Hesslein 1995). These boundary layer effects lead to periphyton that is substantially enriched in ^{13}C relative to phytoplankton, producing a convenient contrast between benthic and pelagic primary producers that can be used to estimate the magnitude of benthic contributions to freshwater (Junger and Planas 1994; Hecky and Hesslein 1995) and marine (Fry 1988; Hobson et al. 1994) food webs.

The $\delta^{15}N$ values of plants generally reflect their nitrogen sources. Plants that take in atmospheric nitrogen have isotope ratios between −2‰ and 2‰, and most plants that derive their nitrogen from soils have ratios varying from −8‰ to 10‰ (Peterson and Fry 1987). Little fractionation of nitrogen isotopes occurs during uptake by plants because nitrogen is often the limiting nutrient in plant growth and all available nitrogen is used (Owens 1987; Peterson and Fry 1987). Phytoplankton also shows little fractionation of nitrogen (Peterson and Fry 1987) and often has $\delta^{15}N$ values between 5‰ and 10‰ (Rau 1981; Yoshioka et al. 1994).

Sulfur isotope ratios of terrestrial plants are usually between 2‰ and 6‰, whereas plankton and seaweed range between 17‰ and 21‰ (Peterson and Fry 1987). Seawater sulfates, which have $\delta^{34}S$ values of about 21‰, are far more enriched than seawater sulfides, with $\delta^{34}S$ values near −10‰ (Peterson and Howarth 1987; Michener and Schell 1994). Benthic systems and marsh plants with a high proportion of sulfide-derived sulfur have lower $\delta^{34}S$ values than those with more sulfate-derived sulfur (Michener and Schell 1994). Therefore, $\delta^{34}S$ is useful for tracing flows of alternative resources in wetland and estuarine food webs that are likely to have substantial contrasts in the importance of sulfide in different food web sources.

The primary determinant of a consumer's isotopic composition is its diet (Haines and Montague 1979; DeNiro and Epstein 1978, 1981). There is some modification of diet isotopic composition, however, because the biological reactions underlying digestion, assimilation, and excretion follow kinetic and thermodynamic laws and can produce biological fractionation. The physiological processes of respiration and excretion release carbon dioxide with a lower $\delta^{13}C$ than the organism's carbon source, leaving the organism with a higher $\delta^{13}C$ in its tissues. Although there is some variation in the amount of carbon fractionation, most organisms show an enrichment, known as a trophic level shift, of nearly 0.7‰ compared with their food sources (DeNiro and Epstein 1978; Haines and Montague 1979; Fry and Arnold 1982; Checkley and Entzeroth 1985; Peterson and Fry 1987; Michener and Schell 1994). There is a large fractionation effect in the production of urea and NH_3 due to preferential excretion of ^{14}N over ^{15}N, and many animals show a trophic level shift in $\delta^{15}N$ of about 3–3.5‰ (DeNiro and Epstein 1981; Checkley and Entzeroth 1985; Owens 1987; Peterson and Fry 1987; Michener and Schell 1994). There is thought to be no significant biofractionation of ^{34}S between diet and consumer (Peterson et al. 1985; Peterson, Howarth, and Garritt 1986; Peterson and Fry 1987; Peterson and Howarth 1987; Fry 1988; Michener and Schell 1994).

Fractionation occurs at smaller scales within consumers, causing different tissues within a single organism to have varying isotope ratios. Fat, brain, muscle tissue, and hair have been shown to have $\delta^{13}C$ and $\delta^{15}N$ values differing from both the food source and one another (DeNiro and Epstein 1978, 1981). This phenomenon is particularly associated with lipids, which often have low $\delta^{13}C$ values (McConnaughey and McRoy 1979; Peterson and Fry 1987; Kling et al. 1992; Rau et al. 1992), and needs to be considered in the design of both the sampling regime and tissue analysis (Gannes et al. 1997). The sensitivity of the stable isotope characteristics of different body tissues to diet composition is related to the turnover rate of those tissues, which can vary from days to years.

As organisms at lower trophic levels expel the lighter isotopes, passing on biomass with a slightly enriched isotope ratio to those organisms consuming them, the change in isotopic composition propagates up the food web. Successively higher levels of consumers also preferentially expel the lighter isotopes, and the difference between primary producer isotope ratios and consumer isotope ratios increases as a function of trophic position in the food web. Thus, stable isotope ratio data may be used to determine both the sources of the organic material and the number of trophic levels in food webs. Although the $\delta^{13}C$ and $\delta^{15}N$ biofractionation patterns hold for many macrofaunal organisms in different taxonomic groups and trophic levels, they do not appear to be generalizable for bacteria (Coffin et al. 1989; Fogel et al. 1989; Macko and Estep 1994; Canfield and Teske 1996; Habicht and Canfield 1996; Hullar et al. 1996).

Food web studies make use of both source and process information, both of which can be gained from stable isotope ratios (Peterson and Fry 1987), especially when multiple isotopes are used together. "Source information" often refers to the primary producers in the system under study, whereas "process information" relates to trophic levels and food web structure. It is essential that organic matter sources have distinct isotope ratios if isotopes are to be used to estimate the relative amounts that each resource contributes to different consumers in a food web (Haines and Montague 1979; Sackett 1989; Michener and Schell 1994; Macko and Ostrom 1994). Thus, stable isotope methods have very limited applications for studying food web interactions in ecosystems that have small variation in the isotope characteristics of different resources.

CASE STUDIES

Waterfowl as Nutrient Vectors between Agricultural Systems and Wetlands

Waterfowl populations have boomed throughout North America in recent decades due to a combination of reductions in hunting pressure, agricultural subsidies during migration and overwintering, and wetland restoration in waterfowl breeding grounds (Ankney 1996; Jefferies et al., chap. 18 in this volume). During migrations and overwintering periods, massive numbers of birds congregate on managed wetlands in areas where there has been extensive loss of natural habitat. At these sites, many waterfowl engage in daily commuting between agricultural fields, where they feed on crops, and wetland systems, where they roost. Roosting aggregations result in substantial deposition of guano, which is rich in nitrogen and phosphorus derived from consumption of agricultural crops (Bildstein et al. 1992; Post et al. 1998). Deposition of nutrients leads to eutrophication of lakes and wetlands (Hutchinson 1950; Brinkhurst and Walsh 1967; Manny et al. 1994; Kitchell et al. 1999) and can be associated with avian cholera and botulism outbreaks (Hartung 1971; Wobeser 1981). Wildlife managers are now concerned that many waterfowl populations are damaging managed wetlands and altering wetland food webs through eutrophication and destruction of vegetation.

Kitchell et al. (1999) evaluated the role of waterfowl as nutrient vectors from agricultural fields to wetland roosting areas at the Bosque del Apache wildlife refuge (BDA) in central New Mexico. Snow goose (*Chen* spp.) densities at the BDA have increased from about 5,000 individuals in 1967 to 60,000 in 1985 and are currently maintained at about 35,000–55,000 (Kitchell et al. 1999). Sandhill crane (*Grus canadensis*) populations have shown similar population trajectories. These birds commute daily between agricultural fields, where they feed extensively on corn and alfalfa, and wetlands, to which they return to roost at night. Post et al. (1998) estimated that nutrient input via guano deposition derived from consumption of agricultural crops accounted for as much as 40% of the nitrogen and 75% of the phosphorus loaded to the primary roosting wetland at the BDA. Kitchell et al. (1999) used nitrogen stable isotope distributions (δ^{15}N) to trace guano-derived N in the food webs of wetlands where snow geese and sandhill cranes roost. The N in guano that was produced from consumption of corn and alfalfa had a δ^{15}N signature that was distinctly depleted in ^{15}N (i.e., low δ^{15}N) compared with the nitrogen loaded to the BDA wetlands from the Rio Grande, which receives substantial inputs of N from sewage. Kitchell et al. (1999) demonstrated that guano-derived N was the major source of N in the food webs of wetland roosting ponds. This important source of nitrogen (in addition to P) has led to eutrophic conditions in wetland ponds where birds roost and is propagated up through the aquatic food webs of the roosting wetlands of the BDA refuge system.

Anadromous Fishes

Anadromous fishes represent important ecological subsidies from marine systems to the freshwater lotic and lentic habitats where they spawn and live as juveniles. Pacific salmon (*Oncorhynchus* spp.), for instance, accumulate more than 95% of their body mass in the marine environment before they migrate back to freshwater rivers, streams, and lakes, where they spawn and die. The nutrients and organic matter in the adult carcasses represent a major subsidy to the freshwater and riparian systems (Willson et al. 1998; Willson et al., chap. 19 in this volume; Naiman et al. 2002). Early evidence for fish-based subsidies was based on mass balance studies that compared inputs from spawning carcasses with other known inputs to aquatic ecosystems (Juday et al. 1932; Donaldson 1967; Krohkin 1967; Durbin et al. 1979). More recently, aquatic ecologists have capitalized on the fact that marine-derived nitrogen has a substantially enriched isotopic composition (i.e., high δ^{15}N) compared with atmospherically derived nitrogen to determine the importance of salmon carcasses to freshwater and riparian food webs (Mathisen et al. 1988; Kline et al. 1990, 1993). Salmon carcasses are consumed by both aquatic and terrestrial predators and scavengers, and they decompose and release inorganic N and P that are assimilated by aquatic and terrestrial primary producers (Ben-David, Hanley, and Schell 1998; Willson et al. 1998; Naiman et al. 2002). Because of the characteristic ^{15}N-enriched composition of marine-derived nitrogen, this nutrient subsidy is easily traced through the aquatic (Kline et al. 1990, 1993; Bilby et al. 1996) and riparian food webs (Ben-David, Hanley, and Schell 1998; Willson et al., chap. 19 in this volume) in areas with dense salmon populations.

A retrospective study of the N isotopes in grizzly bear (*Ursus arctos*) skeletons from throughout the Columbia River drainage in the Pacific Northwest (U.S.A.) demonstrated that salmon represented from 33% to 90% of the diets of bears in this region prior to 1931 (Hilderbrand et al. 1996). Thus, natural stable isotope abundances were used to demonstrate the historical importance of the salmon subsidy to terrestrial ecosystems in this vast ecosystem, and provided an additional and powerful impetus for salmon conservation in this region.

Sewage Inputs to Food Webs

Sewage can act as an anthropogenic subsidy of nutrients and organic carbon to aquatic food webs, even when it is dumped at deep oceanic sites (Van Dover et al. 1992). The isotope ratios of municipal and industrial sewage are distinct from those of naturally occurring organic matter. Waste materials dumped along the coast of New Jersey (Van Dover et al. 1992), California (Rau et al. 1981), and the Baltic Sea (Hansson et al. 1997) accumulate in food webs near the dump sites. Given the evidence that sewage enters these food webs, some questions of interest are how this new source of nutrients and carbon changes the structure of the food webs, and the extent to which transfer between areas by migratory organisms facilitates a corresponding transfer of this waste to otherwise unaffected areas.

Hansson et al. (1997) identified older herring that migrated between polluted and unpolluted sites in the Baltic Sea on the basis of $\delta^{15}N$. The fish from polluted sites were enriched in ^{15}N relative to those from unpolluted sites because the enriched particulate organic matter at the polluted sites was incorporated into all trophic levels. Sewage has enriched $\delta^{15}N$ because denitrification results in preferential loss of ^{14}N from the aquatic N pool.

Rau et al. (1981) found significant differences ($p = .10$) in the $\delta^{13}C$ and $\delta^{15}N$ of Dover sole and ridgeback prawns between a polluted and an unpolluted site off southern California. The average differences in $\delta^{13}C$ and $\delta^{15}N$ between marine organic and sewage-derived detritus were +2‰ and +8‰, respectively. Rau et al. documented isotopic enrichment in tissues of sole and prawns, respectively, of 1.4‰ and 0.7‰ in $\delta^{13}C$, and 4.7‰ and 4.6‰ in $\delta^{15}N$, in unpolluted sites relative to the polluted sites. Because both sole and prawns consume benthic invertebrates, Rau et al. inferred that the difference in the stable isotope ratios resulted from incorporation of sewage-derived detritus into the food web. Casual inspection of the isotopic shifts between the polluted and unpolluted sites leads to an approximate isotopic mass balance of 50% of the diets of sole and prawns being directly or indirectly derived from sewage inputs.

Sewage inputs have also been seen to be incorporated into benthic food webs at a dump site off the coast of New Jersey (Van Dover et al. 1992). In this case there was a large difference between the δ^{34}S of sewage-derived organic material (SDOM) and that of plankton-derived organic material (PDOM) (differences in δ^{34}S were 11.5‰ to 18.7‰). This clear distinction between sulfur isotope signals allowed Van Dover et al. to construct a two-end-member mixing model and estimate the amount of SDOM incorporated into the diets of organisms at this 2,500 m deep site. Of ten species sampled, two had significantly different stable isotope ratios between a reference site and the dump site. Both species were organisms characterized as surface-deposit feeders, one of which was a sea urchin estimated to obtain 35% of its sulfur from SDOM.

Nitrogen stable isotope ratios are usually used to infer the trophic position of organisms within food webs because there is a large shift in δ^{15}N between trophic levels. Because nitrogen is involved in a relatively complex biogeochemical cycle that involves multiple transformations with the potential to fractionate N isotope ratios, using N isotopes to infer source information involves many assumptions (Robinson 2001). Cabana and Rasmussen (1996) showed that before inferring an organism's trophic level based on its δ^{15}N, it is imperative to know the δ^{15}N of the sources at the base of the food web. They examined filter-feeding mussels from different lakes and measured the δ^{15}N of their tissues as a means to estimate the sources of nitrogen input to the ecosystems. Mussels are particle feeders that, because of their relatively large body size and slow tissue turnover rates, integrate plankton isotope characteristics across time. Cabana and Rasmussen found that δ^{15}N in mussel tissue ranged from 1.2‰ to 9.0‰, and that mussel tissue δ^{15}N was positively correlated with human density around the lakes. This enriched N isotope distribution reflected that sewage δ^{15}N was about 15‰ and that other sources had a corresponding value of –5‰ to 5‰. Because the diets of mussels are well characterized (i.e., phytoplankton), Cabana and Rasmussen suggest that these long-lived organisms be used to correct temporally variable primary producer signatures before assigning trophic positions based on δ^{15}N to other organisms in the food web with diets that are less well known.

Hydrothermal Vents

Hydrothermal vents have anomalously high levels of biological production compared with other areas of the deep sea. The mechanisms that support the large vent biomass are poorly understood. Conventional biological techniques are difficult and expensive—if not impossible—to perform at vent sites. Chemical analyses have been key to learning about trophic interactions in these food webs.

Rau (1981) noted that hydrothermal vent worms, clams, and crabs had low $\delta^{15}N$ values that ranged from 1.8‰ to 9.8‰, compared with plankton values of 5‰ to 10‰ and a deep-sea sediment value of 5‰ to 13‰ for organic N. Assuming that hydrothermal vent organisms show a $\delta^{15}N$ trophic level shift of the same general magnitude as other macrofaunal organisms, the animals' signatures were not compatible with phototropically derived material (i.e., produced by plants) being the sole contributor to their diets.

Pond et al. (Pond, Dixon et al. 1997; Pond, Segonzac et al. 1997) used $\delta^{13}C$ and lipid component analyses to study the feeding mechanisms and life history strategies of three species of shrimps inhabiting hydrothermal vents. Two shrimp species, *Rimicaris exoculata* and *Alvinocaris markensis*, collected at the TAG and Snake Pit sites had different fatty acid compositions and different $\delta^{13}C$ ratios associated with those fatty acids (Pond, Dixon et al. 1997). *R. exoculata* had high levels of bacteria-associated monounsaturated fatty acids, while *A. markensis* did not. Furthermore, *A. markensis* had a large fraction of phototrophically associated fatty acids that were less abundant in *R. exoculata*. This information, coupled with $\delta^{13}C$ values of −17‰ to −28‰ for shallow ocean phytoplanktonic material, was used in conjunction with behavioral and morphological observations to characterize *R. exoculata* as dependent on endosymbiotic bacteria and *A. markensis* as detritivorous and at least partially dependent on phototrophically derived organic material (Pond, Dixon et al. 1997).

For the shrimp *Mirocaris fortunata*, lipid composition, lipid $\delta^{13}C$, and pigmentation varied with size and sex (Pond, Segonzac et al. 1997). Pond et al. concluded that *M. fortunata* probably did not rely on sulfur-oxidizing bacteria as a food source. Other bacteria, perhaps associated with vent mussels, and other mussel-derived material were probably important to the diet. Adult shrimp and small shrimp with low pigment levels had $\delta^{13}C$ levels indicative of thiotrophic origins, but overall, *M. fortunata* was considered to be an opportunistic feeder. Its life history strategy may be characterized by different foraging behaviors; the $\delta^{13}C$ of some shrimp fatty acids is indicative of time spent higher in the water column, and dependence on carbon fixed in the marine surface layer, as larvae. Adult shrimp had bimodal $\delta^{13}C$ distributions that could be indicative of different nutritional resources with isotopically distinct characteristics, patchy resources, territorial behavior, feeding preferences, or differences in larval behavior.

FUTURE PROSPECTS

The increased accessibility and affordability of stable isotope analyses have added a powerful tool to the set of techniques used to study food webs (see fig. 3.1; Peterson and Fry 1987; Lajtha and Marshall 1994; Lajtha and Michener 1994). However, most applications of stable isotope techniques have treated data in a qualitative sense, and studies that use the full potential of stable isotope techniques are still exceptional. Opportunities are developing for ecologists to use stable isotopes in more informative ways to better elucidate food web interactions, the importance of material transport by organisms, and how this transport is related to the importance of spatial subsidies. We highlight some of these applications here.

Experimental System-Level Additions of Isotopic Tracers

We have stressed that the usefulness of stable isotopes to trace material flow between spatially segregated components of ecosystems requires that the different subsystems have distinct isotopic signatures. In many cases, spatially separated or functionally different ecosystem components do indeed have distinct natural isotope distributions. However, it is also possible to alter the isotopic composition of certain components of ecosystems experimentally to evaluate flows of materials and energy through food webs. System-level additions of appropriate isotopic tracers can be used either to establish an isotopically distinct component of a food web (e.g., a possible subsidy) or to amplify a natural contrast in isotopic signatures. We briefly describe two examples of whole-system isotopic tracer additions below.

Dissolved organic carbon (DOC) is a major source of carbon to stream ecosystems (Fisher and Likens 1973; Mann and Wetzel 1995). Most DOC in streams originates from terrestrial ecosystems and is incorporated into bacterial production. Some of this production is consumed by organisms at higher trophic levels in food webs. However, the relative importance of bacterial production to the feeding and growth of different consumers in stream food webs is poorly understood. Hall (1995) used sodium acetate enriched in ^{13}C to trace the flows of DOC through the food web of Cold Spring at the Coweeta Hydrologic Laboratory in North Carolina. Hall increased the δ^{13}C of DOC from −26‰ to +100‰ (for about U.S.$130). By artificially altering the ^{13}C signature of DOC, Hall was able to distinguish between inputs of carbon to this system from DOC and those from allochthonous detrital inputs. Addition of the stable isotopic tracer showed that DOC-derived bacterial carbon was not used uniformly by all components of the invertebrate food web, and that in some cases, it was used by organisms that had not been previously recognized as important consumers of bacterial production. In a similar study, Kling (1994) used stable isotope additions to track C and N flows simultaneously through the food webs of an Arctic lake and evaluate the importance of terrestrial C inputs to pelagic food webs.

Hershey et al. (1993) used a ^{15}N tracer addition in the Kuparuk River, Alaska, to resolve the apparent paradox represented by the fact that most lotic invertebrates drift substantially downstream over the course of their lives, yet upstream reaches maintain viable populations of these drifting species. One possible solution to this paradox is that adults fly upstream before spawning, and this upstream dispersal compensates for the net downstream drift observed in the larval aquatic forms (Hershey et al. 1993). Nitrogen fertilizer enriched in ^{15}N was dripped into the Kuparuk River as part of a long-term ecosystem fertilization experiment (Peterson et al. 1993). This system-scale ^{15}N tracer addition resulted in *Baetis* mayflies that were enriched from a δ^{15}N of about 25‰ to almost 250‰. Hershey et al. (1993) were therefore able to track labeled adult *Baetis* and determine the distance that they flew upstream before ovipositing. The estimated upstream flight distance was 1.6–1.9 km, which compared exceedingly well with the estimated 2.1 km downstream drift by larvae.

The two studies described above are examples in which isotopic tracers have been used experimentally to alter the natural distribution of stable isotopes to evaluate the importance of material flow in food webs and ecosystems. Similar techniques using rare element experimental tracers (e.g., rubidium) in agroecosystems have been developed to study colonization and dispersal of terrestrial arthropods (Corbett and Rosenheim 1996a). These other chemical tracers can be used in much the same way that experimental stable isotopic tracers are used to track material transport by organisms. Experimental tracer studies have an auspicious future in food web and ecosystem ecology because they can be used effectively at the spatial and temporal scales relevant to the transport of energy and materials by mobile organisms.

Getting Beyond the $\delta^{15}N$ versus $\delta^{13}C$ Scatterplot

Stable isotope information is more powerful when multiple isotopes are used simultaneously (Peterson et al. 1985; Peterson, Howarth, and Garritt 1986; Peterson and Howarth 1987). Data from multiple isotopes are commonly represented as dual isotope plots (Peterson et al. 1985; Peterson, Howarth, and Garritt 1986; Peterson and Fry 1987; Peterson and Howarth 1987; Sullivan and Moncreiff 1990), which can illustrate qualitatively the flow of nutrients from sources to consumers. There have been few attempts to use the information from such plots to obtain quantitative estimates of feeding relationships (but see Harrigan et al. 1989; Ostrom et al. 1997). Most previous mixing models using stable isotope data have been limited primarily to considering one isotope ratio at a time, even when data from multiple isotope ratios are available (Van Dover et al. 1992; Lajtha and Michener 1994). Furthermore, earlier mixing models have limited the number of sources contributing to the food web or consumer by dividing the sources into no more than two groups, such as polluted or unpolluted (Van Dover et al. 1992) or autochthonous versus allochthonous (Junger and Planas 1994). Theoretically, the number of food web sources that can be differentiated using isotope data is one more than the number of independent isotopes used in the study. For example, to estimate the relative contributions of sources X_1 and X_2 to consumer Y_1, data from a single isotope distribution can be used to solve the following equations to determine the fractional inputs α_1 and α_2:

$$\alpha_1 = \frac{Y_1 - X_2 - f}{X_1 - X_2},$$

$$\alpha_2 = 1 - \alpha_1,$$

where f is the isotopic enrichment due to dietary transfer. For carbon, f is approximately 0.7‰, for nitrogen it is between 3.0 and 3.5, and for sulfur it is assumed to be negligible.

Treseder et al. (1995) used this simple model to estimate that ants (*Philadris* spp.) translocated about 29% of the nitrogen and 39% of the carbon used by a Malaysian forest epiphyte (*Dischidia major*). *Philadris* forage and scavenge organic material and insects away from *Dischidia major* and deposit C and N in its leaves, where young ants are raised. Ant foraging activities can therefore be viewed as an important subsidy to the productivity of *Dischidia major*. This study serves as an excellent example of using natural stable isotope distributions to track the contributions of nutrients and organic matter from the foraging activities of a mobile consumer to the autotrophic production in a simple ecological system.

In many cases, splitting food web sources into two pools is an overly simplistic way of partitioning the trophic base of ecosystems, particularly in systems that have multiple source inputs. However, direct calculation of the relative contributions from more sources than the number of isotopes (plus one) is mathematically impossible. This issue is generally referred to as an underdetermined system of equations (i.e., more unknown parameters than data points).

Two companion models, SOURCE and STEP, have been proposed to use information from multiple isotope ratios simultaneously to obtain estimates of diet composition and organic matter sources to food webs, even when the number of potential foods or organic matter sources is larger than the number of stable isotope ratios that have been measured (Lubetkin 1997; S. C. Lubetkin and C. A. Simenstadt, unpublished data). These linear mixing models were developed to identify the dominant sources of organic matter entering a food web, determine the sequence of trophic steps, and evaluate specific predator-prey linkages. In SOURCE, a consumer's stable isotope ratios are modeled as a weighted average of the organic matter sources its tissues are derived from, plus the cumulative effects of biofractionation. With multiple stable isotope ratios, a system of linear equations is formed, and the relative contribution of each source to the consumer and its trophic level relative to the sources can be estimated through iteration (Lubetkin 1997; S. C. Lubetkin and C. A. Simenstadt, unpublished data). Subsets of the total number of sources are used to estimate source composition based on the subset alone, and these subset mixtures are estimated for all possible combinations of sources. Source composition for the entire food web is then estimated as the center of mass of all biologically feasible subset estimates. This application of linear programming is referred to as a cornerpoint technique (Hillier and Liebermann 1990). STEP works similarly, but in this second model, the consumer's signature is modeled as a weighted average of all its diet components' stable isotope ratio signatures plus a single biofractionation resulting from the consumer's own metabolism. Again, a set of linear equations is formed using the information from multiple stable isotopes simultaneously, and the consumer's diet may be estimated (Lubetkin 1997; S. C. Lubetkin and C. A. Simenstadt, unpublished data.). Thus, SOURCE directly assesses the primary producer origins of consumer tissues and the consumers' trophic levels, and STEP traces more detailed pathways down the food web from each consumer to its diet.

As an example of the application of these models, we used data from Sullivan and Moncreiff (1990) with SOURCE and STEP to generate diet compositions and estimates of the proportions of consumer diets that were derived from different organic matter sources. Sullivan and Moncreiff (1990) used $\delta^{13}C$, $\delta^{15}N$, and $\delta^{34}S$ to analyze the relative inputs of *Spartina, Juncus*, edaphic algae, and phytoplankton to consumers in Graveline Bay Marsh, Mississippi. They concluded that edaphic algae was a large contributor to the diets of consumers in the marsh, either directly or indirectly, and that edaphic algae was an important primary producer deserving of more consideration in other marsh studies. Using dual isotope plots, Sullivan and Moncreiff showed that many of the fifty-six consumer species had isotopic signatures that were intermediate between phytoplankton (as approximated by the zooplankton stable isotope ratios) and edaphic algae isotopic signatures. *Juncus* and *Spartina* were relative outliers on the dual isotope plots, and were therefore assumed to make little contribution to the marsh food web (Sullivan and Moncreiff 1990).

To simplify this example, we chose twelve consumers from the Sullivan and Moncreiff (1990) data set and used SOURCE and STEP to make quantitative estimates of the relative importance of each primary producer to those consumers and to estimate their diets and trophic levels. This analysis confirmed the importance of edaphic algae in this food web, as it contributed strongly to the stable isotopic signatures of several consumers, particularly the bryozoan (*Bugula* spp.), which was estimated to have received 87% of its organic matter from edaphic algae (tables 3.1, 3.2). Phytoplankton was the other main primary producer source in the food web. However, in contrast to Sullivan and Moncreiff (1990), we did find substantial contributions of *Spartina* and *Juncus* (between 9% and 30%) to six of the twelve consumers' diets (table 3.2, fig. 3.2). Our analysis of this data suggests that vascular plants should be considered an important energy source to certain consumers in a marsh food web that is dominated by phytoplankton and edaphic algae primary production.

Like simple isotope mixing models, these multiple isotope models require that the inputs to a food web have distinct isotopic signatures. This criterion is more likely to be met if those inputs come from habitats regulated by different physical or biogeochemical processes. However, sources of organic matter and nutrients are more likely to be distinct in multiple isotope space than in single isotope space.

Table 3.1 SOURCE model analysis of the food web from Graveline Bay Marsh, Mississippi, with estimates of consumer trophic level and the estimated contribution of each organic matter source to consumer isotope composition

Consumer	Spartina	Juncus	Edaphic algae	Phytoplankton[a]	Trophic level[b]
I. recurvum	0.4	1.0	51.0	47.6	1.49
Zooplankton	0	0	0	100.0	2.00
G. demissa	0.7	13.8	55.3	30.2	1.64
Neanthes spp.	15.9	19.6	27.7	36.8	2.05
C. sapidus	16.1	6.6	60.4	16.9	2.01
Bugula spp.	0	0.6	89.8	11.0	2.00
F. majalis	32.0	10.3	31.5	26.2	2.46
S. ocellatus	14.0	11.7	44.6	29.8	2.64
L. brevis	0	4.3	79.6	17.0	2.79
L. xanthurus	5.6	14.8	34.7	44.9	3.06
B. chrysura	2.9	3.6	84.3	9.2	3.11
A. nasuta	0	0.8	91.1	9.1	3.18

NOTE: The rows may not add to 100% due to rounding error.

[a] The phytoplankton stable isotope ratios were estimated by subtracting the assumed effects of a single trophic level shift from the zooplankton stable isotope ratios (0.7‰ for $\delta^{13}C$ and 3.0‰ for $\delta^{15}N$).

[b] Primary producers are a trophic level of 1.

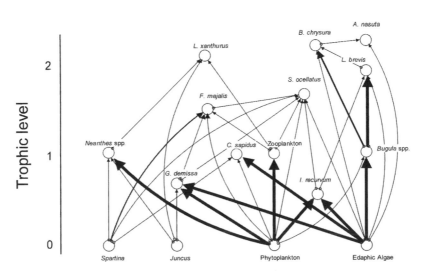

Figure 3.2 A food web diagram for the subset of consumers from Graveline Bay Marsh, Mississippi. Thin lines indicate that the consumer is estimated to derive 8% to 25% of its diet from a source; lines of medium thickness indicate contributions of 25% to 45% of the diet; thick lines represent more than 45% of the diet. (From Sullivan and Moncreiff 1990.)

Table 3.2 STEP model results showing the estimated diets of several consumers from Graveline Bay Marsh, Mississippi

Consumer	Primary producers				Potential foods				Consumers							
	Sp.	Junc.	EA	Phyt.	I. r.	Zoo.	G. d.	N. sp	C. s.	B. sp	F. m.	S. o.	L. a.	L. x.	B. c.	A. n.
I. recurvum	0.4	1.0	51.0	47.6	0	0	0	0	0	0	0	0	0	0	0	0
Zooplankton	0	0	0	100	0	0	0	0	0	0	0	0	0	0	0	0
G. demissa	0.7	13.8	55.3	30.2	0	0	0	0	0	0	0	0	0	0	0	0
Neanthes spp.	18.2	9.3	2.6	54.1	2.3	2.1	1.9	1.6	1.4	1.2	1.2	1.0	0.8	0.9	0.7	0.7
C. sapidus	17.1	2.3	50.0	24.1	1.5	0.8	0.9	0.6	0.7	0.7	0.5	0.4	0.4	0.3	0	0
Bugula spp.	0	0.1	87.5	9.4	2.7	2.3	0.3	0	0	0.5	0	0	0	0	0	0
F. majalis	31.1	2.5	3.6	10.4	6.1	14.9	12.7	3.7	2.2	2.5	2.8	2.7	1.3	1.6	1.0	0.9
S. ocellatus	10.5	2.9	8.4	10.2	8.2	9.9	20.9	5.3	3.0	3.6	8.7	2.2	1.7	1.7	1.5	1.3
L. brevis	0	0.2	20.1	5.3	13.7	3.2	3.1	0.7	0.1	36.6	0.4	0.4	3.0	1.6	3.6	4.3
L. xanthurus	0.6	9.8	4.4	5.9	3.6	21.0	5.1	9.0	1.7	6.3	5.5	5.9	4.2	9.6	3.5	4.1
B. chrysura	0.8	0	18.6	0.2	0.7	0.2	0.5	1.2	5.7	37.2	3.7	3.2	8.1	0.9	10.4	8.6
A. nasuta	0	0	15.3	0	0.5	0	0	0	0.2	54.7	0	0	6.7	0	8.1	15.3

NOTE: The rows may not add to 100% due to rounding error. Estimates less than about 8% indicate mathematically possible combinations of potential foods that may not be biologically likely (Lubetkin 1997).

Abbreviations: Sp., Spartina; Junc., Juncus; EA, edaphic algae; Phyt., phytoplankton; I. r., I. recurvum; Zoo., zooplankton; G. d., G. demissa; N. sp, Neanthes spp.; C. s., C. sapidus; B. sp, Bugula spp.; F. m., F. majalis; S. o., S. ocellatus; L. b., L. brevis; L. x., L. xanthurus; B. c., B. chrysura; A. n., A. nasuta.

CONCLUSION

This chapter is not an extensive review of stable isotope research in ecology. For more detailed discussion of certain techniques and applications of stable isotopes, readers should consult some of the excellent reviews of this subject (e.g., Peterson and Fry 1987; Lajtha and Michener 1994). Although stable isotope research has recently become more commonplace in ecological studies, these data are still poorly integrated into quantitative studies of food webs. Data produced through analysis of stable isotope distributions by multiple isotope mixing models (Lubetkin 1997; Ostrom et al. 1997) are directly applicable to food web and ecosystem models such as Ecosim and Ecospace (Christensen and Pauly 1992; Walters et al. 1997).

Analyses of stable isotope distributions are not a substitute for direct observations of ecological interactions and habitat coupling in food webs. However, stable isotope distributions should be treated as complementary data that can be combined with more conventional data to improve our ability to understand complex ecological processes (Hilborn and Mangel 1997). Stable isotopes also provide a common currency that can be incorporated by both ecosystem ecology and population/community ecology— two subdisciplines of ecology that have been poorly integrated due, in part, to the lack of a common useful currency with which to study ecological processes. One goal of this book is to make progress toward this integration, and stable isotope analyses are a key tool for this purpose.

ACKNOWLEDGMENTS

We thank Gary Polis, Mary Power, and Gary Huxel for the invitation to contribute to this book. Comments by Gary Polis and two anonymous reviewers greatly improved the manuscript. We also acknowledge the inspiration Gary Polis provided to us through his science and friendship. Our research is supported by the National Science Foundation and the Andrew W. Mellon Foundation.

Chapter 4

Cross-Habitat Transport of Nutrients by Omnivorous Fish along a Productivity Gradient: Integrating Watersheds and Reservoir Food Webs

Michael J. Vanni and Jenifer L. Headworth

The exchange of materials across habitats or ecosystems may have profound consequences for the dynamics of food webs (Polis and Hurd 1996a, 1996b; Polis et al. 1996). Organisms can mediate this exchange in several ways. Animals as diverse as plankton, bison, and salmon move and redistribute resources among habitats at a variety of spatiotemporal scales (Kline et al. 1990; Bilby et al. 1996; Vanni 1996; Flecker 1996; Knapp et al. 1999). Animals generally consume particulate nutrients, then egest nutrients in particulate form (feces) and excrete nutrients in dissolved form (urine). Consumption and subsequent excretion of nutrients represents a transformation of potentially limiting resources from unavailable to available form. Thus, animals can mediate directional flow of nutrients from one habitat to another or promote heterogeneity in nutrient availability (Kline et al. 1990; Schindler et al. 1993; Bilby et al. 1996; Vanni 1996; Knapp et al. 1999; Strayer et al. 1999). In lakes, fish and invertebrates can transport nutrients between relatively discrete habitats—for example, between littoral and pelagic areas. This transport can be an important source of nutrients for recipient primary producers (Brabrand et al. 1990; D. E. Schindler et al. 1993, 1996; Vanni 1996; Persson 1997; Schaus et al. 1997; Strayer et al. 1999).

The gizzard shad (*Dorosoma cepedianum*), a common omnivorous fish in fresh waters of eastern North America, can be important in transporting nutrients from sediments to water via excretion (Vanni 1996; Schaus et al. 1997). Gizzard shad larvae are obligate planktivores, while juveniles and adults are omnivores, consuming zooplankton, algae, and sediment detritus. Gizzard shad of all age classes prefer to eat zooplankton if large-bodied, energetically profitable taxa (such as *Daphnia*) are available (Mundahl 1988; Shepherd and Mills 1996). However, when zooplanktivory is not a profitable option, gizzard shad can ingest sediment detritus, a food source that is low in quality but essentially unlimited in abundance, at least over short time scales (Mundahl and Wissing 1987, 1988). Gizzard shad are especially abundant in reservoirs (Stein et al. 1995, 1996). Reservoirs have relatively large watersheds compared with natural lakes and therefore receive large amounts of sediments (Thornton 1990). Furthermore, reservoir zooplankton assemblages are often dominated by small species that are unprofitable food items. Perhaps as a consequence of these two factors, gizzard shad feed heavily on sediment detritus in reservoir ecosystems (e.g., Mundahl and Wissing 1987, 1988; Yako et al. 1996). Benthic invertebrates are rarely found in gizzard shad guts, and these fish apparently assimilate detritus itself as well as microbial flora associated with sediments (Smoot 1999). The detritus assimilated by gizzard shad probably derives from allochthonous sources as well as from phytoplankton that die and sink to the lake bottom. However, the C : N ratio of reservoir sediments suggests that terrestrially derived material contributes substantially to sediment composition (Smoot 1999; M. J. Vanni et al., unpublished data). By excreting nutrients consumed from watershed-derived detritus, gizzard shad may provide a source of available nutrients for phytoplankton and thus link terrestrial landscapes and lake food webs (fig. 4.1).

Variation in diet has large implications for the role of gizzard shad in nutrient flux (fig. 4.2). Their excretion of nutrients previously bound in sediment detritus represents a "new" source of nutrients (Vanni 2002). We refer to this flux through shad as nutrient translocation or transport (Kitchell et al. 1979; Vanni 1996, 2002) because it represents transfer from one habitat to another (benthic to pelagic). In contrast, when gizzard shad excrete nutrients consumed from plankton, they are merely reprocessing nutrients already present in the pelagic habitat, a process we refer to as nutrient recycling (fig. 4.2). Nutrient recycling can be important for sustenance of a given level of phytoplankton primary production, but cannot stimulate "new" primary production as can nutrient transport (Dugdale

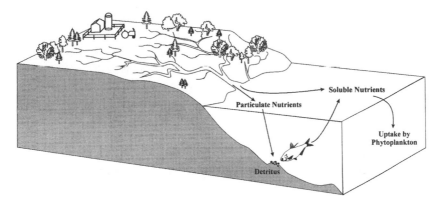

Figure 4.1 Sediment-feeding gizzard shad can serve as a link between watersheds and reservoir food webs and thereby provide a nutrient subsidy for phytoplankton. Nutrients are delivered from watersheds in particulate form (e.g., P attached to silt, clay, sediment; solid arrowheads) or in dissolved form (e.g., as phosphate; open arrowheads). Some particulate nutrients entering the reservoir from the watershed sink to the bottom and become sediment detritus; phytoplankton produced in the water column also sink and enter the sediment detrital pool. Gizzard shad consume sediment detrital nutrients and excrete nutrients in dissolved inorganic form, thereby rendering them available to phytoplankton, a process we refer to as nutrient transport.

and Goering 1967; Caraco et al. 1992). Nutrient transport by gizzard shad can be an important source of new nutrients that can stimulate phytoplankton biomass (Schaus et al. 1997; Schaus and Vanni 2000).

The role played by a particular species in transporting nutrients depends on its abundance. Gizzard shad abundance increases with lake productivity in both natural lakes (Bachmann et al. 1996) and reservoirs (DiCenzo et al. 1996; Michaletz 1997; Bremigan and Stein 2001). This observation suggests that the flux of nutrients through gizzard shad also increases with productivity. However, productive lakes also receive greater inputs of available nutrients from other sources (Carpenter, Caraco et al. 1998). Thus, the relative importance of nutrient transport by gizzard shad (i.e., the proportion of total nutrient flux attributed to shad) may increase or decrease with lake productivity.

In this chapter, we use a modeling approach to explore the hypothesis that the relative importance of nutrient transport by gizzard shad increases with lake productivity, focusing on reservoirs of the eastern United States. By linking watersheds and reservoir food webs, we hope to understand better how reservoirs are affected by two diverse nutrient subsidies. We do not purport to provide a rigorous test of this hypothesis with a model. Rather, our goal is to allow a more quantitative basis for alternative testable

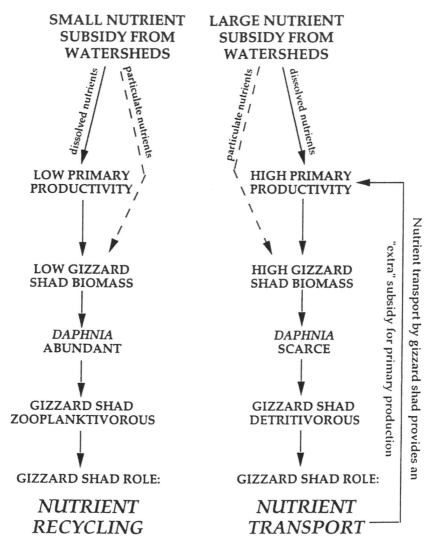

Figure 4.2 Flow chart representing a hypothesized scheme for how nutrient subsidies from watersheds determine the role of gizzard shad in nutrient flux. When nutrient inputs from watersheds are small, primary productivity is low, which ultimately leads to low gizzard shad biomass. Because shad biomass is low, large zooplankton such as *Daphnia* can persist, and shad prefer to feed on them. In so doing, zooplanktivorous shad recycle nutrients within the water column. In contrast, large dissolved nutrient inputs lead to high rates of primary production. In addition, particulate nutrient inputs may directly subsidize the gizzard shad population by providing detritus. Hence gizzard shad biomass is high, leading to a depletion of *Daphnia*. Under these conditions, gizzard shad are mostly detritivorous, and hence transport nutrients from sediments to the water column. This transport provides a further subsidy for phytoplankton primary production.

hypotheses; specifically, whether the importance of nutrient transport by shad increases or decreases with lake productivity.

THE MODEL

Our model explicitly considers how diet (detritivory vs. zooplanktivory), population size structure, and biomass influence nutrient transport and recycling rates along a productivity gradient (fig. 4.3). We gauge the importance of nutrient transport by gizzard shad by quantifying three parameters along a productivity gradient: (1) per capita nutrient transport rates, which can be viewed as an index of per capita interaction strength (Paine 1992; Power et al. 1996); (2) nutrient transport and recycling rates by entire gizzard shad populations (i.e., total flux at the ecosystem scale); and (3) the proportion of phytoplankton production supported by nutrient transport. In most lakes, phosphorus (P) is the limiting nutrient, and the concentration of total P in the water column (hereafter, TP) is highly correlated with algal productivity and biomass (e.g., V. H. Smith 1979, 1998). Thus, TP is used as a surrogate for productivity in many studies (e.g., Hansson 1992; Mazumder 1994) and in our model.

Per Capita Nutrient Recycling and Transport Rates

The mass of phosphorus recycled or transported by an individual gizzard shad (per capita rate) was estimated using a mass balance approach (Kraft 1992; D. E. Schindler et al. 1993, 1996; Vanni 1996). P excretion (X_P, mass of P released in dissolved inorganic form via urine production) for an individual fish was estimated as $X_P = I_P - F_P - G_P$, where I_P, F_P, and G_P are the amounts of P ingested, egested (released as feces), and used for growth, respectively (Kraft 1992; D. E. Schindler et al. 1996). We separately estimated P recycling (i.e., excretion by planktivorous fish, X_{PZ}) and P transport (i.e., excretion by detritivorous, sediment-feeding fish, X_{PS}).

P Recycling

We assumed that a feedback loop exists among gizzard shad biomass, *Daphnia* abundance, and shad feeding behavior. When shad are scarce, *Daphnia* persist, and are the preferred food of shad (see fig. 4.2). When shad are abundant, they depress *Daphnia* abundance and switch to detritus feeding. *Daphnia* abundance generally decreases to near zero when fish biomass is above a threshold of about 20–50 kg ha^{-1} (Post and McQueen

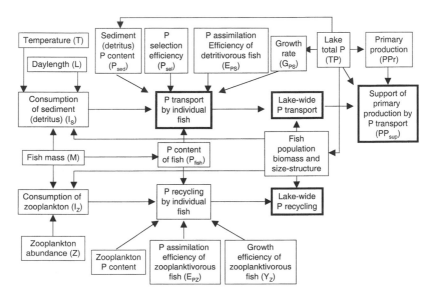

Figure 4.3 Schematic diagram of the model used to estimate nutrient transport and recycling by gizzard shad. Arrows indicate the influence of one factor on another. Parameters encased in heavy boxes are those used to evaluate the importance of nutrient cycling by gizzard shad along a gradient of productivity, and for which output is shown in fig. 4.4.

1987; Mills et al. 1987; Shepherd and Mills 1996; Schaus 1998). Mills et al. (1987) generated the following relationship between *Daphnia* and yellow perch abundance in eutrophic Oneida Lake:

$$Z = 303.8e^{(-0.1B)}, \tag{4.1}$$

where Z is *Daphnia* abundance (l^{-1}) and B is yellow perch biomass (kg ha^{-1}). Shepherd and Mills (1996) also found that a gizzard shad biomass of about 55 kg ha^{-1} could consume all *Daphnia* production in Oneida Lake, and that shad switch from *Daphnia* and other large zooplankton to detritus and algae when *Daphnia* decline. In Acton Lake, a eutrophic reservoir in Ohio, *Daphnia* are present only when shad biomass is less than 35 kg ha^{-1} (Pollard et al. 1998; Schaus et al. 2002). In productive Ohio reservoirs, small species dominate zooplankton assemblages, shad are abundant, and nonlarval gizzard shad are almost entirely detritivorous (Mundahl and Wissing 1987, 1988; Stein et al. 1995, 1996; Yako et al. 1996; Schaus et al. 2002). Thus, in our simulations, we assumed that *Daphnia* declined with increasing gizzard shad abundance according to equation (4.1), substituting shad biomass for yellow perch biomass. We also assumed that consumption of *Daphnia* by an individual gizzard shad (I_Z, g dry mass fish^{-1} d^{-1}) is proportional to *Daphnia* abundance:

$$I_Z = (9.11 \times 10^{-4})MZ, \tag{4.2}$$

where M is fish wet mass (g) and Z is *Daphnia* density (l^{-1}). This equation ($r^2 = .842$, $n = 4$, $p < .025$) was derived from tables 1 and 2 in Shepherd and Mills (1996), setting the intercept to 0 and omitting data from 21–22 July 1992 as an outlier.

Ingestion of phosphorus (I_{PZ}) was obtained by multiplying I_Z by the P content of *Daphnia* bodies (0.015 g P g dry mass^{-1}; Andersen and Hessen 1991). Once P is ingested by zooplanktivorous fish, it is either egested (as feces, F_{PZ}) or assimilated. F_{PZ} is equal to $I_{PZ} (1 - E_{PZ})$; E_{PZ} is the P assimilation efficiency of shad feeding on *Daphnia* and was set to 0.72 in this model (Schindler and Eby 1997). Per capita P excretion by zooplanktivorous fish—that is, P recycling (g P excreted fish^{-1} d^{-1})—was obtained as $X_{PZ} = I_{PZ} - F_{PZ} - G_{PZ}$, where G_{PZ} is P allocated to individual growth, and is obtained as $G_{PZ} = I_Z Y_Z P_{fish}$, where Y_Z is the growth efficiency of shad feeding on zooplankton (0.15 g dry mass growth g dry mass consumed^{-1}; Shepherd and Mills 1996; Roseman et al. 1996) and P_{fish} is the P content of fish bodies. P_{fish} was determined by grinding up whole shad from Acton Lake and quantifying whole-body P content. P content increased with fish size in the following manner:

$$P_{fish} = 0.0215M^{0.101}. \tag{4.3}$$

P Transport

Excretion rates of individual detritivorous fish were obtained by modifying an earlier model (Vanni 1996). Daily detritus ingestion rates (I_S, g dry mass sediment detritus consumed fish^{-1} d^{-1}) were obtained as $I_S = DN$, where D is the dry mass of detritus ingested per gut filling and N is the number of gut fillings per day. We modeled fish of five different sizes (5, 15, 35, 75, and 150 g wet mass) so that we could vary gizzard shad population size structure (see below). D was estimated from fish wet mass (M) and gizzard shad population biomass:

$$D = (-0.0109 + 0.0245M)Q. \tag{4.4}$$

The terms in parentheses were derived from Salvatore et al. (1987) for detritivorous gizzard shad. Q is a scaling factor, set equal to 1 if gizzard shad biomass was greater than 50 kg ha^{-1}, and equal to (gizzard shad biomass/50) if shad biomass was 50 kg or less. This equation assumes that per capita detritus consumption increases linearly with shad population biomass as *Daphnia* abundance declines.

The number of gut fillings per day, N, is equal to L/t_G, where L is the number of hours per day during which gizzard shad feed (equal to day length; Pierce et al. 1981). The gut passage time, t_G (h gut fillings^{-1}), is estimated from temperature, T (°C), and fish wet mass, M:

$$t_G = (76.998T^{-0.977})(0.600M^{0.285}). \tag{4.5}$$

In equation (4.5), the part in the left-hand parentheses is taken from Mundahl (1991), derived from data in Salvatore et al. (1987) for detritivorous gizzard shad. However, Salvatore et al. (1987) measured gut passage time only for fish of about 6 g wet mass. Daily ingestion rates of fish generally scale with body mass as $I = aM^b$, with b commonly near 0.750 (Jobling et al. 1993). Since D scales linearly with body mass (equation [4.4]), we assumed that t_G must scale less than linearly with body size. The part in the right-hand parentheses of equation (4.5) was obtained iteratively, and produces daily ingestion rates (I_S) that scale to body size with $b = 0.750$. We set $L = 16$ h and $T = 25$°C, typical values for an Ohio reservoir in midsummer.

I_S was multiplied by the P content of ingested sediment detritus to obtain the mass of phosphorus ingested via detritivory (I_{PS}, g P fish^{-1} d^{-1}). Detritivorous gizzard shad are selective feeders, consuming detritus with higher carbon and nitrogen contents than are present in bulk sediments (Mundahl and Wissing 1988). We measured P selection efficiency (P_{sel}, P content of gut detritus divided by P content of sediments), and found it to be 5.25, which is similar to the mean N selection efficiency reported by Mundahl and Wissing (1988). Therefore, daily ingestion of P from sediment detritus was estimated as $I_{PS} = I_S P_{sed} P_{sel}$, where P_{sed} is the P content of lake sediments (g P g dry mass sediments^{-1}). P_{sed} generally increases with lake productivity (TP). Using data from Nürnberg et al. (1986) and Nürnberg (1988), we derived the following equation:

$$P_{sed} = 0.00152 + (4.039 \times 10^{-6})TP$$
$$(r^2 = .280, n = 36 \text{ lakes}, p < .001). \tag{4.6}$$

In Nürnberg's studies, P_{sed} was quantified in sediments in anoxic hypolimnia of lakes. However, gizzard shad feed only in oxygenated areas, and we find that P_{sed} in oxic areas is lower than P_{sed} in anoxic areas in Acton Lake (oxic $P_{sed} = 0.4635$ anoxic P_{sed}; Evarts 1997). Thus we multiplied anoxic P_{sed} (from equation [4.6]) by 0.4635 to obtain oxic P_{sed}.

P released as feces (F_{PS}) is equal to I_{PS} $(1 - E_{PS})$, where E_{PS} is P assimilation efficiency for detritivorous shad. E_{PS} was calculated from P contents of detritus taken from foreguts and hindguts of gizzard shad from Acton Lake, using mineral ash as undigested reference matter (Conover 1966). We found that E_{PS} was 0.68, intermediate between that of gizzard shad assimilation efficiencies for C and for N (0.50 and 0.77, respectively; Mundahl and Wissing 1988).

Per capita nutrient transport (g P excreted detritivorous fish^{-1} d^{-1}) was obtained as $X_{PS} = I_{PS} - F_{PS} - G_{PS}$. Phosphorus allocated to growth by detritivorous fish (G_{PS}) was obtained by multiplying per capita growth rate (g dry mass fish^{-1} d^{-1}) times the P content of fish body tissue (P_{fish}, as for planktivorous fish). Per capita growth rates were obtained as the product of fish mass times mass-specific growth rates. Mass-specific growth rates (g g^{-1} d^{-1}) were obtained as $(\ln M_f - \ln M_i)/t$, where M_f and M_i are final and initial wet mass and t is the time period over which growth was measured. For fish 15 g or larger, M_f and M_i were obtained from age-specific changes in length from figure 2 in DiCenzo et al. (1996) and a length-mass regression for Acton Lake shad, assuming that all growth occurred over a 6-month growing season. Specific growth rates for 5 g fish were similarly obtained from detritivorous gizzard shad in Acton Lake (Schaus et al. 2002). We explored two scenarios (table 4.1). In the "constant size structure and growth" scenario, we assumed that for a given fish size, specific growth rate was constant across the productivity (TP) gradient. In the "variable growth and size structure" scenario, we assumed that for a given fish size, specific growth rate declined with TP (derived from fig. 2 in DiCenzo et al. 1996).

Ecosystem-Scale Fluxes

Quantification of nutrient flux through fish at the ecosystem scale requires fish population density or biomass, as well as per capita rates. Surprisingly few studies have quantified whole-lake fish biomass. We obtained gizzard shad biomass data from natural lakes in Florida (Bachmann et al. 1996; R. Bachmann, personal communication) and developed a regression of gizzard shad biomass (kg ha^{-1}) versus TP (μg P l^{-1}): Biomass $= 0.066$TP$^{1.233}$

Table 4.1 Specific growth rates and size structures used in the "constant size structure and growth" and "variable size structure and growth" scenarios

	Specific growth rate (g g^{-1} d^{-1})			Proportion of population biomass		
	Total P (μg P/L)			Total P (μg P/L)		
Fish size class	10	100	200	10	100	200
5 g fish	0.0154	*0.0100*	0.0065	0.05	*0.10*	0.15
15 g fish	0.0092	*0.0060*	0.0039	0.15	*0.20*	0.25
35 g fish	0.0031	*0.0020*	0.0013	0.20	*0.30*	0.40
75 g fish	0.0015	*0.0010*	0.0007	0.35	*0.25*	0.15
50 g fish	0.0015	*0.0010*	0.0007	0.25	*0.15*	0.05

NOTE: For the constant scenario, parameters were held constant at all TP levels and are given in italics. For the variable scenario, parameter values for TP = 10, 100, and 200 are given; values for other TP levels were linearly interpolated.

(n = 60 lakes, r^2 = .238, p < .001). At a given TP, gizzard shad biomass in these lakes is substantially less than that in reservoirs (Miranda 1983; Schaus et al. 1997; Schaus 1998; M. J. Vanni et al., unpublished data), possibly because of the aforementioned watershed-derived subsidies to reservoirs (Jenkins 1982; Adams et al. 1983). Therefore, we adjusted the Florida data as follows. We used the regression above to obtain shad biomass "scaling factors." At each TP, the scaling factor is equal to estimated shad biomass in Florida lakes divided by estimated shad biomass in a Florida lake with TP = 100 μg P l^{-1}. This TP concentration is approximately equal to that in Acton Lake, where we have several years of gizzard shad biomass data (Schaus et al. 1997, 2002), and represents the midpoint of the TP gradient we explore here. To obtain estimated biomass in reservoirs at other TP levels, we then multiplied mean Acton Lake biomass (175 kg ha^{-1}) by these scaling factors at each point along the TP gradient. DiCenzo et al. (1996) and Michaletz (1997) also showed that gizzard shad relative abundance (catch per unit effort) increased along gradients of productivity (chlorophyll) in reservoirs. Therefore, we obtained estimates of gizzard shad biomass from these studies in a similar manner, after estimating TP in their reservoirs with a TP-chlorophyll regression developed for Ohio reservoirs (Chl = 0.775 + 0.308TP, r^2 = .941).

These analyses yielded three separate estimates of how gizzard shad biomass increases with TP in reservoirs (fig. 4.4A). Preliminary data from three Ohio reservoirs agrees well with the Bachmann et al. and DiCenzo et al. estimates, while the Michaletz curve appears to be substantially

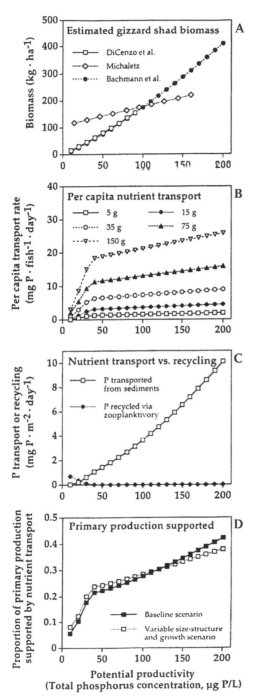

Figure 4.4 Estimated gizzard shad biomass and predicted importance of gizzard shad-mediated nutrient transport along a reservoir productivity gradient (total phosphorus concentration). (A) Estimated gizzard shad biomass, as derived from three sources (Bachmann et al. 1996; DiCenzo et al. 1996; Michaletz 1997). Biomass is plotted only over the range of productivity studied by the cited authors. For model simulations, we used the average of the biomass values estimated by the Bachmann et al. and DiCenzo et al. curves. (B) Per capita P transport by fish of various sizes. (C) P transport and recycling by entire gizzard shad populations (i.e., at the whole-lake scale). (D) Proportion of phytoplankton production supported by P transport. In the "constant growth/ size structure" (baseline) scenario, individual growth rates within a size class and population size structure were constant along the TP gradient, while in the "variable growth/size structure" scenario, these parameters varied with TP. See text and table 4.1 for details.

different. Therefore, we averaged the adjusted Bachmann et al. and DiCenzo et al. estimates at each TP level. The resulting equation, used in our model, is

$$\text{Gizzard shad biomass} = 0.785\text{TP}^{1.175}. \tag{4.7}$$

Nutrient recycling rates at the whole-lake scale were obtained by multiplying the per capita nutrient recycling rate (X_{PZ}) times the number of fish (our model assumes that zooplanktivorous fish of all sizes consume *Daphnia* and grow at the same mass-specific rate). Nutrient transport rates at the whole-lake scale were estimated by multiplying per capita nutrient transport rates (X_{PS}) for a fish of a given mass (5, 15, 35, 75, or 150 g) times the number of fish in that size class and then summing these products. Nutrient transport rates by shad populations should be affected by population size structure because mass-specific P excretion rate declines with fish mass (Schaus et al. 1997). In addition, there is some evidence that size structure varies with productivity (DiCenzo et al. 1996). We thus explored two different scenarios. In the "constant size structure and growth" scenario, the size structure did not vary with lake productivity, and was based on size structure data from DiCenzo et al. (1996) and Acton Lake (table 4.1). In the "variable size structure and growth" scenario, size structure varied with lake productivity, based on DiCenzo et al. (1996) and preliminary data from three Ohio reservoirs (M. J. Vanni et al., unpublished data). Both studies show that size structure shifts to smaller sizes as productivity increases.

Sustenance of Primary Production

We estimated the proportion of phytoplankton primary production supported by nutrient transport by detritivorous gizzard shad. To do so, we first estimated primary production rates (PPr, mg C m^{-3} d^{-1}) using the equation from V. H. Smith (1979):

$$\text{PPr} = -79 + 10.4\text{TP}. \tag{4.8}$$

PPr was multiplied by the phytoplankton cell P:C ratio to obtain the P uptake needed to sustain primary production (mg P m^{-3} d^{-1}). The phytoplankton P:C ratio was set at 0.0133 (mass:mass, equal to molar C:P = 194), which is equal to the approximate mean seston C:P ratio for Acton Lake and is indicative of moderate P limitation (Hecky et al. 1993). Finally, P transport by gizzard shad was divided by P demand by phytoplankton to obtain the fraction of primary production supported by nutrient transport (PP_{sup}). Note that P recycling is not included in this calculation.

Sensitivity Analyses

We conducted two sensitivity analyses to assess how model output is affected by choice of parameter values. For the first analysis, the response variable used to assess model sensitivity was the slope of the PP_{sup} versus TP relationship. This variable can be viewed as an index of how the importance of nutrient transport varies with productivity. The following parameter values were individually decreased and increased by 10%: L, E_{PS}, P_{sel}, and T; the slopes in equations (4.4), (4.6), and (4.8); and the exponents in equations (4.3), (4.5) (both exponents), and (4.7). For the second analysis, we varied (±10%) parameters associated with planktivorous feeding behavior, including the three constants in equations (4.1) and (4.2); the P content of *Daphnia*; E_{PZ}; and G_{PZ}. In this analysis, the response variable was the proportion of total P flux through a gizzard shad population due to planktivorous feeding (i.e., nutrient recycling/[nutrient recycling + nutrient transport]).

RESULTS

Per Capita Nutrient Recycling and Transport Rates

Predicted per capita nutrient transport rates increased with productivity (TP) for fish of all sizes (fig. 4.4B). The relatively steep increase in rates at low TP (<40) reflects both increasing reliance on detritus and increasing sediment P content. At higher TP (≥40), shad are nearly entirely detritivorous; the gentler increase in per capita P transport reflects increasing sediment P content.

Ecosystem-Scale Fluxes

The quantities of P transported versus recycled by entire gizzard shad populations (i.e., at the whole-lake scale) are predicted to vary with TP in opposite fashion (fig. 4.4C). P recycling decreased with increasing TP and exceeded P transport only at TP of 20 or lower. P transport, in contrast, increased exponentially with increasing TP, reflecting increased consumption of detritus by shad, increased sediment P content, and increased shad biomass.

Sustenance of Primary Production

The proportion of primary production supported by gizzard shad-mediated nutrient transport (PP_{sup}), as predicted by this model, increased with increasing TP (fig. 4.4D). The slope of the PP_{sup} versus TP relationship was steeper under the "variable size structure and growth" scenario than under the "constant size structure and growth" scenario (from TP 40 to 200), but the effect was relatively small. Additional model runs, in which either growth or size structure (but not both) varied with TP according to table 4.1, showed that the difference in PP_{sup} between scenarios was due almost entirely to changes in size structure at high TP. At low TP, differences in PP_{sup} between scenarios were due equally to changes in growth and in size structure. This result agrees with models of Schindler and Eby (1997), who found that P excretion rates of fish were relatively insensitive to changes in fish growth rates, and that excretion rates are most affected by variation in growth rates when food P content is low.

Sensitivity Analyses

In the first sensitivity analysis, we assessed how variation in individual parameters changed the slope of the PP_{sup} versus TP relationship (table 4.2). We did this separately for TP 40–200 (i.e., the portion of the TP gradient in which shad were primarily detritivorous) and TP 10–40 (see fig. 4.4C). The model was sensitive to variation in two of eleven parameters. Sensitivity was greater at high TP (≥ 40) than at low TP (≤ 40). In the second sensitivity analysis, we assessed how variation in planktivory parameters affected the proportion of P flux through a gizzard shad population that is attributable to nutrient recycling. The model was sensitive to three of six parameters. We present results only for TP = 40, as sensitivity was maximal for all parameters at this TP. Even though the model was very sensitive to some parameters at TP = 40, nutrient recycling accounted for a very small proportion of total P flux through gizzard shad at this TP level. For example, when the exponent of equation (4.1) was decreased by 10%, nutrient recycling accounted for 5.1% of total P flux through shad, as opposed to 2.9% under the baseline scenario. Furthermore, under all sensitivity

Table 4.2 Results of sensitivity analyses of model parameters

| | Change in slope of TP vs PP$_{sup}$ relationship (%) | | | |
| | High TP (TP > 40 µg/L) | | Low TP (TP < 40 µg/L) | |
Parameter	Parameter decreased 10%	Parameter increased 10%	Parameter decreased 10%	Parameter increased 10%
Hours of feeding per day (L)	−9.7	9.7	−9.5	9.1
Assimilation efficiency of detritivorous fish (E_{PS})	−9.7	9.7	−5.4	5.1
P selection efficiency (P_{sel})	−9.7	9.7	−9.5	9.1
Temperature (T)	−9.5	9.5	−9.3	8.9
Equation 4.4 slope	−10.1	10.1	−5.2	3.7
Equation 4.6 slope	−6.4	6.4	−1.3	1.3
Equation 4.8 slope	9.5	−8.1	−3.9	−5.1
Equation 4.3 exponent	−0.1	0.1	−0.4	2.9
Equation 4.5, first exponent	7.3	−6.8	9	−8.3
Equation 4.5, second exponent	−26.2	35.9	−17.7	28.6
Equation 4.7 exponent	−51.4	118.8	−52.1	27.7

| | Change in proportion of P flux due to planktivorous feeding (at TP = 40 µg/L) (%) | |
	Parameter decreased 10%	Parameter increased 10%
Equation 4.1 coefficient	−9.7	9.7
Equation 4.1 exponent	77.8	−44.3
Equation 4.2 coefficient	−9.7	9.7
Daphnia P content	−17.5	17.4
Assimilation efficiency of planktivorous fish (E_{PZ})	−17.5	17.4
Growth efficiency (G_{PZ})	7.7	−7.8

NOTE: In all analyses, individual parameters were decreased and increased 10% and the percent change in the response variable was measured. In the first analysis, the response variable was the slope of the relationship between TP and the fraction of primary production supported (PP$_{sup}$). This analysis was conducted separately at low (≤40) and high (>40) TP. In the second analysis, the response variable was the fraction of total P flux through gizzard shad populations that is attributable to planktivorous feeding. Results of this analysis are presented only at TP = 40, because sensitivity was maximal at this TP level.

scenarios, P recycling exceeded P transport only when TP was 20 or less, as in the baseline scenario (see fig. 4.4C).

DISCUSSION

Our model suggests that nutrient transport by detritivorous gizzard shad becomes increasingly important as lake productivity increases, in terms of per capita excretion rates, whole-ecosystem nutrient transport rates, and the fraction of primary production supported by this translocation. We caution once again that our results are derived from a model incorporating many data sources, not an explicit experimental test of hypotheses. However, we believe that our results have many implications for food web dynamics and ecosystem management.

Some species exert strong effects on ecosystems simply because they are abundant ("dominant" species). Others ("keystone" species) have strong per capita effects, and their effects on the ecosystem are thus dispro-portional to their abundance (Power et al. 1996). Either way, species may exert their influence through "trophic effects" (via consumption) and "non-trophic effects" (by causing physical state changes in biotic or abiotic materials, which then have subsequent effects on other species; Jones et al. 1994; Polis and Strong 1996). A major challenge of ecology is to develop predictive models of how interaction strength varies with ecological context (Menge et al. 1994; Power et al. 1996). Our model provides a predictive framework within which to explore how the effects of nutrient transport by gizzard shad vary along a productivity gradient, in terms of both per capita and population-level effects.

Our model suggests that the non-trophic effects of shad (nutrient trans-port) increase with productivity at both per capita and whole-ecosystem scales (see fig. 4.4). The increase in rates of nutrient transport by shad pop-ulations with increasing productivity was expected, because shad popula-tion size increases greatly with productivity. Our model also predicts that the proportion of primary production supported by nutrient transport in-creases with increasing lake productivity. This result suggests that gizzard shad-mediated nutrient transport assumes greater importance in highly productive lakes. This prediction suggests, in turn, that watershed inputs and nutrient transport by gizzard shad act synergistically to determine lake productivity (see figs. 4.1 and 4.2). According to this scenario, increases in dissolved nutrient inputs from watersheds lead directly to increased pri-mary production rates. Increased particulate nutrient inputs from watersheds lead to increased detritus for gizzard shad, and ultimately greater gizzard shad biomass. Data collected over a 5-year period show that about 60% of the phosphorus entering Acton Lake from its watershed enters in particulate form (Vanni et al. 2001). By converting particulate nutrients into dissolved form, shad further increase nutrient supply to phytoplankton, leading to even higher primary production rates and higher concentrations of nutrients in the water column.

A synergistic interaction between nutrient subsidies derived from gizzard shad and from watersheds implies a positive feedback loop (see fig. 4.2). Thus it is useful to consider what factors ultimately limit gizzard shad populations. There is some evidence for density-dependent growth rates in reservoirs (that is, growth rates decline in very productive reservoirs; DiCenzo et al. 1996; see table 4.1), and density dependence may ultimately limit shad biomass. In addition, gizzard shad populations may be reduced by low oxygen concentrations during winters with prolonged ice cover. Periods of winter hypoxia are more likely in productive lakes because of higher decomposition rates.

Reductions of both subsidies (shad P transport and P inputs from watersheds) may result in multiplicative improvements in reservoir water quality (i.e., decreased algal biomass). In fact, because watershed nutrient inputs may be the ultimate forcing factor mediating the effects of gizzard shad on lake ecosystems, management strategies for fisheries and watersheds must be unified (Stein et al. 1996; Bremigan and Stein 2001). Note, however, that the effects of reducing nutrient inputs from watersheds may be manifested over relatively long time scales, as it will take time for the existing detrital food supply to be exhausted.

In addition to mediating nutrient flux, gizzard shad also exert trophic effects, both via zooplanktivory and by providing food for piscivores (Stein et al. 1995; Shepherd and Mills 1996). The trophic effects of gizzard shad populations also seem to increase with reservoir productivity (Stein et al. 1996). Additionally, Drenner et al. (1996, 1998) found that the total effects of omnivorous fish (those feeding on benthic as well as planktonic prey, including gizzard shad, carp, and tilapia) increased with productivity in experimental mesocosms and ponds. The pattern of increasing effects of omnivorous fish with increasing lake productivity in warm-water, shallow ecosystems contrasts with the pattern of top-down effects in pelagic habitats of north-temperate, stratified lakes. In the latter, effects of consumers on their prey either diminish with increasing productivity (McQueen et al. 1986; Power et al. 1996; Carpenter et al. 1995), remain constant with productivity (Brett and Goldman 1997), or are maximal at intermediate productivity (Elser and Goldman 1991). Thus, while more studies are needed, there is mounting evidence for a fundamental difference in the relationship between productivity and the effects of fish in stratified, north-temperate lakes and in shallow, warm-water lakes (Drenner et al. 1998).

Our predictions are obviously sensitive to our model assumptions. The modeled relationship between lake productivity and the importance of nutrient transport is quite sensitive to the shape of the TP versus shad biomass relationship, and a good deal of the increase in per capita P transport rate along the TP gradient is due to increased P content of sediments. Thus, future research needs to be directed toward the relationship between lake productivity and these two parameters, as well as shad feeding habits (i.e., the balance of zooplanktivory vs. detritivory). In addition, some parameters were assumed to be constant across the productivity gradient, for lack of data, but may well vary along this gradient. Hence our sensitivity analysis may not provide an accurate depiction of how variation in these parameters affects the way in which the importance of nutrient transport varies with productivity. Such parameters include the P selection efficiency of detritivorous shad and the phytoplankton P : C ratio. Significant progress toward testing the hypothesis that nutrient transport by gizzard shad increases as productivity increases will require assessment of how these parameters vary across productivity gradients.

How general is this model? Our model was developed explicitly for gizzard shad in reservoirs. Reservoirs are the dominant lake type in North America at latitudes below 42°, except for the numerous natural lakes in Florida (Thornton 1990), and gizzard shad often dominate reservoir fish biomass. Thus, even if we restrict our scope to this lake type, nutrient transport by gizzard shad is potentially important in thousands of ecosystems across a wide geographic area. Furthermore, gizzard shad are often abundant in natural lakes and rivers. Finally, nutrient transport by other species of sediment-feeding detritivores may be important in other ecosystems receiving significant inputs of detritus from watersheds. These systems include estuaries (e.g., Deegan 1993; Odum et al. 1995) and some tropical rivers and lakes (Bowen 1983; Lowe-McConnell 1987; Flecker 1996). In these systems, populations of detritivores are often subsidized by terrestrial (or upstream) inputs, and hence their effects on food webs may depend on these subsidies. How external nutrient subsidies and animal-mediated nutrient translocation interact to regulate food web dynamics remains relatively unexplored, but their effects are potentially very important in many ecosystems.

ACKNOWLEDGMENTS

We thank participants in the 1998 INTECOL symposium for discussion of these ideas, and D. J. Berg, R. B. Blair, T. O. Crist, M. J. González, S. J. Harper, A. Hastings, G. A. Polis, D. E. Schindler, and two anonymous reviewers for comments on earlier manuscript drafts. We extend special thanks to Roger Bachmann for providing unpublished data on gizzard shad biomass from Florida lakes. This work was supported by NSF grants DEB 9318452 and 9726877

Chapter 5

Bottom-Up/Top-Down Determination of Rocky Intertidal Shorescape Dynamics

Bruce A. Menge

Understanding how ecological forces structure communities is perhaps the central goal of ecology. This focus has become even more apparent during the past 50 years as research emphases in efforts to understand community dynamics have shifted. These shifts include advancing from predominantly observational to experimental methodologies; from reliance on models stressing equilibrial dynamics to models incorporating non-equilibrial dynamics; from a physically based to a biotically based to an integrated physico-biotic conceptual framework for community dynamics; from simple pairwise to complex multispecies interactive approaches; and from predominantly small-scale to multiple-scale perspectives. Landscape ecology, with its focus on multiple spatial scales, spatially explicit patterns, and cross-scale influences, is one result of these advances. A new direction in landscape ecology focuses on understanding the mechanisms that underlie landscape dynamics, an effort that will require the integration of population and community approaches with those of landscape ecology (Wiens et al. 1993). A particularly promising approach would be to combine the highly successful methods of analysis of community dynamics, such as interaction web experimentation, with larger-scale, spatially explicit quantification of patterns.

Research in marine rocky intertidal habitats has been at the forefront of efforts to understand community dynamics. Their favorable scales in organism size and mobility, their vertically compressed spatial scales, their accessibility, and the relatively rapid pace of their dynamics make these communities amenable to incisive and rigorous scrutiny. The general lessons from investigations of community dynamics in these habitats seem broadly applicable to other habitats, and thus have been highly influential in shaping and directing the conceptual development of community ecology in general (Hairston 1989; Paine 1994; Bertness et al. 2001).

Until recently, marine ecology has emphasized top-down processes (e.g., predation, grazing) as primary determinants of community structure. This view was bolstered by a body of still persuasive studies in a variety of marine intertidal and subtidal habitats (Connell 1961a, 1961b, 1970; Paine 1966, 1974; Dayton 1971; Menge 1976; Estes et al. 1978; Robles 1987, 1997; Barkai and McQuaid 1988; Duran and Castilla 1989). Deviations from top-down control in some of these mostly local-scale studies were generally understood as a predictable consequence of variation in some key physical condition, such as wave turbulence, immersion time, or salinity. For example, high wave exposure can suppress predation, releasing prey from top-down control and leading to increased competition (Menge 1978b; Lubchenco 1986). Increased stress, such as longer exposure to air (Menge 1978a; Bertness et al. 1999) or low salinity (Witman and Grange 1998), can also reduce predation, often leading to increases in both competition and facilitation. Inevitably, however, with an expansion in both geographic scope (i.e., among regions) and spatial and temporal scales (i.e., to "shorescapes" within a region, such as whole rocky benches), researchers detected differences in marine community dynamics that demanded consideration of additional factors. For example, factors such as larval transport, nutrients, and productivity may be relatively uniform at local scales (meters to hundreds of meters), but more variable at larger oceanographic scales (kilometers to hundreds of kilometers).

Growing evidence is consistent with the hypothesis that larger-scale variation in benthic community pattern and regulation is dynamically linked to variation in nearshore oceanographic conditions (Menge 2000a, 2003). For example, in South Africa, the biomass of molluscan grazers and filter feeders increases with increasing benthic algal productivity, thereby altering species interactions both quantitatively and qualitatively (Bustamante et al. 1995a, 1995b; Bustamante and Branch 1996a, 1996b). In the Gulf of Maine, high-flow sites had higher growth of filter feeders, higher recruitment, and lower predation than did low-flow sites, suggesting that delivery of both food and propagules varied with currents (Leonard et al. 1998). These and other studies cited below suggest that to understand benthic community dynamics, we must incorporate additional processes and mechanisms and a greater range of spatial and temporal scales into our investigations. In particular, an intensified focus on the influence of bottom-up processes on species interactions, including top-down effects, and their community and shorescape consequences seems warranted.

CONCEPTUAL FRAMEWORK

Two broad classes of conceptual frameworks for community regulation include "environmental stress" (Connell 1975; Menge and Sutherland 1987; Bertness and Callaway 1994) and "nutrient/productivity" models (Oksanen et al. 1981; Fretwell 1987; Menge 2000a). Environmental stress models (ESMs) assume that species interactions vary predictably along gradients of physical and physiological stress and recruitment. Nutrient/ productivity models (NPMs) assume that species interactions vary predictably along a gradient of productivity and/or nutrients. Both types of models predict increased trophic complexity along these gradients (i.e., with decreased stress or increased productivity). Here, "trophic complexity" refers to a combination of the relative number of trophic linkages per species and food chain length. The two types of models differ in whether or not ecologically significant omnivory occurs. ESMs incorporate an important role for omnivory under conditions of low stress, while NPMs assume that ecologically significant trophic interactions are limited to carnivory and herbivory. Evidence suggests, however, that omnivory is both more frequent and more important than previously thought (Persson et al. 1988, 1996; Menge et al. 1996; Polis 1999), suggesting that the Oksanen-Fretwell models may be a special case of a more general model framework.

During the past decade, community ecologists have increasingly recognized that a broader theoretical framework for the regulation of communities must integrate the complementary ESM and NPM viewpoints (e.g., Persson et al. 1988, 1996; Menge and Olson 1990; Menge 1992, 2000a, 2003; Power 1992b). A first step was to integrate bottom-up and top-down approaches into a more general nutrient/productivity framework. Both theory and empirical evidence (Persson et al. 1988; Diehl and Feissel 2000) suggest that the importance of omnivory can increase with productivity. Building on this idea and the NPM standpoint that predation increases with productivity, Menge et al. (1996) proposed a simple modification that incorporated increasing blurring of the distinctness of trophic levels with increasing productivity. The predictions of this model contrasted with those of the Oksanen et al. (1981) model, but were similar to those of ESMs. The modified model proposed that the basal level was controlled by resource limitation in short food chains and increasingly by consumer pressure in longer food chains (length > 1), as opposed to alternation between competition (food chain lengths of 1 and 3) and consumers (food chain lengths of 2 and 4). The model thus proposed that predation increased in importance with an increase in the magnitude of bottom-up effects.

TESTS IN A MODEL ECOSYSTEM

The modified NPM summarized above was inspired by preliminary studies of the influence of bottom-up oceanographic processes on the structure and dynamics of rocky intertidal interaction webs (defined as the subset of strongly interacting species in a community; Menge and Sutherland 1987) along the Oregon coast (fig. 5.1). Observations had suggested that traditional, top-down-based explanations of control of community structure in rocky intertidal communities were insufficient to explain some major differences in community structure (Menge 1992). Specifically, at wave-exposed sites at Strawberry Hill (hereafter SH), the abundance of macroalgae was low while that of filter feeders and invertebrate predators was high. In contrast, at wave-exposed sites at Boiler Bay (hereafter BB), the abundance of macroalgae was high while that of filter feeders and invertebrate predators was low. Field experiments on predation rates and quantification of recruitment and growth rates of filter feeders at both sites suggested that both consumer effects and prey supply were higher at SH. These differences were not readily explained by differences in wave exposure, since field measurements demonstrated that wave turbulence regimes did not differ between sites (Menge et al. 1996). The different growth rates of sessile filter feeders suggested an alternative interpretation: that bottom-up processes were likely to be stronger at SH.

These data inspired a research program aimed at quantifying the relationship between nearshore oceanographic conditions and rocky intertidal community dynamics. An interaction web-based model was proposed to explain the contrasting community dynamics (fig. 5.2). We initially focused on evaluating the dynamics of the interaction webs at BB and SH, and later expanded the scope of our studies to include a larger geographic range along the Oregon coast (see fig 5.1).

Below, I first summarize the results of studies aimed at testing predictions of the relative magnitudes of links in the interaction webs for BB and

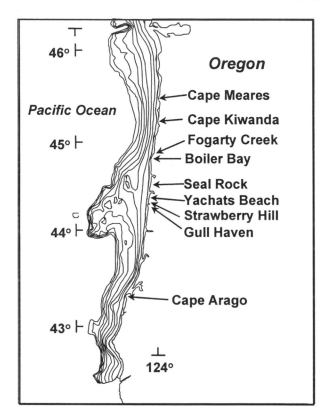

Figure 5.1 The Oregon coast, showing the locations of study sites mentioned in text and the bathymetry of the continental shelf at 20-meter increments.

SH. I then pose the question of whether or not the observed differences between these sites reflect the predicted variation at larger oceanographic scales, and I summarize our current understanding of nearshore oceanography off the central Oregon coast. Finally, I ask whether the oceanographic patterns allow prediction of interaction web dynamics across larger spatial scales and present preliminary results from large-scale predation rate experiments.

INTERACTION WEB DYNAMICS

Our current understanding of the interaction webs in the low intertidal zones at BB and SH is presented diagrammatically in figure 5.2. In 1993, we postulated that the links in the interaction webs were generally stronger at SH and weaker at BB, with the exception of those between macrophytes

BOILER BAY   **STRAWBERRY HILL**

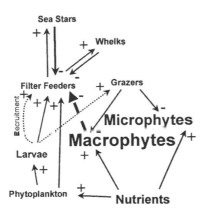
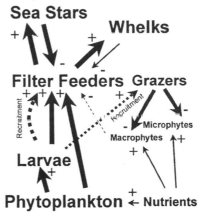

Figure 5.2 Interaction webs at wave-exposed sites at Boiler Bay and Strawberry Hill. The nodes in these webs (font size reflects relative abundances) are shown as functional groupings, not species, although the higher levels consist of relatively few taxa. Arrow thickness reflects the magnitude of the effect, arrows point to affected nodes, and + and – signs indicate whether the effect was positive or negative. Dotted lines indicate recruitment effects, dashed lines (macrophytes to filter feeders) indicate preemptive effects, upward solid arrows indicate bottom-up effects, and downward solid arrows indicate downward trophic effects. Sea stars are *Pisaster ochraceus,* whelks are *Nucella canaliculata* and *N. ostrina,* grazers consist primarily of limpets and chitons (mostly *Lottia* spp. and *Katharina tunicata*), and filter feeders include barnacles (most *Balanus glandula* and *Chthamalus dalli*) and mussels (*Mytilus californianus, M. trossulus*). Macrophytes include many species, but are dominated by laminirians (*Hedophyllum sessile, Lessoniopsis littoralis*), surfgrass (*Phyllospadix scouleri* and *P. torreyi*), and turfy and foliose red algae. Microphytes are primarily algal sporelings and diatoms. Although species composition, diversity and relative abundances differ somewhat between sites, the basic elements of the webs at each site are identical.

and filter-feeding invertebrates. The high cover of benthic algae and low abundance of barnacles and mussels in the low zone at BB led to the hypothesis that macrophytes had a strong negative effect on recruitment of filter feeders via priority effects and subsequent preemption of space. Conversely, we speculated that at SH, filter feeders negatively affected macrophytes through competitive overgrowth, inhibition of recruitment, or both. Although preliminary tests indicated that kelp and surfgrass canopies can indeed reduce barnacle recruitment, more thorough testing is necessary to evaluate these predictions fully, and I consider this interaction no further.

Prior evidence suggested that sea star predation was stronger at SH, and that this difference was fueled by higher recruitment of mussels and higher growth rates of mussels and barnacles (Menge et al. 1994, 1996). The magnitudes and effects of other linkages were more speculative and were regarded as hypotheses to be tested (see fig. 25.6 in Menge et al. 1996). I address these tests after a brief summary of nearshore oceanographic conditions adjacent to these communities.

Oceanographic Conditions

The Oregon coast lies within the California Current (hereafter CC) upwelling ecosystem, which ranges from Washington to southern California and from coastal reefs to 100 km offshore (Smith 1981; Strub et al. 1987a, 1987b, 1991). The CC flows southward, with variations in current speed dependent on wind speed and direction. Upwelling, a consequence of sustained southward coastal winds, occurs intermittently from April to October, but mostly in June through August. Upwelling is defined as current-and wind-driven replacement of offshore-moving surface water by nutrient-rich water welling up from depth along the coast. Shoreward movement of surface waters occurs during downwelling events that alternate with upwelling events. Relaxations (cessations of upwelling) are characterized by decreased wind speed and a shift in wind direction from southward to northeastward. Both upwelling and downwelling events can last from a few days to weeks. For many species, larval delivery to coastal habitats occurs during relaxation and downwelling (Farrell et al. 1991; Shkedy and Roughgarden 1997).

An oceanographic discontinuity associated with a change in the width of the continental shelf (see fig. 5.1) occurs midway between BB and SH along the central Oregon coast (Barth and Smith 1998; Huyer 1983; Kosro et al. 1997; Small and Menzies 1981; Strub et al. 1991; see fig. 5 in Menge et al. 1997b). As it flows southward past BB, the CC veers offshore, following the continental shelf break, as the shelf widens to form Stonewall and Heceta Banks. Sea surface temperatures (from AVHRR satellite imagery) and HF radar current grids indicate that during upwelling, inside the main jet of the CC, nearshore currents off BB are swift and southward, but slow down off SH and can become entrained in a counterclockwise gyre there (Kosro et al. 1997). Water sampling shows that chlorophyll *a* (Chl-*a*) concentration, an index of phytoplankton abundance, is higher over the shelf than beyond it, and is higher within the Heceta-Stonewall Banks gyre than at most sites to the north, including BB. These observations imply that plankton may be rapidly diluted off BB and may be entrained and aggregated off SH.

To test the predictions that water column concentrations of nutrients, phytoplankton, and detritus were higher at SH than at BB, we began shore-based water sampling in 1993. Preliminary data were consistent with this hypothesis. Subsequent samples (ranging from daily to monthly) taken through 1995, however, revealed that while detritus and phytoplankton concentrations and productivity were consistently, and sometimes dramatically, higher at SH, there were no consistent patterns in nutrient concentrations (nitrate, phosphate, silicate) (Menge et al. 1997a, 1997b). Thus, nutrient concentrations were similar in magnitude at both sites, while phytoplankton concentration was greater at SH (see fig. 5.2).

Bottom-Up Links

The initial observation that filter feeders (both barnacles and mussels) grew faster at SH (Menge 1992; Menge et al. 1994, 1997b) has since been confirmed by annually repeated transplant experiments of both mussels (*Mytilus californianus*) and barnacles (*Balanus glandula, Chthamalus dalli*) (B. A. Menge et al., unpublished data; Sanford and Menge 2001). Both mussels and barnacles consistently grow faster at SH. These differences occur in both El Niño and non-El Niño years (B. A. Menge et al., unpublished data). In addition, a molecular index of protein synthetic capacity, RNA: DNA ratio, was higher in SH mussels and was correlated with phytoplankton concentrations, indicating that short-term growth responded rapidly to food abundance (Dahlhoff and Menge 1996; Menge et al. 1997b). Although we have not yet quantified growth patterns for filter feeders such as sponges, tunicates, bryozoans, or hydrozoans, field observations over two decades indicate that these groups are also more abundant at SH (B. A. Menge, personal observations).

These observations are consistent with the hypothesis that the between-site differences in abundance of filter feeders depend at least partly on contrasting regimes of plant-derived particulate food. Other factors must also be at play, however, because field evidence suggests that growth rates of mussels and barnacles remained high through the winter months, when phytoplankton blooms do not occur. For example, the mussel *Mytilus trossulus* settles most densely in autumn. By January, recruits can grow from about 1 to 10 mm in shell length, and by spring, *M. trossulus* at SH can be 20–30 mm in length (B. A. Menge, personal observations). Similarly, barnacles recruiting in autumn grow quickly, particularly at SH, and can reach sexual maturity by winter (Menge 2003). Contrary to expectation, these high growth rates occurred when phytoplankton and detritus concentrations were low, not high.

Two additional factors, consumption of larval zooplankton and more favorable thermal regimes, may help explain the sustained high growth rates of filter feeders during winter (Sanford and Menge 2001). Growth of uncrowded barnacles at both sites confirmed that high growth rates persisted during non-bloom periods. Recruitment rates of barnacles and mussels are consistently high from late summer through November, and mussel recruitment (but not barnacle recruitment) is consistently greater at SH (Sanford and Menge 2001; B. A. Menge, unpublished data). Significantly, the guts of filter feeders commonly contain larvae of mussels, barnacles, and other meroplanktonic invertebrates (Barnes 1959; Crisp and Southward 1961; Zardus et al. 1991; Navarrete and Wieters 2000). Hence, sustained winter growth of filter feeders may depend in part on high concentrations of meroplanktonic larvae delivered to intertidal habitats (Sanford and Menge 2001).

Rocky intertidal thermal regimes also shift in late summer–early autumn. Seasonal drops in daytime air temperature accompany a shift in low tide occurrence from daytime to nighttime (both reducing exposure to potentially stressful high temperatures), and upwelling ceases (reducing exposure to cold water temperatures). Thus, with the onset of autumn, filter feeders are exposed to more moderate and steadier temperatures, both of air and water. Since more moderate temperatures foster higher metabolic and assimilation efficiency (Anderson 1994; Crisp and Bourget 1985; Sanford et al. 1994), sustained autumn and winter growth rates may also depend on more favorable thermal regimes. These observations are consistent with the hypothesis that planktonic larvae are consumed by benthic filter feeders, and that this trophic link is stronger at SH (see fig. 5.2).

A final bottom-up effect postulated to differ between sites is the phytoplankton-larval link (see fig. 5.2). Barnacle and mussel larvae are planktotrophic (feed on plankton), suggesting the hypothesis that larval populations are larger, larvae grow faster, or both where food concentrations are higher (SH). With potential differences in larval transport to shore (Farrell et al. 1991; Menge et al. 1997b; Connolly and Roughgarden 1999b), this putative enhancement of larval populations may thus be a factor underlying higher recruitment rates at SH.

Testing this hypothesis is a focus of ongoing research, but two observations are consistent with this possibility. First, barnacle larvae vary in size at settlement, and size variation can be induced in the laboratory by raising larvae under different food concentrations (R. Emlet, unpublished data). Field outplants of settled larvae demonstrated that larger settlers survived better, suggesting that recruitment density may depend on larval food regime (R. Emlet, unpublished data). Second, a southern California study of the response of mussel larvae (*Mytilus galloprovincialis*) to different food concentrations in the laboratory, samples of mussel larval size at settlement, and growth of outplanted recruits at sites with differing food concentrations indicates that food availability can have strong positive effects on initial size, growth, and survival of larvae (Phillips, in press; Phillips and Gaines, in press). Lipid content, an index of mussel larval condition, was also strongly correlated with food concentration. Although these alternatives require direct testing at our Oregon sites, the evidence is consistent with the hypothesis that high planktonic food concentration enhances larval populations. Since particulate food concentrations are higher at SH, it is reasonable to suggest that larval growth and survival are higher at this site than at BB (see fig. 5.2).

Top-Down Links

Our earlier experiments demonstrated strong trophic links between sea stars and filter feeders, but other links were hypothetical (Menge et al. 1996). We postulated that ecologically significant links connected whelks and filter feeders, sea stars and whelks, sea stars and grazers, and grazers and benthic plants. Further, preliminary results suggested that all trophic links were stronger at SH. For example, low plant abundance at SH could be explained by stronger grazing pressure. Increases in whelk abundance and size resulting from sea star deletions could be explained by either release from sea star predation ("intra-guild" predation; Polis and Holt 1992; Polis et al. 1989) or release from exploitation competition with sea stars. The smaller size of grazers at SH (a pattern not shown in fig. 5.2) could reflect stronger size-selective predation on large grazers by sea stars at this site. We address these alternatives below.

Grazing Pressure

The effect of molluscan grazers (limpets and chitons) on microalgal abundance was tested using herbivore exclosure experiments at each site (Menge 2000a). Grazers were excluded using 20 × 30 cm barriers of copper-based anti-fouling paint (over which grazers are reluctant to crawl; see Cubit 1984; Paine 1992). Controls included two treatments allowing access by grazers: partial paint barriers (+*paint*, +*grazers*) and marked plots (–*paint*, +*grazers*). Rock surfaces were cleared and sterilized with NaOH (lye) to initiate the experiments, which lasted for 2 to 3 months each summer.

To test the hypothesis that grazing pressure on macrophytes differed between sites, we transplanted to low zone habitats 3 × 10 cm strips of *Mazzaella splendens*, a common bladed red alga known to be favored by limpets and chitons (Gaines 1984). Algal strips were held next to the substratum using Plexiglas clamps fastened to the rock with screws. We qualitatively categorized damage to the algae during 3-day deployments (none = 0, minor = 1, major = 2, severe = 3).

Results suggested that grazer effects on both microalgae and macroalgae were greater at SH (Menge 2000a; Menge et al. 1997b; B. A. Menge, unpublished data). This difference is consistent with the observed lower abundance of algae at SH. It is uncertain, however, whether differences in grazing pressure can fully explain between-site differences in algal cover. Most of the dominant macrophytes appear little affected by grazers (e.g., surfgrass, B. A. Menge et al., unpublished data; laminarians and algal turfs, B. A. Menge, personal observations), and clear associations between overall algal abundance and herbivore abundance at different rocky shores are not evident. Alternative views, such as determination of the grazing regime, and thus the nature of the top-down effects of herbivores, by bottom-up processes (e.g., nutrients) or by limpet recruitment, are under consideration (e.g., T. Freidenburg et al., unpublished data). Evidence consistent with a nutrient hypothesis comes from recent studies of the effects of nutrients and herbivores on tide pool algal assemblages, which showed that nutrient addition led to increased algal abundance and productivity in wave-sheltered areas, but that herbivores did not respond to nutrients (Nielsen 2000).

Whelk Effects on Filter Feeder Abundance

The effects of whelks (*Nucella ostrina, N. canaliculata*) on the survival of a major prey, the mussel *M. trossulus*, in relation to both the joint and separate effects of the sea star *Pisaster*, were evaluated experimentally (Navarrete and Menge 1996). These studies monitored the survival of transplanted mussels in the presence and absence of both sea stars and whelks (both manually manipulated) at both sites. Results indicated that, although sea stars had an overwhelmingly larger effect, and much larger per capita effects, on mussel survival, whelks substantially reduced mussel survival in the absence (but not the presence) of sea stars. We concluded that whelk predation had a significant effect on mussel survival, but one that was much smaller than that of the keystone predator *Pisaster*. Interestingly, the intensities of whelk predation on mussels at BB and SH were similar (see fig. 5.2). In a separate study of the effect in Oregon of predation by the whelk *N. canaliculata* on the zone-forming, competitively dominant mussel *M. californianus*, whelks did not consume a single individual of the California mussel (E. Sanford et al., unpublished data). Overall, then, whelk effects on mussels seem generally weak (see fig. 5.2).

Sea Star Effects on Whelks

Prior results had suggested that at SH (but not DD), whelk abundance was higher, and whelk shell length was larger, in the absence than in the presence of sea stars (Menge et al. 1994). The non-response by whelks at BB was most likely due to the usual absence at this site of dense patches of mussel prey in the low zone and to sparse *Pisaster* populations. Since differences in whelk size and abundance could result from either direct predation by sea stars on whelks or depletion of mussels, their joint prey, by the more voracious sea stars, we tested the combined effects of food and sea stars on whelks (Navarrete et al. 2000). We quantified the responses in abundance and shell length of *Nucella ostrina* (formerly *N. emarginata;* Marko 1998) and *N. canaliculata* to the presence and absence both of sea stars (manually manipulated) and mussels (existing patches were either left untouched or removed by scraping).

The responses by whelks evidently resulted from exploitation competition with sea stars (Navarrete et al. 2000). Regardless of sea star presence or absence, whelks were abundant and larger in the presence of mussels, and sparse and smaller in the absence of mussels. Further, in *+sea star +mussel* plots, whelk abundance dropped sharply once sea stars had eliminated the mussels, but in *−sea star +mussel* plots, where mussels persisted longer, whelk abundance remained high. Since whelks have nondispersive, crawl-away larvae, I infer that whelk abundance is determined largely by the bottom-up effect of food supply (filter-feeder productivity). Sea star abundance is likely to be affected by both recruitment, which is higher at SH (B. A. Menge et al., unpublished data), and food supply.

Sea Star Effects on Limpets

Sea stars consume limpets and chitons, and surveys suggested that limpets were smaller (but more abundant) at SH (Menge 2000a, 2003; B. A. Menge, personal observations). To test the possibility that sea star predation affected limpet size and abundance, we followed changes in limpets in the presence and absence of sea stars at both sites. No effect was detected; limpet densities and sizes were indistinguishable in +*sea star* and −*sea star* plots (B. A. Menge, unpublished data). Although further investigation is needed, I tentatively conclude that limpet and probably chiton populations are little affected by top-down processes.

INTERACTION WEB DYNAMICS AT THE SHORESCAPE SCALE

How general are these dynamics? Are similar community patterns and processes observed at sites other than these two? Does ecologically significant variation in coastal oceanography occur, and if so, does it predictably affect communities on rocky shores? Recent research has begun to consider variation in oceanography and community dynamics both within an upwelling ecosystem (the California Current; Menge 2000a; Menge et al. 1997a, 1997b) and between upwelling and non-upwelling ecosystems (the South Island of New Zealand; Menge et al. 1999, 2002). We are accumulating evidence that bottom-up forces are broadly important determinants of community structure, species interactions, and their consequences. Our research has focused on rocky intertidal habitats in the Pacific Ocean (Menge et al. 2002), but there are similarities between our results and those obtained in other habitats (Menge 2000a). These habitats include marine subtidal and intertidal regions (e.g., Bustamante et al. 1995a; Bustamante and Branch 1996b; Witman et al., chap. 9 in this volume; S. D. Gaines, personal communication) as well as nonmarine habitats (Polis and Hurd 1996a; Polis et al. 1998). These similarities suggest that the lessons learned are broadly applicable.

Generality of Between-Site Oceanographic Differences

Water sampling from the shore at nineteen sites (including BB and SH) along the Oregon coast showed that chlorophyll *a* was low north of Newport, and with the exception of consistently low values at Cape Arago, high from Newport southward to Cape Blanco (Menge et al. 1997b). More recent sampling indicates that Chl-*a* is low south of Cape Blanco as far as Fort Bragg, California (B. A. Menge et al., unpublished data). Thus, the phytoplankton abundance differences observed between BB and SH reflect oceanographic variation occurring at a much larger spatial scale (>300 km of coastline).

Do Oceanographic Conditions Predict Onshore Community Dynamics?

To determine whether larger-scale nearshore oceanographic conditions can forecast rocky intertidal community dynamics, we carried out three studies.

First, to determine whether mussel growth varied in relation to phytoplankton concentration, we performed transplant experiments at eight study sites that spanned the region of low to high to low Chl-*a* concentration (Cape Meares to Cape Arago; see fig. 5.1), following previously established methods (Menge et al. 1994). Second, to determine whether the supply of mussel recruits varied in relation to phytoplankton concentration and/or mussel growth, we quantified mussel recruitment at the same eight sites using standard collectors (e.g. Menge et al. 1994, 1997b, 1999; Connolly and Roughgarden 1998; Leonard et al. 1998). Third, to determine whether predation intensity reflected oceanographic conditions at larger spatial scales, we conducted predation rate experiments at four sites, two with low planktonic Chl-*a* (Fogarty Creek, BB) and two with high planktonic Chl-*a* (SH, Gull Haven). The experiments had the same design as those performed earlier (Menge et al. 1994; Navarrete and Menge 1996): we translocated clumps of mussels to low intertidal regions and followed their survival in the presence and absence of sea stars.

Relations among Chlorophyll a *and Mussel Growth and Recruitment*
Mussel recruitment was greater at sites of higher phytoplankton concentration (higher Chl-*a*; fig. 5.3C) and, we assume, reflects the abundance of larvae delivered to a site. Mussel growth rates increased with both Chl-*a* concentration and mussel recruit density (fig. 5.3A, B). As suggested above, phytoplankton and larvae represent two potentially important sources of particulate food for mussels. Hence, the bottom-up effects on mussel growth, and perhaps on recruitment, documented at our two core sites may represent phenomena occurring at much larger spatial scales. In this case, the eight sites used were spread along an oceanographic gradient of phytoplankton productivity spanning more than 250 km.

Predation Rate Experiment
As expected, the intensity of predation on mussels by sea stars was low at BB and Fogarty Creek, the additional site of low phytoplankton productivity, and high at SH and Gull Haven, the additional site of high phytoplankton productivity (fig. 5.4). Predation by sea stars on mussels therefore appears to be more intense at sites of high phytoplankton productivity, and less intense at sites of low phytoplankton productivity, ranging over more than 80 km of coast. Although the number of sites in this study was still relatively small, these data are consistent with the hypothesis that the magnitude of top-down effects— in this case, sea star predation—reflects variation in bottom-up oceanographic effects as expressed through effects on the predator's food supply.

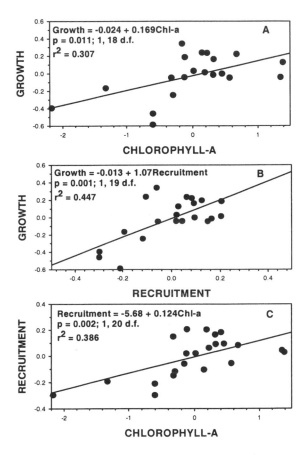

Figure 5.3 Scatterplots and regressions between growth rates (mm yr⁻¹) and recruitment of mussels (*Mytilus californianus*) and concentrations of chlorophyll a, an index of phytoplankton productivity, at eight sites over 3 years. Each symbol represents one site for 1 year. Sites were Cape Meares, Cape Kiwanda, Fogarty Creek, Boiler Bay, Seal Rock, Yachats Beach, Strawberry Hill, and Cape Arago (see fig. 5.1). Annual growth rates were quantified from June of one year to June of the next, for 1996–1997, 1997–1998, and 1998–1999. Recruitment is the sum of mean monthly recruitment for July to late summer or early autumn (usually September) and mean monthly recruitment from spring (April or May) to June of the next year. Chlorophyll a was quantified each July and August each summer except 1996, when only one multi-site survey was available. Because years differed in absolute values of each variable, but still showed similar within-year patterns, the data were standardized prior to analysis. Data were ln-transformed and standardized to the annual mean, as $(xi - m)/m$, where xi is the ith value of the variable and m is the annual mean of the variable. This procedure results in standardized values ranging above and below 0. Linear regressions were done using these standardized values, with a critical $p = .05$.

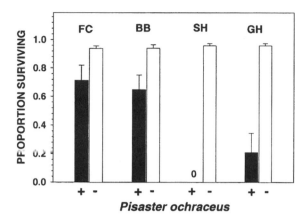

Figure 5.4 Results of experiments to quantify the rates of removal of transplanted mussel prey (*M. californianus*) by the sea star Pisaster ochraceus at four sites, Fogarty Creek, Boiler Bay, Strawberry Hill, and Gull Haven (in order from north to south). (See Menge et al. 1994 and Navarrete and Menge 1996 for detailed explanations of methods.) Data are the proportion of mussels surviving from mid-June to early September in clumps of thirty mussels each in +*sea star* and –sea star plots (*n* = 6 per site). Results of two-way ANOVA: Site, *F* = 6.34, *p* = .001; Pisaster, *F* = 21.5, *p* < .0001; Site × Pisaster, *F* = 2.29, *p* = 0.093; d.f. 3, 1, 3, and 40.

DISCUSSION

Ecologists have long known that movement of organisms and materials across community boundaries (e.g., "trophic subsidies" or "spatial flows") can be substantial and can have major consequences for community structure (Polis, Anderson, and Holt 1997). Only relatively recently, however, have investigators begun to address how these linkages influence the structure and dynamics of interaction webs (Bustamante et al. 1995a; Menge et al. 1997b; Polis 1999). Although incorporation of multiple processes and multiple scales in time and space adds great complexity to studies of community dynamics, significant advances have been made, particularly in freshwater and terrestrial habitats (Carpenter and Kitchell 1993; Polis et al. 1998; Polis 1999). Progress in marine habitats has lagged, in part due to technical limitations, but research directed toward these issues has increased dramatically in recent years (Menge 2003).

When we explicitly consider the role of nearshore oceanography, we advance our understanding of the striking variation in intertidal communities noted long ago on local scales (Lewis 1964; Stephenson and Stephenson 1972). Although mysteries remain, results to date imply that filter-feeder population density and biomass may be a joint function of rates of delivery of both propagules and particulate food (e.g., fig. 5.3). In turn, these rates can depend on both mesoscale (tens to hundreds of kilometers) and macroscale (hundreds to thousands of kilometers) variation in nearshore oceanographic conditions. Such variation can occur within a particular upwelling or downwelling regime, as well as between upwelling and downwelling regimes (Menge et al. 2002).

In the California Current upwelling ecosystem, our studies, along with others (Farrell et al. 1991; Connolly and Roughgarden 1998, 1999a, 1999b), suggest that rocky intertidal communities vary predictably with nearshore oceanographic conditions. Earlier work suggested that recruitment density of key species varied among sites as a consequence of oceanographic mechanisms that diluted or concentrated larvae (Roughgarden et al. 1991; Connolly and Roughgarden 1998). The evidence presented here indicates that these effects can be complemented by trophic flows of plankton to filter feeders and perhaps of phytoplankton to zooplanktonic larvae. High food inputs may underlie high predation rates and strong top-down effects on filter feeders (see fig. 5.4; Menge 2000a), with potentially major indirect consequences for abundance and species interactions among macrophytes and herbivores.

One key mystery is the relationship between nutrients and benthic algae. Nutrients did not vary meaningfully between BB and SH (Menge et al. 1997a, 1997b). Recent studies, however, suggest that nutrient supplementation can increase algal biomass and productivity (at small, tide pool scales; Nielsen 2000). Further, nutrient variation over larger spatial scales and longer temporal scales may underlie macroscale variation in macrophyte biomass, both in Oregon (T. Freidenburg et al., unpublished data) and in Chile (K. Nielsen et al., unpublished data).

Our results, and similar data from New Zealand (Menge et al. 1999, 2002), suggest that predation effects on community structure are strongest where bottom-up inputs are highest. Thus, as predicted by our modification of nutrient/productivity models, predation effects appear to increase with increased productivity of prey. Although these results did not involve an increase in omnivory (e.g., the sea stars and whelks studied increased in abundance and effect with increased production, but did not change diet composition to include plants or feeding at different trophic levels), even larger magnitudes of production, or the occurrence of high production over longer periods of time during the annual cycle, might encourage the invasion of omnivores.

Rocky Intertidal Shorescape Dynamics: Integrating Community
and Landscape Ecology

Although rocky intertidal habitats have served for decades as a crucible for
conceptual advancement in community ecology, the approaches of landscape ecology have been employed minimally in these habitats. Yet the factors making rocky shores so amenable to incisive investigation at the
population and community levels also should enhance study at the landscape level. Shorescapes are dominated by sessile animals and macrophytes, and the investigation-friendly sizes and generally high densities of
these organisms make it relatively easy to quantify patterns on a spatially
explicit basis. The ease of manipulating "stands" of these sedentary biota,
and of evaluating their interactions with their generally sluggish consumers, should favor the integration of the more traditionally mechanistic
approaches of community ecology with the pattern-quantifying approaches
of landscape ecology in rocky intertidal shorescapes. Landscape ecology
also considers phenomena across multiple scales in space and time, and
the addition of these layers of complexity to community investigations on
rocky shores has also lagged.

How might such linkages affect rocky intertidal shorescapes? Evidence

The discovery that signals originating at large, oceanographic scales
are readily detectable among sites in rocky intertidal communities provides
a major impetus toward integrating population and community marine
ecology with landscape ecological approaches. Although rocky intertidal
habitats and their biota can be isolated by intervening sandy or muddy
beaches, their interconnections through planktonic larvae, nutrients, phytoplankton, and detritus provide the linkage with seemingly isolated benthic dynamics. Research into the dynamics of these coupled systems is still
in its infancy. Qualitatively comparable linkages have been detected in
hard-bottom habitats in at least seven geographic regions (Alaska, Duggins
et al. 1989; Oregon, Menge et al. 1997b; southern California, S. D. Gaines,
personal communication; New England, Leonard et al. 1998; South
Africa, Bustamante and Branch 1996a, 1996b; New Zealand, Menge et al.
1999, B. A. Menge et al., unpublished data; and Chile, Broitman et al.
2001), indicating that such linkages are broadly significant in nearshore
ecosystem dynamics.

How might such linkages affect rocky intertidal shorescapes? Evidence
cited in this chapter suggests that on the Oregon coast, across spatial scales
of more than 300 km and temporal scales of several years, variation in
upwelling and, presumably, nutrient concentration in the nearshore can

affect macrophyte abundance, size, and growth (T. Freidenburg et al., unpublished data). We are currently investigating the effects of such variation on patch dynamics and mosaic structure on both local and regional scales. A similar link between upwelling intensity and macrophyte abundance has been suggested in Chile (Broitman et al. 2001).

Mussel beds have been the subject of much classic research on disturbance and its consequences for community structure (Levin and Paine 1974; Paine and Levin 1981; Sousa 1984). Recent evidence from our work in Oregon suggests that in addition to wave turbulence, the dynamics of mussel beds at the shorescape scale are responsive to nearshore oceanographic gradients (concentrations of particulate food, including phytoplankton, detritus, and perhaps zooplankton) and shore geomorphology (Guichard et al., in press; P. Halpin et al., unpublished data). This conclusion is based on a combination of spatially explicit annual surveys of mussel beds at four sites (Fogarty Creek, BB, Yachats Beach, SH) along 70 km of coast with quantification of Chl-*a* concentration, mussel recruitment, and mussel growth rates. In general, it appears that rates of recovery from disturbance are at least twice as fast where both mussel recruitment and growth rates are high. Further, the disturbance regime seems dependent on substratum heterogeneity. Specifically, at SH, where the rock underlying the mussel bed is fragmented by surge channels and tide pools extending below the lower edge of the bed, the sizes of the largest disturbances are small relative to those at sites of more homogeneous substratum (BB, Fogarty Creek). This may be because the area of disturbance is limited by the fragmented state of the mussel bed. Further, recovery from disturbance is relatively fast at SH, probably because of the smaller maximum disturbance size and because the processes driving recovery, recruitment, and individual growth are greater than at Fogarty Creek or BB. While these results are still short-term (7 years), they appear to expand the classic wave disturbance-based interpretation of the dynamics of mussel beds to include a greater range of potentially important mechanisms.

Future Directions

Future research is likely to move in at least three general directions. Most importantly, much remains to be learned about the mechanisms that underlie benthic-pelagic linkages. Because of the large scales involved, the diverse array of talents and expertise required, and the necessity of ever more powerful remote sensing techniques, this effort will require substantial investment of funds, personnel, and technology, as well as strongly multidisciplinary approaches. Extension of these rocky shore-nearshore pelagic studies to the investigation of linkages to other marine habitats, such as sandy beaches, estuaries, and deeper subtidal areas, and to adjacent terrestrial habitats is a second obvious important direction for future study. A third direction will be to integrate ESM and NPM theory. Our discovery of strong bottom-up signals in a system long regarded as one in which environmental stresses, such as wave action and desiccation, were the major physical contexts suggests the need for an integrated theoretical perspective. Steps in this direction have begun (Guichard et al., in press), and further research is in progress. The prospects are great for dramatically improved insights into the dynamics of rocky intertidal shorescapes as the current scope of research matures and widens into new areas.

ACKNOWLEDGMENTS

I thank T. Freidenburg, G. Huxel, G. Polis, and two anonymous reviewers for comments on the manuscript. Research support was provided by grants from the Andrew Mellon Foundation, the David and Lucile Packard Foundation, and NSF, and by an endowment to Oregon State University by the Wayne and Gladys Valley Foundation. This chapter is contribution number 6 from PISCO, the Partnership for Interdisciplinary Studies of Coastal Oceans: A Long-Term Ecological Consortium funded by the David and Lucile Packard Foundation.

Chapter 6

Allochthonous Nutrient and Food Inputs: Consequences for Temporal Stability

Wendy B. Anderson and Gary A. Polis

In practice and in theory, allochthonous resources can produce a variety of dynamics in recipient populations. Under some conditions, these donor-controlled resources can stabilize the dynamics of recipient populations with a mechanism analogous to the existence of refuges for prey—a constant food supply is provided, and this supply cannot be overexploited by the consumer. In other cases, the same inputs can destabilize interactions between recipients and their local resource if spatially subsidized consumers depress their prey toward extinction (Polis, Anderson, and Holt 1997; Huxel et al., chap. 26 in this volume). This chapter addresses the effects of allochthonous inputs on ecosystem stability using a real system, desert islands in the Gulf of California, as a model to assess some theoretical predictions.

Often, disparate conclusions concerning food web stability are created by different uses of the term "stability" (Pimm 1984, 1992). Here we are interested in the numerical stability of populations—that is, the amplitude of numerical fluctuations through time. We assume that increased amplitude enhances the probability of stochastic events causing extinction of small populations. We do not consider other components of stability such as resistance, resilience, and persistence (Pimm 1984, 1992). We examine the effects of allochthonous nutrients and food on the temporal stability of primary productivity and on the size of consumer populations at trophic positions throughout the food web.

Ecologists agree that the dynamics of natural systems are driven by multiple factors (Polis 1999). Focusing on one factor at a time increases our understanding of its importance, but we cannot fully assess the role of any factor when it is studied in isolation from the context of whole communities. Although this chapter begins with a focus on allochthonous nutrient inputs, we conclude by considering the effects of nutrients in concert with the other allochthonous resources that affect island systems.

THEORETICAL PREDICTIONS

Existing theory shows that the flux of allochthonous nutrients can produce a variety of outcomes (De Angelis 1992; Polis, Hurd et al. 1997; see DeAngelis and Mulholland, chap. 2, Huxel et al., chap. 26, and Power et al., chap. 24 in this volume). DeAngelis (1992) reviewed various theoretical outcomes of nutrient input to a system. Under some food web configurations (i.e., chains; see Oksanen et al. 1981), large nutrient inputs can destabilize autotroph-herbivore interactions; enhanced productivity can stimulate population growth among higher consumers, which then can produce either a temporary "trophic cascade" or limit-cycle behavior. This "paradox of enrichment" (sensu Rosenzweig 1971) pattern can lead to reduction of plant populations to a level that either approaches extinction or is below the threshold needed to support the herbivore population. This potential destabilization is due to temporary increases in both the quantity and quality (i.e., nutrient content) of plants. However, smaller amounts of nutrients can stabilize an autotroph-herbivore interaction if the rate of nutrient input is above the threshold for herbivore survival, but below the threshold for supporting higher consumers.

Large pulses of nutrients can also allow the persistence and coexistence of plant species that might otherwise be excluded by competition (Huston and DeAngelis 1994). In temporally heterogeneous environments, resource limitation frequently shifts among resources, which prevents any small group of species from achieving competitive dominance. Such resource shifts, and hence competitive outcomes, are contrary to the findings of traditional competition models. These shifts may alter plant-herbivore interactions via stabilization of primary productivity.

Like DeAngelis (1992), who predicted that there is a limited range of nutrient inputs at which plant-herbivore interactions are stable, Huxel and

McCann (1998) predicted that small to moderate inputs of allochthonous prey would stabilize local herbivore-predator interactions. At high input levels and with strong predator preference for the allochthonous prey, the local herbivore-predator interaction becomes decoupled and thus destabilized.

Seasonal or annual shifts in resource availability, as well as many small inputs into multiple trophic levels, lead to complex food webs with weak to moderate links among interacting species. Polis and Strong (1996) argued strongly for the stability of complex food webs. McCann et al. (1998) subsequently provided theoretical support for their argument that showed, contrary to traditional views of food web stability (Polis 1998), that such complexity in food webs would actually enhance the stability and persistence of communities. The empirical data in the remainder of this chapter are presented within the context of these theories.

ISLAND SYSTEMS IN THE GULF OF CALIFORNIA

Arid desert islands in the Midriff region of the Gulf of California are an excellent model system in which to study the effects of allochthonous resources on recipient ecosystem dynamics (Polis et al., chap. 14 in this volume). These islands are characterized by high temporal and spatial heterogeneity in nutrient availability and in primary and secondary productivity. Primary productivity on these islands is typically limited by precipitation (Polis, Hurd et al. 1997). However, on some islands during wet El Niño years, nutrient availability becomes the major factor that limits productivity.

Sources of "background" nutrients on all islands include weathering of parent material, decomposition of terrestrial organic matter inland or marine detritus at the shoreline, and allochthonous inputs from sea spray and dust arriving via wind and precipitation. These background nutrient levels are minimal; on islands without additional nutrient input, nutrients limit plant growth when water is not the limiting factor (Anderson and Polis 1999; G. A. Polis and W. B. Anderson, unpublished data).

Some islands receive additional nutrients in the form of seabird guano. Seven of the fourteen islands in the local archipelago are roosting and/or nesting sites for seabirds such as gulls, pelicans, boobies, and cormorants. These islands are 13–76% covered by seabird guano (Anderson and Polis 1999), a substance rich in phosphorus, nitrogen, and important cations such as calcium and magnesium (Hutchinson 1950).

PLANT RESPONSES TO NUTRIENTS IN WET AND DRY YEARS

Most nutrients are available to plants only during pulses of rain, even on islands with high nutrient levels. Rains dissolve and mobilize nutrients, stimulate plant germination, and thus facilitate growth (fig. 6.1). However, in wet years, plant response varies substantially between bird and non-bird islands, although all islands receive the same amount of rain (Anderson and Polis 1999; G. A. Polis and W. B. Anderson, unpublished data). Plant cover in wet years is substantially higher (44.5% vs. 16.3%) on bird than on non-bird islands (Anderson and Polis 1999). This relationship is nonlinear because areas of extremely high guano cover can be chemically "burned" or so physically disturbed by seabird activity that few or no plants survive (fig. 6.2).

Annual net primary productivity, estimated by annually harvested standing biomass of herbaceous and sub-shrub plants, responds to precipitation on all islands, but the strength of that response is mediated by nutrient availability (fig. 6.3). Of the four islands for which we have long-term data, the two bird islands with substantial guano cover show more extreme increases in plant biomass in wet years than do the two non-bird islands.

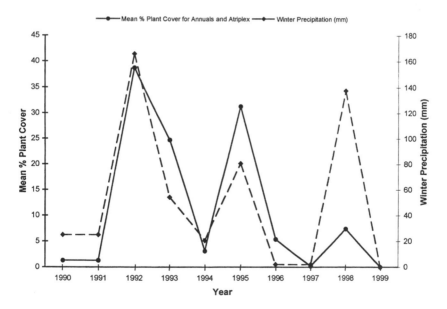

Figure 6.1 Precipitation and plant cover from 1990 to 1998 on twelve Midriff islands of the Gulf of California near Bahia de los Angeles. Plant cover includes all annual plants and drought-sensitive biennials and perennials such as *Atriplex barclayana*.

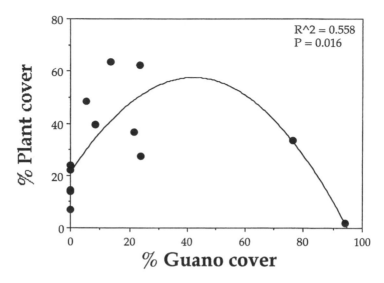

Figure 6.2 Plant cover in 1995, a wet year, as a function of guano cover on twelve islands in the Bahia de los Angeles area. Plant cover increases with low to moderate levels of guano cover, but decreases at extremely high levels.

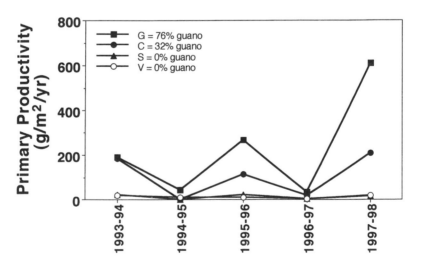

Figure 6.3 Plant productivity from 1993 to 1998 on two bird islands (fertilized by guano) and two non-bird islands in the Midriff region of the Gulf of California. Primary productivity is estimated as the aboveground biomass of annual and biennial plants. Productivity is much more variable on bird than on non-bird islands. G = Gemelos West, 76% guano cover; C = Coronadito, 21% guano cover; V = Ventana, 0% guano cover; S = Smith, 0% guano cover.

In fact, the plant biomass on each of the four islands varies in direct correlation with soil nutrient availability in wet years. In contrast, in dry years (precipitation <20 mm), no relationship exists between soil nutrient content and plant biomass, primarily because virtually no primary production occurs in typical dry years. Hence we see a pattern of great annual fluctuations in plant biomass on these islands, with pulses of plant growth followed by long periods of little to no growth. Precipitation-induced increases in plant biomass are greater on bird (68.2–104% increases) than on non-bird islands (19.2–34.3% increases). In essence, rain-induced, nutrient-driven pulses of high growth produce greater fluctuations in biomass, and thus destabilize plant productivity, on bird islands. In contrast, wet and dry year productivity is lower and more stable on non-bird islands.

Tilman and Downing (1994) reported similar results from Cedar Creek Natural History Area, Minnesota: biomass in patches with increased nutrients varied more between years than in patches without nutrients. Patches with nutrients experienced growth spurts of large, competitively dominant species in normal years; in drought years, these patches were more similar in species composition and biomass to unfertilized patches.

Species composition, and possibly interspecific plant competition, are the most likely explanations of the biomass response to nutrients on our study islands as well. Bird islands are generally dominated (>95% of plant biomass) during wet years by two or three species of nitrogen-loving, fast-growing annuals, including *Atriplex barclayana, Chenopodium murale,* and *Amaranthus watsonii*. Only *Atriplex* is somewhat drought-tolerant; *Chenopodium* and *Amaranthus* die quickly upon the return to hot, arid conditions. We speculate that in wet years, the dominant species on bird islands competitively exclude slower-growing species. Such biomass-mediated competitive effects thus directly destabilize plant productivity on bird islands as species are lost. The lower diversity on fertilized bird islands leads to fewer species with alternative strategies for dealing with inevitable droughts. Thus, the lack of these species leads to greater fluctuations in productivity.

Non-bird islands support a diverse array of smaller, slower-growing plants, sometimes up to eleven different species per square meter in wet years (W. B. Anderson and G. A. Polis, unpublished data). Plant communities on these islands include small grasses (e.g., *Aristida californica*) with low nitrogen requirements and nitrogen-fixing legumes (e.g., *Lupinus* spp.), which provide further evidence that soils on non-bird islands are nitrogen-limited. Very few plants on any island persist beyond the wet period, although some biennials persist on non-bird islands. Only *Atriplex barclayana,* found on all islands, persists from wet into dry seasons or years.

Plant nutrient quality varies along with plant biomass among islands (Anderson and Polis 1999). Herbaceous species on bird islands have N concentrations 2.6 times higher than those on non-bird islands. Further, even the same species, *Atriplex,* contains up to 2.4 times more N in its tissues on bird islands than on non-bird islands. These differences become important as we consider how resource quantity and quality influence herbivore population dynamics.

HERBIVORE RESPONSES TO NUTRIENTS IN WET AND DRY YEARS

Herbivore growth and fecundity apparently are limited more often by N, P, and other nutrients than by energy from organic C (White 1993; Polis and Strong 1996). Therefore, diets rich in N and P should lead to more abundant herbivore populations. As we have just seen, bird islands have larger quantities and higher quality of plant food than non-bird islands in wet years (Polis, Hurd et al. 1997; Anderson and Polis 1999; Polis et al., chap. 14 in this volume). As expected, the population dynamics of herbivorous insects track plant dynamics both temporally and spatially. In dry years, herbivore abundance is quite low on all islands. In wet years, herbivore abundance increases, but not equally on all islands—increases are particularly large on bird islands. In dry years, herbivore abundance is lower on bird islands than on non-bird islands. This pattern may be due to the greater abundance of biennial and perennial plants on non-bird islands. Thus, annual fluctuations of herbivorous insects mirror spatial and temporal changes in plant productivity and exhibit the greatest amplitude (i.e., the lowest stability) on bird islands (fig. 6.4).

We suggest that, although herbivore dynamics are driven by resource quality and availability, the effect of herbivores on plant biomass and community dynamics is generally not biologically significant. Although individual plants or species may sustain losses to herbivores, the plant community is ultimately and primarily regulated by bottom-up factors (e.g., rain, nutrients). Top-down consumption by herbivores appears minimal, even during the short period when annual plants are present. Our observations suggest that the crashes in annual plant populations after rains are driven almost totally by resource availability (returning aridity), not by "paradox of enrichment" (Rosenzweig 1968) herbivory, as modeled by DeAngelis (1992). Likewise, we strongly believe that declines in herbivore

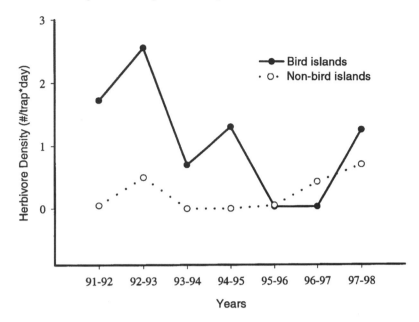

Figure 6.4 Abundance of insect herbivores from 1991 to 1998 on bird and non-bird islands in the Midriff region of the Gulf of California. Herbivore abundance was estimated by the number of herbivorous insects captured on our sticky traps per trap-day. Productivity is much more variable on bird than on non-bird islands.

populations are generally caused by the drought-induced deaths of plants, not by top-down pressure from their predators.

THE EFFECTS OF ALLOCHTHONOUS INPUTS ON CONSUMERS

Processes in the plant and herbivore compartments of a spatially subsidized system are but one of several external factors that could affect the dynamics of higher trophic levels on our study islands. In addition to nutrient inputs via seabird guano, two other major conduits bring marine subsidies to species at trophic positions throughout the food web (fig. 6.5; Polis et al., chap. 14 in this volume). Abundant marine foods that wash ashore form the first major conduit. Detritivores and scavengers directly consume shore drift (algal detritus and marine carrion), and predators eat marine and supralittoral prey. The densities of the diverse consumers that use this shoreline input are greatest on small islands, which have the highest coast-to-inland ratios (Polis and Hurd 1995, 1996a, 1996b). Seabird "products" other than guano form the second major conduit. Myriad consumers eat bird tissue, including scavengers and detritivores on dead chicks, adults,

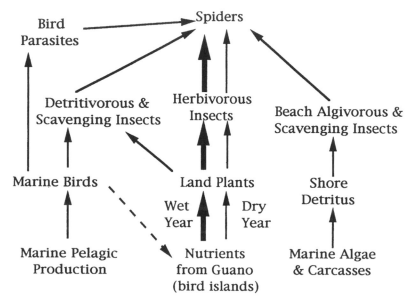

Figure 6.5 Schematic food web on islands in the Midriff region of the Gulf of California. There are three sources of organic carbon: marine shore detritus, seabirds, and primary productivity by land plants. Shifts in the importance of plant productivity and herbivore abundance occur in wet years, while marine shore detritus and seabirds are relatively constant resources.

feathers, and egg tissue and parasites on living birds (Polis and Hurd 1995, 1996b; Anderson and Polis 1998).

Thus the food web on any particular island is quite complex, with consumers using three sources of prey: plants, the coast, and seabird tissue (Sánchez-Piñero and Polis 2000; Stapp et al. 1999; Polis et al., chap. 14 in this volume). Moreover, allochthonous input into each of these three pathways varies profoundly between drought and El Niño years, between small and large islands, and between bird and non-bird islands (see above; Polis, Hurd et al. 1997; Polis et al., chap. 14 in this volume). Such variation undoubtedly affects the dynamics of consumers and food webs in several ways. However, we suspect that the likely overall effect of multiple links of resources to consumers is to dampen the effects of any one resource on a consumer population (Polis and Strong 1996; see models in McCann et al. 1998; Polis 1998; Huxel 1999; Huxel et al., chap. 26 in this volume).

One confounding factor for our analysis is that most of the islands receiving significant guano input are also the smallest islands; that is, those concurrently receiving the most shore drift and seabird products per unit area. Thus small bird islands receive the most marine inputs and also show the highest (guano-enhanced) primary productivity. This complexity makes it difficult, but not impossible, to separate the effects of nutrient inputs via the plant-herbivore pathway from those of inputs via the detritivore pathway when examining the dynamics of higher trophic levels (for approaches, see Sánchez-Piñero and Polis 2000; Stapp et al. 1999).

EFFECTS OF INPUTS ON HIGHER-LEVEL CONSUMERS: OBSERVED DYNAMICS

How do these multiple inputs and variable in situ dynamics affect the dynamics of higher-level consumers? Here we look at the population dynamics of web-building spiders, a group of predators we have studied for 9 years on twenty-one islands in our study archipelago. Web-building spider dynamics show the most relative stability on islands with the highest temporal variation in in situ terrestrial primary productivity (fig. 6.6; Polis et al. 1998). Populations on small (mostly bird) islands exhibit significantly

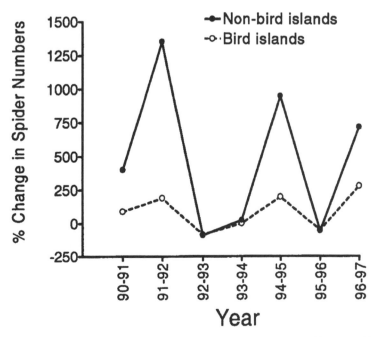

Figure 6.6 Changes in spider densities from 1990 to 1997 on small bird and large non-bird islands in the Midriff region of the Gulf of California. Spider densities are much more variable on large non-bird islands, which are less influenced by inputs of marine resources than are small bird islands with multiple pathways of resource inputs from the ocean and land.

less change from year to year than do populations on large non-bird islands, which show significantly larger changes between wet and dry years.

At first examination, these dynamics are contrary to expectations based on our previous findings that the temporal variation in plant biomass and herbivore abundance was much greater on small bird islands than on large islands not influenced by birds. Why? Our working hypothesis underlying these observed differences in population stability is that the baseline of resource availability differs systematically between small and large islands. On all small islands, the effect of marine materials that arrive via shore drift on the island's "energy budget" is proportionally much (from 2 to more than 100 times) greater than on large islands (see Polis and Hurd 1996b, especially figs. 2 and 4). Thus, populations of spiders (and many other higher-level consumers; G. A. Polis, unpublished data) on small islands are heavily subsidized by marine-based foods from the shore and by seabird tissues. Such input provides a relatively high but constant baseline of total available resources. Conversely, the diets of spiders in the inland regions of large islands contain relatively few or no marine-derived foods (Polis and Hurd 1995; Anderson and Polis 1998); these spiders are more strongly dependent on terrestrial-based foods that propagate through the in situ primary productivity compartment of the web. This baseline of total resource availability is generally lower per unit area, but also relatively more variable, being almost entirely dependent on annual precipitation (see fig. 6.1). Thus spider resource availability on large islands is much more variable through time compared with the more constant (marine-derived) resources for spiders on small islands. We suggest that this baseline of resource availability is one of the major factors dampening fluctuations of spiders and other higher-level consumers on small islands.

These findings prompt us to suggest an important and possibly systematic difference in the dynamics of consumers as a function of their trophic level in a food web. Herbivores and many detritivores eat only plant tissue. Thus, fluctuations in populations of these lower-level consumers are driven predominately by differences in primary productivity. On the other hand, fluctuations in populations of higher-level consumers (i.e., predators) are driven by resources from multiple compartments. Thus, predators (e.g., spiders) eat prey derived from primary productivity, shore drift, and bird tissue. We would expect that fluctuations in populations of these higher-level consumers would be an integrated product of their multiple sources of food.

This logic underlies the observed differences in the way that populations of herbivores and predators fluctuate in our system. Since primary productivity is more variable on small (bird) islands, we expect populations of species that use just this resource (i.e., herbivores) to be more variable, and that is exactly what we observe. Since secondary productivity (i.e., herbivores plus detritivores and scavengers on shore drift and seabird carcasses) is less variable on small islands, we expect populations of species that integrate across resource compartments (i.e., spiders and other predators) to be less variable. That is also what we observe. These findings are quite consistent with the fact that the relatively more constant marine resources form a larger proportion of the resource base available to higher-level consumers on small than on large islands.

We note, in closing this section, that our logic uses "source omnivory" as a general framework to explain differences in the relative stability of consumers at different positions in the web and on different islands. Here, "source omnivory" is omnivory across compartments (plants, shore input, seabird input). This use differs from the usual sense of "omnivory," which is either across taxonomic groups ("resource omnivory," i.e., number of species eaten within one "taxon," e.g., plants) or trophic groups ("trophic omnivory," i.e., number of different food types eaten, e.g., plants, detritus, arthropods) (Levine 1980; Polis 1991). Source omnivory allows populations with multiple allochthonous inputs (e.g., on small bird islands) to be relatively more stable than populations of these same species in areas without multiple inputs (e.g., large islands and mainland). Many have argued that broad, "multi-channel" omnivory should generally act to stabilize the populations of individual consumers and entire communities (see Polis and Strong 1996; McCann et al. 1998; Polis 1998).

OTHER EFFECTS OF ALLOCHTHONOUS RESOURCES ON STABILITY

Our explanations for stability are largely "bottom-up," in the sense that we focus on the effects of differences in resource supply (nutrients, prey) among sites. Obviously, many other factors influence the stability of a population or community. These include several factors that we cannot explore in this short chapter, such as life history characteristics, species composition, species diversity, and consumer-resource interactions.

Increases in primary and secondary productivity fostered by allochthonous marine inputs also probably affect species diversity and top-down effects mediated through the food web. If increased productivity leads to higher consumer diversity and increases food web complexity and the number of "weak links," we might expect the system to be more stable, at least based on recent models by Huxel and McCann (1998; McCann et al.

1998; see Polis 1998). On the other hand, stability and diversity could be decreased if productivity increases the populations of "strong link" species. Thus, high plant productivity associated with heavy El Niño rains provides the nectar and pollen required for foraging and egg production by spider wasps (Pompilidae). Populations of these parasitoids then erupt from very low levels to destabilize spider populations by reducing spider densities by 60–95% on our study islands (Polis et al. 1998).

CONCLUSIONS

Plants and Herbivores

Allochthonous nutrient inputs in the form of guano cause greater fluctuations in plant productivity and herbivore population dynamics than in areas not subsidized with these nutrients. These nutrients, although ever present, are available for plant uptake only during irregular rainy periods (e.g., El Niño events, winter rains), and hence stimulate plant response only in pulses. Herbivore populations probably respond not only to increased plant availability, but also to the enhanced plant nutrient quality that characterizes species and populations on islands with guano (White 1993; Anderson and Polis 1999). These patterns support the generalizations by DeAngelis (1992) that large pulses of allochthonous nutrient inputs can destabilize plant and herbivore dynamics and interactions.

In dry years, water availability limits plant growth and consequent herbivore response across all islands. In wet years, nutrient availability probably limits plant growth on islands without guano, while we speculate that competition for space or light resources limits plant growth on islands with guano. These temporal shifts in resource limitation among years do not seem to promote coexistence of several species within islands, as Huston and DeAngelis (1994) predicted, because only very few species can germinate or survive during the drought periods. On the other hand, the variation in resource limitation among islands creates variation in dominant species among islands, which increases regional species diversity during wet periods.

Higher Trophic Positions

Nutrient subsidies from guano through plants and herbivores exert variable effects on the dynamics of consumers at most higher trophic positions (e.g., spiders), depending on the importance of allochthonous prey resources and the complexity of the rest of the web. Typically, in wet years, populations of annual spiders increase due to the enhancement of prey availability through the plant-herbivore channel, but these increases are only in the range of 1.3–3.0 times. Such responses are much less than the one to two order of magnitude increases observed in plants and their herbivores. On smaller bird islands, spider dynamics are somewhat decoupled from the plant-herbivore pathway because of the multiple marine inputs that arrive through various pathways (see fig. 6.5; Polis et al., chap. 14 in this volume). These marine- and detritus-based pathways represent a more stable resource for spiders. A combination of plant-based and marine- or detritus-based resources also creates a more complex food web. The relative stability of spider populations on small bird islands supports Huxel and McCann's (1998; McCann et al. 1998) suggestion that more complex food webs generally have greater stability.

In contrast, the dynamics of spiders and other higher-level consumers on large islands without guano exhibit different patterns. Here, marine-based resources are generally less important, and dynamics are more strongly linked to within-island fluctuations of plants and herbivores. However, such fluctuations on these islands without nutrient subsidies are relatively low, as are both productivity and the densities of spiders and other higher-level consumers.

In conclusion, in our island system, allochthonous nutrient inputs tend to destabilize plant and herbivore dynamics. However, allochthonous nutrient inputs are often accompanied by inputs of resources used by consumers at multiple trophic positions throughout the web. At the population level, such "source omnivory" stabilizes the populations of many consumers. At the community level, such food subsidies make food webs more complex and probably more stable.

ACKNOWLEDGMENTS

The authors wish to acknowledge the National Science Foundation for support to GAP (DEB 9527888) and the Mexican Ministry of the Environment for Research Permit DOO 700 (2)-2341. We thank S. D. Hurd, F. Sanchez-Piñero, C. T. Jackson, J. D. Barnes, and A. Beld for help in the field and laboratory. We also thank D. A. Wait, D. L. DeAngelis, and another anonymous reviewer for suggestions for improving the manuscript.

Chapter 7

Implications of System Openness for Local Community Structure and Ecosystem Function

Robert D. Holt

There is growing evidence from a wide range of ecological systems that local population, community, and ecosystem dynamics can be dramatically influenced by fluxes of organisms and materials among spatially separated habitats (Holt 1993; Polis, Anderson, and Holt 1997; Power and Rainey 2000). An appreciation of the effect of such spatial coupling has led to a recent reevaluation of many traditional issues in ecology. For instance, weak coupling of different habitats may be an important source of stability in complex communities (Huxel and McCann 1998). This phenomenon has potentially important implications for the diversity-stability relationship (Polis 1998). Moreover, the spatial scale and behavioral details of consumer movement can alter the "bottom-up" effects of enrichment in food chains (Oksanen et al. 1995; Nisbet et al. 1997), and predator "spillover" can greatly depress local prey populations, in effect magnifying "top-down" effects in some habitats (e.g., Estes et al. 1998; Ekerholm et al. 2001).

The quantitative effect of spatial linkages on local dynamics, however, should vary greatly among habitats, for several reasons. First, the world is heterogeneous at almost all spatial scales for factors that affect population growth and interspecific interactions (Williamson 1981). Some habitats with high productivity can exert substantial effects on other habitats without a correspondingly large reciprocal effect. Spatial fluxes can also exhibit strong directionality due to the action of prevailing winds, currents, or gravity (e.g., on passive dispersal down a montane slope). Because of such asymmetries, in open systems one can often pragmatically identify "source" and "recipient" habitats. The recipient habitat may have a species composition generated by spatial fluxes that is dramatically different from that expected based on local conditions alone. Second, systems may be closed with respect to one set of components, yet operationally open for others. For instance, an endemic consumer species on an oceanic island may have demographically self-contained population dynamics, so that changes in its numbers are driven solely by in situ births and deaths. Yet its birth rate may be heightened by allochthonous resource inputs, and its death rate by migratory predators (Polis, Anderson, and Holt 1997). Habitats can vary greatly in which particular components are strongly coupled by spatial fluxes to the external environment.

In this chapter, I consider two general conceptual issues that should be considered whenever one considers the dynamics of open habitats, yet which have been almost entirely ignored in the ecological literature. The first has to do with the traditional focus in community theory on mechanisms of local coexistence (e.g., Kotler and Brown 1988; Chesson and Huntly 1997). In open systems, species tending toward exclusion in a local community (e.g., due to interactions with resident competitors or predators) can persist, and even be abundant, due to recurrent immigration. A consideration of the factors controlling abundance in such species leads to a focus on *rates* of competitive exclusion, and on temporal variability in such rates, rather than just the qualitative phenomenon of exclusion versus coexistence. The second general issue I examine is the relationship between rules of community structure and ecosystem function. I will show that the exact way in which a local system is open to external subsidies and exports can profoundly influence how community interactions map onto ecosystem properties such as productivity and total biomass. I illustrate both issues with simple models that capture the flavor of more complex models.

THE CENTRAL PARADIGM OF CLASSIC COMMUNITY ECOLOGY

A key organizing theme in community ecology is that local communities are restricted subsets of a regional species pool (e.g., Weiher and Keddy 1995). Over a sufficiently long time scale, all local communities arise from a suite of successful invasions from larger spatial arenas (Ricklefs and Schluter 1993; Holt 1993; Zobel 1997; Huston 1999; Loreau and Mouquet 1999). One general organizing factor that can generate local community structure is the exclusion of species by interactions with resident species such as predators and competitors. This insight defines a basic protocol or paradigm in community ecology, which is to articulate how local community structure arises from patterns in the success and failure of invasions, as species are "tested" against the template of local environmental conditions and interactions with resident species. If an invading species increases when rare, it is potentially a permanent community member. By contrast, if an invading species declines when rare, it is excluded. A vast body of community ecology can be boiled down to elaborations of this conceptual protocol.

More formally, let us assume that the population dynamics of species i is described by a continuous-time differential equation model. Let N_i be the density of invading species i, when it is sufficiently rare that any direct density dependence can be ignored, and let $\mathbf{N_r}$ be a vector describing the densities of resident species and resource pools in the community. The instantaneous per capita growth rate of species i is f_i (in general, a function of the abundances of resident species). The equation

$$\frac{dN_i}{dt} = N_i f_i(\mathbf{N_r}), \quad i \neq j \tag{7.1}$$

can represent a wide variety of assumptions about interactions between the invader and resident species, including resource competition, interference, and attacks by resident natural enemies. If f_i is a linear function of densities of resident species with constant coefficients, the model is the familiar Lotka-Volterra model. More generally, the growth rate of the invader will be a nonlinear function of the resident species' abundances (e.g., Abrams and Roth 1994), and the model coefficients may vary through time as well (e.g., Dunson and Travis 1991; Chesson and Huntly 1997). If the environment is constant and the resident community has settled into a stable set of abundances, the invader will have a constant per capita growth rate. If $f_i < 0$, the invader declines toward zero abundance in the focal habitat. Otherwise, the invader increases and eventually is likely to influence the preexisting community.

Iterating over repeated invasions by species drawn from the regional species pool generates a local community of cohabiting species, defined in part by their autecological requirements and in part by mechanisms of coexistence (e.g., niche partitioning, keystone predation). The rules that determine which species persist in the local community and which are excluded can be thought of as "sorting" or assembly rules. For instance, in simple pairwise exploitative competition for a single limiting resource, the consumer that can persist at the lower resource level tends to displace the less efficient consumer (Tilman 1990). Over repeated invasions, this rule leads to a sorting of species, such that the consumer species in the regional pool that depresses local resources the most eventually dominates.

MODIFYING THE CENTRAL PARADIGM IN OPEN COMMUNITIES

This familiar conceptual protocol rests on a qualitative assumption: potential community members are assumed to be excluded, and thus absent from the community, whenever their per capita growth rates when rare are negative. This assumption can be reasonable, for instance on distant oceanic islands where colonization is a rare, sporadic event. In effect, the usual protocol assumes that communities are closed except for occasional bouts of colonization by nonresident species. But for many local communities in continental settings, spatial couplings among habitats lead to recurrent or chronic invasions (one of the mechanisms implicit in island biogeographic theory; MacArthur and Wilson 1967; Holt 1992; Rosenzweig 1995). A species excluded by the resident community could nonetheless be regularly present, and even be locally abundant, because of dispersal.

A simple extension of model (7.1) reveals that the magnitude of this effect depends on the interplay of the rate of input from external sources and the rate of local exclusion (Holt 1993). We add a term I representing allochthonous inputs of a given invading species (for simplicity, we drop the index i):

$$\frac{dN}{dt} = f(t)N + I. \tag{7.2}$$

If this species has a constant, negative growth rate—that is, $f(t) = f < 0$—the excluded species equilibrates at a standing density of

$$N^* = \frac{I}{|f|}. \tag{7.3}$$

The denominator is the absolute value of the rate of exclusion of the invading species. Equation (7.3) implies that an excluded species can be locally abundant if it enjoys a low rate of exclusion or high allochthonous input.

Conversely, an excluded species will be rare (and typically missed in standard field sampling) if it is both strongly excluded and has a low rate of input from external sources. The rate of input, I, is governed by local properties of the source habitat (e.g., productivity), landscape attributes (e.g., movement rates from source into recipient habitat), and properties of the recipient habitat (e.g., edge permeability) (Polis, Anderson, and Holt 1997). The rate of exclusion, $|f|$, depends on properties of both the invader and the recipient habitat and on resident community structure. In open systems, for any model of interacting species that does not predict coexistence, one needs to know not just the mere fact of exclusion, but the *rate* of exclusion. The use of community models to quantify rates of exclusion (rather than simply exclusion vs. coexistence, as in closed communities) has largely been ignored in theoretical community ecology.

One assumption leading to equation (7.3) is that the excluded species has a constant, negative growth rate. More generally, growth rates will vary, driven by temporal variation in the external environment or in the densities or activities of resident community members. In a closed community, exclusion will still occur if the long-term average growth rate is negative. However, in an open community, further analysis reveals that temporal variation tends to enhance (sometimes quite substantially) the average abundance of species persisting because of immigration, particularly if direct intraspecific density dependence is weak (Gonzalez and Holt 2002; Holt et al., in press). Qualitatively, immigration sustains local populations through bad times, permitting populations to capitalize on runs of good times and potentially increase to high numbers, even if such times are insufficient to permit sustained persistence without immigration.

Equation (7.3) does not directly express feedbacks that arise through the effects of the invader on resident species' abundances. To assess such feedbacks, it is necessary to examine models with explicit mechanisms of potential exclusion. One can take any standard model of a community module (sensu Holt 1997a), incorporate an input term, and then evaluate how spatial subsidies modify local community structure and dynamics. Here I present two examples.

Exploitative Competition with External Inputs

The most familiar community module may be two consumers competing exploitatively for a single shared limiting resource (Tilman 1982, 1990). As is well known both theoretically and empirically (Grover 1997), given such pairwise competition, if the system is closed to immigration and settles down to an equilibrium, the equilibrium will be dominated by the consumer species that persists at the lower resource level; the other species declines to extinction.

Incorporating immigration by the superior consumer just hastens the fate of the inferior one. By contrast, incorporating regular immigration by the inferior consumer from another habitat where it persists (possibly because it is a superior exploiter of resources there) permits it to persist in the recipient habitat as well. The following model shows that the abundance of the excluded species in the focal community depends on the interplay of input rates and local rates of exclusion. Moreover, sufficiently high input rates can force the exclusion of the superior local competitor.

For simplicity, we measure abundances on scales such that each unit of consumed resource is converted into an equivalent number of consumers. Let $g_i(R)$ be the birth rate of consumer i on the single limiting resource (of abundance R), and m_i a density-independent rate of local mortality and emigration of consumer i. In general, one expects $g_i(R)$ to increase monotonically with R, approaching an asymptote at high resource levels. The net growth rate of species i is $F_i(R)$. Without resources, there should be no births (i.e., $g_i(0) = 0$). The renewal dynamics of the resource are denoted by G (which may be a function of R). With these assumptions, the model is

$$\frac{dN_1}{dt} = N_1[g_1(R) - m_1] = N_1 f_1(R)$$

$$\frac{dN_2}{dt} = N_2[g_2(R) - m_2] + I = N_2 f_2(R) + I \tag{7.4}$$

$$\frac{dR}{dt} = G - N_1 g_1(R) - N_2 f_2(R).$$

We assume that the superior local competitor, species 1, does not have external inputs. The value of R at which species i has a zero local growth rate (viz., the R such that $g_i(R) = m_i$) is $R_i^* = g_i^{-1}(m_i)$. If $R_1^* < R_2^*$, then species 2 should be competitively excluded by species 1 when they co-occur in a closed community (Tilman 1990; Holt et al. 1994; Grover 1997).

If both species persist at equilibrium in an open community, we have

$$R^* = R_1^*,$$

$$N_2^* = \frac{I}{|f_2(R_1^*)|}, \text{ and} \tag{7.5}$$

$$N_1^* = \frac{1}{m_1}[G - N_2^* g_2(R_1^*)].$$

The ambient level of the resource is set by the superior competitor. The abundance of the inferior competitor is its input rate divided by its rate of exclusion at the ambient level of resources-exactly as in the simpler single-species model of equation (7.2) above.

A locally inferior competitor can thus be abundant in an open community if it is weakly excluded and enjoys high rates of input from external sources. Although the system potentially includes strong nonlinearities (e.g., in resource uptake rates, or in G), the relationship between the equilibrial abundances of the competitors and the rate of spatial subsidy of the inferior competitor turns out to be very simple: the abundance of the inferior competitor increases linearly with its rate of input, I, whereas the abundance of the locally superior competitor declines linearly with inputs of the inferior competitor.

A sufficiently high rate of input implies exclusion of the locally superior competitor. The rate of input needed to exclude the superior competitor is

$$I > G\frac{|f_2(R_1^*)|}{g_2(R_1^*)} = G\left[\frac{m_2}{g_2(R_1^*)} - 1\right] \tag{7.6}$$

Inspection of this simple expression reveals that reversal of local competitive dominance arising from allochthonous inputs of a locally inferior competitor is facilitated by three circumstances: (1) factors associated with a low rate of exclusion (e.g., low m_2); (2) a habitat with a low rate of in situ renewal of the shared limiting resource; or (3) a high rate of immigration of the inferior competitor. A very effective resident competitor depletes resources to a low level, which makes it harder for it to be displaced by a subsidized inferior competitor. Conversely, if the two competing species have similar competitive abilities, low input rates can tip the balance toward an inferior resident competitor.

The quantitative effect of spatial subsidies on the competitive exclusion of species that are locally competitive dominants can be modulated by local factors. For example, assume that an invading prey species is inferior at consuming the shared resource, but enjoys a lower rate of mortality due to resident generalist predators (acting as density-independent mortality agents) than does a resident competitor. In this case, competitive exclusion should be weakened because ambient resource levels should be higher, increasing the birth rate for the invader. Along with the invader's assumed lower death rate, this implies that the intrinsically superior competitor for resources will be more vulnerable to exclusion at lower rates of subsidized inputs of the inferior competitor. At a fixed rate of spatial subsidy (I), species specialized for habitats with lower productivity (G) are more likely to be vulnerable to competitive exclusion by spatially subsidized species. As a special case, assume that the basal resource grows logistically (so that $G = rR(1 - R/K)$) and that consumption rates (and consumer birth rates) are linear functions of resource abundance (i.e., $g_i(R) = b_i R$). If resource levels are depressed by consumption well below K, the condition for exclusion of the superior competitor due to spatial subsidy of the inferior competitor is

$$I > r\left(\frac{m_2}{b_2} - \frac{m_1}{b_1}\right).$$

Spatial subsidies are particularly likely to reverse competitive dominance if the consumers are roughly equivalent (as measured by a small magnitude for the term in parentheses) or the resource has low recruitment rates.

Apparent Competition with External Inputs

Similar phenomena arise in systems in which exclusion is driven by shared predation leading to apparent competition (Holt 1984, 1997a; Holt and Lawton 1994; Huxel and McCann 1998). Consider the following model, in which an effective generalist predator keeps all its prey at low densities and in which the predator's functional response to each prey is linear. For simplicity, the system in the absence of external subsidies is assumed to be stabilized by interference competition among predators, with no direct density dependence in the prey. The model is

$$\frac{dP}{dt} = P\left(\sum_i a_i R_i - m - qP\right)$$

$$\frac{dR_1}{dt} = R_1(r_1 - a_1 P) \tag{7.7}$$

$$\frac{dR_2}{dt} = R_2(r_2 - a_2 P) + I.$$

Here, m is the density-independent mortality rate of the predator, and q gauges direct density dependence due to predator interference competition. The per capita attack rate by the predator on prey species i is a_i (prey densities are scaled so that prey consumption and predator births are in equivalent units). The r_i are the prey species' intrinsic growth rates, and I is the rate of spatial subsidy for prey species 2. We assume that $r_1/a_1 > r_2/a_2$, which implies that in a closed community, prey species 1 will sustain enough predators to eliminate prey species 2 via apparent competition (Holt and Lawton 1994). With immigration, species 2 persists. The equilibrium with all species present is

$$P^* = \frac{r_1}{a_1},$$

$$R_2^* = \frac{I}{|r_2 - a_2 P^*|},$$

$$R_1^* = \frac{1}{a_1}\left[m + qP^* - \frac{a_2 I}{|r_2 - a_2 P^*|} \right]. \tag{7.8}$$

As in the model of exploitative competition, the abundance of the inferior prey species is governed by the interplay of the immigration rate and a rate of local exclusion (the denominator in the expression for R_2^*). The abundance of the superior prey species is depressed by the inferior prey, and the superior prey species is eliminated at high inputs of the allochthonous inferior prey. Given that the two prey coexist, there is a linear relationship between their abundances and the rate of spatial subsidy for the inferior prey. Elimination of the non-subsidized prey occurs if

$$I \geq \frac{1}{a_2}(m + qP^*)|r_2 - a_2 P^*| = \left(m + \frac{qr_1}{a_1} \right)\left(\frac{r_1}{a_1} - \frac{r_2}{a_2} \right). \tag{7.9}$$

A subsidized prey species that is only weakly excluded by the resident prey-predator community via apparent competition (as measured by relative r/a ratios) can readily exclude a non-subsidized prey. Such exclusion is more likely if the predator is very effective at catching each prey type (high a's) or if the habitat is low in productivity (low r's). Conversely, changes in parameters that depress the predator's growth rate (increased m) or dampen its numerical response (increased q) make it harder for a subsidized prey to exclude a non-subsidized prey.

The above model assumes that the system is spatially closed for the predator and open for one (but not both) prey species. If instead the system is closed for both prey species, but open for the predator (with an input term I' added to the expression for dP/dt and $I = 0$ in model [7.7]), prey coexistence is impossible: the prey species with the higher r/a dominates, excluding the alternative prey. Moreover, if the predator input rate is sufficiently great, then both prey species will be excluded (Holt 1984; see also Holt 1996b; Polis, Anderson, and Holt 1997; Huxel and McCann 1998).

The two models suggest that spatial subsidies can substantially influence community structure, and they give guidelines as to where such effects might be sought. In particular, both models suggest that spatial subsidies should have a particularly marked effect in low-productivity habitats and when considering species that are roughly equivalent. Species that are specialized for low-productivity habitats are particularly vulnerable to exclusion by "spillover" of species from other habitats.

THE INFLUENCE OF COMMUNITY SORTING ON ECOSYSTEM PROCESSES: PROFOUND CONSEQUENCES OF SYSTEM OPENNESS

A basic challenge in ecology today is linking community dynamics with ecosystem processes (Jones and Lawton 1995; Loreau 1998; Kinzig et al. 2002). In the previous section, I examined how system openness within a given trophic level influences species coexistence via spatial subsidies that modify the outcome of "sorting" by local interactions such as exploitative and apparent competition. In this section, I use very simple models with detrital feedbacks to demonstrate that allowing system compartments to be open also controls how species sorting rules affect local ecosystem function. Holt and Loreau (2002) analyze much more complex, realistic ecosystem models and arrive at essentially the same conclusions derived below for simple models.

My basic approach is to splice compartment models of local ecosystem processes (DeAngelis 1992) to species sorting rules. The ecosystem models describe how nutrients flow among various compartments in a community defined by food chain interactions. Figure 7.1 shows three systems, differing only in whether or not there is spatial coupling between the local system and an external environment, and if so, the compartment that is involved in the coupling. The abiotic compartments of the local system include a pool of a limiting resource, of abundance R, and detrital pools generated by each trophic level (D denotes detritus produced by producer mortality; D_1, that produced by the herbivore). For each detrital pool, a constant

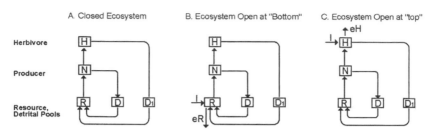

Figure 7.1 Ecosystems with a defined trophic structure can be coupled spatially to the external world in various ways. (A) Closed ecosystem (no spatial coupling). (B) Ecosystem with fluxes only at the basal resource level. (C) Ecosystem with fluxes (immigration and emigration) only at the top consumer level.

rate of decomposition frees resources. The biotic compartments include a producer population, of abundance N, fed upon by a herbivore, of abundance H. Figure 7.1A depicts a closed ecosystem; figure 7.1B, a system in which the basal resource compartment is spatially open, importing and exporting nutrients; and figure 7.1C, a system in which the top consumer alone is linked to the external world.

We imagine that at the producer trophic level, there are a large number of species potentially available in the regional species pool, but that the local community has just a single top consumer species. The attributes of the producers, as expressed in the local environment, imply that just a single species competitively dominates in pairwise competition. With occasional colonization at very low densities, one expects to observe species sorting, as species superior at persisting in the local environment invade and supplant other species. At the producer trophic level, the factors that determine local superiority include both exploitative and apparent competition (Holt et al. 1994; Liebold 1996; Grover 1997). With respect to ecosystem processes such as local nutrient cycling and spatial fluxes in nutrient pools, biotic colonization can be trivial. Except during transient phases of invasion and exclusion, the local system is thus a simple unbranched food chain. The question we address is how species sorting driven by species replacement at the producer trophic level maps onto changes in the ecosystem attributes of primary productivity and total biomass.

Closed Ecosystem

We start with the closed ecosystem in figure 7.1A. The ecosystem model is defined by linear functional responses between the producer and the limiting resource and between the herbivore and the producer (scaled by attack rates a and a_1, respectively); by constant density-independent death rates for both the producer and the herbivore (m and m_1, respectively); and by constant rates of detrital regeneration of the nutrient (at decomposition rates of d and d_1, corresponding to the detrital pools generated by the producer and the consumer, respectively). The abundances of all compartments are measured in terms of nutrient content. In the closed system, we assume that the total pool of nutrients is Q. These assumptions lead to the following model:

$$\frac{dR}{dt} = dD + d_1 D_1 - aRN$$

$$\frac{dN}{dt} = aRN - mN - a_1 NH$$

$$\frac{dH}{dt} = a_1 NH - m_1 H \tag{7.10}$$

$$\frac{dD}{dt} = mN - dD$$

$$\frac{dD_1}{dt} = m_1 H - d_1 D_1.$$

Model (7.10) implies the following equilibrial abundances:

$$N^* = \frac{m_1}{a_1}$$

$$D^* = \frac{m}{d} N^*$$

$$H^* = \frac{Q - \dfrac{m}{a} - \dfrac{m_1}{a_1}\left(1 + \dfrac{m}{d}\right)}{1 + \dfrac{a_1}{a} + \dfrac{m_1}{d}}$$

$$D_1^* = \frac{m_1}{d_1} H^*$$

$$R^* = \frac{m}{a} + \frac{a_1}{a} H^*.$$

Note that biomasses (living and dead) and free nutrients sum to Q, the total nutrient pool present in the closed system. Gross primary production, W, defined as the rate of production of new tissue by the producer, is aR^*N^* at equilibrium.

How is primary production influenced by producer sorting? Consider the producer isocline in a space defined by herbivore and resource abundances (Holt et al. 1994; Liebold 1996; Grover 1997). In the above model, this isocline is a straight line with positive slope, intersecting the R-axis at the R^* value required to sustain the producer in the absence of herbivory (fig. 7.2A). Different producer species in the regional species pool can vary in any one (or more) of three parameters defining producer dynamics: m, a

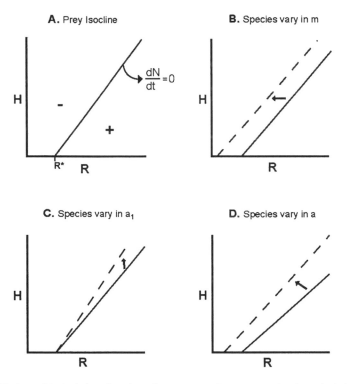

Figure 7.2 A graphic depiction of sorting rules among producers competing for a single limiting resource and sharing a common herbivore. (A) Each producer species has a zero growth isocline in a phase plane with axes of resource density (R) and herbivore abundance (H). The positive slope reflects how increasing resource abundance is required to offset losses to the herbivore. For producer coexistence, the isoclines must cross (due to trade-offs; see Holt et al. 1994). The examples considered in the text do not have crossing isoclines, so single species should dominate in pairwise competition. (B) Producers differ only in resistance to density-independent mortality, m. At any given herbivore abundance, a species with lower m can persist at lower resource abundances. (C) Producers differ only in resistance to herbivory, a_1. All species have the same R^* (intercept on R-slope); species that are more vulnerable to herbivory have shallower isoclines. (D) Producers differ only in assimilation rate, a. More efficient consumers can persist at lower resource abundances and withstand more consumption; the isocline shifts to the left and tilts upward. (Adapted from Holt et al. 1994; Liebold 1996; Grover 1997.)

basic mortality rate (fig. 7.2B); *a*, a resource assimilation rate (fig. 7.2D); and a_1, vulnerability to herbivory (fig. 7.2C). We assume that the species pool varies with respect to just one parameter at a time; thus, the isoclines of alternative producers do not intersect. This implies that in pairwise competition in the local environment, just one producer species persists, due to the combined effects of exploitative competition for the shared limiting resource and apparent competition via the shared resident top consumer (Holt et al. 1994; Liebold 1996; Grover 1997).

If species colonize at random (relative to their local abilities), the first species present is not likely to be the one that ultimately dominates. With continued colonization, the community should evolve by species replacement, with sorting toward producer species with lower intrinsic death rates (*m*), higher assimilation rates (*a*), or lower vulnerabilities to herbivory (a_1). Substitution of equilibrial values for N^* and R^* into the expression for primary production, *W*, permits one to examine how the dynamics of species sorting at the producer level influences primary production (Holt and Loreau 2002). Rather than presenting algebraic details, here I simply sketch the results.

1. Species sorting by exploitative efficiency. If producer species differ only in resource assimilation rates, production increases as more efficient producers replace less efficient producers. The direct effect of increased assimilation by the producer is to depress the resource supply. However, assimilated nutrients are passed on to the herbivore population and recycled through the detrital pool, leading indirectly to increased production at the producer level.

2. Sorting by resistance to mortality factors other than herbivory. If producer species differ in *m*, the intrinsic death rate, the dominant producer should be the one with the lowest mortality rate in the local environment. If $d > d_1 m_1/(d_1 + m_1)$, then sorting lowers production, because lowering the basic mortality rate for producers delays the reentry of nutrients into the detrital pool. This inequality is satisfied if the decomposition rate of detritus from producers at least equals that of dead herbivores, or if herbivores have a sufficiently low death rate. If this inequality is reversed, species sorting enhances primary production because of more rapid nutrient recycling.

3. Sorting by resistance to herbivory. More complex patterns are possible here, with species sorting either increasing or decreasing production. Given sufficiently high *Q*, prey sorting increases production. With lower nutrient pools, prey sorting can depress production.

For instance, the latter situation arises if producers have low intrinsic mortality rates, or if decomposition rates for producer detritus are low.

Ecosystem with Resource Imports and Exports

We now change the assumption of a closed ecosystem. Consider an ecosystem open to nutrient flux at the "bottom," but otherwise closed (fig. 7.1B). The model is unchanged, except that the resource equation becomes

$$dR/dt = I - eR + dD + d_1 D_1 - aNR.$$

Assume that the system settles into a steady state. At equilibrium, from conservation of mass, nutrient imports into the entire system must match exports. Since coupling with the external world is entirely through the resource compartment, the imports and exports of this compartment must balance; hence $I = eR^*$, or $R^* = I/e$. In effect, coupling the resource compartment to the rest of the world constrains that system variable. With R^* fixed, the other compartments have the same abundances as before, except that $H^* = (aR^* - m)/a_1$.

Once again, we imagine that a set of producer species is sorted by pairwise competitive encounters. Gross primary production is $aR^* N^* = (am_1 I)/(a_1 e)$. This expression leads to the following conclusions:

1. Prey species sorting by assimilation rate always increases primary production, matching the closed ecosystem pattern.

2. Prey species sorting by resistance to mortality factors other than the herbivore has *no* effect on primary production, in sharp contrast to the closed ecosystem, in which such sorting could either increase or decrease production. The basic reason for this is that system openness leads to a nutrient pool that is no longer dependent on flows through the decomposer system.

3. Prey species sorting by resistance to the herbivore always increases primary production. Again, this conclusion differs from that drawn for the closed ecosystem model, in which invasion by species able to resist the herbivore could either increase or decrease the long-term productivity of the system.

Species sorting can now also influence biomass and, in general, tends to increase the total biomass (living plus detrital pools) of the system (sorting by m increases total biomass if $1/d < 1m_1 + 1d_1$). In effect, species sorting permits the system to capture and retain more of the influx of nutrients before those nutrients are lost to the external world.

Ecosystem with Herbivore Imports and Exports

Finally, we assume that the herbivore compartment is coupled to the external world, whereas the rest of the ecosystem is closed (fig. 7.1C). The herbivore equation becomes

$$dH/dt = I - eH + a_1NH - m_1H,$$

and the rest of the model is as in equation (7.10). At equilibrium, nutrient inputs into the entire system should match outputs, so $H^* = I/e$. The equilibrial abundances of the other components are as before, except that standing abundance of the resource is now $R^* = (a_1I/e + m)/a$. Gross primary production is $m_1(I/e + m/a_1)$. Species sorting in the producer level now has the following implications:

1. Sorting by assimilation rate has *no* effect on primary production. This conclusion sharply contrasts with those from the other scenarios.

2. Sorting by resistance to mortality factors other than the herbivore now uniformly *reduces* primary production. This conclusion also differs from that for the closed system and the system open at the base. The reason for this difference is that with lower density-independent mortality, the flux of nutrients into the free nutrient pool is reduced, leading to less nutrient being available for use by the producer.

3. Sorting by resistance to the herbivore always increases primary production. This is comparable to what happens in the system that is open at the base, but differs from the closed system. Once again, modifying where the ecosystem is open versus closed has radically changed how the community process of species sorting maps onto ecosystem functioning.

With respect to total biomass (living and dead), producer sorting by assimilation rate has no effect, sorting by resistance to herbivory increases total biomass, and sorting by intrinsic mortality decreases biomass. These outcomes of species sorting differ sharply from those expected in an otherwise similar system that is open at the resource level, rather than the consumer level.

The above models are quite simple, but can be made more complex in various ways without changing the basic message (Holt and Loreau 2002; R. D. Holt and M. Loreau, unpublished data). For instance, the import and export terms for the system with an open resource compartment could be replaced by $I - e(R)$, where $e(R)$ is a nonlinear function increasing with R. If the basal resource is the only open compartment in the local ecosystem, then there will be some R^* at which imports equal exports; this defined resource level then constrains the remainder of the system. Species sorting could also occur at the top consumer level, or at both levels simultaneously. Preliminary study of a wide range of models suggests that the following qualitative message is robust: the relationship between community dynamics and ecosystem processes is sensitive to whether—and if so, how—a system is open to external subsidies and exports.

The above conclusions all assume that species sorting arises from rare colonization episodes, so that there is no local coexistence. If there is recurrent immigration, then we need to fuse ecosystem models with models for local interspecific interactions, comparable to equations (7.4) and (7.7) above. This strategy permits one to examine the effects of local biodiversity on ecosystem function in open systems. As discussed in more detail elsewhere (Holt and Loreau 2002; M. Loreau and R. D. Holt, unpublished data), whether or not local biodiversity enhances an ecosystem process depends in a detailed manner on which functional parameter determines local dominance, and on the pattern of spatial openness in the ecosystem. Finally, most natural systems are temporally variable as well as spatially open, and such variation can be expressed in both the magnitude of subsidies and the strength of local population and community processes. An important direction for future work is to examine subsidies in nonequilibrial, temporally varying systems (see Sears et al., chap. 23 in this volume; Holt 2002, Holt et al., in press).

CONCLUSIONS

The ideas presented above underscore several important implications of community and ecosystem openness that warrant further attention from theoretical and empirical ecologists.

The central paradigm of community ecology must be modified to accommodate spatial coupling. Equation (7.2) describes the dynamics of species that are excluded from a local community, yet maintained by recurrent subsidies from an external source. This model highlights the importance of understanding quantitative *rates* of exclusion. A general methodological conclusion is that if rules of local community structure arise from patterns of colonization and local extinction, such rules are likely to be difficult to discern when communities are open. This should be particularly true in habitats with high rates of allochthonous inputs and within which exclusion by competition or predation occurs weakly. The effects of spatial subsidies on population size can be particularly pronounced when the local environment also varies temporally (Gonzalez and Holt 2002). It should thus be particularly difficult to detect rules of local community organization in systems that are both open and temporally variable.

Spatial subsidies can control the results of local competition. Models (7.4) and (7.7) for exploitative and apparent competition with spatial subsidies show that local competitive interactions can be readily reversed by allochthonous inputs of subordinate species.

System openness influences the mapping of community dynamics onto ecosystem processes. The ecosystem models presented above demonstrate that the influence of species sorting on ecosystem properties such as primary production or total biomass is altered by system openness. Moreover, the implications of species-level "sorting rules" (defining local community membership) for ecosystem processes can be strongly influenced by which components of a system are spatially coupled with external sources. This conclusion suggests that across ecological systems, one should not expect a one-to-one mapping of rules determining community assembly onto ecosystem effects.

System openness influences species redundancy. A topic of growing concern is the functional redundancy of species (Lawton and Brown 1993). The above models shed light on this issue. For instance, in a system with an open resource compartment, producers that differ only in their resistance to mortality factors other than the herbivore all have the same gross primary production, and so by this measure are functionally redundant. However, this redundancy is not really a function of *species* properties, but of species properties as expressed in a particular *system*. If the resource compartment is instead closed (and the herbivore compartment either closed or open), then these same producer species exhibit different productivities at equilibrium and so are no longer functionally redundant.

System openness alters the importance of local decomposition processes. In the closed ecosystem model, primary production is influenced by rates of decomposition from the detrital pools. In the two open ecosystem scenarios, however, these rates do not appear at all in the expression defining primary production at equilibrium. In open systems, local primary production may be largely dominated by coupling with the external environment.

The specific results presented above are, of course, tied to specific models. I suggest, however, that the qualitative insights drawn from these models are applicable to a much wider range of models (see also Holt and

Loreau 2002). More broadly, I suspect that as population, community, and ecosystem ecology become more tightly integrated into a holistic theory of ecological systems in the coming years, a fundamental theme that will tie different strands of ecological thought together is an abiding concern with the issues of system openness, spatial subsidies, and asymmetries in a heterogeneous world.

ACKNOWLEDGMENTS

I acknowledge research support from the National Science Foundation and the University of Florida Foundation, and the supportive intellectual environment provided by the NERC Center for Population Biology, Imperial College at Silwood Park, United Kingdom, during the initial development of these ideas. I thank Michel Loreau and Michael Barfield for insights and assistance. Finally, I thank my departed friend G. A. Polis for his invitation to contribute to this volume and for the many deep insights about ecological systems—and life—he shared with me over the years.

Chapter 8

Migratory Neotropical Fishes Subsidize Food Webs of Oligotrophic Blackwater Rivers

Kirk O. Winemiller and David B. Jepsen

Motivated in part by growing interest in the role of biodiversity in ecosystem health, ecologists have turned their attention to the influence of population dynamics and species interactions on ecosystem processes (Jones and Lawton 1995; Vanni and DeRuiter 1996). Interest in this issue also has been stimulated by recent studies showing that consumers can have pronounced direct and indirect effects on energy and nutrient fluxes, which in turn influence productivity and trophic dynamics in other portions of the food web. For example, detritivorous fish feeding in the littoral zone of lakes have been shown to transport and release nutrients (via excretion) to pelagic areas, which then promote phytoplankton growth in support of the grazer food web (Carpenter et al. 1992; Vanni 1996). Similarly, caiman of the Amazon River feed in productive lagoons, then move to unproductive lagoons, where their excretion releases nutrients (Fittkau 1973). (Reviews of consumer regulation of nutrient dynamics can be found in Kitchell et al. 1979 and Polis, Anderson, and Holt 1997.) Researchers have demonstrated top-down (consumer) and bottom-up (producer) control in grazer food webs of aquatic systems without addressing how phytoplankton production and upper-level trophic dynamics are linked to longer-term nutrient cycling in these webs (Carpenter et al. 1992). Additionally, there may be energy and nutrient conduits into food webs that have gone undetected using traditional research paradigms.

Pathways that import energy and nutrients from one ecosystem to another have been called "food web spatial subsidies" (Polis and Hurd 1996a). These subsidies can be crucial to our understanding of community and ecosystem dynamics and processes (Polis et al. 1996; Polis, Anderson, and Holt 1997). For example, ecosystems can be linked by food web spatial subsidies resulting from animal migration. Anadromous salmon import nutrients from the ocean into oligotrophic streams of Alaska and the Pacific Northwest (Northcote 1988; Kline et al. 1990; Deegan 1993; Schuldt and Hershey 1995; Willson and Halupka 1995; Bilby et al. 1996; Willson et al. 1998). Marine-derived production assimilated by salmon is consumed directly by a variety of terrestrial carnivores, and decomposition of salmon carcasses enriches riparian landscapes (Willson et al., chap. 19 in this volume).

This chapter examines how migratory fishes transfer production from nutrient-rich whitewater river ecosystems to predators of nutrient-poor blackwater ecosystems in South America. Blackwater rivers of the Amazon and Orinoco basins have high fish species richness, low primary production, and high piscivore abundance. During the wet season, large schools of algivorous/detritivorous fishes (*Semaprochilodus* species) migrate downstream and exit the river to spawn and feed on the productive floodplains of the Amazon and Orinoco. During the early dry season, massive schools of juvenile detritivores enter and ascend blackwater rivers. During their long upstream migrations, these fish are consumed by resident blackwater piscivores. Stomach contents data from a population of *Cichla temensis* in Venezuela revealed heavy feeding on detritivores during the period of their upstream migration. Reproduction by *Cichla* follows this period of intense feeding on *Semaprochilodus,* and fecundity is undoubtedly influenced by this nutritional source. Preliminary estimation of this nutritional subsidy in a Venezuelan blackwater river suggests that migratory detritivorous fishes of the family Prochilodontidae create trophic linkages between river ecosystems throughout South America.

TROPICAL FLOODPLAIN RIVERS AND LANDSCAPE HETEROGENEITY

Neotropical floodplain rivers are dynamic ecosystems that support high species richness and important fisheries (Lowe-McConnell 1987; Welcomme 1989). In contrast to the regulated rivers common in temperate regions, most Neotropical rivers still retain their natural flow, so that floodplains are periodically inundated (Junk et al. 1989; Sparks 1995). Increased demand for animal protein has affected fish populations and community structure in many tropical areas (Welcomme 1989), and fisheries management must deal with socioeconomic pressures to alter river discharge and exploit natural resources. We are only beginning to understand the influence of changes in producer and consumer biomass on tropical river food webs and primary and secondary production and the relative importance of alternative producer pools to fish production and biodiversity (Winemiller 1996).

Tropical rivers are classified based on their optical qualities, which reflect the combined influence of geochemistry, soils, and vegetation characteristics (Sioli 1975). Whitewater systems are turbid, nutrient-rich rivers that carry large loads of suspended material (mostly clays) and have low transparencies in the range of 0.1–0.5 m (Secchi disk), with pH readings of 6.2–7.2 (Sioli 1975). The abundant inorganic nutrients and lentic conditions in marginal habitats promote extensive development of aquatic macrophytes (Forsberg et al. 1988). The ecology of these "floating meadows" and their role in fish production in the Amazon was described by Howard-Williams and Junk (1977).

Clearwater rivers are colorless and more or less transparent, with Secchi readings of 1.1–4.3 m and pH in the range of 4.5–7.8. The high transparency of clearwater rivers is due to a small suspended sediment load and a lack of the humic compounds that stain other systems. Rivers with this profile usually drain latosols (Sioli 1975).

Blackwater rivers have few suspended solids and a low pH (4.0–5.5). These nutrient-poor systems represent the other end of a continuum from productive whitewater systems. The transparency of black waters (1.3–2.3 m) is reduced by the high levels of humic acids that stain the water. Blackwater rivers usually originate in bleached sand soils where podzolization occurs. Sioli (1975) noted that the absence of suspended matter is explained by the almost complete lack of clay particles in the top horizon of sandy catchment areas. In South America, blackwater and whitewater river drainages may be present in the same region, reflecting the irregular distribution of podzols (Lewis et al. 1995). In contrast to whitewater and clearwater rivers, blackwater rivers typically support very little macrophyte growth, presumably due to low nutrient availability.

The net primary productivity of a whitewater floodplain lake in the Central Amazon was estimated at about 0.8 g C $m^{-2} d^{-1}$ (Schmidt 1973). Aquatic macrophytes dominate the producer biomass in white waters (Sioli 1975). Consumers have been shown to accumulate carbon from C_3 plants (*Eicchornia* spp. and a diverse array of plants from other families), but very little carbon from C_4 plants (mostly *Paspalum repens* and other

Graminaceae), which presumably passes through the microbial loop. Phytoplankton and C_3 macrophytes are the most nutritious primary production sources in the Amazon floodplain and are more important in supporting fish biomass than their standing biomass would indicate. Although C_4 macrophytes constitute a major portion of the primary producer biomass in many systems, they seem to be a minor source of energy for aquatic consumers (Hamilton et al. 1992; Forsberg et al. 1993). There is little information on production rates of attached algae (periphyton) in tropical rivers, but their importance as a source of carbon for aquatic fauna has been demonstrated in stable isotope studies (Hamilton et al. 1992).

Systems with little apparent in situ primary production can nevertheless support impressive fish biomass. Many nutrient-poor blackwater rivers in South America support important fisheries, but the basal production supporting this secondary and tertiary production is not apparent. Model simulations have shown that even at low light intensities, small standing crops of periphyton can support a relatively large consumer biomass (McIntire 1973). The consumer webs of these oligotrophic systems often are augmented, to varying degrees, by terrestrial allochthonous sources (Goulding et al. 1988).

FOOD WEB SUBSIDIES FROM MIGRATORY FISHES

Fishes of the Neotropical characiform family Prochilodontidae have complex feeding and spawning migrations that coincide with seasonal changes in water levels in the tropics (Goulding 1980; Ribeiro and Petrere 1990; Barbarino et al. 1998; Winemiller and Jepsen 1998). Prochilodontids support major commercial and subsistence fisheries throughout the Amazon, Orinoco, and Paraná river basins. Prior research has shown that *Semaprochilodus* species, abundant algivore/detritivores of blackwater rivers of the Amazon and Orinoco basins, migrate downstream to the floodplains of the whitewater Amazon or Orinoco rivers, where they spawn during the high-water period (Ribeiro 1983; Goulding et al. 1988; Vazzoler et al. 1989). The productive floodplain habitats of whitewater systems are nursery areas for young *Semaprochilodus*. During the period of falling water, spawned-out adults and young-of-the-year (YOY) migrate up to several hundred kilometers back into blackwater rivers, where the YOY are preyed upon by large predatory fishes, especially *Cichla temensis* (Perciformes, Cichlidae), an important diurnal piscivore of Neotropical blackwater rivers (Jepsen et al. 1997; Winemiller et al. 1997). For *Semaprochilodus*, the period from zygote to first migration to blackwater is 3–6 months. Most adults probably spend

3–4 months feeding on the Orinoco floodplain each year and live 2–3 years (based on data for the better-studied *Prochilodus mariae*; Lilyestrom 1983; Barbarino et al. 1998). *Semaprochilodus* spawning migrations are an ecological strategy that exploits the high productivity of flooded whitewater ecosystems to enhance the survival and growth of early life stages (Ribeiro 1983; Vazzoler et al. 1989). Presumably, YOY and adults migrate to blackwater tributaries during the falling-water period in order to avoid stranding in drying floodplain pools and the extremely high densities of competitors and predators in the channels and permanent lagoons of the whitewater system.

The Río Cinaruco is a blackwater tributary (ca. 400 km) of the Orinoco in Venezuela's llanos region (fig. 8.1). It is characterized by high fish species richness (>250 species), low aquatic primary production, and low invertebrate abundance, but high piscivore abundance. The Cinaruco has lower pH, alkalinity, turbidity, and dissolved nutrient concentrations than do whitewater tributaries of the Orinoco (Lewis 1988; Winemiller et al. 1997). Primary production and aquatic macrophyte and phytoplankton densities are much lower in the blackwater Cinaruco than in the whitewater Orinoco (Lewis 1988). Terrestrial leaf litter tends to be high in refractory organic matter and extremely slow to decompose in the acidic environment of blackwater rivers (information summarized by Goulding et al. 1988). In the lower Río Negro in Brazil (a blackwater tributary of the Amazon), *Semaprochilodus* species have been observed feeding on fine detritus from the surface of submerged vegetation in the flooded forest. This fine detritus is believed to be derived from decomposition of terrestrial litter during the annual flood period. Stomach contents of Río Cinaruco *Semaprochilodus* collected during the low-water period indicate that diatoms growing over the sand substrate were a more important food resource than fine detritus.

Juvenile *Semaprochilodus kneri* were estimated to provide nearly 50% of the annual biomass ingested by large *Cichla temensis* (>40 cm) of the Cinaruco (fig. 8.2). These large *Cichla* constitute about 60% of the adult *Cichla* population during the falling-water and low-water periods (Jepsen et al. 1997; Winemiller et al. 1997), constituting a much greater percentage in terms of population biomass. Gonadal maturation in *Cichla* follows the falling-water period, when feeding on *Semaprochilodus* and other fishes is most intense. Body condition and fat stores of *Cichla* increase significantly during this falling-water period (Jepsen et al. 1997). The migratory detritivores probably provide *Cichla* and other large piscivores (the piranha, *Serrasalmus manueli*, payara, *Hydrolycus scomberoides*, and river dolphin, *Inia geoffrensis*) of nutrient-poor rivers with a significant nutritional subsidy that allows them to achieve significantly greater growth, fecundity,

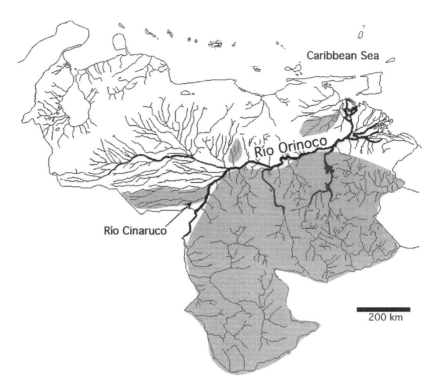

Figure 8.1 Map showing the Río Orinoco and its blackwater tributary, the Río Cinaruco, in western Venezuela. Shaded regions are dominated by blackwater and clearwater rivers. The region north of the Río Orinoco drains the nutrient-rich terrain of the Andes Mountains, Coastal Range, and alluvial plains (llanos); the region south of the Orinoco drains the nutrient-poor terrain of the ancient Guyana Shield formation.

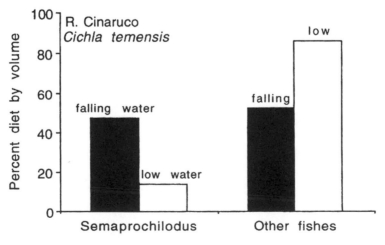

Figure 8.2 Volumetric proportions of YOY *Semaprochilodus* and other fishes in the diet of large *Cichla temensis* (>40 cm) in the Río Cinaruco during the falling-water period and low-water periods of 1993–1994 (sample N = 381).

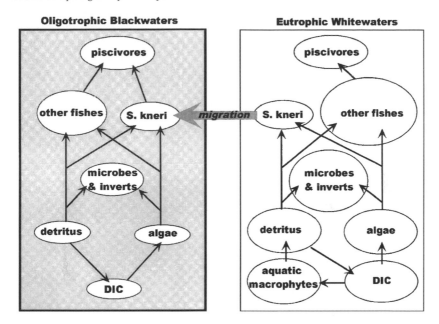

Figure 8.3 Schematic of the food web subsidy to blackwater piscivores from migratory fishes from whitewater floodplains. Compartment sizes represent the relative production of each web component (kg ha^{-1} yr^{-1}); arrows represent trophic pathways. DIC = dissolved inorganic carbon. Detritus is presumed to be derived from both aquatic and terrestrial sources; the absolute contribution of terrestrial sources is not assumed to be different in the two systems.

recruitment, and population densities than would be supported by in situ production alone. Based on the mark-recapture method, Taphorn and Barbarino (1993) conservatively estimated *C. temensis* density at approximately 71 adults ha^{-1} in a lagoon of the Cinaruco.

As the low-water season progresses, the density of YOY fish declines, and the frequency of *Semaprochilodus* in *Cichla* diets declines markedly (see fig. 8.2). During the latter phase of the low-water period, virtually all *Cichla* have empty stomachs, and gonadal maturation in preparation for spawning occurs just prior to the beginning of the annual floods (Jepsen et al. 1997; D. B. Jepsen and K. O. Winemiller, unpublished data).

There are two species of *Semaprochilodus* in the Cinaruco, with *S. kneri* far more abundant than *S. laticeps*. No other fishes of the Cinaruco undertake seasonal movements of the size, regularity, and spatial scale (several hundred kilometers each year) of the *Semaprochilodus* migrations. When these abundant fishes migrate back into nutrient-poor blackwater rivers, they provide blackwater piscivores with a significant nutritional subsidy (fig. 8.3).

Using preliminary data from the Cinaruco, we now explore how stable isotope data can complement stomach contents data in estimates of the nutritional subsidy of migratory fishes to resident piscivores of the blackwater river. Young *Semaprochilodus* that immigrate into the blackwater river have assimilated food resources with carbon isotopic signatures different from those of the resources in the blackwater river. These stable isotopic signatures permit estimation of the proportions of whitewater-derived biomass and blackwater-derived biomass assimilated by the resident piscivores that feed on migratory fishes.

STABLE ISOTOPES AND TROPHIC ECOLOGY

Food web interactions have traditionally been inferred from direct feeding observations or analyses of stomach contents. Yet these techniques provide relatively short-term assessments of consumption, and they provide no information on assimilation and limited information on rates of energy and nutrient acquisition at broader temporal and spatial scales (see also Schindler and Lubetkin, chap. 3 in this volume). In some cases, ingested material may represent refractory material of low nutritive value to consumers. Conversely, labile material that is assimilated rapidly may be underestimated. Stomach contents data do not integrate an individual's diet over time; thus, large amounts of dietary data are needed to gain insights about population responses to long-term fluxes in food resources.

A useful approach to linking flows of nutrients and energy with consumption is analysis of stable isotope ratios (Polis et al., chap. 14, and Schindler and Lubetkin, chap. 3 in this volume). Carbon ($^{13}C/^{12}C$), nitrogen ($^{15}N/^{14}N$), and other isotopes exist in material pools, and their ratios can be measured with great precision using mass spectrometry. (These ratios, expressed as $\delta^{13}C$, $\delta^{15}N$, and so forth, are actually the isotope ratio of the sample divided by the ratio of a standard reference material.) Isotope ratios provide information on sources of organic matter important to consumers as well as insights about how materials are processed within trophic networks (Peterson and Fry 1987; Fry 1988; Harrigan et al. 1989; Hobson and Welch 1992; Yoshioka et al. 1994; Hansson et al. 1997; Anderson and Polis 1998; Polis et al., chap. 14, and Schindler and Lubetkin, chap. 3 in this volume).

Carbon isotope ratios in different primary producers reflect differential fractionation of the heavier isotope member relative to the lighter member during fixation of inorganic carbon (Rounick and Winterbourne 1986). For example, C_4 plants fix CO_2 via the Hatch-Slack photosynthetic pathway, resulting in $^{13}C/^{12}C$ ratios of $-23‰$ to $-9‰$, whereas terrestrial C_3 plants using the Calvin photosynthetic pathway fix carbon at ratios of $-32‰$ to $-25‰$. After fixation by autotrophs, there is little fractionation of carbon at successive trophic transfers (ca. 1‰), so when producers have divergent isotopic signatures, it is possible to determine the relative contribution of each to consumers in the food web (reviewed by Fry and Sherr 1984; Peterson and Fry 1987; Lajtha and Michener 1994). In aquatic systems, primary producers have a range of carbon isotope ratios that reflect not only variation in sources of dissolved inorganic carbon (DIC), but also hydrological and geochemical conditions during photosynthesis (Hecky and Hesslein 1995).

Carbon isotope analysis has also been used to discriminate the trophic importance of different detrital sources in aquatic systems (McArthur and Moorhead 1996). Autochthonous organic matter (phytoplankton and macrophytes) can be distinguished from allochthonous (terrestrial) organic matter because terrestrial plants typically have a more negative carbon isotopic signature ($-28‰$) than phytoplankton $-20‰$ to $-25‰$) or submersed aquatic plants ($-12‰$ to $-30‰$) (Boutton 1991). Detritus from aquatic macrophytes has long been assumed to support most estuarine and many freshwater food chains, but isotope studies have demonstrated that phytoplankton and periphytic algae are more important than their abundance suggests (Araujo-Lima et al. 1986; Hamilton et al. 1992; Forsberg et al. 1993).

The ability to distinguish various sources of organic matter can be increased by examining several isotopes (McArthur and Moorhead 1996). Nitrogen isotopes are differentially fractionated by food web processes. ^{14}N is excreted by animals more efficiently than ^{15}N; therefore, as food webs become more complex (i.e., more trophic levels), ^{15}N accumulates at higher trophic levels. An approximate $3–4‰$ enrichment with each successive transfer has been used to assign trophic levels to species and enumerate trophic links within food webs (Minagawa and Wada 1984; Vander Zanden et al. 1997; Vander Zanden and Rasmussen, in press). In many food webs, $\delta^{15}N$ increases $+10‰$ to $+15‰$ from autotrophs to top consumers, depending on the number of trophic transfers (Peterson and Fry 1987). In combination, $\delta^{13}C$ and $\delta^{15}N$ signatures can resolve sources of organic matter and food web structure, and they have been used successfully in a number of aquatic studies (Kline et al. 1990; Hesslein et al. 1991; Hamilton et al. 1992; Bunn and Boon 1993; Forsberg et al. 1993; Kidd et al. 1995; Bootsma et al. 1996; Schlacher and Wooldridge 1996; Schindler and Lubetkin, chap. 3 in this volume).

Many fish species undergo diet shifts with age (Werner and Gilliam 1984). Evaluation of isotope ratios from different size classes within a species can indicate the life stages at which these shifts occur (Gu et al. 1997; Hentschel 1998). Most fish live in habitats where the availability of different resource pools changes seasonally, and such changes are reflected in isotopic signatures from tissues with short turnover times. Stable isotope analysis can indicate the relative importance of seasonal nutritional inputs (Goering et al. 1990), as well as ontogenetic diet shifts (Hesslein et al. 1991; Hobson 1993) and the nutritional origins of organisms that migrate between ecosystems (Hesslein et al. 1991). Many fish living in tropical floodplain river systems accumulate fat stores during the flood period (Junk 1985; Goulding et al. 1988; Winemiller 1989; D. B. Jepsen and K. O. Winemiller, unpublished data). However, it is not known whether this growth is simply a function of increased foraging opportunities in the aquatic habitat, or whether inundated terrestrial sources represent a significant input to the energy budget on a seasonal basis. Liver tissue turns over faster than muscle tissue, and our preliminary data show a mean difference (liver-muscle) of 2.5‰ for ^{13}C in Cinaruco *Semaprochilodus kneri* ($n = 9$) and 3.2‰ in large *Cichla temensis* ($n = 5$). In each case, the shifts are consistently in the predicted direction (see below) based on the size class and the time of year the fish were collected (e.g., liver is more negative than muscle for immigrant *Semaprochilodus* following several months of residence in the blackwater river).

ESTIMATION OF NUTRITIONAL SUBSIDY USING STABLE ISOTOPES

The isotopic signatures of dissolved inorganic carbon (DIC) and particulate organic carbon (POC) differ between whitewater and blackwater rivers. In a study of isotope geochemistry in the Orinoco basin in Venezuela, Tan and Edmond (1993) reported DIC signatures ranging from −8‰ to −12‰ for whitewater rivers draining nutrient-rich soils of the Orinoco basin, and values between −12‰ and −23‰ for blackwater rivers. They inferred that the high negative values for black waters reflect a large fractionation from biogenic processes, whereas the less negative values of white waters reflect values close to atmospheric CO_2 or carbonate minerals. The main Orinoco channel shows a gradual transition in DIC δ^{13}C from its headwaters, which receive black waters (−20.1‰), to its whitewater lower reaches (−11.3 to −16.0‰). POC δ^{13}C ranges from −34.8‰ for Venezuelan blackwater rivers to −24.1‰ for whitewater rivers. Provided that consumers of autochthonous production reflect these values, such differences provide a basis for comparing isotopic signatures among various consumers from rivers with different chemical characteristics (Jepsen and Winemiller 2002).

Cichla temensis, S. kneri, and other common fishes and major food web components were collected monthly from the Río Cinaruco in the vicinity of Laguna Larga from October 1993 through June 1994 (Jepsen et al. 1997), and again during January–March 1997 and January 1998, for stable isotope analysis of muscle and liver tissues. Fish were measured and weighed, then muscle and liver tissues (5 g) were removed and preserved in salt. In the laboratory, the tissues were rinsed in deionized water, dried in an oven at 60°C for 48 hr, then ground into a powder (<250 μm) with a mortar and pestle. Powder samples were weighed with a precision balance and then sealed in tin capsules, labeled, and sent to the isotope laboratory at the University of Georgia (Institute of Ecology), where standard procedures were used for isotope analysis.

The isotopic signatures of comparable primary producers and detritus are different in blackwater and whitewater systems (Hamilton et al. 1992; K. O. Winemiller and D. B. Jepsen, unpublished data). Muscle tissue samples taken from a wide variety of fishes from the blackwater Río Cinaruco and the whitewater Río Apure (a major Orinoco tributary ca. 150 km north of the Cinaruco) clearly indicate that fish assemblages of these two ecosystems have different mean δ^{13}C values (Cinaruco –30‰ [$n = 117$], Apure –23‰ [$n = 55$]) (fig. 8.4). Since fish feeding on detritus and periphyton in whitewater ecosystems have more positive carbon isotopic signatures than those feeding in blackwater systems, a shift in isotopic signatures should be associated with the body sizes differentiating juveniles (first-year migrants) from adult *Semaprochilodus* returning to blackwater. A study of migratory *Semaprochilodus insignis* from the Río Negro clearly demonstrated marked ontogenetic shifts in the carbon signature that coincide with early dietary shifts and migration from whitewater to blackwater (fig. 8.5).

To examine how isotopes record ontogenetic diet shifts, muscle tissue from different size classes of *Semaprochilodus* and *Cichla* was examined. Based on Fernandez's (1993) study in Brazil, shifts (on the order of 5–6‰) in δ^{13}C were predicted to coincide with the migration of YOY *Semaprochilodus* into the blackwater Cinaruco (ca. 10–13 cm standard length [SL]). Preliminary data for Cinaruco *Semaprochilodus kneri* show the same isotopic pattern as *S. insignis,* although δ^{13}C values are not as negative at each size interval (i.e., ca. –27‰ for 13–14 cm *S. kneri* vs. –30‰ for 11.5–12.5 cm *S. insignis,* and ca. –32‰ vs. –36‰ for adult *S. kneri* and *S. insignis,* respectively) (fig. 8.6). The Río Negro has a more extreme black-

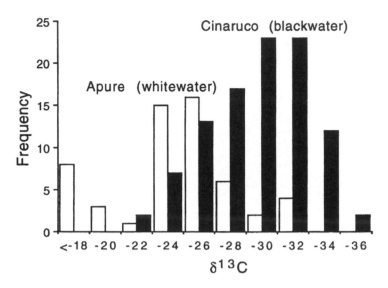

Figure 8.4 Frequency distributions of $\delta^{13}C$ values for fish from the whitewater Río Apure (open bars) and the blackwater Río Cinaruco (solid bars). (Apure mode = −24.8‰, Cinaruco mode = −30.5‰.)

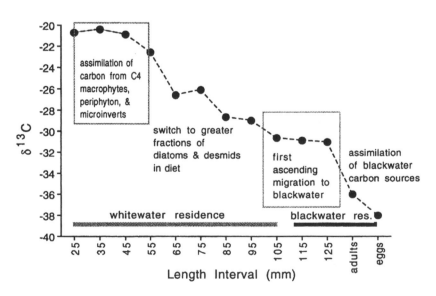

Figure 8.5 Shifts in $\delta^{13}C$ with increasing total length in migratory *Semaprochilodus insignis* of the blackwater Rio Negro and the whitewater Río Amazonas in Brazil. The time from first-feeding larva to the first ascending migration is approximately 4 months. The trend in this system is similar to that shown by Río Cinaruco *S. kneri* in fig. 8.6. (Data from Fernandez 1993 and Forsberg et al. 1993.)

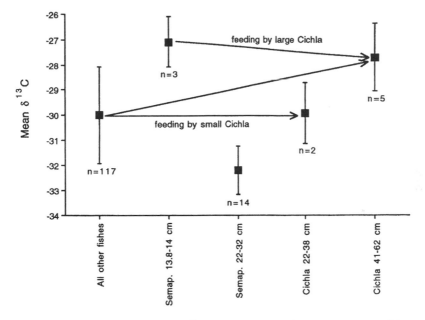

Figure 8.6 Mean values (bars = 1 SE) of $\delta^{13}C$ for a wide variety of blackwater Cinaruco fishes, small YOY *Semaprochilodus kneri* with recent residence in the Cinaruco (SL < 15 cm), large (age 1+) migratory *Semaprochilodus kneri* (SL > 15 cm) in their second or third year of residence in the Cinaruco, small piscivorous *Cichla temensis* (SL < 40 cm), and large *Cichla temensis* (SL > 40 cm). Arrows show pathways for a mixing model of piscivore carbon isotope ratios (assuming $\delta^{13}C$ shifts < 1‰ during digestion and assimilation).

water geochemistry than the Río Cinaruco, which apparently causes a greater shift toward lighter ^{13}C ratios in its biota.

An end-member mixing model (equation [8.1]) was used to evaluate the contributions of whitewater and blackwater carbon sources (W, B) to the biomass assimilated by *Cichla* (Forsberg et al. 1993):

$$\% W = \left[1 - \left(\frac{\delta^{13}C_{consumer} - \delta^{13}C_w}{\delta^{13}C_B - \delta^{13}C_w} \right) \right] \times 100. \qquad (8.1)$$

In the case of carbon, the model had end members that consisted of imported whitewater carbon from YOY *Semaprochilodus* prey at −27‰ versus the weighted mean value for in situ blackwater prey at −28.4‰. The weighted mean for in situ prey was based on findings from analyses of stomach contents and isotope ratios of prey taxa. There were two prey groups that caused mean ^{13}C for consumed in situ prey to be lower than the overall average for the blackwater fish assemblage: the benthivorous cichlid *Geophagus surinamensis* (−27.3‰), which feeds primarily on benthic infauna, which may feed heavily on organic matter derived from terrestrial plants; and midwater characids, *Brycon* spp. (−25.2‰), which feed on fruits and seeds from terrestrial plants plus smaller fractions of terrestrial insects.

The $\delta^{13}C$ values for Cinaruco *S. kneri* of different sizes indicate carbon assimilation histories consistent with their migration from whitewater to blackwater. Muscle tissue samples from larger *Semaprochilodus* have more negative $\delta^{13}C$ signatures that reflect longer histories of consumption and assimilation of blackwater food resources and proportionally less assimilation of whitewater resources (see fig. 8.6). Again, this pattern is the same as that described by Fernandez (1993) for Río Negro-Río Amazonas *S. insignis* (see fig.8.4), in which YOY individuals showed a rapid decline in $\delta^{13}C$ at approximately 12.5 cm SL (the size of entry into black water).

The results of stomach contents analysis (expressed as volumetric proportions) indicated that 45% of the prey ingested by *Cichla temensis* larger than 40 cm SL during the falling-water period were YOY *Semaprochilodus*. This major nutritional input is reflected in the $\delta^{13}C$ signatures (showing more positive, whitewater-like values) of muscle tissue from *Cichla* large enough (>40 cm) to exploit YOY *Semaprochilodus* (8.5–15.5 cm). Smaller *Cichla* (<40 cm) have more negative $\delta^{13}C$ signatures, reflecting feeding on a variety of smaller, nonmigratory fishes that reflect assimilation of in situ sources of organic matter (see fig. 8.6).

Tissue stable isotopic signatures of carbon support the contention that *Semaprochilodus* is the major nutritional component for large *Cichla,* and the assimilation estimate (46%) closely matches the (falling-water period) consumption estimate from stomach contents analysis. The great importance of this nutritional input is reinforced by a more negative $\delta^{13}C$ of liver tissue from adult *Cichla* captured during the mid- and late falling-water period, which also corresponds to the period of peak condition and gonadal maturation in preparation for the major spawning period (April–May).

Ratios of stable isotopes of nitrogen ($\delta^{15}N$) reflect inter- and intraspecific variation in trophic level (with a 3–4‰ $\delta^{15}N$ shift with each successive consumption/assimilation step in a food chain) and have the potential to improve statistical resolution of different carbon sources (e.g., $\delta^{15}N$ for equivalent primary producers can differ by as much as 3‰ between black water and clear water; Jepsen and Winemiller 2002). *Semaprochilodus* revealed little size-based variation in $\delta^{15}N$ (fig. 8.7), indicating that juveniles and adults feed at similar trophic levels. Mean $\delta^{15}N$ for large *Cichla*

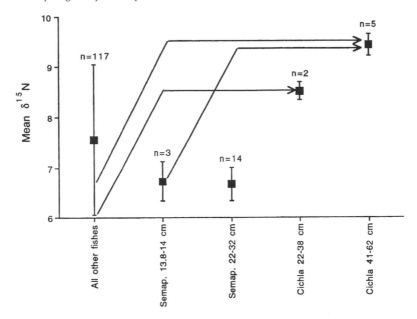

Figure 8.7 Mean values (bars = 1 SE) of $\delta^{15}N$ for a variety of blackwater Cinaruco fishes, small YOY *Semaprochilodus kneri* (SL < 15 cm), large migratory *Semaprochilodus kneri* (SL > 15 cm), small piscivorous *Cichla temensis* (SL < 40 cm), and large *Cichla temensis* (SL > 40 cm). Arrows show pathways for a mixing model of piscivore nitrogen isotope ratios, with nitrogen ratio assumed to increase 3‰ per trophic level. Consequently, both large and small *Cichla* are assumed to feed on a subset of the resident prey fishes with $\delta^{15}N$ lower than the group mean.

was about 3‰ greater than the mean for *Semaprochilodus* and 2‰ greater than the mean for all Cinaruco fishes. This nitrogen pattern indicates the high trophic position of large *Cichla*, the importance of *Semaprochilodus* in their diet, and selective feeding on resident fishes lower than the mean trophic level. $\delta^{15}N$ for small *Cichla* was only approximately 1‰ greater than the mean for all Cinaruco fishes, an indication that these smaller predators feed mostly on small fishes from lower trophic strata (i.e., detritivorous and omnivorous characiforms; Jepsen et al. 1997; Winemiller et al. 1997).

POTENTIAL INDIRECT EFFECTS OF THE SUBSIDY

The Río Cinaruco supports at least 250 fish species from 32 families. Most of these species are small (<10 cm SL) and feed at lower trophic strata (detritivores, algivores, omnivores, invertivores). *Cichla temensis* is an abundant, voracious predator that consumes a diverse array of characiform, siluriform, and perciform fishes (Jepsen et al. 1997; Winemiller et al.

1997), and thus may have the potential to influence the population dynamics of prey. There are other piscivores in the Cinaruco, but none are as abundant within such a wide array of habitats as *C. temensis*. Nocturnal siluriforms are the dominant large piscivores of tropical whitewater rivers, but are uncommon in the blackwater Cinaruco. After *C. temensis,* the most abundant large diurnal piscivores of the Cinaruco are *Cichla orinocensis* (butterfly peacock) and *Serrasalmus manueli* (piranha). *C. orinocensis* is most common in lagoons and shallow shoreline habitats (Winemiller et al. 1997), whereas *S. manueli* is most common in deeper waters, especially in the river channel. Because *S. manueli* is not a gape-limited predator (but rather removes pieces of flesh from prey), it has the potential to exploit migratory prochilodontids, but this has not yet been investigated. *Cichla orinocensis* is a gape-limited predator (ingests prey whole) and rarely grows to 40 cm SL in this system (Winemiller et al. 1997), so feeding on migratory prochilodontids by this species is probably uncommon.

Because *S. manueli* and *C. orinocensis* are abundant and consume a variety of prey taxa, they could influence the population dynamics of prey taxa within their respective habitats under appropriate conditions. Yet because *C. temensis* can consume a greater range of prey sizes and occurs in a wider range of habitats, it should have greater potential to exert top-down effects on the food web. This potential for top-down effects should be increased by the nutritional subsidy it receives from the abundant migratory fishes that import production from a distant landscape.

Could migratory fishes subsidize blackwater river food webs via other mechanisms? It seems unlikely that migratory fishes import significant quantities of dissolved inorganic nutrients into the blackwater system via excretion of material assimilated from the whitewater system. This probably occurs during the initial period of migration, but, given that migrants continue to feed during their migrations, the net import of nutrients is probably insignificant.

Additionally, migratory fishes might have an indirect effect on resident prey populations if they reduce the availability of benthic algae and high-quality detritus in the oligotrophic blackwater system. This hypothesis is being tested with exclosure experiments. Visual observations of foraging fish and stomach contents analysis indicate that *S. kneri* feeds mostly on diatoms and fine detritus taken from sand substrates. Standing biomass of diatoms and fine detritus is extremely low, but no quantitative estimates of benthic production in these systems have yet been made. The large shoals of *S. kneri* that are observed grazing along shallow sandbank areas during the low-water period reduce the availability of benthic algae and detritus

(K.O. Winemiller, unpublished data) for small resident algivores and detritivores (25 species from 5 fish families). Manipulative research and quantitative models are needed to demonstrate the degree to which migratory fishes might enhance diversity via subsidization of piscivores and suppression of competitively superior grazer populations, or reduce diversity via competition for limited benthic primary production.

CONCLUSIONS AND FUTURE RESEARCH

Migratory prochilodontids are major food web components of oligotrophic blackwater ecosystems in South America, and thus may provide major food web subsidies derived from nutrient-rich whitewater floodplains. Based on analysis of predator stomach contents, juvenile *Semaprochilodus* were estimated to provide approximately 45% of the annual biomass ingested by large *Cichla*. Preliminary data for tissue stable isotopic signatures of carbon support the contention that *Semaprochilodus* is the major nutritional component for large *Cichla,* and the assimilation estimate (46%) closely matches the consumption estimate. Moreover, a major fraction (perhaps >50%) of *Cichla* egg production (potential fitness) should be derived from nutritional inputs from whitewater production. This prediction could be explored using information from a mixing model that estimates biomass assimilation using information from stomach contents analysis and stable isotope data from mature oocytes.

We are collecting larger diet and tissue samples to facilitate statistical inferences and to improve estimates of assimilation of whitewater sources of basal production. In addition to comparing fish of different sizes during different periods, we will exploit the spatial variation in the connectivity of the river hydroscape. We should observe a significant difference between large piscivores from lagoons that are isolated from the river channel and those from connected lagoons that receive immigrating detritivores. Individuals from isolated lagoons should have more negative δ^{13}C signatures (indicating assimilation of in situ food sources only) than conspecifics from lagoons connected to the river channel (indicating heavy feeding on immigrant YOY *Semaprochilodus*).

Quantitative estimation of food web subsidies from migratory fishes is important for the conservation and management of Neotropical river fisheries. The ongoing construction of dams in South America will have negative consequences not only for migratory fish populations (Barthem et al. 1991; Barbarino et al. 1998), but also for predator populations in the oligotrophic rivers that receive migratory fishes.

ACKNOWLEDGMENTS

We thank D. Taphorn, A. Barbarino Duque, G. Webb, C. Lofgren, D. Rodríguez Olarte, and H. López Fernández for providing valuable assistance during fieldwork in Venezuela. Funding was provided by grants from the National Geographic Society (nos. 5609-96, 6074-97), Texas Agricultural Experiment Station, and Texas A&M Office of Research (KOW); and a Tom Slick Doctoral Research Fellowship and grants from the Sigma Xi Scientific Research Society and Texas A&M Office of Graduate Studies (DBJ).

Chapter 9

Benthic-Pelagic Linkages in Subtidal Communities: Influence of Food Subsidy by Internal Waves

Jon D. Witman, Mark R. Patterson, and Salvatore J. Genovese

A fundamental goal of ecology is to understand the mechanisms controlling spatiotemporal patterns of population and community structure. As in other environments, the structure of marine communities reflects the integration of ecological and evolutionary processes operating at different spatial and temporal scales. For many years, marine ecologists working on bottom-dwelling (benthic) communities in intertidal or subtidal habitats focused on processes occurring within a single ecosystem and on small spatial scales (i.e., less than a meter to hundreds of meters; local habitat scale). These studies and many others demonstrated the importance of local processes such as competition (Connell 1961b; Dayton 1971; Peterson 1982), predation (Paine 1966; Menge 1976; Lubchenco 1978), disturbance (Dayton 1971; Woodin 1978), mutualism (Glynn 1976; Vance 1978), and larval recruitment (Underwood et al. 1983; Gaines and Roughgarden 1985) and were instrumental in building the foundation of marine community theory.

Since then, there has been a growing body of empirical work indicating that the transfer of food, nutrients, and consumers across habitats and ecosystems plays a large role in the dynamics of marine, terrestrial, and aquatic food webs (Polis and Hurd 1996b; Polis, Anderson, and Holt 1997; Menge, chap. 5 in this volume). Although cross-habitat and cross-ecosystem effects are an integral part of energy flow models (Moloney 1992), they have been largely ignored in conceptual and mathematical models of marine food webs, populations, and communities until recently (Polis and Winemiller 1996; DeAngelis and Mulholland, chap. 2, Huxel et al., chap. 26, and Holt, chap. 7 in this volume).

The marine subtidal zone covers 71% of the globe (Sverdrup et al. 1970). By virtue of its size and its "openness" to fluxes of organic matter and consumers, there are a multitude of cross-habitat exchanges in the marine subtidal zone. For example, major conduits of organic matter and nutrients from terrestrial or intertidal marine habitats to the subtidal zone include terrestrial-subtidal nutrient runoff in estuarine plumes (Day et al. 1989), flows of nutrients from seabird guano (Beckley and Branch 1992; Smith and Johnson 1995), and leaching of N and P from tropical bedrock (Littler et al. 1991). Subtidal sources of kelp detritus are vital to consumers in adjacent intertidal habitats (Stuart et al. 1982; Duggins et al. 1989; Bustamante et al. 1995a) and in the deep sea (Vetter 1994). Human nutrient subsidies to subtidal habitats by way of sewage and runoff from fertilized land must be added to the list of important cross-ecosystem exchanges (see Power et al., chap. 24 in this volume). Striking examples of changes in subtidal community structure brought about by such anthropogenic transfers have been documented in polluted tropical bays, where elevated phosphate levels from sewage have led to increased abundance of macroalgae, subsequently increasing coral mortality (Smith et al. 1981).

In this chapter we review the some of the evidence that cross-habitat or cross-ecosystem linkages influence population and community dynamics in the subtidal zone. Focusing on coupling between pelagic (water column) and benthic ecosystems as a major type of cross-ecosystem exchange, we address the question, Does benthic-pelagic coupling have ecological consequences for individuals, populations, and communities in hard-bottom subtidal habitats? Given that allochthonous food and nutrient supply results from benthic-pelagic coupling, one might predict that food webs would be regulated by bottom-up forces (Hunter and Price 1992) in such situations. If bottom-up forces do not prevail, then it would be informative to consider the factors limiting or obscuring their influence (Power 1992b). Sites of high allochthonous food and nutrient supply may be regulated by joint bottom-up and top-down (predation) forces (Oksanen et al. 1981; Menge 1992; Bustamante et al. 1995b; Menge et al. 1997a; Menge, chap. 5 in this volume).

Oceanographic processes primarily control the links between benthic and pelagic ecosystems, so we stress physical factors at the outset. We then present some of our work on internal tidal phenomena, suggesting that benthic-pelagic linkages in food and nutrient supply are particularly important in topographically raised areas of the sea floor in temperate regions. The subtidal habitats considered here are hard-substrate habitats dominated by plants and animals on the sea floor of rocky ledges, pinnacles, banks and sills, and on the vertical walls of fjords and coral reefs. Our contribution is skewed toward coupling mechanisms associated with internal tidal phenomena (waves, solitons, and bores) because they have far-reaching ecological effects and because we are familiar with them. A soliton is a solitary internal wave, while an internal bore is a breaking internal wave (Mann and Lazier 1991; Herman 1992).

BENTHIC-PELAGIC COUPLING

On a holistic level, more regard should be paid to the physical boundaries between systems and the magnitude and temporal scales of imports and exports.
—Smetacek 1984

Bottom-dwelling populations are usually dependent on the pelagic environment for food and nutrients and for the dispersal of their larvae or propagules. The supply of particulate organic matter to the benthos has long been recognized as a key factor regulating benthic community structure, biomass, and metabolism in soft-bottom habitats (Hargrave 1973; Grassle and Morse-Porteous 1987; reviewed in Gooday and Turley 1990; Graf 1992). Despite Smetacek's (1984) comment and the long history of research emphasizing the importance of allochthonous nutrient inputs to the benthos, it appears that the resurgence of interest in larval recruitment as a process generating pattern on the bottom (Gaines and Roughgarden 1985; Butman 1987; Young 1987) was responsible for motivating large numbers of ecologists to consider how extrinsic events in the water column might influence the structure of bottom-dwelling communities.

Mechanisms of benthic-pelagic coupling can be grouped into the following four categories: (1) sedimentation, (2) freshwater runoff, (3) upwelling (which is ultimately driven by the interaction of wind, large-scale currents, and topography), and (4) mechanisms associated with tidal mixing (internal tidal phenomena and fronts; Longhurst 1981; Mann and Lazier 1991). Sedimentation of organic matter to the deep sea plays a key role in structuring deep-sea communities. It imparts a patch dynamic to the benthic landscape, consisting of ephemeral, organic-rich patches in a background matrix of less enriched sediments (C. R. Smith 1985; Grassle and Morse-Porteous 1987). In addition, a steady, albeit patchy, rain of wood and whale carcasses helps maintain the abundance and diversity of deep-sea communities and may select for the evolution of opportunistic life histories (Turner 1973; C. R. Smith 1985). Sedimentation of phytoplankton to the bottom is common, particularly after spring phytoplankton blooms, on the continental shelf and deeper, where it drives increased benthic metabolism and biomass (Hargrave 1978; Graf 1992). Freshwater runoff from fjord watersheds is a strong cross-ecosystem signal that can regulate predation in the shallow rocky subtidal zone (Witman and Grange 1998). Wind-driven upwelling is predictable on the western margins of continents (Richards 1981; Mann and Lazier 1991), where subtidal to intertidal transfers of nutrients and phytoplankton can increase the growth and abundance of intertidal algae and invertebrates (Bosman et al. 1987; Bustamante et al. 1995b; Menge, chap. 5 in this volume).

Internal waves are a form of tidal mixing that is globally important to benthic-pelagic coupling. Internal waves commonly interact with the bottom in areas of shallow ocean topography (e.g., ledges, seamounts, pinnacles, sills, banks, edges of continental shelves, and coral reefs) and thus probably play important, largely unrecognized roles in structuring benthic communities at local (meters to a few kilometers) to landscape (20–200 km; Mittlebach et al. 2001) spatial scales. As mechanisms of cross-ecosystem (pelagic-benthic) transfer of organic matter and larvae, internal waves can affect benthic communities by causing rapid changes in food and nutrient availability and in temperature and by transporting larvae or propagules to the shoreline (Shanks 1983, 1986, 1995; Pineda 1991, 1995, 1999). They have the potential to influence the structure of food webs, interaction strengths, recruitment rates, patch dynamics, rates of competitive overgrowth, distribution, and diversity in benthic communities. After a brief overview of the physical oceanography of internal waves, we focus on the ecological effects of food and nutrient subsidies by internal wave phenomena.

INTERNAL WAVES

The three basic conditions required for internal waves to form are (1) a stratified water column (i.e., by density), (2) an abrupt change in bottom topography, and (3) tidal energy. Because these conditions are common

(Baines 1974; LeFevre 1986; Shanks 1995), internal waves are widespread in coastal and offshore regions of the world's oceans. Internal waves may be especially common in temperate oceans due to the more distinct density stratification of the water column in temperate than in tropical oceans. Sets of internal waves, referred to as wave "trains" or "packets," can be generated by the fission of a single nonlinear wave (Herman 1992). Internal waves are generated when the tide forces a shallow thermocline over the edge of a topographically raised area of sea floor, causing the flow to be supercritical (i.e., the tidal advection speed exceeds the phase speed of the greatest internal gravity wave mode; Apel et al. 1985). This process produces an internal bore, or hydraulic "jump," which creates a prominent depression of the thermocline structure over the topographic high. When the ebbing currents slacken, the depression, which was held stationary over the raised area of the bottom in the maximum ebb, is released to propagate away into stratified waters, where it can separate into a rank-ordered series of solitons, with the highest-amplitude wave traveling the fastest in a packet (Apel et al. 1985). Internal waves also form when the thermocline and pycnocline (an abrupt density transition in the water column) are disturbed by freshwater runoff from rivers (Zeldis and Jillett 1982). These phenomena are called "internal" waves because they have a minimal signature on the ocean surface, consisting of alternating bands of slicks and ripples, and thus are largely within, or "internal" to, the water column. With wave amplitudes of tens of meters and wavelengths over a kilometer (LeFevre 1986), internal waves are often larger than surface waves. The bands of slicks and ripples typical of internal waves are visible on the ocean surface from space (Apel et al. 1975; LaViolette et al. 1990) and have been detected in all the world's oceans (NASA images).

The biological consequences of internal waves include the mixing of nutrients above and below the pycnocline and large vertical displacements of concentrated phytoplankton constituting subsurface chlorophyll maximum layers (SCMs; Denman 1977; Holligan et al. 1985). Using innovative acoustic techniques, Haury et al. (1979) showed that packets of semidiurnal internal waves cause rapid vertical displacements of the SCM and the zooplankton associated with it in Massachusetts Bay. Zimmerman and Kremer (1984) argued that upwelling driven by internal waves provided pulses of nutrients that were critical to the survival of *Macrocystis* kelp off California during the 1983 El Niño. Similarly, internal waves propagating under ice cover influence algal production in highly stratified sounds (Ingram et al. 1989).

ECOLOGICAL EFFECTS OF INTERNAL WAVES IN THE TROPICS

Internal waves are known from the tropics (Holloway 1987; Wolanski and Pickard 1983), but the significance of internal wave coupling to the ecology of coral reefs was not appreciated until Leichter et al. (1996) discovered that broken internal waves (internal bores) commonly impinge on coral reefs in the Florida Keys, U.S.A. To a certain extent, the internal wave regime described for coral reefs (Leichter et al. 1996) is a reverse image of internal wave phenomena on rocky ledges in the Gulf of Maine (Witman et al. 1993; see below). Instead of downward transport of the warm, phytoplankton-rich water of the SCM to rocky-bottom communities (Witman et al. 1993), internal tidal bores in the Florida Keys slosh cool water that is rich in nutrients, phytoplankton, and zooplankton up the reef slope (Leichter et al. 1996). These internal bores cause dramatic short-term (1–240 min) variation in concentrations of chlorophyll *a*, water velocity, temperature, and salinity (Leichter et al. 1996). Environmental variation associated with internal waves was most pronounced at a depth of 35 m and decreases with depth up to 6 m. Since corals are known to be nutrient-limited (Muscatine and D'Elia 1978), the delivery of nitrate-rich deep water by internal tidal bores may be important for coral growth. Leichter et al. (1996) hypothesized that spatial variation in the nutrient regime created by internal bores influences a broad spectrum of ecological processes, including spatial competition between corals and algae and the growth of corals, sponges, and planktivorous fish. Recent tests of some of these hypotheses indicated that growth rates of a branched scleractinian coral, *Madracis mirabilis,* were higher at the depth of internal wave influence than on the shallow reef (Leichter et al. 1998). Dense concentrations of copepods, larval fish, and crabs were associated with the run-up of cool water from the internal tidal bores, supporting the hypothesis that the internal bores influence zooplankton abundance.

Recent studies in tropical subtidal rock wall habitats of the Galápagos Islands indicate strong benthic-pelagic coupling (upwelling, downwelling) that apparently derives from breaking internal waves (Witman and Smith 2003). In this system, large increases in epifaunal diversity, growth, and abundance of sessile suspension-feeding invertebrates (barnacles, sponges, ascidians) occurred at a site with strong localized upwelling (Witman and Smith 2003). There was some evidence of a linkage between bottom-up and top-down effects, as an estimated 37% of the barnacle biomass produced in a year was consumed by predatory whelks (Witman and Smith 2003). The generality of these effects is currently being tested at replicate upwelling and non-upwelling sites in the Galápagos.

CASE STUDY OF THE ECOLOGICAL EFFECTS OF INTERNAL WAVES IN ROCKY SUBTIDAL COMMUNITIES OF THE GULF OF MAINE

We have been investigating links between the water column and rocky subtidal habitats on Cashes Ledge, a prominent topographic feature of the central Gulf of Maine (fig. 9.1). The ledge rises steeply from 200–300 m deep basins on the eastern and western flanks of the Gulf of Maine to within 12 m of the surface at Ammen Rock (fig. 9.2). The 40 km long ridge trends parallel to the coast and is serrated by many widely spaced pinnacles and knolls with average depths of 28–36 m. The site is characterized by dense populations of sessile invertebrates—particularly suspension feeders, which are dependent on the water column for food (Witman and Sebens 1988; Leichter and Witman 1997). The circulation pattern around Cashes Ledge is a gyrelike offshoot of the eastern Maine coastal current (Brooks 1985), with predictable, northward-flowing surface and bottom currents on the flood tide that reverse to a southward flow on the ebb. The interactions between the ledge and the oceanic flow field cause accelerated flow over the tops of the pinnacles (Witman and Dayton 2001), generate local internal waves (lee waves) on the ebb tide, and are affected by soliton packets propagating from the south that are probably generated at Georges Bank (Witman et al. 1993). Rapid shifts in near-bottom flow speed have been observed at Ammen Rock Pinnacle (ARP) in phase with passing internal waves (fig. 9.3; Genovese 1996; Leichter and Witman 1997; Genovese and Witman 1999). Maximum amplitudes of 27 m and an average period of 10.6 minutes characterized internal waves at ARP (Witman et al. 1993).

The depression of the thermocline and SCM by internal waves causes large, predictable pulses of warm, phytoplankton-rich water to the bottom. For example, bottom temperatures increased by as much as 8.8°C in 10 minutes as chlorophyll *a* concentrations increased two- to threefold (Witman et al. 1993). Chlorophyll *a* concentrations determined by the acetone extraction method were significantly higher in internal wave than in non-wave samples (J. D. Witman, unpublished data). Time series conductivity temperature depth (CTD) casts indicated that the greatest depth to which the thermocline is displaced by internal waves was 40 m (see fig. 9.2).

Simultaneous independent plankton pumps (SIPPs) were designed and fabricated to quantify the nature and magnitude of allochthonous food input to rocky subtidal habitats by internal waves (M. R. Patterson and J. D. Witman, unpublished data). The SIPPs contain a microcontroller and precision temperature sensor so that they can be triggered by the elevated

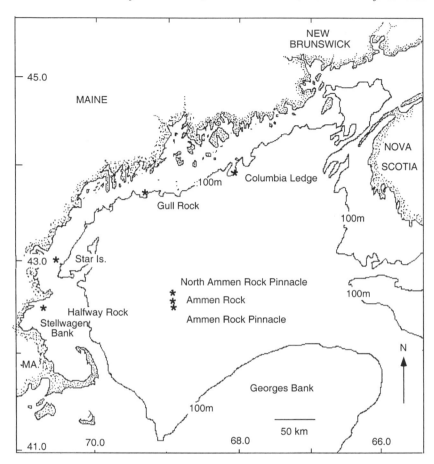

Figure 9.1 Location of study sites. Internal waves are common at all three sites on Cashes Ledge (Ammen Rock, North Ammen Rock Pinnacle, and Ammen Rock Pinnacle; referred to as internal wave sites in text), on Stellwagen Bank, and on Georges Bank. We have not detected internal waves or bores at Halfway Rock, Star Island, Gull Rock, or Columbia Ledge (referred to as non-wave sites in text).

temperature signal of arriving internal waves. Up to eight replicate SIPPs were deployed across the top of ARP (28–33 m depth) to sample food supply by internal waves during cruises in July 1994, July and September 1995, and August 1996. Concentrations of phytoplankton (centric diatoms, *Peridinium, Ceratium*), picoplankton, ultraplankton, and zooplankton (copepod nauplii) were found to be significantly elevated during internal wave events compared with non-wave events sampled immediately after the internal wave passed by (J. D. Witman and M. R. Patterson, unpublished data). Not all internal waves had similar effects on food supply to the

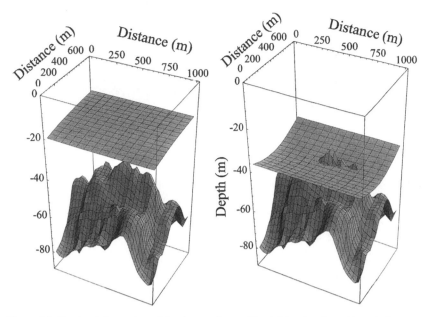

Figure 9.2 Simulated depression of the thermocline and food-rich subsurface chlorophyll maximum layer (SCM; gridded layer above pinnacle on left) by an internal wave at Ammen Rock Pinnacle (ARP), Cashes Ledge, showing the large scale of this cross-ecosystem mechanism. The pinnacle is approximately 1 km by 760 m and drops off steeply on the north slope (shown) to 80 m depth (vertical scale exaggerated). The actual bottom topography, which was reconstructed from sonar soundings, is depicted. Internal wave effects (increased phytoplankton and zooplankton supply to the bottom, rapid temperature increases) predictably extend to 40 m depth. These two snapshots are separated in time by periods as short as 7 minutes.

bottom, however, as there was significant variation between waves on some days.

Individual Growth Responses

Growth across Depth Gradients of Internal Wave Influence
Because the internal wave signal was strong at the top of Cashes Ledge and diminished sharply with depth, we used depth as a proxy for the internal wave effect in experiments. To test the general null hypothesis that the growth of suspension-feeding invertebrates did not differ between depths in the internal wave zone (28, 35 m) and those in the non-wave zone (50 m), we conducted a reciprocal transplant experiment at two pinnacles on Cashes Ledge with two species of sponges, an encrusting bryozoan, a mussel, a sea anemone, and a solitary ascidian on cruises between 1989 and 1992. These species were selected for the transplant

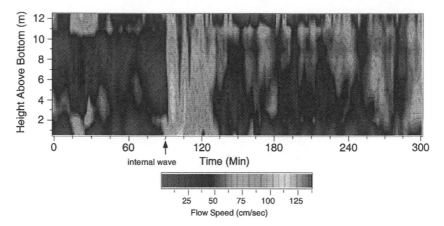

Figure 9.3 Profile of water flow speeds at ARP from 0.5 m off the bottom to a height of 12 m above the bottom (in the water column) obtained by a bottom-mounted Acoustic Doppler Current Profiler on 15–16 July 1995. Strong benthic-pelagic coupling created by a packet of internal waves is indicated by the 5 closely spaced light gray bands after the arrow. Simultaneous temperature and fluorometer data (not shown) indicated that each band was synchronous with a vertical displacement of the thermocline and SCM. Note that flow speeds increased by a factor of 2 or more during internal wave passage, and that high internal wave flow extended from 12 m above the bottom to the bottom during these events.

experiment to ensure that a range of suspension-feeding modes utilizing the wide spectrum of food resources associated with the internal waves were represented. The suspension feeders in this community included active suspension-feeding sponges feeding on picoplankton (Pile et al. 1996), mussels, bryozoans, and ascidians feeding on phytoplankton, and passive suspension-feeding anemones consuming zooplankton (Sebens and Koehl 1984).

We collected organisms from their original habitat (depth) at Ammen Rock Pinnacle and North Ammen Rock Pinnacle (ARP, NARP), measured their original size, and transplanted them to experimental substrata (slate tiles or plastic mesh). Bryozoans colonized the tiles a year prior to the experiment. Tiles were then moved up or down the slope to one of three depth stations (28, 35, or 50 m), with controls established by returning suspension feeders to their original habitat. Mussels were transplanted in dome-shaped mesh cages attached flat to the bottom. Sea anemones (*Metridium senile*) escaped through the mesh of transplant cages, so the hypothesis was tested by measuring changes in the basal area of anemones in permanent quadrats on rock walls between 35 and 28 m depth. Submersibles were used to set up the 50 m transplants, while the shallower experiments were set up using scuba gear. The experimental design included two plots at each depth separated by 5 to 15 m horizontal distance to avoid pseudoreplication in testing for the depth effect (Hurlbert 1984). Control and treatment organisms were treated identically. Storms modified the experimental design by eliminating some of the shallow (28 m) experiments at ARP, so a completely balanced design with replicates for all depths and sites was not possible. Experiments affected by the storm were repeated.

Growth responses of suspension feeders in the internal wave zone were striking, as shown in a representative plot of horse mussel (*Modiolus modiolus*) growth (fig. 9.4). ANCOVA revealed significant differences in the slopes of growth regressions in and below the zone of internal wave influence at ARP (table 9.1). Mussels on the peak (28 m) and slope (35 m) had significantly faster growth rates, as measured by changes in shell length, than those moved downslope from either depth to the 50 m station. In addition, the treatment group transplanted from 28 to 35 m grew significantly faster than the group moved down to 50 m. There was no difference in growth rate between the control groups at 28 and 35 m depths. However, the control mussels at 28 m had significantly faster growth rates than those transplanted from 28 to 35 m. The main depth-specific growth differences at the replicate site (NARP) were similar to those at ARP; that is, mussels grew faster in the internal wave zone (28, 35 m sites) than below it (50 m; table 9.1).

The encrusting bryozoan *Parasmittina jeffreysi* showed large growth differences in colony area between the 35 m and 50 m depth stations over one year at the one site where the experiment was conducted (ARP). ANOVA indicated a significant effect of depth on *Parasmittina* growth ($F = 12.13$[***], d.f. 3, 189, data log $x + 1$ transformed). Post hoc comparisons with Fisher's least significant difference test indicated that the mean growth of the 35 m control group (1.6 cm^{-2} area) was significantly greater than the mean growth of the 50 m controls (0.9 cm^{-2} area, $p \leq .05$). Treatment groups moved upslope from 50 to 35 m showed the largest increase in mean colony growth (2.4 cm^2 per year; J. D. Witman and S. J. Genovese, unpublished data).

A summary of the growth experiments is shown in figure 9.5, since it is not possible to present a detailed treatment of the growth responses of all species in this review. Growth rates were significantly higher at the 28 m and 35 m depths influenced by internal waves than deeper (50 m) for the horse mussel *M. modiolus* (both sites) and the sponge *Myxilla fimbriata*, so the general null hypothesis can be rejected. Because the flux of food particles decreases with depth, even between the 28 m and 35 m sites (Genovese and Witman 1999), which on the north slope of ARP

Table 9.1 Results of analysis of covariance (ANCOVA) and Tukey multiple comparisons tests comparing slopes of regression lines of growth in mussel transplant experiments

Group	Regression equation	F	ANCOVA d.f.	ANCOVA F	Multiple comparisons
Ammen Rock Pinnacle					
28 m C	$y = 0.696 + 0.751x$	240.3***	3, 100	17.0***	28 m C vs. 35 m C, ns
35 m C	$y = 0.423 + 0.800x$	405.5***			28 m C > 28–35 m T*
28–35 m T	$y = 0.509 + 0.649x$	696.3***			28 m C > 28–50 m T*
28–50 m T	$y = 0.229 + 0.584x$	449.4***			28–35 m T > 28–50 m T***
					35 m C > 28–50 m T***
					35 m C vs. 28–35 m T***
North Ammen Rock Pinnacle					
28 m C	$y = 0.296 + 0.861x$	247.0***	4, 104	6.13***	28 m C vs. 35 m C, ns
35 m C	$y = 0.393 + 0.811x$	881.0***			28 m C vs. 28–35 m T, ns
28–35 m T	$y = 0.376 + 0.716x$	1,013.6***			28 m C vs. 28–35 m T, ns
28–50 m T	$y = 0.122 + 0.639x$	238.3***			28 m C > 28–50 m T**
35–28 m T	$y = 0.395 + 0.805x$	685.0**			35 m C > 28–50 m T***
					35 m C vs. 28–35 m T, ns
					35 m C vs. 35–28 m T, ns
					28–35 m T > 28–50 m T***
					35–28 m T > 28–50 m T***
					28–35 m T vs. 35–28 m T, ns

NOTE: See figure 9.4. Regression equation represents log of shell length (mm) in 1991 (y) on log of shell length (mm) in 1990 (x). The slope is the growth rate (in shell length) of the treatment or control mussels. Multiple comparisons tests indicate which groups differed significantly ($p < .05$), T = a treatment group, i.e., moved up or downslope; C = a control group, i.e., transplanted back to original depth. For all regressions d.f. = 1, 20 except for ARP 28–50 T (1, 40) and NARP 28–50 T (1, 30) groups. Groups separated by an inequality sign were significantly different, with the sign indicating the direction. m = depth in meters.

*$p < .05$; **$p < .025$; ***$p < .0001$; ns = nonsignificant

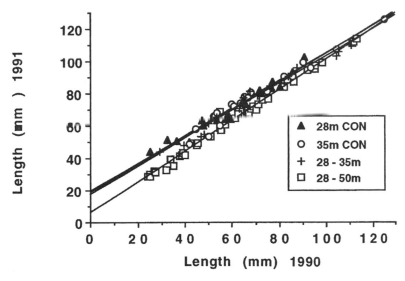

Figure 9.4 Results of one of the transplant experiments testing the null hypothesis of no difference between the growth of sessile invertebrates in the internal wave zone (28, 35 m depths) and below it (50 m) at Ammen Rock Pinnacle. One-year growth in shell length is depicted for the horse mussel, *Modiolus modiolus*. A comparison of the slopes of regression lines indicated slower growth rates for mussels moved below the internal wave zone (to 50 m) than for those in the internal wave zone (see table 9.1).

were about 75 m apart, we predicted that the passive suspension-feeding anemones in the more energetic internal wave regime (28 m depth) would have the fastest growth rates. This prediction was supported by significantly higher growth at the shallowest depth (28 m) compared with the 35 m depth at ARP for the anemone *M. senile,* the sponge *Halichondria panicea,* and the ascidian *Ascidia callosa* (J. D. Witman and S. J. Genovese, unpublished data).

Because internal waves in the central Gulf of Maine increase water temperature, affecting the respiration and pumping rates of suspension feeders while simultaneously increasing food availability, the effects of wave events are difficult to predict. Even so, a common result of the transplant experiments was that suspension feeders grew more in the internal wave zone than below it (50 m). To better resolve the biological effects of internal waves, we developed a bioenergetic model to predict the influence of internal waves on the scope for growth of an active (*Modiolus modiolus*) and a passive (*Metridium senile*) suspension feeder (M. R. Patterson and J. D. Witman, unpublished data). The model indicated that the number of

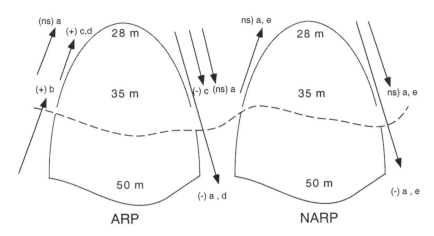

Figure 9.5 Summary of major results of individual growth experiments across depth gradients at Ammen Rock Pinnacle (ARP) and North Ammen Rock Pinnacle (NARP). The curved horizontal line shows the typical lower depth of the internal wave zone (~ 40 m depth). (+) indicates a significant increase in growth; (–), a significant decrease in growth; (ns), no significant difference. Species: a = *Modiolus*, b = *Parasmittina*, c = *Halichondria*, d = *Ascidia*, e = *Myxilla*.

internal waves passing by the pinnacle per day would increase energy gain proportionally in both species; however, the passive suspension-feeding *Metridium* is more dependent on the allochthonous food source supplied by internal waves than the active suspension-feeding mussel *Modiolus* (M. R. Patterson and J. D. Witman, unpublished data).

Within depths influenced by internal waves, there were differences in the growth of mussels (*Mytilus edulis*) and sponges (*Halichondria panicea*) related to small-scale topographic effects on flow (Leichter and Witman 1997). Active suspension feeders (mussels) grew faster on the tops of 1.5 m high rock walls, where flow speed was greatest, while the maximum growth rates of facultatively active suspension feeders (sponges) were greater in low flow at the bases of rock walls.

Growth Differences between Sites with and without Internal Waves
Horse mussels, sea anemones, and encrusting bryozoans have been transplanted between coastal and offshore rocky subtidal sites in the Gulf of Maine to test for growth differences between the offshore food supply regime mediated by internal waves and that in the coastal zone where internal waves are absent (Lesser et al. 1994; Genovese and Witman 1999; J. D. Witman, S. J. Genovese, and K. P. Sebens, unpublished data). In con-

trast to the "vertical" growth experiments described above, which were conducted along a depth gradient at internal wave sites, "horizontal" distance across the Gulf of Maine shelf was the factor of interest and depth was constant in these experiments. Lesser et al. (1994) documented a higher scope for growth and growth rates for the passive suspension-feeding anemone *Metridium senile* in the higher flow regime of one internal wave site (ARP). Conversely, *Modiolus modiolus,* an active suspension feeder, grew faster in the lower-flow regime of the coastal zone. Whether or not these results are representative of a true regional effect is unknown because of the lack of site replication. Genovese and Witman (1999) found a greater flux of chlorophyll *a* offshore, but no significant differences in growth rates of the bryozoan *P. jeffreysi,* transplanted on tiles and in natural communities, between two coastal sites without internal waves and two internal wave sites offshore. This result was apparently due to the detrimental effects of spatial competition and high flow speed on bryozoan growth offshore.

Population Growth Responses

Can one extrapolate from the results of manipulative growth experiments conducted with individual organisms over short time periods to natural population and community dynamics on longer temporal scales?

We predicted that rates of population growth would track the growth rates of individual suspension feeders; that is, that population growth would be greatest at the depths and sites where internal waves occur. This hypothesis was tested by monitoring changes in the abundance of sea anemones, sponges, and bryozoans in permanent quadrats on rock walls at internal wave sites and non-wave sites (see fig. 9.1). Population growth was estimated from the abundances summed over all quadrats for each year (t) as N_t/N_{t-1}, which is the finite rate of increase, λ, and thus an estimate of the annual rate of population growth, since the time interval is one year. Populations are increasing when λ is greater than 1.0 and decreasing when λ is less than 1.0 (Caswell 1989). Long-term data from two internal wave sites and four non-wave sites have not yet been completely analyzed, but preliminary results indicate that the population growth of sea anemones (*Metridium senile*) at one of the internal wave sites (ARP) was highly variable, decreasing abruptly from 1986 to 1987 and from 1989 to 1990 and recovering somewhat between 1987 and 1989, but with λ remaining below 1.0 from 1990 to 1994 (fig. 9.6A). Severe predation on anemones by the nudibranch *Aeolidia papillosa* was responsible for the

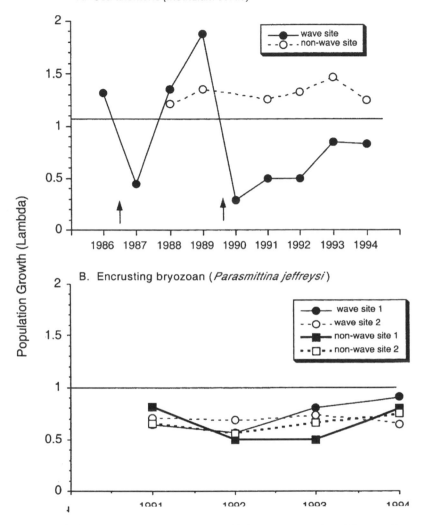

Figure 9.6 (A) Changes in population growth rate (λ) for sea anemone (*Metridium senile*) populations in permanent quadrats (30 m depth) at a site with internal waves (Ammen Rock Pinnacle) and a non-wave site (Columbia Ledge). Populations are increasing when λ is greater than 1.0. Note the large fluctuations in λ at the internal wave site, which were caused by predation from nudibranchs (arrows). Anemone population growth remained positive at the site lacking internal waves. (B) Population growth of the encrusting bryozoan *Parasmittina jeffreysi* in monitored quadrats at sites with and without internal waves (wave sites 1 and 2 are on the flanks of Ammen Rock Pinnacle, while non-wave site 1 is at Halfway Rock and non-wave site 2 is at Star Island). Note that bryozoan populations declined at all sites. (Data from Genovese 1996.)

sharp reductions in population growth at ARP (Witman 1998). *A. papillosa* is a voracious subtidal predator of sea anemones in temperate-boreal regions (Harris 1976). It has pelagic veliger larvae that settle and grow rapidly to 3–4 cm body length in several months, at which time it can consume 50–100% of its body weight in each feeding bout (Harris and Howe 1979). In the present study, the same *Aeolidia* predation also devastated *Metridium* populations at the other internal wave site (NARP; J. D. Witman and S. J. Genovese, unpublished data). In contrast, anemone populations at a site without internal waves grew rapidly from 1987 to 1994 (fig. 9.6A). These results suggest that a bottom-up effect of internal waves on the individual and population growth rates of sea anemones may have triggered a top-down regulation by predatory nudibranchs.

In contrast to the sea anemones, the population growth of the encrusting bryozoan *Parasmittina* was unaffected by predation (Genovese 1996; Genovese and Witman 1999). There was no trend of greater population growth at internal wave sites than at non-wave sites, despite the higher concentration and flux of chlorophyll *a*-containing food particles at internal wave sites (fig. 9.6B; Genovese 1996; Genovese and Witman 1999). For example, population growth rates of *Parasmittina* (λ) were less than 1.0 at two non-wave sites and at one wave site during 1990–1994. Population declines were attributed to the negative effects of interspecific overgrowth competition by other sessile invertebrates and to comparatively higher physical disturbance at one internal wave site (Genovese 1996). High sedimentation decreased colony survivorship at the coastal sites lacking internal wave activity (Genovese 1996; Genovese and Witman 1999).

FOOD WEB LINKAGES

The results of our observations and experiments are summarized in a food web for a site with internal wave activity in the Gulf of Maine (fig. 9.7). We postulate that food subsidy by internal waves has effects cutting across benthic-pelagic ecosystems and on multiple trophic levels at high-exchange sites. Bottom-up effects are often short-term, however, and they may be suppressed by top-down control (predation), competition, or physical disturbance.

The supply of planktonic food to suspension feeders is subsidized by internal waves, with individual growth effects seen for sponges, mussels, bryozoans, ascidians, and sea anemones. Calanoid copepods are known to track and graze SCM layers in the Gulf of Maine (Townsend et al. 1984), and

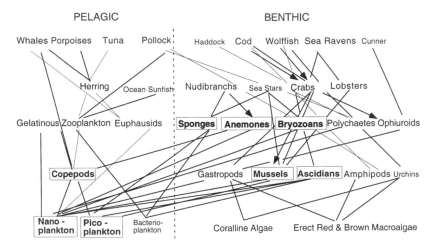

Figure 9.7 Cross-ecosystem connectedness food web from an internal wave site (ARP) in the Gulf of Maine, showing joint bottom-up and top-down regulation. The abundances or individual growth rates of groups enclosed in boxes and in boldface type are increased by the benthic-pelagic coupling driven by internal waves. Downward arrows show top-down control (e.g., anemones regulated by nudibranch predation). Web connections are based on gut contents of predators sampled on site (Witman and Sebens 1992; J. D. Witman and K. P. Sebens, unpublished data) and from Bigelow (1926). Groups in small type are less abundant.

they are abundant at the internal wave sites on Cashes Ledge. Copepods and euphausiids are an important food source for gelatinous zooplankton, fish, and whales (Bigelow 1926).

Several lines of evidence indicate that strong linkages between bottom-up and top-down forces result from the cross-ecosystem effects of internal waves. These findings include (1) regulation of anemone population growth by nudibranch predation at internal wave sites where the growth rates of individual anemones are greatest; (2) selective predation on *Mycale lingua,* one of the sponges at ARP utilizing picoplankton supplied by internal waves (Pile 1996; Pile et al. 1996), by the sea star *Henricia sanguinolenta* (Shellenbarger 1994); and (3) high predation by wolffish (*Anarhichas lupus*) on horse mussels (*Modiolus modiolus*) transplanted from the coastal zone to the depths at ARP, where mussel growth is significantly enhanced by internal waves (see figs. 9.4 and 9.5; Witman and Sebens 1992). Moreover, Witman and Sebens (1992) hypothesized that other mobile and sessile invertebrate species at internal wave sites offshore in the Gulf of Maine are controlled by cod (*Gadus morhua*) predation. These links are between cod and crabs, between cod and tubicolous polychaetes and amphipods, and between cod and brittle stars (see fig. 9.7).

A remarkable study (Hunt et al. 1998) demonstrated the potential effects of internal waves on food webs in the Aleutian Islands. Internal waves generated from underwater sills between islands in the Aleutian chain influenced zooplankton aggregations, in turn affecting the periodicity of feeding by seabirds (auklets) at the surface of the ocean.

Our results support the prediction that predation increases with ecosystem productivity (Oksanen et al. 1981; Menge et al. 1996), given the high predation on anemones and sponges at internal wave sites. Herbivory can similarly regulate the bottom-up signal of accumulating plant biomass (Hairston et al. 1960). Interpretation of the high predation by fish observed at internal wave sites as a support for the ecosystem productivity model is confounded by overfishing of these key predators (cod, wolffish) at the non-wave sites in the coastal zone (Witman and Sebens 1992). Since many subtidal food webs bear the stamp of overfishing at high trophic levels (Dayton et al. 1995), it may be especially difficult to test the relation between productivity and predation, or other hypotheses about joint bottom-up and top-down control, in subtidal food webs containing harvested species (Steneck and Carlton 2001). In some cases, however, subtidal populations of predators controlled by bottom-up forces have few predators at higher trophic levels (Witman et al. 2003).

In summary, an important conclusion from these studies is that the effects of internal waves on individual growth rates do not necessarily translate into increased growth rates of populations, nor do they necessarily predict increased abundances of species at the community level. This is partly due to high removal of individuals by disturbance or predation. Predation may be higher in subtidal than in intertidal habitats (Witman and Dayton 2001) and thus may commonly mask bottom-up effects in the subtidal zone unless prey populations are tracked frequently enough to reveal top-down effects (Witman et al. 2003).

SUMMARY AND FUTURE DIRECTIONS

It is now established that internal wave phenomena deliver warm, shallow water to bottom-dwelling communities in the rocky subtidal zone and deep, cool water enriched in phytoplankton, nutrients, and/or zooplankton to coral reefs. Sites of such high cross-habitat or cross-ecosystem exchange between the bottom and the water column can be predicted in part by abrupt changes in the topography of the bottom.

The spatial range of this benthic-pelagic cross-ecosystem coupling is very large, extending from spatial scales of tens of kilometers in terms of packets of internal waves at the surface to variation on the scale of tens of meters and centimeters on the bottom. Internal waves provide an unparalleled opportunity to investigate the spatial and temporal scaling of physical-ecological linkages from space (e.g., remote sensing of packets of internal waves at the surface) to the bottom of the ocean. Since the conditions for the propagation and intensity of internal wave activity depend on the temperature stratification of the water column, this important cross-ecosystem coupling will be sensitive to changes in climate.

Cross-ecosystem coupling by internal waves has many potential ecological effects. Few of these effects have been investigated to date, but increased growth rates in individual organisms using food and or nutrients supplied by internal waves is the most common response, although it may not always occur. The effects of internal wave-mediated food and nutrient regimes on the reproductive output of benthic organisms have not been investigated. The effects of the large, short-term temperature fluctuations associated with internal waves on the ecology and physiology of benthic organisms needs to be investigated further. For example, it is possible that the rapid temperature changes associated with the onset of internal waves may act as a stimulus to spawning in benthic organisms over large spatial scales, particularly at the beginning of the stratified season (e.g., late spring in temperate oceans). Whether these rapid temperature fluctuations are stressful to organisms living in internal wave regimes has not been investigated. Following the early lead of Zimmerman and Kremer (1984), more work needs to be conducted on the responses of benthic plants to enhanced nutrient supply driven by internal waves or other oceanographic benthic-pelagic coupling phenomena. This nutrient-plant linkage may be an unrecognized bottom-up linkage influencing marine grazers. There is enough preliminary evidence that internal wave effects extend up to vertebrates to warrant future research on the effects of internal waves at higher trophic levels (i.e., fish, marine mammals, and birds). Investigation of the influence of internal waves on community structure and dynamics (e.g., interaction strengths, competition, distribution, and diversity) is in its infancy, although this is a promising area of future research.

Perhaps not surprisingly, it is not always possible to extrapolate from the results of manipulative growth experiments conducted with individual organisms over short time periods to natural population and community dynamics. The positive effects of increased food supply on organism growth at the individual level do not necessarily result in increased growth of populations. In some cases, as when predators quickly remove prey populations stimulated by bottom-up effects, these observations may reflect linkage between bottom-up and top-down control. Our work in the Gulf of Maine and in the Galápagos Islands suggests that cross-ecosystem coupling by internal waves links bottom-up and top-down forces at some rocky subtidal sites.

ACKNOWLEDGMENTS

We are grateful to the National Science Foundation, the National Undersea Research Program at the University of Connecticut (Avery Point), and the Andrew Mellon Foundation for supporting this research.

Chapter 10

Effect of Landscape Boundaries on the Flux of Nutrients, Detritus, and Organisms

M. L. Cadenasso, S. T. A. Pickett, and K. C. Weathers

BOUNDARIES AND FOOD WEBS IN MOSAIC LANDSCAPES

The goal of this book is to place food webs in a landscape context and to explore how this placement may enhance our understanding of web structure and dynamics. The merging of landscape and food web ecology requires that the spatial heterogeneity of the system in which the food web resides be considered. An important component of heterogeneity is landscape boundaries, but this component has been little studied. The role of boundaries can be understood from the disciplinary perspectives of landscape ecology and food web theory. Landscape ecology recognizes that boundaries are functional components of mosaic landscapes. Food web theory has not yet explored the role of boundaries explicitly, but boundaries are implicitly incorporated into studies of cross-system subsidies to food webs. We will begin by briefly outlining the understanding of the role of landscape boundaries from these two disciplinary perspectives.

Landscape Ecology Perspective

Landscape ecology is the disciplinary home for the study of spatial heterogeneity. Taking a landscape approach to food webs requires superimposing a spatially explicit template over other factors already incorporated into food web studies. Spatial heterogeneity can be described by recognizing patches of the earth's surface that differ in composition, structure, or the properties of some attribute such as vegetation, soil, parent material, or water. Landscapes are mosaics of such patches connected by fluxes of material, organisms, and energy (Forman and Godron 1986). These mosaics necessarily contain boundaries. To understand the function of mosaic landscapes at any scale, ecologists need to understand how patches, boundaries, and fluxes interact (Wiens et al. 1985; Forman 1995; Pickett and Cadenasso 1995; Weathers et al. 2000b).

Food Web Theory Perspective

The study of food webs has evolved from describing links and feedbacks between organisms to placing the food web within its environmental context (Winemiller and Polis 1996). This advance required recognizing influences on web structure and dynamics that arise from outside the collection of member organisms, their life histories, and their interactions with one another. Because organisms occupy patches differing in biotic and abiotic characteristics and processes, the trophic web may also vary by patch (Polis et al. 1996). Therefore, a mosaic landscape can consist of different patches, each characterized by a different food web.

We now recognize that in addition to the patch-specific drivers of food web structure, processes in adjacent patches can affect food web structure and dynamics in a particular patch (Polis, Anderson, and Holt 1997). Several chapters in this volume contribute to this critical advance in food web theory by addressing between-system fluxes of materials that function to subsidize the food web of the receiving system. For example, fluxes of subsidies between streams and adjacent riparian and upland forests (Power et al., chap. 15, and Willson et al., chap. 19 in this volume), between marine and island systems (Polis et al., chap. 14 in this volume), and between marine and freshwater systems (Riley et al., chap. 16 in this volume) have all been discussed. These studies demonstrate that fluxes of food web subsidies link patches in mosaic landscapes, and this functional link between patches is an important enhancement to the study of both food webs and landscape ecology.

Though these chapters have considered the links between adjacent patches provided by the flux of subsidies, we have yet to consider what *controls* the flow of subsidies between patches. Therefore, the field of food web ecology is poised to examine the control of fluxes between landscape patches. Fluxes between systems must cross system boundaries, and boundaries may regulate the exchanges of subsidies between patches. Consequently, boundaries may influence the trophic structure and dynamics of food webs. This chapter will focus on the role of boundaries in regulating the flow of food web subsidies between systems. We will explore the structure and function of landscape boundaries and, using data from our research on forest edges, discuss how boundaries control the flux of potential food web subsidies such as nutrients, detritus, and organisms. Key features of the recipient or donor system also influence cross-system subsidies (Witman et al., chap. 22 in this volume). Our goal, however, is to assess how the understanding of food webs can be enhanced by accounting for the function of boundaries in the landscape.

OVERVIEW OF BOUNDARY STRUCTURE AND FUNCTION

Boundaries are physical and compositional discontinuities in the landscape. For instance, boundaries between fields and forests represent large shifts in both the physical structure and the composition of the community. In contrast, a boundary that is primarily compositional is represented by boundaries between forested riparian zones and adjacent upland forests. Boundaries between terrestrial and aquatic systems are extreme shifts in both structure and composition, as are terrestrial-atmospheric boundaries. The discontinuities of structural boundaries may occur at various spatial scales and can be characterized and investigated at the scale of kilometers or at the scale of meters, whichever is appropriate for the phenomenon in question (Cadenasso and Pickett 2000).

Boundaries are created and maintained by a variety of processes that can influence their structure. A boundary may be very abrupt or gradual, depending on the spatial extent over which the transition occurs and the way the boundary is formed. Boundaries that are maintained by humans, such as some types of forest edges, are frequently abrupt. Other structurally abrupt boundaries may be formed by sudden changes in underlying physical geomorphology, such as mountain ranges and transitions from land to water, or by natural disturbances such as fires, large forest blowdowns, or lava flows. Abrupt boundaries are often characterized by steep gradients of abiotic and biotic change. More gradual boundaries may occur, however, because of a relaxation of the process that originally created them or because the underlying causal gradient is shallow or diffuse. For example, boundaries between forests and adjacent open fields in the northeastern United States may change from abrupt to gradual if mowing or some other form of management is discontinued and trees become established in the field. Gradients of soil moisture extending out from lowland streams are frequently gradual and patchy. These gradients may be manifested in the aboveground vegetation by a gradual transition from a dominance of riparian species to a dominance of upland forest species. There is a reciprocal and dynamic interaction between the structure of a boundary and the steepness of abiotic and biotic gradients across it.

We have suggested that, rather than thinking of boundaries first or primarily as structures, they should be thought of as functional components of mosaic landscapes (Pickett and Cadenasso 1995). What is unknown is whether physical discontinuities in structure are concomitant with important functional discontinuities. Boundaries are expected to be hotspots of ecological change because they are the primary site for interactions between two systems. Organisms, materials, and energy can all move across boundaries, and boundaries may regulate the magnitudes of exchanges between habitats (Wiens 1992; Forman 1995; Pickett and Cadenasso 1995; Cadenasso 1998; Cadenasso and Pickett 2001). Rates of ecosystem inputs (Potts 1978; Hasselrot and Grennfelt 1987; Draaijers et al. 1988; Weathers et al. 1992, 1995, 2000a, 2000b, 2001) and processes (Cadenasso 1998; Weathers et al. 2001) may also change at boundaries. However, there are few studies that examine the control of fluxes by boundaries, and the assertion that boundaries regulate fluxes between ecosystems remains an important, largely untested hypothesis (Wiens 1995; but see Cadenasso and Pickett 2000, 2001).

Up to this point, we have explored how understanding the role of boundaries could enhance the development of landscape and food web ecology and how both of these disciplinary perspectives are prepared to incorporate the function of boundaries explicitly into their frameworks. We have provided an overview of how boundaries may be formed, maintained, and changed, and we have suggested that boundaries function to regulate the fluxes of materials, energy, and organisms between landscape patches. Now we will demonstrate how the function of boundaries can be experimentally tested, and we will employ a conceptual model we have developed to summarize our thinking and organize our research (Pickett and Cadenasso 1995). The model consists of three components: any two systems or patches and the boundary between them. Fluxes between the two systems or patches must cross the boundary, and the boundary may inhibit the flux, enhance the flux, or be neutral to the flux. Regulation of fluxes by the boundary may be affected by the structure and dimensionality of the boundary, by the type of flux, and by the types of systems or patches in contact at the boundary (Cadenasso and Pickett 2000, 2001). In this discussion we have been using the inclusive term "boundary" to encompass transitions between all types of systems and to facilitate the search for a generalizable effect of boundaries on food webs. However, we tested our conceptual model using a specific type of boundary—the boundary between deciduous forests and adjacent old fields—commonly referred to as a forest edge. All of the research presented here has been done in the Hudson Valley, New York, U.S.A.

EXAMPLES OF BOUNDARY STRUCTURE AT FOREST EDGES

Quantitative research on forest edges has focused on describing gradients of change in abiotic and biotic variables with distance from the edge (table 10.1; reviewed by Murcia 1995). Deviations from these trends are usually due to specific conditions at a study site, such as the presence of a young forest (Williams-Linera 1990; Bierregaard et al. 1992; Kapos et al. 1997), disturbance of the forest edge (Bagnall 1979; Palik and Murphy 1990), or creation of a new edge (Bruner 1977; Ranney et al. 1981; Chen et al. 1992). Deviations from the expected general trends can also be due to differences in the way in which the data are collected, such as the use of different definitions of the "zero" point for transects traversing the edge zones (Brothers and Spingarn 1992; Matlack 1993) or differences in the relative distances into the forest that a study extends (Cadenasso et al. 1997). Research on forest edges, to date, has been largely static and descriptive. An exception to this general approach has been research on avian nest predation and parasitism, which has assessed how the forest edge interacts with the activity of nest predators and, as a result, how the edge functions to regulate nest predation (Andrén et al. 1985; Andrén and Angelstam 1988; Andrén 1995).

The static and descriptive approach is, of course, a necessary first step and provides a strong foundation for the dynamic and experimental research that will be necessary to further the understanding of edge function. This foundation also provides a platform for the integration of research on community and ecosystem processes, abiotic and biotic factors, and the structure and function of boundaries. Based on assessment of the empirical data and on the conceptual analysis of edges, there have been increasingly frequent and well-reasoned calls for research on the function of forest edges (Wiens et al. 1985, 1993; Wiens 1992, 1995; Angelstam 1992; Forman and Moore 1992; Forman 1995; Murcia 1995). The empirical studies have been very slow in coming, however. To elucidate the functional role of edges in the landscape empirically, ecologists need to determine how edges interact with fluxes of organisms, materials, and energy (Pickett and

Table 10.1 Abiotic and biotic gradients across forest edges

Change with distance into forest	Gradients			
	Biotic	Sources	Abiotic	Sources
Increase	Average basal area of trees	López deCasenave et al. 1995	Humidity Soil moisture	Matlack 1993; Chen et al. 1995 Kapos 1989; Camargo and Kapos 1995
Decrease	Species richness	Gysel 1951; Wales 1972; Ranney et al. 1981; Brothers and Spingarn 1992; Fox et al. 1997	Light	Geiger 1965; Kapos 1989; Matlack 1993; Cadenasso et al. 1997
	Total basal area of trees	Wales 1972; Ranney et al. 1981; Williams-Linera 1990	Wind	Chen et al. 1995
	Density of woody stems	Gysel 1951; Wales 1972; Bruner 1977; Ranney et al. 1981; Whitney and Runkle 1981; Williams-Linera 1990	Soil temperature	Cadenasso et al. 1997
			Maximum air temperature	Kapos 1989; Williams-Linera 1990; Matlack 1993; Cadenasso et al. 1997

Cadenasso 1995). In the following sections we will summarize an experiment we conducted to determine how the forest edge mediates fluxes of nutrients, detritus, and organisms. Central to this experiment is the link between edge structure and edge function.

FOREST EDGES AS REGULATORS OF FOOD WEB SUBSIDIES

The structure of the forest edge is linked to the function of the edge as a mediator of the fluxes traversing it (Cadenasso and Pickett 2000, 2001; Weathers et al. 2001). If these fluxes can subsidize or influence food webs, then the architecture of the edge may significantly affect the trophic structure of food webs. Forest architecture is composed of the biomass of the vegetation, the layering of the canopy strata, the branching of trees and shrubs, and the life forms of the plants. The architecture of the edge differs from the architecture of the forest interior due to differences in the density and size of stems and in the architecture of individual trees (see table 10.1). Viewed from outside the forest, the forest edge appears to be a curtain or wall of vegetation, made up of lateral branches of the canopy trees, small trees, and shrubs. Vines are also frequent at the forest edge, but in the forest interior they are abundant only in large canopy gaps.

Architecture may be the general means by which edges influence food web structure through, among other things, its influence on flux mediation. Many factors affect the architecture of edges, such as time since creation and feedbacks between abiotic changes and biotic responses. The emphasis here is not on how the architecture of an edge changes through time, but rather on how a given architecture may influence fluxes that are potential subsidies to food webs. We will discuss three fluxes that cross forest edges, are influenced by edge architecture, and can be food web subsidies. These fluxes include nutrients, detritus, and organisms.

Nutrient Subsidies

The abundance of nutrients can affect the base of the food web. If there are spatial patterns of variation in nutrient input and availability across a landscape, then those spatial patterns may generate a corresponding spatial structure in food webs. Essential nutrients, such as nitrogen (N), delivered to terrestrial ecosystems from the atmosphere are delivered via wet and dry deposition. Wet deposition of nutrients occurs via rain, snow, cloud or fog water, and rime ice, while gases and particles are dry-deposited (Lindberg et al. 1986; Likens et al. 1990; Lovett 1994). Following the deposition of nutrients to the canopy, many chemical exchanges and transformations can take place, including foliar uptake and leaching (Lindberg and Lovett 1992). Therefore, the material that is delivered to and washed from the canopy surface during rainfall and deposited on the forest floor as through-fall can be a composite of wet and dry deposition as well as canopy processing. The architecture of the forest canopy, as the receptor surface, influences the spatial pattern of atmospheric nutrient deposition; other influences include wind speed and direction and canopy wetness (Weathers et al. 1992, 1995, 2000a; Lovett 1994). Atmospheric deposition of nutrients across landscapes, at the scale of tens to hundreds of kilometers, is especially likely to be sensitive to the location and architecture of edges (Weathers et al. 1992, 1995, 2000a, 2000b, 2001).

We quantified spatial differences in nutrient input to the forest floor by collecting throughfall at the edge and in the interior of two replicate low-elevation deciduous forest stands (sites 1 and 2). We found greater inputs of dissolved inorganic nitrogen (DIN) below the canopy at the edge than in the forest interior (fig. 10.1). This result was consistent throughout the growing season and was observed at both sites. Bulk precipitation collected in the open field adjacent to site 1 indicated that, for the entire growing season, DIN flux in the open was equivalent to that received in the forest interior (fig. 10.1). The below-canopy DIN flux was, on average, 49% higher at the edge than in the forest interior and the adjacent field (Weathers et al. 2001). Such spatial differences, if they resonate through the soil bacterial and fungal compartments of an ecosystem or through the primary producers may ultimately ramify throughout entire food webs (Cadenasso 1998).

Differences in the architecture of the forest canopy between the edge and the interior may provide an explanation for the differences in below-canopy nutrient input we observed. Edges may be the first point of contact between the forest and air moving across this landscape. Air masses moving across adjacent open areas may be disrupted when they contact forest edges; frequently, the air eddies, and gases and particles contained in the air are deposited on leaf surfaces at the edge (Draaijers et al. 1988; de Jong and Klaasen 1997). The wall of vegetation at the edge provides a much greater surface area to filter or capture the deposited gases and particles than the adjacent open field or interior forest canopy. Leaf area index (LAI), a measure of leaf surface area per unit ground area, is thought to be greater at the edge than in the interior. Wiman and Ågren (1985), Beier (1991), and Erisman and Draaijers (1995) proposed that the greater LAI

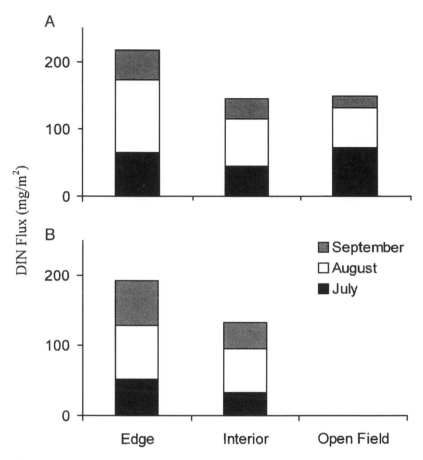

Figure 10.1 Dissolved inorganic nitrogen flux (DIN; mg m^{-2}) in throughfall during the 1995 growing season. Throughfall was collected by event at two forest edges and their interiors (A, site 1; B, site 2). Bulk precipitation was collected in the open field adjacent to site 1. Samples were bulked by month for analysis and were analyzed using a Danaus DX500 ion chromatograph in the Institute of Ecosystem Studies analytical laboratory. Stacked bars represent DIN flux during July, August, and September. It did not rain in June. Both edges are abrupt, adjacent to maintained open fields, and have a northeastern aspect. The edge was defined by 80% canopy cover, and the interior was 25 m from the edge. At site 1, DIN flux was 53% greater at the edge, and at site 2, it was 45% greater. Data were analyzed for each site using a one-tailed t test to compare the DIN flux at the edge with that in the interior at each site across the entire growing season. The enhancement of DIN flux at the edge was significant at both sites (site 1, $p = .0337$; site 2, $p = .0093$). (From Weathers et al. 2001.)

contributes to patterns of enhanced deposition to edges. However, few studies have adequately documented a greater LAI at edges or demonstrated that a higher LAI enhances deposition. Two components of edge architecture—the physical discontinuity that can disrupt air flow and the vegetational wall that comprises the receptor surfaces for horizontally driven droplets, gases, and particles—both contribute to the spatial pattern of greater below-canopy fluxes at the edge than in the forest interior.

The enhanced below-canopy flux of DIN at forest edges may cascade throughout the biotic community. Inorganic N can be readily taken up and used by organisms (Eviner and Chapin 1997). Shallow-rooted plants, such as seedlings and herbs, may quickly incorporate inorganic N into their tissues, increasing the N content in plants at the edge compared with those in the interior of the forest (Thimonier et al. 1992). Increased nutrient content in plants could lead to herbivores selecting for seedlings with greater tissue nutrient content, and further, higher tissue nutrient content could affect decomposition rates and soil nutrient pools and fluxes. Research is needed to determine whether the enhanced below-canopy nutrient fluxes found at forest edges have a significant effect on food web dynamics at the edge or in adjacent areas.

Detrital Subsidies

The architecture of the edge also influences the spatial pattern of detritus deposition over a landscape by causing differences in the density of potential sources of detritus and by changing the pattern of wind speed and direction that may determine where detritus is deposited. Detritus can be an important source of nutrient availability to the biotic system. The architecture of the edge may influence nutrient availability by (1) influencing the amount of litter entering the system, (2) influencing the quality of the litter substrate for decomposition, and (3) influencing the abiotic conditions for decomposers in the edge zone. We explore each of these three influences in turn.

The higher vegetation density of forest edges may result in greater amounts of litter being produced in the edge zone than in the interior (Bierregaard et al. 1992). Detrital inputs to the edge and interior zones can differ in amount, as well as in quality, depending on the different relative amounts of grass, forb, and woody species present. The relative abundances of plant life forms are indirectly affected by the architecture of the edge because architecture influences gradients of microenvironmental variables (Cadenasso et al. 1997), which may influence the growth and performance of life forms differentially.

Litter deposited at the edge from vegetation growing there may be of higher quality than litter from the interior of the forest because of the enhanced nutrient input to this zone (Cadenasso 1998; Weathers et al. 2001). Though no research, to our knowledge, has examined litter quality at forest edges, information does exist on litter quality in relation to forest stand nutrient status. For example, Tietema et al. (1993) compared N-limited and N-saturated oak forests in the Netherlands and found that those stands that received the highest N deposition, and consequently the highest throughfall flux of N, had both increased concentrations of N and reduced lignin in oak litter. Higher litter quality resulted in faster decomposition rates. In addition, the N content of litterfall on a N-rich forest floor was proportionally higher than on a N-poor forest floor in the taiga forests of interior Alaska (Flanagan and Van Cleve 1983).

Decomposition of organic material is performed by the community of microorganisms in the soil. The rate at which the microbial community decomposes litter and turns over nutrients depends, in part, upon the quality of the litter substrate (Nadelhoffer et al. 1983; Prescott et al. 1993) and the abiotic conditions where decomposition takes place (Moore 1981; Van Cleve et al. 1981). Higher light conditions and warmer soil and air temperatures at the edge than in the forest interior (Cadenasso et al. 1997) may support greater microbial activity there.

Whether these influences of forest edge architecture on the amount and distribution, quality, and decomposition rate of plant litter reverberate through food webs is an open question (Cadenasso 1998). Therefore, future research must determine whether any changes in food web structure across forest edges are responses to resource and stress effects on primary producers, responses to the differential availability and performance of the soil decomposer compartment, or responses to previous land use history.

Organismal Subsidies

The architecture of a forest edge may influence fluxes of organisms traversing it. Animals using multiple habitats may move between them or may concentrate either near or away from edges. For example, deer use both forest and nonforest habitats and may concentrate at the forest edge or traverse it frequently. Other organisms may avoid forest edges because of the increased exposure to predators there relative to the closed-canopy forest, and birds of prey may perch on edges because of the increased visibility of potential prey. In addition, plants may disperse across forest edges by a variety of means—wind, water, or animals. If organisms respond to the architecture of the forest edge in a spatially explicit way, then the food webs in which these organisms interact should likewise be spatially structured. A straightforward way to test whether the architecture of the forest edge influences the food web structure of edge communities is to alter the architecture and quantify the animal and plant responses to the experimental alteration.

We performed such an experiment to determine how mammalian herbivores (Cadenasso and Pickett 2000) and wind-dispersed seeds (Cadenasso and Pickett 2001) would respond to an alteration in the architecture of the edge vegetation. The architecture of forest edges at two sites was altered by removing all vegetation lower than half the canopy height, including the lateral branches of the canopy trees, entire small trees, and shrubs. This alteration (the thinned treatment) mimics the architecture of an edge newly created either by cutting or by a natural disturbance such as windthrow. At each site, one half of the edge was left with an intact wall of vegetation as the control (the intact treatment).

Using this experimental design, we tested the link between edge structure and edge regulation of the foraging activities of mammalian herbivores. In our system, the two mammalian herbivores are white-tailed deer (*Odocoileus virginianus*) and meadow voles (*Microtus pennsylvanicus*). Deer move freely across forest edges each day; voles inhabit fields adjacent to forests, but forage on forest edges. In each of the two edge treatments, we planted three species of tree seedlings in cages and in paired open plots. The seedlings were all grown from seed and were of equivalent size when planted. No seedling planted inside a cage was damaged by herbivory. The cages excluded herbivores and were used to verify that any damage to seedlings planted in the open plots could be attributed to a specific herbivore. Deer and voles damage seedlings in characteristic ways, allowing for accurate attribution of damage. The seedlings were monitored weekly for the first 10 weeks and every 2 weeks after that until leaf drop (see Cadenasso and Pickett 2000 for methodological details).

In the intact treatment, 40% of the uncaged seedlings were damaged by herbivores, and voles caused 96% of that damage. In contrast, deer were the primary herbivores in the thinned treatment; voles clipped only 2.8% of the seedlings there (fig. 10.2; Cadenasso and Pickett 2000). The foraging activity of the herbivores was significantly different between the two edge structures, and the herbivores also showed distinct preferences for specific tree species (Cadenasso and Pickett 2000). The combination of the effect of edge architecture on herbivore activity and the feeding preferences of the herbivores demonstrated a shift in the strength of interactions in the trophic structure. The long-term effects of this shift on forest structure and composition remain an open question.

Figure 10.2 Percentage of planted tree seedlings damaged by herbivores in the intact and the thinned edge architectural treatments. Twelve seedlings of each of three species were planted in paired caged and open plots. Only seedlings planted in the open plots are represented in this graph. Solid bars represent seedlings damaged by deer, and open bars represent seedlings damaged by voles. Each bar is an average of five open plots ±1 SE. Significantly more seedlings were damaged in the intact treatment than in the thinned treatment (MANOVA; $p = .0005$). This difference is attributable to variations in vole herbivory with edge architecture (ANOVA; $p = .0001$). (From Cadenasso and Pickett 2000.)

   Dispersal of organisms from the surrounding landscape into the forest interior may alter the composition, structure, and interactions of the biotic community, and consequently, the trophic structure and dynamics of the system. We quantified the flux of wind-dispersed seeds traversing the thinned and intact edge treatments to determine how the structure of the edge mediated that flux (Cadenasso and Pickett 2001). Seed traps were placed on transects parallel to the edge and spaced 3 m apart at distances from 5 to 50 m into the forest. Traps were in the field from 15 September to 16 December 1996 to encompass the entire period of fall dispersal and leaf drop (see Cadenasso and Pickett 2001 for methodological details). All seeds of plant species not found in the forest were identified and counted. Fifteen hundred seeds were trapped during the experiment, and 83% of them were from three species of *Solidago*. The remaining 17% were from eleven taxa. Across the entire dispersal period, significantly more seeds crossed the thinned edge than the intact edge (fig. 10.3). The significant interaction of distance from edge with edge structure indicates that the distance seeds disperse into the forest is dependent on the structure of the edge (Cadenasso and Pickett 2001).

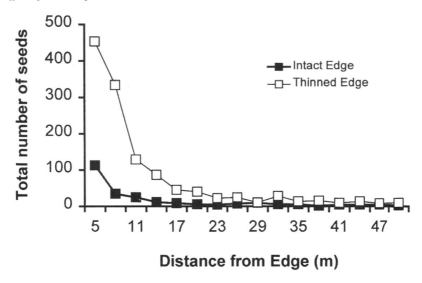

Distance from Edge (m)

Figure 10.3 Total number of seeds trapped in the intact and thinned edge treatments during the entire dispersal period of 15 September to 16 December 1996. Twelve seed traps were randomly placed along transects parallel to the forest edge at each distance from the edge. Only seeds of open field species not found in the forest are included in this figure. Though the traps were in the field continually during the dispersal period, they were replaced periodically. Each sample period was analyzed using a two-way analyses of variance with distance from edge and edge architectural treatment as the dependent variables. For all sample periods, significantly more seeds crossed the thinned edge treatment than the intact edge treatment (all $p < .0003$), and the interaction term was also significant (all $p = .0001$) for all periods, indicating that how far into the forest seeds dispersed was dependent on the architecture of the edge. (From Cadenasso and Pickett 2001.)

FUTURE DIRECTIONS

The research outlined above has indicated that the architecture of forest edges affects five different processes that can influence food web structure: gradients of abiotic factors; the below-canopy flux of atmospherically deposited nutrients; the amount, distribution, substrate quality, and decomposition rate of plant litter; the activity of mammalian herbivores; and the flux of wind-dispersed seeds. Research is needed to determine the extent to which such effects of boundaries actually contribute to differences in food web structure and dynamics across landscapes. Other specific questions concerning the effects of edge environments on food web structure include the following: At what spatial scales do differences in edge architecture result in different food web structures? Are these scales determined by the grain of response of certain key species in the food web? Do food web structures and dynamics respond more strongly to physical environmental differences (i.e., regulator or stress factors) or to differences in nutrients and biomass (i.e., resources)? Do shifts in trophic interactions in response to structural changes across edges constitute simple substitutions of trophically equivalent species or qualitative shifts in food web structure, and hence in complexity and energy flow through food webs? These are just a sample of the types of questions that address the direct effects of the environmental and architectural shifts across forest edges on food webs. Synthesis of the abiotic and biotic gradients documented across edges, the differential nutrient input to edges versus interiors, herbivore responses to edges of different architectures, and the different permeabilities of edges to seed dispersal will provide a strong foundation for hypothesis generation and future research.

Understanding the ecology of food webs in landscapes requires an evaluation of the spatial variation in factors potentially influencing food web structure and dynamics. Our focus in this chapter has been on the importance of biotic and abiotic spatial patterns across forest edges as a tractable and illustrative case of boundaries in general. Structural and functional boundaries in the landscape represent areas where biotic and abiotic factors change and where interactions between two habitats are expected to occur.

In summary, the composition and complexity of food webs may well respond to the structural discontinuities and gradients that appear at landscape boundaries and edges. There is a growing body of evidence that points to alterations of architecture, and consequently of spatial gradients in environmental resources and stress factors that can influence food webs, across edges. Although neither the spatial pattern of food webs across landscape boundaries nor their response to those boundaries has yet been the subject of focused and extensive research, our data suggest that the potential influences of boundaries are large and merit serious attention.

ACKNOWLEDGMENTS

Research support from the National Science Foundation, DEB 9307252, and the Andrew W. Mellon foundation is appreciated. This chapter is a contribution to the program of the Institute of Ecosystem Studies.

Chapter 11

The Variation of Lake Food Webs across the Landscape and Its Effect on Contaminant Dynamics

Joseph B. Rasmussen and M. Jake Vander Zanden

Persistent contaminants such as organochlorines, methyl mercury, and radiocesium have become widespread atmospheric pollutants and constitute important allochthonous inputs to aquatic ecosystems (Eisenreich et al. 1981; Patton et al. 1989). Even in areas remote from industrial activities and other anthropogenic sources, such as the open sea and extreme polar regions, dangerously high concentrations of these substances can be found in biota (Bidleman et al. 1989; Kidd et al. 1998). In addition to being widely distributed by atmospheric processes, these substances are concentrated within watersheds and transported along streams and rivers to lakes, estuaries, and oceans. Within food webs, trophic transfer can result in dangerously high exposures for high-level consumers, even though input concentrations may be barely detectable. At the base of the aquatic food chain, phytoplankton bioaccumulate these substances, and in so doing, concentrate them by several orders of magnitude. This bioaccumulation represents an important avenue of contamination for animal consumers, and because significant biomagnification usually occurs at each trophic link, concentrations can build up through the food web and pose significant health risks to both humans and wildlife, especially if food chains consist of many trophic levels (Oliver and Niimi 1988; Rowan and Rasmussen 1992, 1994).

Ever since these risks began to be recognized, we have also been impressed by the enormous between-system variation in contaminant levels in top predators that exists across the landscape, even within relatively homogeneous regions such as the Canadian Shield (Rasmussen et al. 1990; Vander Zanden and Rasmussen 1996). Generic approaches to fisheries and environmental management tend to focus on "average" conditions and will naturally tend to ignore exposure risks associated with this variation unless the processes that generate it are understood and the likelihood of spatial transport is recognized. For example, lakes and oceans with long food chains can have highly contaminated top piscivores that migrate long distances, thus serving as important transport vectors on the landscape (Scrudato and McDowell 1989; Eggold et al. 1996). This is a problem of importance to ecosystem management, and one that is readily incorporated into a broad landscape framework (Turner 1989).

Both the processes that generate variation and the transport processes that broadcast it pose daunting problems for scientific study because of the large spatial scales involved and the obvious limits to experimentation. These problems have led to an emphasis on comparative methods that focus on among-system variation (Cabana et al. 1994; Vander Zanden and Rasmussen 1996) and the use of widely disseminated geochemical tracers (e.g., stable isotopes and radioisotopes) to study trophic transfer within food webs and transport processes on the landscape (Cabana and Rasmussen 1994, 1996; Vander Zanden, Casselman, and Rasmussen 1999; Vander Zanden, Shuter et al. 1999; Trudel and Rasmussen 2001). Tracer methodologies provide cost-effective measures that can be used to characterize local food webs in a manner that is not only accurate and replicable, but also time-integrative, and therefore very useful in large-scale comparative studies. In this chapter we will outline how such approaches have been used to examine variation among lake food webs, its relationship to contaminant biomagnification, and the transfer of biomagnified contaminants across the landscape by migrating fish.

UNDERLYING FACTORS CONTRIBUTING TO THE SPATIAL VARIATION OF LAKE FOOD WEBS

The variation among lake food webs in eastern Canada appears to depend mainly on the mix of species that have had the opportunity to colonize the postglacial landscape (Carter et al. 1980; Roff et al. 1981; Dadswell 1974; Sprules and Bowerman 1988). Since glaciers covered the entire region as little as 10,000 years B.P., it is not speciation and endemism that result in lake-specific differences, but rather biogeographic factors affecting postglacial dispersal. Of importance in this regard are the physical relationship of a lake to major proglacial lake basins and the sea, the drainage patterns that were reestablished on the landscape after glacial retreat, and the "watershed capture" events that accompanied continental rebound and allowed species from southern and eastern drainages to move northward (Martin and Chapman 1965; Dadswell 1974). Similarly, Hershey et al. (1999) have linked food web structure in North Slope Alaskan food webs to geomorphological history and biogeographic factors operating on the postglacial landscape. The biogeographic distributions of a great many of the fish and invertebrate species that constitute lake food webs are far from equilibrium, and many important species are confined to lakes situated along drainages that served as major avenues of postglacial dispersal. The extent of this "biogeographic disequilibrium" has neither been completely documented—especially in regard to invertebrate taxa—nor has its importance to food web structure been generally appreciated.

Superimposed on the postglacial lottery has been the rapidly accelerating onslaught of modern-day introductions, many of which are unintentional, or at least poorly recorded, with the consequence that we sometimes know little more about them than about the postglacial invasions (Evans and Loftus 1987). In many other cases, these introductions have been intentional efforts to create or enhance fisheries. The Great Lakes have received the greatest influx of exotic introductions. Among the most noteworthy of these invaders are the Pacific salmonids, which have not only dramatically altered the lake food webs, but have also established spawning migrations that span considerable distances inland (Crawford 2001). These introductions, as well as subsequent lake management protocols, have also contributed to the variation and uniqueness of individual lake food webs in the region, although introductions, if widespread enough, can ultimately lead to broadscale uniformity.

PELAGIC FOOD WEBS

The pelagic food webs of eastern Canadian lakes generally contain lake trout (*Salvelinus namaycush*) if the lakes are deep enough for stable thermal stratification and the hypolimnia have sufficient levels of dissolved oxygen during the summer months (Goddard et al. 1987). Lake trout are native to the area, although they have been stocked in some headwater lakes that lacked populations naturally. These fish are generally confined to the hypolimnion during the summer months, where they feed on pelagic forage

fishes (e.g., rainbow smelt, cisco, lake whitefish, sculpins), benthic invertebrates, or zooplankton (including *Mysis*) (Trippel and Beamish 1993;
Vander Zanden and Rasmussen 1996). During the winter months lake
trout feed actively throughout the lake and often feed extensively on littoral
fishes and invertebrates (Martin 1954). In lakes lacking pelagic forage fishes
(class 1 lakes), lake trout typically consume zooplankton and benthic invertebrates during summer and a broader spectrum of foods during the
winter months (Martin 1952, 1954; Vander Zanden and Rasmussen 1996).
These lakes are typically small headwater lakes that were isolated from the
main drainage channels that were used as dispersal corridors following
glaciation. Lakes situated along major drainages and glacial spillways typically have a richer pelagic community characterized by multiple species of
pelagic forage fishes. Lakes that fall within the boundaries of former
proglacial lakes typically contain glacial relict crustaceans, such as *Mysis relicta* and *Diporeia hoyi* (Dadswell 1974). Although *Mysis relicta* has been no-
toriously introduced to lakes throughout the world (for example, Lake
Tahoe, California-Nevada; Flathead Lake, Montana), populations of *Mysis*
in eastern Canada are strictly native, glacial relict populations (Dadswell
1974). Lakes containing these species have the longest and most complex
food webs, and we refer to them as class 3 lakes (Rasmussen et al. 1990;
Vander Zanden and Rasmussen 1996). Class 2 lakes are those that include
pelagic forage fishes, but no megazooplankton.

We have compared the trophic positions of lake trout from class 1, 2,
and 3 lakes using both stable isotope measures (Cabana and Rasmussen
1994; Vander Zanden, Casselman, and Rasmussen 1999; Vander Zanden,
Shuter et al. 1999) and stomach contents analyses (Vander Zanden and
Rasmussen 1996). Both methods reveal that trophic position increases significantly with lake class. An important factor contributing to the between-
system variation in food web structure, and to the variation in the trophic
positions of most common fish species, is the generalist nature of these
consumers. Indeed, most lake trout feed on a broad spectrum of prey
species, sizes, and functional groups, and most include prey that vary
widely in trophic position within their diet (Vander Zanden and Rasmussen 1996). Thus omnivory is an important general feature of lake food
webs, and lakes with similar species composition can differ considerably in
trophic structure (Vander Zanden, Shuter et al. 1999).

While the distinction between class 1, 2, and 3 lakes is based largely on
zoogeographic history, pelagic communities have often been influenced by
introductions of lake trout, brook trout, Atlantic salmon, walleye, lake whitefish, cisco, alewife, and rainbow smelt (Evans and Loftus 1987; Scott and

Crossman 1973). The pelagic communities of the Laurentian Great Lakes have been altered the most through introductions of forage fishes such as rainbow smelt, alewife, and emerald shiner, top predators such as chinook salmon, coho salmon, pink salmon, rainbow trout, and brown trout, and hybrid species such as the splake (Martin and Baldwin 1960; Fraser 1980; Crawford 2001). These invaders not only have completely restructured the Great Lakes food webs, but also, by virtue of their extensive spawning runs, have invaded most of the major waterways draining into the lakes.

Although lake trout are still present in most of the Great Lakes basins, their numbers are often far below historical levels due to a combination of overfishing, the effects of marine lampreys, and competition and predation from exotic species (Christie et al. 1987). Furthermore, the spawning stock of lake trout in each of the lakes is a fraction of what was formerly present, indicating a loss of genetic diversity.

Although the rainbow smelt is native to some lakes in this area, usually those that were at one time part of Champlain Sea (a marine intrusion during the last glaciation), this species has been stocked into a great many lakes in Ontario and Quebec, usually illegally, and has subsequently colonized much of the region (Evans and Loftus 1987). This species feeds heavily on young-of-the-year fish and can become highly cannibalistic in certain lakes (Vander Zanden and Rasmussen 1996; Evans and Loftus 1987). This behavior is rarely encountered in coregonids. Consequently, the trophic position of lake trout is typically higher in lakes containing rainbow smelt, and this higher trophic position corresponds with increased levels of contaminant bioaccumulation in lake trout (Vander Zanden and Rasmussen 1996).

While large lakes typically have a well-developed pelagic food web, including piscivorous fishes such as lake trout, walleye, and burbot, medium-sized and smaller lakes, especially those situated higher in drainage systems (e.g., headwater lakes), often lack a "true" offshore fish community. In such lakes a wide variety of littoral species make part-time use of the pelagic zone.

An important factor that introduces variation into food webs in eastern Canada is the rapid rate at which exotic species are invading and spreading throughout the region. Some of these invasions are intentional and authorized—historically, introductions of smallmouth bass have been carried out throughout the region with the aim of enhancing the sport fishery (Vander Zanden, Casselman, and Rasmussen 1999). Other species are expanding by virtue of their remarkable capacities for dispersal; the northern pike and rock bass appear to be examples. The most significant factor in the spread of littoral fish species throughout the region, however, appears to be the "bait bucket brigade," whereby fisherman purchase what they generally refer to as "minnows," but which actually are a mixture of sunfish, rock bass, bullheads, cyprinids, and perch, and at the end of their day of fishing dump the bucket into the lake they are fishing (Litvak and Mandrak 1993). Although policies are being introduced to deal with this problem in Ontario and Quebec, other states and provinces do not adequately restrict the use of live bait and are at risk of future species invasions emanating from the bait bucket. Although the public is becoming gradually more aware of the problems that species introductions can present, introductions of this type will continue.

FOOD WEB VARIATION AND FISHERIES MANAGEMENT

The tremendous variation in food web structure has important implications for fisheries management. Consider that variation in the length of food chains leading to piscivores often spans nearly two entire trophic levels. From a trophic-dynamic perspective, assuming ecological efficiencies of 10%, each additional trophic level should correspond to an order of magnitude decrease in fisheries production. A trophic-dynamic approach to predicting fisheries production has not been well developed in freshwater systems, possibly because of the extreme variation of trophic positions and the amount of omnivory in these food webs. For lake trout, two studies observed that yield was independent of community composition, which reflects the number of trophic levels (Kerr and Martin 1970; Goddard et al. 1987). Furthermore, many factors other than the amount of available energy will affect the potential fisheries yield. For example, Kerr and Martin (1970) concluded that piscivorous lake trout achieved greater foraging efficiency than planktivorous lake trout populations, resulting in similar levels of production for the two populations.

The life history characteristics of lake trout (maximum size, age and size at maturity, reproductive output, natural mortality rates, and growth rates at various life stages) are highly variable across lakes. Sustainable yield and sensitivity to exploitation should vary as a function of the life history pattern of the population. This variation makes managing lake trout (as a species) a difficult prospect, particularly considering that lake trout occur in thousands of lakes across Canada and that they are commonly affected by overexploitation. One approach would be to uncover some correlates of life history patterns and use these surrogate variables as a basis for management. Early studies reported that life history was influenced primarily by differences in diet (Martin 1966). Shuter et al. (1998) found that lake area and total dissolved solids correlated with lake trout life history patterns. They proposed that these environmental correlates be used to develop fisheries exploitation guidelines, particularly in situations in which lake-specific data are not available.

Martin (1966) reported that the growth, size, reproduction, and life history of lake trout were driven by the relative contributions of zooplankton and fish to lake trout diet. This finding is a strong indication that food chain structure can play an important role in shaping lake trout life history. We found that lake area and fish species richness were close determinants of food chain length, explaining more variation in food chain length than the lake class variable (Vander Zanden, Shuter et al. 1999). This close relationship between food chain length and lake area suggests that lake area, although correlated with life history patterns, serves as a surrogate for trophic structure. Further studies should examine the role of trophic structure in influencing fish life history and its implications for fisheries management.

THE APPLICATION OF FOOD WEB APPROACHES TO CONTAMINANT STUDIES

Many persistent contaminants are atmospherically dispersed and deposited on the landscape through precipitation and dry deposition (Blais et al. 1998). The pattern of distribution is often complex, and the long-range dispersal of these contaminants often obscures any relationship between the pattern of deposition and their original source. Persistent contaminants of this kind are bioaccumulated by fish and other biota, and concern over human health risks has led to outright consumption bans and the closure of fisheries, or more often, to consumption advisories on many species of fish and wildlife. Contaminants of interest in this regard include PCBs, DDT and other organochlorine pesticides, methyl mercury, and ^{137}Cs. Surveys carried out by federal and provincial agencies during the 1970s revealed not only major differences in the degree of contamination among fish species, but more interestingly, major differences among lakes in contamination levels of the same species that could not be explained by local pollution sources (Crawford and Brunato 1978).

Although the cycling of all of these contaminants can be influenced in important and complex ways by biogeochemical processes, trophic transfer is the major pathway of exposure to these contaminants. Indeed, for Great Lakes fish, in which a great many organochlorines and other contaminants have been measured, it could be easily shown that levels of these contaminants were many times higher than the levels predicted by bioconcentration models (through direct uptake from water) based on laboratory contaminant exposures (Oliver and Niimi 1988; Rowan and Rasmussen 1992). The higher the trophic position of the fish, the greater was the discrepancy, and for large piscivorous salmonids, the levels of contaminants in tissues could be up to a hundred times as much as could be accounted for by direct uptake from water (Oliver and Niimi 1988). For this reason, we hypothesized that a great deal of the variation in contamination levels, both among species and among systems, was related to the trophic structure of the food web and that the trophic position of a fish was a reflection of the amount of biomagnification occurring in the food web leading to that fish (Rasmussen et al. 1990). In this way, we felt that ecological approaches oriented around the food web could make important contributions to the understanding of exposure risk to humans and wildlife—an area traditionally dominated by chemists and toxicologists. In addition, the study of contaminants in fish and wildlife could potentially provide us with possible tracers useful in the study of food webs.

Detailed analyses of food webs using stomach contents analysis require the sacrifice of a great many fish, making this approach costly and labor-intensive. Thus, for most ecosystems, there are too few food web data available to allow convincing comparative studies to be carried out. In order to test the hypothesis that food web structure—and more specifically, the length of the food chain—was an important determinant of contaminant levels, we needed to develop efficient surrogate variables that could be measured in many lakes and, at the same time, were convincingly related to food web structure.

The most extensive fish contaminant surveys had been carried out by the Ontario Ministry of the Environment during the 1970s, and the most complete data were available for PCBs and methyl mercury in lake trout (Crawford and Brunato 1978). Although no food web data were available for the eighty or so lakes in which trout had been examined for contaminants, a database on fish species composition had been compiled by the Ontario Ministry of Natural Resources. We were thus able to assign the lakes for which we had lake trout contaminant data to class 1, 2, or 3, reflecting food chain length based on pelagic community composition (Rasmussen et al. 1990). Using lake class as a predictor variable in multiple regression models, we demonstrated that food chain length was a strong determinant of PCB and mercury concentrations in lake trout (Rasmussen et al. 1990; Cabana et al. 1994). Other significant predictors of lake trout PCB concentrations were the lipid content of the trout and the distance north from major Great Lakes cities.

We were confident that this simple lake classification scheme reflected broad-scale patterns in food chain length: class 1 lakes should have the shortest food chains because they lacked pelagic forage fishes and *Mysis*, while class 3 lakes should have the longest food chains because both of these trophic levels were present. Yet this simple classification based on the concept of discrete trophic levels certainly overlooked many intricacies of trophic structure. To address this problem, we examined the food web structure of these systems in a series of stable isotope and dietary studies.

The $\delta^{15}N$ values measured in lake trout, pelagic forage fishes, and *Mysis* from a sample of lakes in each food web class indicated that although $\delta^{15}N$ increased with lake class, the increase between classes was typically less than a full trophic level equivalent (Cabana and Rasmussen 1994), suggesting high levels of omnivory in the food chain (Cabana and Rasmussen 1994). This study also demonstrated that $\delta^{15}N$ was a strong predictor of mercury concentrations in lake trout—in fact, a much stronger one than lake class. The isotopic analysis also revealed significant among-lake variation in $\delta^{15}N$ values in lake trout within each class of lakes, although it was uncertain whether this variation reflected underlying among-lake differences in $\delta^{15}N$ at the base of the food web (Cabana and Rasmussen 1996; Vander Zanden and Rasmussen 1999b), or whether it was a result of actual among-lake differences in the food webs leading to lake trout. We have since developed the means to correct for baseline variation at the among-lake (Cabana and Rasmussen 1996; Vander Zanden et al. 1997) and within-lake levels (Vander Zanden and Rasmussen 1999b). Development of this baseline correction permitted us to examine broad-scale patterns in trophic structure and food chain length at the individual lake level (Vander Zanden, Shuter et al. 1999); indeed, both baseline and trophic differences contributed to variation in $\delta^{15}N$ signatures among lake trout and other biota. This work demonstrated that lakes with similar community composition can exhibit dramatic differences in trophic structure and that other variables, such as fish species richness and lake area, provide better predictors of food chain length than simple measures of food chain length based on the number of trophic levels (Vander Zanden, Shuter et al. 1999).

We also carried out another comparison of class 1, 2, and 3 food webs based on published and unpublished lake trout and pelagic forage fish diet data (Vander Zanden and Rasmussen 1996). In this study, we calculated trophic positions for more than two hundred lake trout and pelagic forage fish populations. Trophic position is a continuous variable calculated from the weighted mean trophic position of stomach contents. To do this calculation, we had to make the approximation that invertebrates as zooplankton and noncarnivorous benthos were primary consumers (trophic position = 2). Our dietary food web reconstruction corroborated the stable isotope results in that lake trout ranged in trophic position from 3.0 to 4.6, lake trout from class 3 lakes had the highest trophic position (mean = 4.4), those from class 1 lakes had the lowest trophic position (mean = 3.6), and those from class 2 lakes had intermediate values (mean = 3.9). Despite this qualitative correspondence with lake class, omnivory was prevalent at certain food web linkages. For example, lake trout from class 3 lakes exhibited a trophic position significantly lower than the value of 5.0 predicted from the lake class variable. Although class 3 lake trout were almost completely piscivorous, the forage fishes that they were utilizing were highly omnivorous, with *Mysis* rarely dominating their diets. Stable isotope analysis showed that *Mysis* were also quite omnivorous and that only the adults were completely carnivorous (Braunstrator et al. 2000).

Dietary and stable isotope studies indicated that class 1 lake trout, although highly variable in their trophic position, were typically much more piscivorous than we had initially expected, given the absence of pelagic forage fishes in these lakes (Vander Zanden and Rasmussen 1996). In fact, lake trout from such lakes can rely heavily on littoral fishes, particularly during the cold-water period of the year (Vander Zanden and Rasmussen 1996). Subsequently, we were able to determine that the extent to which lake trout used littoral forage fishes was mediated by the presence of introduced smallmouth bass and rock bass (Vander Zanden, Casselman, and Rasmussen 1999). This bass-induced trophic shift on the part of lake trout is hypothesized to lead to reduced contaminant burdens in lake trout as a result of their depressed trophic position.

Lipids have long been known to be highly correlated with PCB levels in fish, and this has often led to confusion about the nature of the biomagnification process in fish, since many chemical toxicologists have argued that the relationship to lipids was evidence that the contaminants were being taken up directly from the water (Hamelink et al. 1971). While our analyses always showed that trophic position was strongly linked to lake trout PCB levels, even when lipid-adjusted (or lipid-partialed) data were used (Rasmussen et al. 1990; Vander Zanden and Rasmussen 1996), the lipid patterns were still very interesting, since the lipid content of the flesh also strongly increased with trophic position. Many of the lipids found in fish tissues contain polyunsaturated fatty acids, which are synthesized only by plants and algae. These fatty acids are biomagnified through the food chain, stored in body tissues, and ultimately play an important role in reproduction. The relationship between lipid content of lake trout and trophic position suggests that there may be an important food web dimension to lipid metabolism that has been little studied by aquatic ecologists. Although other persistent contaminants do not typically exhibit biomagnification factors as high as those of the highly lipophilic organochlorines (Vander Zanden and Rasmussen 1996), clear enrichment patterns in relation to food chain length could be seen for methyl mercury and for radiocesium (Cabana et al. 1994; Rowan and Rasmussen 1994).

POTENTIAL FOR TRANSFER OF BIOMAGNIFIED CONTAMINANTS ACROSS THE LANDSCAPE

Wild salmonids grow very large, reach high trophic positions, have high lipid concentrations, and thus bioaccumulate very large body burdens and concentrations of organic contaminants. Thus Great Lakes salmonids are among the most contaminated fish in the world (Rowan and Rasmussen 1992; Oliver and Niimi 1988). Many of the introduced Pacific salmonids have established major runs in river systems leading to the Great Lakes. Many fish and invertebrates in these river systems become contaminated when they consume the eggs of these migrating salmonids (Merna 1986). In addition, since most Pacific salmonids die immediately after spawning, and their carcasses decompose and are consumed by insect larvae and crustaceans, salmonid spawning runs have significantly increased the loadings of Mirex, PCBs, and other organochlorines in these tributaries (Scrudato and McDowell 1989; Eggold et al. 1996). There is also evidence that Mirex transported by introduced salmonids enters the terrestrial food web when blowfly larvae consume rotting carcasses (Johnson and Ringler 1979).

On the west coast of North America, salmonids complete their life cycles over enormous distances and home very precisely to their place of origin. These fishes biomagnify contaminants that they accumulate through the Pacific ocean food web before returning to spawn and die in freshwater systems far inland. While documentation of organochlorine levels in Rocky Mountain lakes is very sketchy, there are some lakes in which levels of contamination in the biota and sediments (Kidd et al. 1995) greatly exceed the levels (by orders of magnitude) expected from atmospheric inputs and runoff, which are generally very low in the Northwest. Although this ecosystem transfer link has yet to be clearly demonstrated, it is reasonable to speculate that systems supporting large populations of anadromous salmonids high in trophic position should have significantly enriched contaminant budgets. Many of the more important organochlorines have bioconcentration factors on the same order as key nutrients such as N and P (up to 10^6), and it is well known that the decomposing carcasses of Pacific salmonids are so numerous and large that they often contribute greatly to the nutrient budgets and isotopic signatures of West Coast lakes and rivers (Cederholm et al. 1999; Milner et al. 2000; Gresh et al. 2000; MacAvoy et al. 2000).

THE IMPORTANCE OF BIOENERGETIC VARIABLES TO BIOMAGNIFICATION OF PERSISTENT CONTAMINANTS

Because we were able to develop precise first-order elimination models for cesium (Rowan and Rasmussen 1995) and methyl mercury (Trudel and Rasmussen 1997), and because their uptake and elimination are not complicated by a complex interplay with the lipid pool, we pursued mass balance studies on these contaminants with the aim of testing biomagnification models and evaluating the importance of the bioenergetic budget to biomagnification (Trudel and Rasmussen 2001). Most discussions of biomagnification are chemically oriented, and the biomagnification factor is generally considered to be solely an attribute of the contaminant in question, ignoring biological variables. Our modeling studies were very successful in demonstrating that the activity budget of fish was an important determinant of the biomagnification factor (Trudel and Rasmussen 2001) and that the energy spent on activity varied by twofold or more among systems for the same species (Rowan and Rasmussen 1996; Trudel et al. 2000). These models show that fish with high consumption rates coupled to high activity costs build their contaminant burdens much more rapidly than they grow, since intake greatly exceeds the rate of elimination.

The precision that was possible in measuring and modeling the mass balance of cesium and methyl mercury made it possible to invert the models and calculate feeding rates from the increase in body burden over an age class (adjusted for elimination using a temperature-specific allometric model). This approach allowed us to work out the in situ bioenergetic budgets for a wide spectrum of fish and to study variation among systems in the energy they spent on activity (Rowan and Rasmussen 1996; Trudel et al. 2001). Just as the growth curves of two widespread species, the yellow perch and the lake trout, were highly variable, so was the activity component of the energy budget. This variation appears to be explainable by ecological variables such as the transparency of the water and the relative size of the prey being consumed (Boisclair and Rasmussen 1996; Pazzia et al. 2002; Trudel and Rasmussen 2001).

Using this radiotracer method, we have shown that metabolic and activity costs can be highly site-specific for fish and that these costs are sensitive to environmental variables. In certain lakes, lake trout feed almost exclusively on invertebrates, and these fish grow much more slowly than piscivorous populations (Martin 1952, 1966). Recent work shows that this growth deficiency is not related to below-normal food consumption, but rather to greatly elevated metabolic costs, which lead to reduced growth efficiency (Pazzia et al. 2002).

Even different subpopulations of a species within a lake can exhibit significantly different activity costs. Dwarf lake whitefish (zooplanktivores) have daily rations similar to those to normal lake whitefish (zoobenthivores), but expend far more energy (Trudel et al. 2001). This increased energy expenditure appears to result from the need to assemble the daily ration from a large number of individually consumed small prey (zooplankton). Thus, bioenergetically, the energy budgets of dwarf whitefish resemble those of stunted yellow perch or lake trout, in that the energy investment required for a fish to obtain a normal ration is much greater when the ration is composed of many small prey than when it is made up of a few larger ones. Dwarf whitefish have tissue concentrations of mercury several times higher than those of normal whitefish, although they feed at similar trophic positions on prey with similar levels of mercury contamination (Trudel et al. 2001).

These studies on the effect of bioenergetic factors on biomagnification show that variation among food webs in the degree of biomagnification is not simply a result of differences in food chain length (trophic position) from system to system, but also results from considerable among-system variation in energetic conversion efficiencies, which ultimately determine the tissue concentrations of contaminants in organisms.

We have also used this tracer approach to estimate the bioenergetic costs of living in a metal-contaminated environment. Yellow perch inhabiting metal-contaminated lakes in the Noranda region of northern Quebec have significantly lower growth efficiencies and higher activity costs than perch from reference lakes. These effects are probably a combination of direct metal toxicity effects on fish and the indirect effects of metal contamination on communities of benthic invertebrates. Metal-contaminated lakes have very few prey-sized fish and also lack many of the macroinvertebrates that yellow perch normally consume during their transition to piscivory (Sherwood et al. 2000).

Thus, while it is indeed unfortunate that our environment is polluted with hazardous compounds that poison wildlife and force us to restrict consumption of wild fish and game, we at least can make the best of a bad situation by utilizing these substances as tracers to learn more about the bioenergetic and trophic variation of wild populations.

CONCLUSIONS

It is important to recognize that persistent contaminants such as organochlorines, mercury, and radiocesium are widespread atmospheric pollutants and that they constitute potentially important allochthonous inputs to ecosystems, putting both humans and wildlife at risk, even in parts of the world very remote from industrial activities and anthropogenic point sources. In order to draft policies that adequately deal with the risks to humans and wildlife that persistent contaminants pose, it is necessary that we understand the factors that magnify their concentrations and how those factors vary across the landscape. Food web processes play an important but poorly recognized role in biomagnification, and an impressive amount of variation exists among food webs in the degree to which persistent contaminants are biomagnified. Much of this variation results from the zoogeographic processes that have occurred on the landscape in the past together with the species invasions that are occurring at present.

Management policies cannot be uniquely constructed for each lake or river system, but rather need to operate within management plans derived for larger regional scales. Generic approaches to fisheries and environmental management tend to focus on "average" conditions, and they will naturally tend to ignore exposure risks associated with variation unless the processes that generate it are understood and the likelihood of spatial transport is recognized. Humans and wildlife that consume fish can be exposed to serious risks if they concentrate their foraging in high-risk areas. It is also important to recognize that lakes and oceans with long food chains can have highly contaminated top piscivores that migrate long distances, thus serving as important transport vectors on the landscape.

ACKNOWLEDGMENTS

This work has been supported by NSERC strategic grants and NSERC operating grants to JBR.

PART II.

FOOD WEB DYNAMICS ACROSS THE LAND-WATER INTERFACE

Chapter 12

Food Web Subsidies at the Land-Water Ecotone

M. Jake Vander Zanden and Diane M. Sanzone

Consideration of the connections between seemingly discrete systems has deep roots in ecology. In fact, recognition of the interrelatedness of systems is often perceived as one of the hallmarks of the "ecological" worldview (Oelschaeger 1991). Yet ecological studies traditionally emphasize the dynamics occurring within individual habitats or ecosystems. The linkages among habitats have drawn much less attention. Real-world ecological landscapes are heterogeneous mosaics of disparate habitats that are energetically and dynamically linked at multiple ecological scales (Turner 1989). Understanding these linkages requires a shift in focus from the internal dynamics of a predefined study system to the dynamics of a heterogeneous landscape consisting of interacting habitat elements. The chapters presented in part 2 take such an approach in their studies of food web dynamics across the land-water interface.

Research conducted at the land-water interface has traditionally focused on the unidirectional flow of energy, nutrients, and organisms from terrestrial habitats to adjacent aquatic habitats (Hasler 1975; McClelland and Valiela 1998; Smith 1998). The productivity of headwater streams, for example, can be dominated by terrestrial (allochthonous) organic matter and arthropod inputs (Wallace et al. 1997; Nakano et al. 1999), while higher-order river reaches can be jointly fueled by instream primary production, nutrients and organic matter from upstream river reaches (Vannote et al. 1980), and land-derived materials from riparian zones and floodplains (Goulding 1980). Streams act as conduits of land-derived nutrients and organic matter, while lakes and estuaries typically serve as sinks for nutrients and contaminants within watersheds (Hasler 1975; McClelland and Valiela 1998). In fact, anthropogenic nutrient and contaminant inputs originating from the land have long been recognized as a primary source of eutrophication and contamination of inland waters (Hasler 1975; Vollenweider 1968).

In contrast to the well-known examples of trophic linkages from land to water, exchanges from water to land are poorly understood, and their consequences for the dynamics of populations and ecosystems are even less well known (but see Jansson 1988; Willson and Halupka 1995; Polis and Hurd 1996b; Polis, Anderson, and Holt 1997; Ben-David, Hanley, and Schell 1998). As the chapters in part 2 illustrate, our unidirectional view of land-water coupling must be revised in light of growing evidence that aquatic productivity can influence the dynamics of terrestrial ecosystems. Since terrestrial inputs also contribute to sustaining aquatic productivity (Peterson, Hobbie, and Corliss 1986), trophic connections across the land-water interface are most accurately represented as a cyclic process rather than a unidirectional flow. With this in mind, the chapters presented here address two main topics: first, the extent to which nutrients, materials, and organisms cross the land-water interface, particularly in the direction from water to land, and second, the consequences of these allochthonous resources for populations and food webs dynamics in the receiving system.

To what extent do allochthonous resources support river productivity, and does the export of this material back to terrestrial consumers affect terrestrial food webs? Power et al. (chap. 15) examine this question in the Eel River watershed by looking at broad-scale patterns in aquatic insect emergence and the consequences of this aquatic subsidy for a range of terrestrial consumers. Emergence is highly variable in space and time, but highest in the presence of floating algal mats. Once insects have emerged, their densities decline rapidly with distance away from the river. Terrestrial consumers (lycosid spiders, lizards, and bats) are more abundant along the stream margin, and their densities are correlated with densities of aquatic insects. Bats play a particularly important role in the transport of aquatic nutrients and materials back to terrestrial systems, providing a clear example of how rivers can influence watershed ecosystems far upslope from regions traditionally classified as riparian zones.

Henschel (chap. 13) also examines the magnitude and consequences of the transport of riverine aquatic insects to terrestrial consumers. He combines observational studies and predator removal experiments to demonstrate that spider populations inhabiting riparian corridors are subsidized by emergent aquatic insects. This subsidy allows the spiders to depress populations of terrestrial herbivores. Such a reduction in terrestrial herbivore populations may translate to reduced levels of herbivory in the system, hence initiating a trophic cascade.

Polis and colleagues (chap. 14) review the energetic and dynamic importance of marine inputs to coastal ecosystems. Marine subsidies are delivered to coastal systems in many forms, including algal wrack, carrion, seabird- and marine mammal-derived detritus, and windblown materials. Wrack and carrion, and the invertebrate communities associated with them, benefit a broad range of terrestrial consumers, such as coyotes, lizards, and rodents. Meanwhile, seabirds supply guano and foraging scraps to the system, which augment plant, arthropod, scavenger, and detritivore production. Resource-driven "bottom-up" effects support elevated populations of consumers at multiple trophic levels. These augmented consumers are more likely to exert top-down control of in situ prey.

Riley and colleagues (chap. 16) draw comparisons between land-water interactions in two very different systems: the Taieri River in New Zealand and the Ythan estuary in Scotland. In the Taieri River, nutrient loading has led to increased productivity, lengthened food chains, and dampened trophic cascades, while transporting surplus algal biomass downstream to subsidize food webs. Meanwhile, in the Ythan estuary, increases in organic material and nutrient loading have resulted in increased biomass of macroalgae and deposit feeders, although decades of increasing nutrient enrichment have had little effect on the overall dynamics of the system.

The movement of energy, nutrients, prey, and consumers across habitat boundaries can be energetically and dynamically important to the receiving ecosystem. The consequences of such movement can range from effects on population dynamics to influences on broader ecosystem processes. Another recurring theme in each of the chapters in part 2 is that allochthonous resource subsidies typically vary in space and time; in fact, the spatial and temporal details of a resource subsidy can dictate whether, and to what extent, that subsidy affects the dynamics of the receiving system. Allochthonous subsidies need not be energetically important in the long term to enhance and concentrate populations of consumers in the receiving system. Minor allochthonous subsidies delivered during critically low food periods can maintain consumer populations at much higher levels than in the absence of the subsidy. In turn, a subsidized consumer may be more likely to exert top-down control of in situ prey (see Henschel, chap. 13).

Where multiple trophic levels are subsidized (as in Power et al., chap. 15, and Polis et al., chap. 14), the dynamic consequences of these inputs become greatly confounded.

Identifying trophic connections across ecotones is particularly important in disturbed or fragmented landscapes, where human interruption of these connections can resonate well beyond the sphere of the perturbed system. The quantity and quality of aquatic and marine resources available at the land-water interface creates an enormous potential to concentrate terrestrial consumers, producing "hotspots"—or, in this case, "hotstrips"—of animal abundance, activity, and species richness along coasts, rivers, estuaries, and lakeshores. Destruction and alteration of these ecotones can interfere with the transport of marine and aquatic resources to terrestrial and coastal consumers, with unknown, but potentially dramatic, consequences for terrestrial communities.

However we decide to delineate the boundaries of our study systems, greater recognition of the trophic connections across them has clear and implicit implications for our understanding of human effects on both natural and managed ecosystems. For example, the movement of persistent contaminants such as methyl mercury, DDT, and PCBs is expected to shadow the movement of energy across such habitat boundaries (Crawford 2001). Another outcome of a better understanding of trophic subsidies across habitat boundaries may be that we are better able to predict interactions between exotic and native species. Exotic species that effectively outcompete and replace native species may achieve success because they possess a heightened ability to utilize allochthonous inputs, or perhaps other introduced species, as a resource (Savidge 1987). Finally, the goal of selecting and protecting natural areas is inherently a landscape-level endeavor. By broadening the arena of food web ecology to consider trophic interactions across ecotones and at the landscape level, our efforts will become increasingly relevant to conservation and resource management decisions.

Chapter 13

Subsidized Predation along River Shores Affects Terrestrial Herbivore and Plant Success

Joh R. Henschel

Rivers, streams, lakes, and pools represent pockets or belts of water on land that affect their surroundings in numerous ways. Not only do they represent a resource for terrestrial organisms, thus affecting their distribution and movements, but they are also places of origin for many aquatic animals that subsequently move over land. Shorelines are habitats where animals often concentrate, and this concentration can have consequences for the communities occurring there. For instance, adult forms of aquatic insects, such as midges or caddisflies, cross over or settle on shorelines, where they occur at densities that vary over time and space. Many terrestrial predators, such as spiders, thrive on insects of aquatic origin in addition to their staple food of terrestrial arthropods. For such predators, aquatic insects represent a subsidy. This chapter examines how subsidized predators affect the shoreline food web.

Subsidies are donor-controlled flows of resources from external systems or habitats that augment population densities of consumers in recipient habitats, potentially allowing them to exert higher effect strength on their alternative in situ resources (Power, Sabo et al. 1998; Power et al., chap. 15 in this volume). In other words, rich subsidies can result in increases in the population density of consumers beyond levels that the local resource base would be capable of sustaining and can enable those consumers, in turn, to change the local resource base without necessarily incurring negative feedback effects (Polis, Anderson, and Holt 1997). This effect of subsidies was revealed by Polis and Hurd (1995) for spiders feeding on littoral flies along ocean shores. Moreover, if the subsidized consumers are polyphagous, they can depress local resources without incurring negative population effects, as was demonstrated by Bustamante et al. (1995a) with algal-feeding limpets on a rocky shore. Such effects of allochthonous resources on autochthonous resources via a common consumer can be regarded as a form of apparent competition (Holt 1984).

Increased consumer pressure caused by subsidized predators operates according to two principles. First, resident prey populations are more susceptible to depression by predators than are migratory prey within the habitat under consideration (e.g., Sinclair and Norton-Griffiths 1979). Second, predators are more capable of depressing a prey species when being sustained by alternative prey (e.g., Jeffries 1988; Holt and Lawton 1993). This effect operates on a long-term basis, involving a numerical response of predator populations to sustained enrichment (Readshaw 1973; Lawton et al. 1975; Spiller 1992).

Spiders are good candidates for study as shoreline predators because of their conspicuous abundance along rivers (Jackson and Fisher 1986; Jadranka 1992; Malt 1995; Williams et al. 1995; Henschel, Stumpf, and Mahsberg 1996; Power et al., chap. 15 in this volume) and ocean coasts (e.g., Schoener and Toft 1983; Spiller 1986; Schoener and Spiller 1995; Polis and Hurd 1995, 1996a, 1996b). Even at sites away from shores, it has been shown that spiders can concentrate in resource-rich habitats (Greenstone 1978; Janetos 1982; Morse and Fritz 1982) and that these polyphagous predators can control insect prey populations (Kajak 1978; Riechert and Lockley 1984; Nyffeler and Benz 1987; Riechert and Bishop 1990; Wise 1993).

This chapter reports on a study in the Würzburg riparian forest along the Main River in southern Germany (hereafter referred to as the Würzburg study). The hypothesis elucidated was that abundant flying insects of aquatic origin subsidize spiders on river shores, thus increasing the capacity of these predators to depress terrestrial insect populations. Furthermore, it was suggested that, as a result of this depression, shore plants should be less damaged by herbivorous insects than plants farther from the river. This was demonstrated by comparing arthropod communities in the herb layer at various distances from the river.

The greater stinging nettle, *Urtica dioica* L. (Urticaceae), with its characteristic fauna (Davis 1983, 1989; Henschel, Stumpf, and Mahsberg 1996;

Sommaggio et al. 1995) served as standardized microhabitat so as to reduce the number of variables affecting community composition and facilitate comparison. Stinging nettles respond to herbivory by increasing defense and regrowth; that is, they grow new leaves and side branches that often bear higher densities of stinging trichomes (Van der Meijden et al. 1988; Pullin and Gilbert 1989; Mutikainen et al. 1994). Regrowth therefore served as an index of the relative degree of herbivory at different sites.

The Würzburg study also involved an experiment to establish whether abundant shore spiders depressed terrestrial insect populations. We physically removed arthropod predators from some experimental plots and monitored the herbivore population for changes. We predicted that terrestrial insects should increase at spider removal sites compared with control sites (where spiders were not removed), and that this response should not be as strong at sites away from the shore, where spiders are less abundant as a result of lower allochthonous resource input.

METHODS

Fieldwork was conducted during 1994 and 1995 in the Würzburg riparian forest (Wasserschutzgebiet Würzburg Mergentheimerstraße: 49°37'N; 09°57'E), an 18 ha area along the western shore of the 70 m wide Main River within the town of Würzburg, Bavaria, Germany (Henschel et al. 2001). The riparian willow-poplar forest was 900 m long and 60–100 m wide and lay 1–3 m above the river level. Tree canopies covered about half of the forest floor, which was dominated by stinging nettles (*Urtica dioica*). The nettles served as a standard substrate for investigating terrestrial arthropod assemblages without conflicting effects of herbivory by large vertebrates, which avoid stinging nettles. Twenty-seven sample sites were selected at 1, 30, and 60 m distances from the shore. These sites were arranged in nine rows extending from the river, with sample sites in each row having similar exposure to sun and wind.

Over 300 taxa of arthropods occurred in the nettles. Beetles (Coleoptera), bugs (Heteroptera), and leafhoppers (Homoptera) were abundant, forming characteristic assemblages (Richards 1948; Davis 1983, 1989; Henschel, Stumpf, and Mahsberg 1996). Aquatic insects, mostly midges (Nematocera, Chironomidae) and caddisflies (Trichoptera), were abundant and diverse (51 species; W. D. Schmidt, unpublished data, Government of Lower Frankonia). Spiders, comprising 72 taxa, represented two-thirds of the abundance of insectivorous arthropods, with the remainder comprising a total of 52 taxa of harvestmen (Opiliones), pseudoscorpions (Pseudoscorpiones), mites (Acari), beetles (Coleoptera), bugs (Heteroptera), earwigs (Dermaptera), dragonflies (Odonata), scorpionflies (Mecoptera), lacewings (Neuroptera), thrips (Thysanoptera), and katydids (Tettigoniidae). Of these other insectivores, only the harvestmen, representing nearly 10%, were important polyphagous predators, while the remainder were either uncommon or stenophagous, and none appeared to have a keystone role.

Several habitat variables were measured at each sample site on the day or night before sampling to determine whether there were environmental gradients among sites, to standardize arthropod samples, and to assess plant damage. These variables included microclimate (temperature, humidity, light), general plant characteristics (canopy cover, distance to trees, herb composition), and nettle structure (density, height, number of leaves and branches). Nettle regrowth was determined by counting small new leaves and branches that emerged along the stem well below the apical bud, where primary leaves arise. No significant gradient of temperature or humidity was detectable in the relatively narrow study area (Henschel et al. 2001). Nettle density was a good proxy for nearly all environmental variables, and it affected arthropod abundance and biomass (Henschel et al. 2001). Although nettle density varied between plots (range = 30–101 nettles m^{-2}, CV = 27.6%), it was normally distributed and did not vary with distance from the shore. Abundance and biomass of terrestrial arthropods were expressed both per square meter and per nettle to remove the effect of nettle density.

Arthropods were sampled with sticky traps and by beating nettles. One set of samples was taken at each of twenty-seven sample sites during May, June, August, and October. Beating samples were collected during climatically stable nights, when most web-building spiders foraged. About 1 m^3 volume of nettles (n = 20–60 plants) was beaten with a stick over a tray for 1 minute, and arthropods were captured by D-vac. Sticky traps were made of clear plastic plates of 10 × 10 cm coated on both sides with insect glue. These traps were placed vertically with the base 10 cm above nettles over sampling periods of 14 days each. To determine nocturnal densities, another set of sticky traps was similarly placed for about 4.5 hours each on 27 nights during the study period.

The experiment involved removing predators over a 10-week period between May and August 1995, after which arthropod abundance was measured. The experimental plots consisted of strips of nettles of 1 × 2 m, surrounded by 1 m wide paths kept clear of vegetation as a hurdle for immigrating predators. Thirty-six plots were situated in pairs. One plot of each pair was designated for predator removal, while the fauna was not manipulated in the control plot. Each pair of shore plots (0–5 m from water) was matched with a pair of plots 30 m distant from the shore. Each treatment plot was searched weekly for about 20 minutes at night. Individual spiders, harvestmen, and other predatory insects were removed from the herb layer with a small electric D-vac (Henschel 1995), while ground predators were ignored. Of 4,931 predators removed, 78.2% were spiders. At the end of the experiment, predator removal efficiency was tested by first removing predators as usual, then collecting arthropods by beating (as above). Predator manipulation removed 61% of spider biomass and 48% of other predators.

Feeding arthropod predators were observed at night on 779 occasions. Prey capture rates by spiders were estimated by comparing proportions of feeding and non-feeding spiders.

During the course of this study, 102,095 arthropods were collected, preserved in 75% ethanol, identified, and measured (±0.01 mm) by microscope. From these body length measurements, biomass (dry mass) was calculated using equations of Rogers et al. (1976) and Henschel, Mahsberg, and Stumpf (1996) for insects, spiders, and harvestmen.

RESULTS

Arthropod Distribution

Winged aquatic insects were significantly more abundant at the shore than farther away (fig. 13.1; Henschel et al. 2001). At the shore, midges and caddisflies were 2.5–4.5 times more abundant and had 4.5–7.4 times more biomass than at 30 m from the shore. This gradient was most strongly evident at night, when aquatic insects as well as spiders were most active.

Predators found in the nettles comprised mostly spiders (65% of abundance), with the remainder being bugs (17%), harvestmen (9%), and other taxa (10%). Only spiders showed significant differences in abundance and biomass with distance from the shore, having 2.3–4.0 times the abundance and 2.5–6.9 times the biomass at the shore than at 30 m (see fig. 13.1). This effect was clearly expressed for the most important species, the long-jawed spider *Tetragnatha montana* (42% of spider biomass) and the sac spiders *Clubiona lutescens* and *C. phragmites* (24%). By contrast, the dwarf spider *Gongylidium rufipes* (Linyphiidae), which

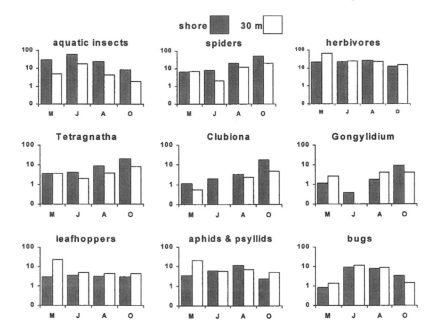

Figure 13.1 Mean standardized biomass (log scale) of arthropods in nettles at the shore (solid bars) and 30 m (open bars) from the Main River during the months of May, June, August, and October. Biomasses of aquatic insects are expressed in mg m^{-2} day captured on sticky traps; biomasses of other insects are expressed in mg m^{-2} of nettles.

formed 18% of the biomass, did not show significant differences with distance. There were no trends in abundance and biomass between 30 m and 60 m for any spiders.

Nettle regrowth, indicative of insect damage to leaves, was 43% less at the shore than at 30 m (see fig. 13.3 below). Abundances and biomasses of terrestrial herbivorous insects were up to 6.4 times lower at the shore than at 30 m (see fig. 13.1). This significant difference was mainly due to Homoptera, which represented two-thirds of the herbivore population. For the leafhoppers (Cicadellidae), the most common terrestrial herbivores, a significant reduction at the shore compared with 30 m was evident for one or another life stage throughout the study period as the insects reproduced and developed. There were fewer adults and nymphs at the shore in May, fewer adults in June, fewer nymphs in August, and fewer adults and nymphs again in October. There were no trends in herbivore populations between 30 m and 60 m.

As a result of these differences in predator and herbivore biomass, the overall predator : herbivore ratio was 1.8 at the shore and 0.7 at 30 m.

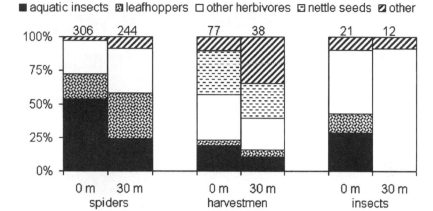

Figure 13.2 Proportions of prey items in the diets of spiders, harvestmen, and predatory insects at the shore and 30 m away. The category "other herbivores" includes aphids, psyllids, bugs, and beetles. Numbers above columns indicate sample size.

Predator Diet

Spiders, harvestmen, and assassin bugs (Nabidae) were observed consuming prey of terrestrial as well as of aquatic origin (fig. 13.2). Spiders accounted for 70% of the observations and had significant differences in diet between sites. Shore spiders fed primarily on aquatic insects (53%) and secondarily on leafhoppers (19%), while at 30 m, aquatic insects (24%) were of secondary importance to leafhoppers (34%). These proportions did not differ from those of the herbivores found in the sticky traps, indicating that spider diet tracked prey availability. These patterns were clearly evident for the two most common spiders, *Tetragnatha* and *Clubiona*. From spider abundance and the proportion of spiders seen feeding, Henschel et al. (2001) estimated that the *T. montana* population at the shore captured 4.4 times as much aquatic prey and 1.5 times as much terrestrial prey as the population at 30 m. For *Clubiona*, these two ratios were 6.4 and 3.7, respectively. This finding indicated that these spiders exerted higher predation pressure on terrestrial insects at the shore.

Experimental Predator Removal

The biomass of spiders was 43–57% lower in predator removal plots than in control plots (fig. 13.3). There were no significant differences for other predators. Leafhopper abundance and biomass increased by 24–26% in

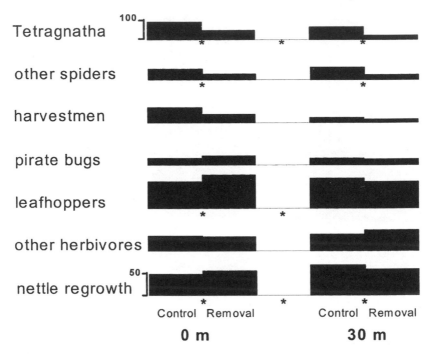

Figure 13.3 Differences in mean biomass of arthropods and in nettle regrowth between control and predator removal plots at the shore and 30 m distant from it. Differences for *Tetragnatha*, other spiders, and harvestmen were caused by the treatment, while the rest are potential response variables. Asterisks indicate that differences between treatments (below graphs) or between shore and 30 m plots (below center) were significant according to Wilcox test ($p < .05$; $n = 9$ pairs). The scale for arthropods indicates biomass, expressed as mg m^{-2}, and for nettle regrowth, the number of secondary leaves per nettle.

shore treatment plots compared with control plots, but there were no significant differences between treatment and control plots at 30 m. Nettle regrowth was significantly increased by 18% in shore treatment plots compared with control plots, but this pattern was not evident at 30 m (fig. 13.3).

DISCUSSION

This study involved a complex food web (fig. 13.4), and only part of the assemblage was investigated. There was a sequence of correlations across trophic levels at the Main River shore (Henschel et al. 2001). Midges were most abundant at the shore, as were polyphagous spiders, while the herbivorous leafhopper population was reduced, as was nettle plant damage. The experiment demonstrated that shore spiders, leafhoppers, and plant damage were sequentially linked; that is, there was an apparent top-down

Trophic link: ➡ increased by subsidy ⇨ reduced by subsidy

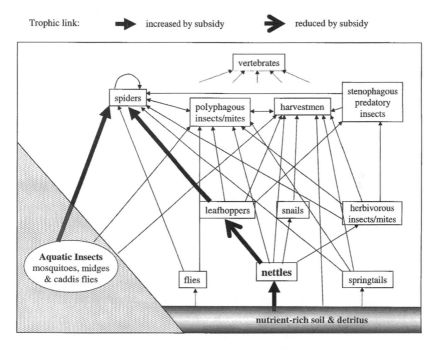

Figure 13.4 Food web in the herb layer of nettles at the river shore based on observations and other information on the diet. The thick arrows indicate the effects of allochthonous aquatic insects—the subsidy—as determined in the present study. The "reduction" of nettles indicates reduced regrowth of less damaged nettles.

trophic cascade (Polis and Hurd 1996a; see figs. 13.3 and 13.4). Spiders ate leafhoppers, so fewer of these herbivores remained to damage nettles.

There were strong indications that highly abundant aquatic insects were the reason why shore spiders were so abundant. Midges declined with distance from the shore, their point of origin and eventual destination. The spiders *Tetragnatha* and *Clubiona* also declined with distance from the shore, as did the proportion of midges in their diets. These spatial changes in the spiders were evidently not due to other environmental factors, as there was no gradient of microclimate or nettle density across the low-lying study area. By contrast, dwarf spiders (*Gongylidium rufipes*), which fed on leafhoppers rather than on aquatic insects, were sometimes less abundant at the shore, where leafhoppers were less abundant (the same was observed for pirate bugs [Anthocoridæ]; see fig. 13.3). These observations support the suggestion that polyphagous spiders were most abundant at the shore because it was near a rich source of prey—namely, allochthonous aquatic insects that they consumed when not consuming terrestrial prey. In other words, there was a bottom-up effect of aquatic insects on spiders.

The experiment demonstrated the top-down effects of the subsidy. Predator removal is a well-established method of examining the effects of predation (e.g., Clarke and Grant 1968; Eickwort 1977; Henschel 1986; Spiller and Schoener 1988, 1990, 1994; Dial and Roughgarden 1995; Holland and Thomas 1997; Lang et al. 1999). The experiment in this study demonstrated that the top-down effect occurred only at the shore, where the spiders *Tetragnatha* and *Clubiona* were abundant and exerted high predation pressure on leafhoppers.

Furthermore, the effects of spiders on leafhopper populations affected nettles (see fig. 13.3). Stronger nettle regrowth, a response to increased herbivory (Van der Meijden et al. 1988; Pullin and Gilbert 1989; Mutikainen et al. 1994), occurred in those shore plots where spiders had been removed and herbivorous leafhoppers had increased. By comparison, nettles in control plots at the shore were less damaged by herbivores. Both leafhoppers and nettle damage increased away from the shore. Other nettle characteristics did not differ consistently with distance from the shore, and there was no evidence of a gradient in the ability of nettles to defend themselves against herbivores. The experiment indicated that reduced nettle damage at the shore was a result of reduced herbivory by leafhoppers (see figs. 13.3 and 13.4).

Distinct ecosystems, such as water and land, become intertwined by subsidies. Spider predation incorporates river productivity into the terrestrial ecosystem, thus enriching it. Without this biomass trap, many adult aquatic insects would merely pass across land to return to the river. The allochthonous input thus exerts its effects only via spiders as retention agents. Conversely, terrestrial arthropods often fall into rivers, where they represent a subsidy to consumers in the river (e.g., Winemiller 1996; Riley et al., chap. 16 in this volume). Reduction of terrestrial herbivores by spiders reduces this allochthonous resource for the aquatic food web. Thus, consumers in adjacent habitats—in this case, water and land—may have indirect effects on each other's food webs. In the river, for example, the degree of predation by fish on aquatic insects could affect the availability of this prey to spiders, while the degree of predation by spiders on terrestrial insects could affect the numbers of insects falling into the river to become fish food. Fish and spiders could be indirectly connected by subsidies in both directions. These examples demonstrate that rivers comprise more than an aquatic component and should be viewed as having a terrestrial component as well. The flux between water and land is an essential process that characterizes both components, making it a fundamental part of the landscape ecology of rivers.

A somewhat surprising result was the rather small spatial scale over which aquatic insects and their trophic effects declined. The 70 m wide Main River is a fairly substantial water body with a low-water discharge rate on the order of 100 m^3 s^{-1}. The biomass of midges and caddisflies declined by over three-fourths within 30 m from the shore across constant micro-habitat conditions and no longer had much effect on the spider population at that distance. This pattern did not appear to be very different from that found by Henschel, Stumpf, and Mahsberg (1996) along streams less than 10 m wide in southern Germany. This observation poses the question of whether there is a relationship between the size of a water body and the spatial scale over which aquatic insects are important subsidies in terrestrial food webs. Does size matter, or is there a fairly constant effect along shores of freshwater bodies (perhaps dependent on the behavior of the particular species or other factors, such as suburban lights)?

I would expect large bodies of water with relatively less shoreline per unit volume (or surface area) to have higher densities of emerging aquatic insects on their shores. In addition to the size of the water body, I would expect its productivity and trophic turnover rate (which tends to be higher in water than on land; Cyr and Pace 1993; Chase 2000) to determine the rate at which aquatic insects emerge and cross land. Furthermore, the behavior of particular species of aquatic insects in crossing land, as well as the nature of terrestrial habitats, should also affect their density at some distance away from water. Thus, further study, incorporating the factors of water body size and productivity and the behavior of allochthonous aquatic insects on land, as well as the nature of the terrestrial habitat and food web, is required to answer this question and to facilitate generalization about the importance of allochthonous connectivity between river and riparian food webs.

ACKNOWLEDGMENTS

I am most grateful to the late Gary Polis for suggesting this study. It formed part of my fellowship at the Alexander-von-Humboldt Foundation, hosted by Eduard Linsenmair and Dieter Mahsberg of Zoology III at the University of Würzburg. Many other helpful staff, students, and associates are thanked for their sterling efforts, especially Inge Henschel and Helmut Stumpf. I especially thank Gary Huxel, Gary Polis, Mary Power, and John Sabo for very helpful suggestions on the manuscript.

Chapter 14

Trophic Flows from Water to Land: Marine Input Affects Food Webs of Islands and Coastal Ecosystems Worldwide

Gary A. Polis, Francisco Sánchez-Piñero, Paul T. Stapp,
Wendy B. Anderson, and Michael D. Rose

At some scale, all habitats can be viewed as functional islands, surrounded by other habitats. Streams, rivers, marshes, ponds, and lakes are all isolated by land. Likewise, land masses, from small atolls to large islands to continents, are insular, surrounded by the oceans that cover 70.8% of Earth's surface. In terms of their biogeography, land masses and freshwater bodies act as islands, containing endemic species and species limited in their abilities to disperse to other islands. Islands of all types are also open to outside influences, which include climate modification, colonization, dispersal, and reception of organic resources from other habitats. Indeed, a major goal of this book is to show that a focal habitat is often greatly affected by the flow of nutrients, detritus, prey, and consumers from other habitats.

We have long recognized that lakes, streams, and other freshwater systems are deeply influenced by the surrounding terrestrial landscape. Gravity assures that a substantial portion of terrestrial nutrients and the carbon fixed by plants eventually works its way into aquatic systems. These products enter bodies of water as dissolved material in runoff and dust, particulate organic carbon in detritus, and other resources such as insect prey. Such "allochthonous flux" is often a major driver of the dynamics, structure, and function of freshwater communities. This "watershed perspective," championed early by Bormann and Likens (1967, 1979), is well recognized (Wetzel 1983; Ward 1989).

Likewise, oceanographers have long studied how "terrigenous input" affects marine primary and secondary productivity at the land-sea inter-face. Nutrients, particulate and dissolved organic carbon, and living prey enter the ocean directly from land and indirectly via rivers and estuaries (Barnes and Hughes 1988). These inputs contribute immensely to the marine coastal fringe, making it one of the most productive habitats in the ocean, and indeed, on the planet.

Nutrients, organic energy, and food derived from aquatic systems influence land communities as well. We are increasingly aware that the export of freshwater nutrients and materials can greatly influence the dynamics of land communities (see Henschel, chap. 13, and Power et al., chap. 15 in this volume; Polis, Hurd et al. 1997). The extent of the influence of such input is potentially extraordinary, particularly since much of the land's surface is closely juxtaposed to reticulated stream and river systems. Recently, considerable attention has been directed at the effects of the ocean on coastal and insular terrestrial systems. Willson et al. (chap. 19 in this volume) give a specific example of the effects of marine input (from anadromous fishes) on land consumers. Anderson and Polis (chap. 6 in this volume) show how allochthonous marine input affects the stability of terrestrial systems. In this chapter, we document the diverse and often substantial influence of marine input on the structure and dynamics of populations, communities, and food webs in coastal and island ecosystems worldwide.

TROPHIC FLOW ACROSS THE COASTAL ECOTONE

The importance of marine input is potentially great because a large area of land is juxtaposed to the sea. The ocean-land interface forms a major ecosystem, the coastal ecotone, which occupies about 8% of Earth's surface along an estimated 594,000 km of coastline (Polis and Hurd 1996b). Productivity differences across this ecotone are often large. Primary productivity in coastal habitats is quite variable, ranging from 3 to 3,500 g m^{-2} yr^{-1} dry mass (Whittaker 1975; Lieth 1978; Polis and Hurd 1996b). In contrast, waters along the shallow littoral fringe are almost always very productive, reaching 3,000–4,000 g m^{-2} yr^{-1} or more in extensive coastlines bordered by kelp forests, estuaries, and reefs. Coastal areas contribute disproportionately to marine productivity (Ryther 1962; Lieth 1978): continental shelves represent 7.2% of marine surface area, but contribute 16.9% of

annual marine primary productivity; estuaries and reefs account for only
0.55% of surface area, but 7.3% of total primary productivity. Macroalgal
beds cover only 0.028% of the ocean surface, but contribute an estimated
1.8% of total marine primary productivity.

Marine nutrients and carbon enter terrestrial food webs via three major
conduits. First, coastal areas receive organic matter via shore drift of carrion and detrital algae. Shore drift varies from tens to thousands of kilograms per meter of shoreline per year (Polis and Hurd 1996b). Two
beaches bordering the Benguela Current ecosystem, characterized by
dense kelp beds, received 2,179 kg and 1,200–1,800 kg m^{-1} yr^{-1} of kelp
(Koop and Field 1980; Stenton-Dozey and Griffiths 1983). Polis and Hurd
(1996b) conservatively estimated that about 28 kg m^{-1} yr^{-1} of macroalgae
entered the shores of the Gulf of California. Carrion and stranded algae
are converted into large populations of specialized intertidal and supralittoral detritivores, scavengers, and predators that inhabit sandy beaches
and rocky shores. These consumers include isopods, Orchestoidea amphipods, beetles, flies, spiders, and scorpions (e.g., Koepcke and Koepcke
1952; Hayes 1974; Cheng 1976; Griffiths and Stenton-Dozey 1981;
Chelazzi and Vannini 1988; Koop and Lucas 1983; Brusca 1980; Brown
and McLachlan 1990). Their populations can reach astounding densities;
for example, *Psamathobledius punctatissimus* occurs at densities up to 2,260
beetles m^{-2} (Griffiths and Griffiths 1983). *Vaejovis littoralis,* a littoral specialist, occurs at densities of 8–12 scorpions m^{-2}, the highest density of
any scorpion (Due and Polis 1985). Shore drift thus fuels a food web that
is based in the supralittoral zone and extends inland to fuel the high numbers of diverse animal species that thrive on coasts and small islands.

Second, many animals feed in the sea, but also use the land. Pinnipeds,
sea turtles, and seabirds import organic material and nutrients to the terrestrial web through their carcasses, reproductive by-products (placentae,
eggshells), food scraps, egg mortality, feathers, and waste products such as
guano (Heatwole 1971; Burger et al. 1978; Burger 1985; Williams et al.
1978; Polis and Hurd 1996b; Bouchard and Bjorndal 1998). Worldwide,
seabirds transfer 104–105 tons of P in the form of guano to land annually
(Hutchinson 1950). For example, seabirds that feed on massive schools of
Peruvian anchovetta deposit millions of tons of guano on islands, which
reaches depths of up to 29 m (Murphy 1981). Input by marine vertebrates
increases the abundance of many terrestrial plant and animal species and
forms the base of a productive food web on coasts and islands (see below;
Anderson and Polis, chap. 6 in this volume).

Third, a surprising amount of marine carbon and nutrients is transported inland via windblown sea foam and spray. These marine materials especially affect areas near the shore, but can travel great distances as dust and aerosols. The potential effect of sea foam is understudied, but the continual input of such nutrients may substantially alter the soils and primary productivity of systems from the coast to hundreds of kilometers inland.

What Determines the Degree and Amount of Marine Input?

Witman et al. (chap. 22 in this volume) discuss the general factors that control the production and input of allochthonous materials across systems. Here, we discuss features of the ocean and land that determine the arrival, magnitude, penetration, and ecological importance of marine inputs. For example, only locations free of effective predators, such as small or distant islands and isolated beaches, host marine vertebrates. Moreover, pinnipeds and seabirds primarily aggregate in large colonies near productive areas that provide abundant fish and other prey. Sea turtles and some seabirds nest in historically determined locations.

The exact amount of material that drifts ashore is a function of several factors. Marine factors include the amount of offshore productivity, its form (e.g., kelp vs. phytoplankton vs. zooxanthellae in corals), and the physical factors that set productivity (e.g., upwelling, bottom depth, strength of thermoclines, mixing), and determine how much of that productivity washes ashore (e.g., currents, winds, fronts). Terrestrial factors regulate how much and how far marine productivity penetrates inland: coastal topography (e.g., beach slope, cliffs), location (e.g., windward vs. leeward), efficiency of the shore biota (plants, predators, scavengers, detritivores) in converting marine input to terrestrial tissue, and consumer mobility (restricted to coast vs. widely roaming).

The ratio of "edge" to "interior" (i.e., the perimeter-to-area ratio, P/A) is a major determinant of allochthonous input to any habitat (Witman et al., chap. 22 in this volume; Forman and Godron 1986; Turner 1989; Polis, Hurd et al. 1997). P/A ratios are paramount to understanding the effects of marine inputs on terrestrial systems (Polis and Hurd 1996b). On islands, the amount of drift is a function of shore perimeter; the relative impact of that drift is a function of island P/A. Thus, although both small and large islands should receive, on average, the same mass of drift per unit area of shoreline, small islands receive much more marine biomass per total unit area than large islands and continents.

What Is the Relative Importance of Marine Input to Island Productivity?

Recognition of the effect of P/A permits us to examine the relative contributions of marine inputs to the energy budgets of terrestrial ecosystems. Polis and Hurd (1996b) and Polis, Anderson, and Holt (1997) compared the importance of marine input (MI) via shore drift with that of in situ terrestrial productivity (TP) by plants for nineteen Gulf of California desert islands. In dry years, the sixteen smallest islands received 1–22 times more biomass from MI than from TP. In wet (El Niño) years, MI/TP ratios declined because high growth by land plants increased the absolute and relative importance of TP. In dry years, only one island had an MI/TP ratio of less than 1, and three other islands had an MI/TP ratio of less than 2; in wet years, eight of twenty islands had MI/TP ratios of less than 1, and seven more islands had ratios of less than 2.

The Gulf of California study was conducted in an area of high MI and low TP. A remaining goal is to quantify the relative contribution of MI in other habitats, even those with high TP. In general, we suggest that MI will always contribute "much" to small islands and the immediate coast worldwide. Specifically, MI/TP ratios should always be substantial under two conditions.

First, we suggest that coastal areas and small islands worldwide with high offshore macrophyte (e.g., kelp, sea grass) productivity should be strongly influenced by ocean input, regardless of their TP. If the mass of macrophytes entering the shore were 10 times greater than in the Gulf of California, then even if TP were 1,000 g m^{-2} yr^{-1}, we calculate that islands of sizes similar to those of the study site would have similar MI/TP ratios. (This level of TP is equivalent to that in a warm temperate mixed forest and twice that of either a boreal forest or a temperate grassland; Lieth 1978.)

Second, even in areas where marine productivity is not high, MI can be significant where substrate or climatic conditions reduce TP. Examples include unvegetated coral cays and rock islands, subpolar and desert islands, islands entirely or partially lacking plant cover due to volcanic activity, islands unsuitable for plants due to extreme guano deposition, and continental edges along deserts (e.g., southwestern Africa and South America, Australia, Middle East) and (sub)polar shores (see Heatwole 1971; Polis and Hurd 1996b). In these systems, even small amounts of MI should be relatively important to the land system's energy budget.

EFFECTS OF MARINE INPUT ON TERRESTRIAL CONSUMERS

To determine the effect of marine input on adjacent land systems, one must measure the transfer of nutrients and energy across the coastal eco-tone. When transfers are mediated by organisms, traditional techniques can be used to evaluate connections between coastal and inland populations (e.g., direct observations, radiotelemetry, mark-recapture studies). These techniques are usually supported by diet analyses to show how consumption of marine foods varies with distance from the ocean or other marine sources (e.g., seabird colonies). More recently, researchers have employed stable isotope analysis of consumer tissues to quantify the flow of energy and nutrients among systems (see Schindler and Lubetkin, chap. 3, and Willson et al., chap. 19 in this volume). The marked differences in stable isotopic signatures of carbon, nitrogen, and sulfur between marine and terrestrial resources (Peterson and Fry 1987) make stable isotope analysis particularly suitable to the study of trophic connections at the ocean-land interface (e.g., Hobson and Sealy 1974; Ramsay and Hobson 1991; Angerbjörn et al. 1994; Pond et al. 1995; Anderson and Polis 1998).

For terrestrial consumers that use marine resources, marine inputs represent a resource subsidy that contributes to higher population densities than can be supported by in situ terrestrial resources alone. Thus, consumers in areas with high marine inputs (small islands, coastal areas off kelp beds, areas near marine vertebrate colonies) reach higher abundances than in similar areas without such inputs. Here we review the effects of marine inputs on terrestrial consumers, both worldwide and with special reference to our ongoing studies on desert islands in the Gulf of California (Polis and Hurd 1996b; Polis, Hurd et al. 1997; Polis, Anderson, and Holt 1997; Polis et al. 1998; Rose and Polis 1998). Because nothing is known of the trophic effects of marine dust and foam on consumers, we consider only coastal shoreline effects and the contributions of seabirds.

Coastal Shoreline Effects

Worldwide, diverse taxa are subsidized by marine input via shore drift (de-trital algae and carrion). Our Gulf of California studies show that supralit-toral arthropods are orders of magnitude more abundant than terrestrial arthropods of adjacent inland regions (10 times for ground arthropods to more than 100 times for flying insects) (Due and Polis 1985; Polis and Hurd 1996b; G. A. Polis et al., unpublished data; fig. 14.1). These high

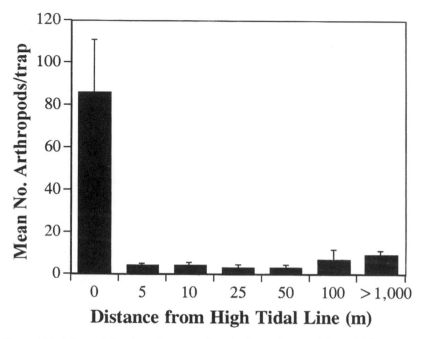

Figure 14.1 Arthropod abundance (mean number of arthropods per trap) from pitfall trapping along a transect from the high tide line (0 m) to the supralittoral (5, 10 m) and inland (25, 50, 100, > 1,000 m) zones along the Gulf of California. Arthropod abundance was about 10 times higher immediately above the high tide line than in the supralittoral and inland zones (F = 5.76, p = .0001, d.f. = 6, 56.71; Welch ANOVA).

numbers of intertidal and supralittoral arthropods subsidize populations of entirely terrestrial arthropods in and above the supralittoral zone. These terrestrial arthropods include web-building spiders (which feed on kelp flies; Spiller 1992; Polis and Hurd 1995, 1996a, 1996b), scorpions, and centipedes (which eat crustaceans and insects; Due and Polis 1985; A. D. Due and G. A. Polis, unpublished data). For example, orb-web spider densities are more than 6 times higher in supralittoral zones than farther inland in Baja California (table 14.1), as well as in Namibia (Polis and Hurd 1996a).

These high densities of coastal invertebrates, combined with the abundance of marine carrion and algal wrack, attract a variety of vertebrates that prey upon coastal invertebrates or scavenge marine carcasses. Intertidal prey are common in the diets of insular skinks, geckos, and other lizards worldwide (G. A. Polis et al., unpublished data). One species of gecko in the Philippines even specializes on crabs. Omnivorous small mammals such as mice, rats, and shrews eat the abundant arthropod prey in both rocky intertidal and sandy beach habitats (Koepcke and Koepcke 1952;

Table 14.1 Abundances of supralittoral and terrestrial taxa on islands and coastal zones of the Gulf of California

Taxa	Shore	Inland	Seabird islands	Non-bird islands	Mainland	Within colonies	Outside colonies	Source
Orb-web spiders (no./m³)	0.155 ± 0.400	0.025 ± 0.165	25.1 ± 2.11	6.15 ± 3.92	0.09 ± 0.01	14.3 ± 1.5	1.8 ± 1.5	Polis and Hurd 1996a
Flying insects (mm/trap/hr)	175.8 ± 2.0	1.6 ± 2.5	12.20 ± 1.65	5.46 ± 1.45	0.90 ± 0.30	12.3 ± 1.0	4.5 ± 1.8	Polis and Hurd 1996a
Tenebrionidae[a] (no./trap)	0.62 ± 0.15	1.39 ± 0.36	20.73 ± 5.62	5.37 ± 2.71	1.3 ± 0.3	41.28 ± 5.89	8.14 ± 1.67	Sánchez-Piñero and Polis 2000; . F Sánchez-Piñero and G. A. Polis, unpublished data
Ground arthropods (no./trap)	43.89 ± 15.75	7.57 ± 1.51	37.89 ± 21.56	5.34 ± 1.97	3.05 ± 0.98	14.27 ± 3.53	4.00 ± 1.17	Sánchez- Piñero and Polis 2000; F. Sánchez-Piñero and G. A. Polis, unpublished data
Lizards (no./hr/person)	30.15 ± 1.16	9.98 ± 1.08	72.49 ± 16.54	14.78 ± 3.61	4.48 ± 1.23	177.55 ± 21.83	33.98 ± 4.12	G. A. Polis et al., unpublished data
Peromyscus G. A. (no./100 trap nights)	48.15 ± 7.77	30.11 ± 6.95	50.69 ± 5.28	31.83 ± 8.94	2.15 ± 0.45	54.00 ± 14.00	34.00 ± 24.00	P. Stapp and G. A. Polis, in press
Chaetodipus (no./100 trap nights)	14.17 ± 4.84	26.04 ± 6.75	—	21.65 ± 5.69	14.17 ± 4.84	—	—	P. Stapp and G. A. Polis, in press
All rodents (no./100 trap nights)	59.47 ± 8.88	47.11 ± 7.34	50.69 ± 5.29	44.82 ± 8.40	21.09 ± 17.30	54.00 ± 14.00	34.00 ± 24.00	P. Stapp and G. A. Polis, in press
Canis latrans (no. tracks/night)	14.70 ± 3.33	3.18 ± 0.89	—	—	—	—	—	Rose and Polis 1998

[a] Not including supralittoral *Phaleria* spp.

Thomas 1971; Osborne and Sheppe 1971; Gleeson and van Rensburg 1982; Navarrete and Castilla 1993; Teferi and Herman 1995). Terrestrial songbirds that feed on arthropods also commonly forage along the coast (Sealy 1974; Egger 1979; Wolinski 1980; Burger 1985; Zann et al. 1990). Large carnivorous mammals and birds (vultures, condors, corvids, eagles) eat carcasses of fish, seabirds, and marine mammals or prey on intertidal invertebrates (Koepcke and Koepcke 1952; Zabel and Taggart 1989; Brown and McLachlan 1990; Jimenez et al. 1991; see Rose and Polis 1998). Even large herbivores such as red deer (Clutton-Brock and Albon 1989), cattle (Hall and Moore 1986), and sheep (Paterson and Coleman 1982) feed on seaweed in the intertidal zone when terrestrial plants are scarce.

These marine subsidies are often translated into high local consumer densities. Most of the world's most dense populations of reptiles occur on islands (G. A. Polis et al., unpublished data). One of the highest lizard densities is that of *Leiolopisma suteri,* a skink that primarily eats intertidal amphipods in beach habitats of New Zealand (Towns 1975). In many coastal areas, deer mice (*Peromyscus* spp.) are more abundant along beaches than farther inland (Thomas 1971; McCabe and Cowan 1945; but see Marinelli and Millar 1989). Similarly, significantly more shrews are captured in rocky intertidal zones than in adjacent vegetated habitats (Stewart et al. 1989; Spencer-Booth 1963). Along the mid-Atlantic seaboard, migrating songbirds are significantly more abundant in coastal scrub and on barrier islands than in interior habitats, in part due to abundant cover and food resources (McCann et al. 1993). Carcasses of marine mammals, seabirds, and fish support higher densities of carnivorous mammals and birds than can be supported by terrestrial resources alone (Bridgeford 1985; Avery et al. 1987; see Rose and Polis 1998).

Similarly, our studies of coastal and insular populations of vertebrates in Baja California demonstrate that a wide range of terrestrial species take advantage of rich shoreline resources. We recorded over twenty species of resident and migratory land birds foraging for coastal arthropods (e.g., hummingbirds, sparrows, flycatchers, wrens; Polis and Hurd 1996b). Abundant *Uta* lizards feed heavily on intertidal isopods and seabird parasites (Wilcox 1980; Grismer 1994; G. A. Polis et al., unpublished data). Some establish feeding territories within carcasses of beached pinnipeds and feed on abundant carrion flies and beetles. Arthropodivorous lizard densities tend to be higher (96.2 ± 16.2 lizards observed per hour) on Gulf of California islands with high shore and seabird input than on other islands and the mainland (19.1 ± 3.0 and 5.0 ± 1.6 lizards per hour), and are

1.8–37 times higher on shore than in inland regions of the same island (G. A. Polis et al., unpublished data; table 14.1).

Omnivorous mice (*Peromyscus*) also inhabit many islands in the Gulf of California and eat marine-based resources. Small islands with seabird colonies support high densities of these mice (see below). On islands without seabirds and on the mainland, *Peromyscus* are typically 2–8 times more abundant along the shore than farther inland (table 14.1). In contrast, granivorous mice (*Chaetodipus*), which occupy fewer islands as well as the adjacent mainland, show no differences in abundance between shoreline and inland areas or are 2–4 times more numerous inland. Stable isotope analyses of *Peromyscus* tissues (Stapp et al. 1999; Stapp and Polis, in press) indicate that coastal mice are subsidized heavily by marine-based resources.

Marine foods also support dense coastal populations of mammalian carnivores. Coastal coyotes (*Canis latrans*) are subsidized by diverse marine resources (Rose and Polis 1998). Consequently, coyotes were 4–33 times more frequent on the coast than inland. Over 69% of all coastal coyote scats contained marine items (compared with fewer than 1% inland), resulting in a more even and diverse diet composition than in inland regions. Marine birds (in 16.5% of scats), fish (12%), and crustaceans (17.7%) constituted a significant portion of the coastal diet, whereas rodents (48%) and rabbits (15.9%) were the major foods inland. Higher foraging activity by predators such as coyotes may result in higher predation risk for terrestrial prey near coastal shores. Preliminary seed tray foraging experiments suggest that this may be the case: granivorous pocket mice (*Chaetodipus*) near shore spent relatively less time foraging in exposed seed trays than in those under protective cover, and they harvested fewer seeds overall than did individuals farther inland (P. T. Stapp, unpublished data).

Seabird Effects

The overwhelming impact of nesting and roosting seabirds on the nutrient budgets of islands is well documented (see Anderson and Polis 1999; Anderson and Polis, chap. 6 in this volume). Marine input via seabirds and other marine vertebrates fuels arthropod communities on islands and in coastal areas in tropical, temperate, and subpolar regions. Marine vertebrates subsidize diverse trophic groups of arthropods in three different ways. First, parasites (e.g., ticks and parasitic flies) feed on vertebrate hosts (Heatwole 1971; Duffy 1983, 1991; Duffy and Campos de Duffy 1986;

Boulinier and Danchin 1996). Second, by providing animal tissue (carcasses, food scraps, etc.), seabirds enhance scavenger populations (e.g., Tenebrionidae, Dermestidae) (Hutchinson 1950; Heatwole 1971; Heatwole et al. 1981; Polis and Hurd 1996b). Third, seabird guano, through its fertilizing effects on plant productivity and quality, enhances populations of herbivores and detritivores (e.g., oribatid mites, Lepidoptera, and scolytid beetles; Onuf et al. 1977; Lindeboom 1984; Mizutani and Wada 1988; Ryan and Watkins 1989).

In turn, abundant arthropod communities around marine vertebrate colonies support a diversity of secondary consumers. Off New Zealand, tuatara (*Sphenodon punctatus*), three skink species, and four gecko species feed on prey that arise directly or indirectly from seabirds (East et al. 1995).

Our trapping on Gulf of California islands shows that arthropod abundances are two orders of magnitude greater on seabird colony islands than on those without nesting colonies. Arthropod primary consumers (detritivores, scavengers, herbivores, and parasites) are more abundant on islands where seabirds breed or roost, and abundances are higher within than outside colonies (table 14.1). These bottom-up effects reticulate through the food web to produce similar patterns of abundance in secondary arthropod predators (e.g., scorpions, spiders, centipedes). Similarly, arthropodivorous and scavenging vertebrates respond to both seabird carrion and the increased abundance of arthropods. Arthropodivorous lizards are more abundant on bird islands (G. A. Polis et al., unpublished data). On the subset of these islands large enough to support rodent populations, *Peromyscus* densities are 1.5–4 times higher than on non-bird islands, and 18–21 times higher than on the adjacent peninsula. Moreover, *Peromyscus* are uniformly dense on bird islands, regardless of distance from shore.

Seabird effects are evident even at the microhabitat scale. *Peromyscus* are captured frequently near stick and debris nests constructed by pelicans. Lizard, tenebrionid beetle, and spider densities are 5–20 times greater within than outside of seabird colonies (Polis and Hurd 1996b; G. A. Polis et al., unpublished data; Sánchez-Piñero and Polis 2000).

Our work on detritivorous tenebrionid beetles (Sánchez-Piñero and Polis 2000) serves to illustrate the complex but critical role of seabirds on Gulf of California islands, and probably worldwide. Tenebrionids are about 5 times more abundant on islands where seabirds roost and nest than on non-bird islands. Large populations are produced by two distinct pathways, one indirect and one direct. First, on islands where birds roost but do not nest, bird carrion is rare, and beetle abundance is an indirect function of seabirds. Here, guano increases nutrient concentrations in the soil (Anderson and Polis 1999 and chap. 6 in this volume). In wet years, these nutrients enhance plant growth and quality, and productivity is very high (fig. 14.2A). In dry years, seabird guano produces no significant effect on either plants or beetles (fig. 14.2B); however, beetle populations are still high because beetles eat plant detritus from previous wet years (Polis, Hurd et al. 1997; Sánchez-Piñero and Polis 2000).

Second, on nesting islands, seabirds primarily affect beetle populations directly through high carrion availability and secondarily through the guano-plant detritus pathway. The carrion pathway predominates because of the large amount of seabird tissue available on breeding islands and because, on islands with large nesting colonies, excess guano may "scorch" vegetation, depressing productivity even in wet years. Thus, beetle abundance is a direct function of the number of nesting seabirds and carrion availability (fig. 14.2C). The interesting point is that the same factor (here, seabirds) produces similar consequences (i.e., high beetle populations), but via two distinct mechanistic pathways (via guano and plants or via carrion) in the food webs.

DYNAMICS OF MARINE INPUT: TROPHIC EFFECTS ON POPULATIONS AND COMMUNITIES

The effects of marine inputs on food webs can be divided into direct and indirect and bottom-up and top-down effects (Huxel et al., chap. 26 in this volume). Changes in resource availability directly alter the abundances of recipient species. Thus, marine resources directly enhance a variety of species. Predators from arachnids to songbirds eat marine prey; scavengers from flies to carnivores eat marine carrion and detritus; and parasites eat tissues of marine birds and mammals. Moreover, marine aerosols and guano directly increase plant productivity via fertilization (Anderson and Polis, chap. 6 in this volume).

Marine inputs also operate indirectly through the food web to influence species one or more links away from the recipient species. Bottom-up indirect effects originate at the base of the web, from which changes in productivity work their way up to increase secondary productivity of consumers throughout the community. Thus, marine materials flow through three food web channels to enrich terrestrial consumers: algae and animals along the shore, marine vertebrates that come to land, and fertilized plants. A common observation is that marine input generally supports high num-

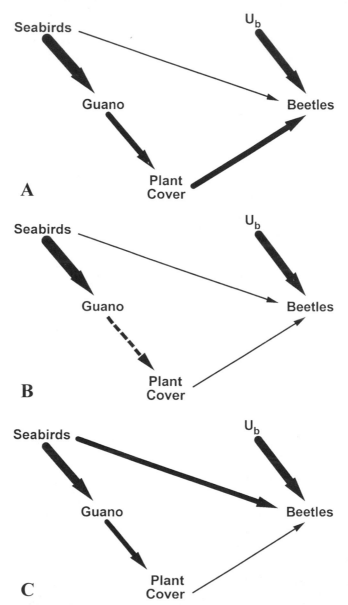

Figure 14.2 Direct and indirect effects of seabirds on tenebrionid beetles on seabird roosting (A and B) and nesting (C) islands of the Gulf of California after a wet year (1995) and a dry year (1996). On roosting islands during the wet year (A), bird effects occurred indirectly: guano enhanced plant productivity, which increased beetle abundance. However, during the dry year (B), seabirds did not show any significant effect on beetle populations on roosting islands, either directly or indirectly. (Beetle abundance remained high on these islands compared with non-bird islands because plant detritus had accumulated from pulses of productivity in the previous wet year.) On nesting islands (C), direct effects from seabird carrion most strongly affected beetles; this effect was consistent in both years.

bers of species at many trophic positions in coastal communities and near bird and mammal colonies (see below; Polis and Hurd 1996b; Polis, Anderson, and Holt 1997).

Top-down food web effects are all indirect, arising from consumers whose numbers are directly increased by marine input. These subsidized consumers also eat in situ resources, with several possible outcomes (Polis, Anderson, and Holt 1997). Most commonly, local resources are depressed in an interaction analogous to apparent competition. How often this occurs is unknown, although we suspect that such depression is quite common. For example, lizards and omnivorous rodents subsidized by marine detritus or seabird input may depress populations of ground spiders and darkling beetles (Tenebrionidae) on Gulf of California islands. Abundances of beetles and *Peromyscus* mice are strongly negatively correlated (F. Sánchez-Piñero and P. T. Stapp, unpublished data). Densities of spiders and beetles increase 8 times and 2 times, respectively, in cages excluding vertebrate predators compared with densities in control cages.

Occasionally, "apparent" trophic cascades occur: a subsidized consumer depresses its prey, and this allows the prey's resource to increase (Polis and Hurd 1996a). Thus, along the shores of the Skeleton Coast in Africa, abundant coastal spiders eat many marine detritivorous flies and suppress insect herbivores; plant damage is significantly less on plants protected by subsidized spiders than on unprotected plants (Polis and Hurd 1996a). Such cascades are apparent because the energy sustaining high consumer densities comes not from in situ productivity (as usually modeled), but from outside the focal habitat.

Note that the food web effects of allochthonous input are largely donor-controlled; that is, consumers benefit from allochthonous resources, but do not affect the resource renewal rate (Polis et al. 1996; Polis, Hurd et al. 1997; Polis and Strong 1996; Persson et al. 1996). Donor control occurs whenever a resource population is spatially partitioned into subpopulations that occupy different (micro-) habitats, only one of which can be accessed by consumers. Donor control then, on occasion, enables recipient control. This phenomenon is exemplified by the Skeleton Coast spiders (i.e., donor control of the spiders by marine flies whose renewal rate in the sea is not influenced by the spiders) that depress insect herbivores (i.e., recipient control of herbivores by the spiders).

Marine input can affect other aspects of populations, food webs, and communities. Anderson and Polis (chap. 6 in this volume) explore the relationship between marine input and population and community stability. Allochthonous resources often act to stabilize recipient populations by serving as food refuges, forming a floor below which populations do not drop (Polis, Anderson, and Holt 1997). Thus, marine input allows populations to persist even when in situ productivity fluctuates greatly. For example, in Baja California, terrestrial primary productivity can be very low (10 g m^{-2} yr^{-1}), but many consumers maintain healthy populations. In other cases, marine input can be destabilizing. Subsidized predators can depress local prey to very low levels, thus increasing their probability of extinction. Similarly, input of nutrients from seabird guano can destabilize plant communities (Anderson and Polis, chap. 6 in this volume).

High-density populations of marine-subsidized consumers can affect inland populations in other ways. First, because of the high reproductive output permitted by abundant and protein-rich marine resources, shoreline populations may serve as a source of dispersers to more interior (sink?) habitats, increasing abundances overall.

Second, we expect that the large contribution of marine input is an important, if not universal, mechanism underlying the high densities of many consumers on islands compared with mainland areas. A variety of other, but not mutually exclusive, explanations have been proposed for this phenomenon (e.g., competitive or predatory release; MacArthur et al. 1972; Case et al. 1979).

Third, marine input may affect competitive interactions. Subsidized populations of omnivorous species also eat terrestrial resources, reducing the amount of resources available to species or individuals that use only terrestrial foods. Thus marine input may favor generalists at the expense of specialists, especially in areas where marine inputs are high, such as coastlines and small islands. Earlier, we showed that subsidized mice (*Peromyscus*) depressed ground spiders and darkling beetles. These same mice also eat seeds, the main resource of granivorous mice (*Chaetodipus*). Such consumption may intensify competitive interactions. On one of the few Gulf islands to support both species, *Chaetodipus* numbers decreased by 50% during an El Niño-related increase in *Peromyscus* abundance (P.T. Stapp, unpublished data). This result was surprising because El Niño-associated precipitation presumably increased the abundance of seeds, *Chaetodipus*'s preferred food. *Peromyscus* may either monopolize such pulses of seeds more effectively or interfere with *Chaetodipus* foraging. In either case, the ability of *Peromyscus* to use marine resources probably permits it to persist during periods of seed scarcity, and probably contributes to its greater success on islands compared with *Chaetodipus*.

Finally, marine inputs probably affect community diversity, particularly on islands. Although much theory addresses how various factors, including primary productivity, set diversity levels (Huston 1994; Rosenzweig 1995), to our knowledge, no theory addresses how allochthonous input affects diversity. Most obviously, altered probabilities of persistence or extinction change the number of consumer species at a site. The availability of marine input also relaxes "assembly rules" for generalist consumer species that colonize islands (Holt et al. 1999). Predators, scavengers, and detritivores can live on islands with few or no land plants and herbivores. For example, centipedes, spiders, scorpions, and lizards live on barren desert islands and coral cays (Heatwole 1971; Polis and Hurd 1996b; Polis, Anderson, and Holt 1997). The first successful colonists on Krakatoa were predators (spiders and lizards) that persisted on prey from the supralittoral food web (Thornton and New 1988). Overall, we are unsure how marine input affects consumer diversity. In some cases, allochthonous resources may depress diversity. Those Gulf of California islands that receive the most seabird guano are dominated by a few species of highly productive plants (e.g., *Amaranthus, Chenopodium*), whereas islands without such fertilizing input are more diverse, but less productive (G. A. Polis and W. B. Anderson, unpublished data).

MARINE INPUT AND HUMAN ECOLOGY

We close by addressing how the sea affects humans and how humans affect coastal systems. The rich bounty of the sea has attracted humans to coastal areas throughout history. Excavation of middens and stable isotope analysis of human bones show that dense populations of prehistoric humans used marine resources worldwide (e.g., Schoeninger et al. 1983). For example, the Chumash of southern California used molluscs, urchins, fish, and other near-shore resources to rank among the most dense populations of pre-Columbian native Americans.

Modern-day humans likewise use many marine resources and live in great numbers near the coast. About 60% of humans—3.8 billion people—now live within 100 miles of the sea (Hinrichson 1997). Within three decades, a projected 75% (6.3 billion) of the world's population could reside in coastal areas. Many factors attract humans to the coast (e.g., aesthetics, clement weather, ease of transportation). However, humans also rely heavily on marine resources. Invertebrates, algae, guano fertilizer, and 100 million tons of fish are harvested annually by about 200 million people, yielding $70 billion (Botsford et al. 1997). Protein is particularly limiting to humans, along with all other animals on this planet. The sea now provides 21% of all protein for humans (Botsford et al. 1997). Such subsidies that concentrate high densities of humans along the coast undoubtedly contribute to the degradation of coastal regions worldwide.

Humans and marine subsidies are interwoven in other ways. In many places, marine subsidies have exacerbated the harmful effects of introduced species. Humans have introduced a variety of "beneficial" domestic animals to islands worldwide (Atkinson 1994), including grazers (sheep, goats, rabbits, cattle), omnivores (pigs and boars), and some predators (domestic cats, mongooses, foxes). Other species have accompanied humans by accident, most notably many species of rats. The results of all these introductions have been almost uniformly catastrophic to the native biota. In many cases, the harmful effects of invasive species are highly subsidized by marine resources (see Power et al., chap. 24 in this volume). Typically, an introduced animal increases its population by using marine resources, either shore material or marine birds. These subsidized exotics then depress populations of local endemic species, sometimes to the point of extinction.

OVERVIEW AND CONCLUSIONS

In this chapter, we have shown that flow between the sea and the land occurs in both directions. Islands and coastal areas worldwide are deeply influenced by their juxtaposition to the ocean. Marine nutrients from birds and sea foam fertilize plants and increase productivity. Marine prey and detritus power dense populations of diverse coastal consumers. Marine vertebrates, particularly seabirds, provide a rich source of food that alters the communities of entire islands. Allochthonous marine inputs not only affect recipient species, but percolate and ramify through the food web to govern the dynamics of most species on islands and in coastal areas.

We conclude by stressing that it is not possible to understand either the demography of coastal and island species or the structure and dynamics of coastal communities and ecosystems without the inclusion of the great energetic and nutrient effects of the ocean. Land and water, although existing separately and easily recognized as distinct biological communities, are very real extensions of each other. This insight must govern our research, and our conservation and management efforts, in the coastal ecotone.

Chapter 15

River-to-Watershed Subsidies in an Old-Growth Conifer Forest

Mary E. Power, William E. Rainey, Michael S. Parker, John L. Sabo,
Adrianna Smyth, Sapna Khandwala, Jacques C. Finlay, F. Camille McNeely,
Kevin Marsee, and Clarissa Anderson

Ecologists have often assumed that material exchange between rivers and terrestrial watersheds is highly asymmetric, with watersheds typically feeding their rivers. There are three reasons to expect this asymmetry. First, watershed land area is considerably greater than that inundated by streams and rivers. Second, terrestrial plant biomass typically dwarfs that of aquatic primary producers. Third, gravity pulls material down slopes. Despite these factors, the amount of energy supplied to food webs by aquatic primary producers can be surprisingly large because of the high biomass-specific productivity and the excellent nutritional quality of common aquatic producers such as diatoms (Boyd 1973; Hanson et al. 1985; Fuller and Mackay 1981). It is therefore interesting to investigate, for various types of ecosystems, how much algae and other aquatic producers support secondary production in rivers, and whether export of river production has any measurable effect on watershed consumers and ecosystem processes.

Ecologists working in desert streams first pointed out that higher trophic levels in rivers could depend primarily on algal rather than terrestrial plant production (Minshall 1978) and that exports from desert rivers could also fuel watershed consumers (Jackson and Fisher 1986). In Sycamore Creek in the Sonoran Desert of Arizona, Jackson and Fisher

(1986) estimated that 97% of the aquatic insect emergence (22.4 grams carbon m^{-2} yr^{-1}) was exported to the watershed, where it fed consumers including ants, birds, and bats. In a prairie stream with higher adjacent terrestrial plant biomass, Gray (1989a, 1996b) documented consumption of aquatic insects by a variety of birds and mammals. Mayer and Likens (1987) studied a heavily shaded New Hampshire stream, Bear Brook. The beech and maple trees surrounding the brook have soft leaves and produce relatively edible litter. Mayer and Likens found that algae, while less than 1% of the standing stock of organic manner in the stream, made up more than 50% of the ingested material in the guts of a common caddisfly species, *Neophylax aniqua*. From published assimilation data, they estimated that at least 75% of this caddisfly's growth was fueled by algae rather than by the more refractory terrestrial detritus. Their results suggested that even in streams that are darkly shaded by deciduous trees with relatively high-quality litter, common aquatic insects may be largely built of algal carbon. More recently, a survey of stable carbon isotope data from studies of seventy temperate rivers suggested that in drainages of greater than 5–10 km^2, algae contributed significantly to the carbon of all lotic consumers except for shredders (Finlay 2001). Members of one guild, scrapers (which includes caddisflies), retained their algal signal even upstream in smaller, darker headwater streams (Finlay 2001).

Do emerging aquatic insects transport significant amounts of algal carbon into terrestrial food webs? Clearly, the degree to which aquatic consumers are fueled by algal versus terrestrial plant production, as well as the predominant direction of trophic exchange between rivers and watersheds, varies among species and sites. We have very little quantitative information as yet about these exchanges (but see Nakano and Murakami 2001). The increasing availability to ecologists of stable isotope and other tracer analyses will greatly facilitate the study of resource fluxes among habitats, once local sources of variation in tracer signatures are sufficiently understood (Finlay et al. 1999, 2002; Cabana and Rasmussen 1996; Ben-David, Hanley, and Schell 1998).

We have begun to study the river-to-forest exchange at the South Fork Eel River of northern California. The watershed supports a mixed old-growth conifer and deciduous forest with some chaparral along ridgetops. This vegetation is not particularly edible (Harris 1984). During the biologically active low-flow summer season, sunny reaches of the South Fork Eel support rapid algal growth, including that of nutritious epilithic algae on rocks and epiphytic algae coating seasonally lush blooms of the filamentous green macroalga *Cladophora glomerata* (Power 1990a, 1990b). Algal standing crops vary dramatically from year to year, depending on whether bed-scouring winter floods have occurred (Power 1992a, 1995; Wootton et al. 1996; and see discussion below). Because insect production and emergence rates are strongly influenced by algae, we wanted to examine whether year-to-year variations in algae and insect emergence affected watershed consumers. This question can be expanded into five, following Polis, Anderson, and Holt (1997):

1. What are the patterns of spatial and temporal variation in aquatic insect emergence?

2. Do watershed consumers track this variation? If not, what constrains them?

3. Does this subsidy influence consumer performance (feeding rates, somatic growth, reproduction, survival, recruitment)?

4. Do subsidies to particular consumers alter their effects on watershed communities or ecosystem properties?

5. How do subsidies and their effects change across landscape gradients or thresholds?

Here, we discuss the application of these questions to the South Fork Eel. First, we describe sources of spatial and temporal variation in insect emergence from the river and its lateral penetration into the watershed. Then we present preliminary observations on how different consumers, which forage over a range of spatial scales, respond to this resource subsidy. Finally, we discuss hypotheses, as yet largely untested in our system, about how subsidies to these consumers might indirectly affect watershed food webs, and about how such effects and interactions may vary across the landscape.

THE STUDY SITE

Our primary study site encompasses about 10 km of the South Fork Eel River and its tributaries (fig. 15.1) within or adjacent to the 3,200 ha Angelo Coast Range Reserve of the University of California Natural Reserve System in Mendocino County, California (39°44′N, 123°39′W). Though timber extraction and cattle grazing are ongoing in the South Fork drainage basin upstream from our site, the reserve and several entire tributary watersheds have been protected from logging since its purchase by Heath and Marjorie Angelo in the 1930s. Where it enters the reserve, the South Fork Eel drains a watershed of 130 km². The river flows north, parallel to the Pacific coast, for another 160 km before reaching the sea just south of Arcata, California. Tectonic uplift at the mouth of the river (5 mm per year) is more rapid than

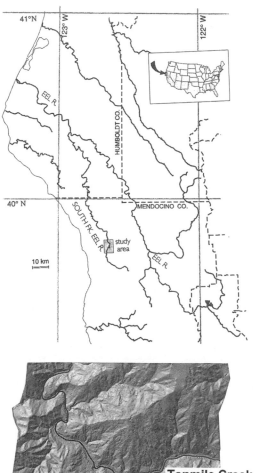

A

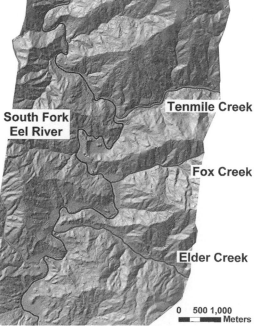

B

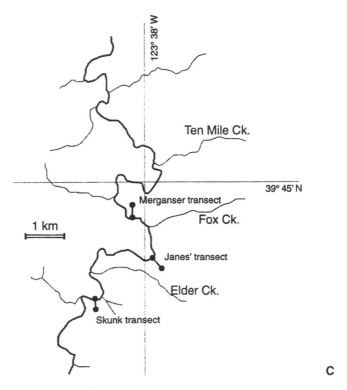

123° 38' W

Ten Mile Ck.

39° 45' N

Merganser transect

Fox Ck.

1 km

Janes' transect

Elder Ck.

Skunk transect

C

Figure 15.1 (A) Location map of the Angelo Coast Range Reserve study site. (B) Digital elevation data from airborne laser altimetry, showing the topography of the South Fork Eel watershed. (C) Locations of sampling transects and experimental studies.

uplift at our study site (1 mm per year). As a consequence, episodic waves of incision have propagated upstream over several thousand years, causing the river to cut deeply into its bedrock channel. The South Fork Eel is therefore canyon-bound, and is flanked by a series of terraces (abandoned river floodplains) of different ages (Seidl and Dietrich 1992), visible as white areas along the mainstem and tributary channels in figure 15.1B. The river is gravel-bedded, with areas of coarse boulders and intermittently protruding bedrock. The mean gradient of the South Fork Eel is 0.005, with discharge ranging from about 0.5 m^3 s^{-1} during summer base flow to up to 56 m^3 s^{-1} during recorded flood peaks (Power 1990b; USGS monitoring data, 1960–1970). Winter floods scour an active channel much wider than the wetted channel during the biologically active summer season. Therefore, despite the incision of the river and the steep surrounding terrain, most reaches of the mainstem receive direct sunlight for 6 to 8 hours a day

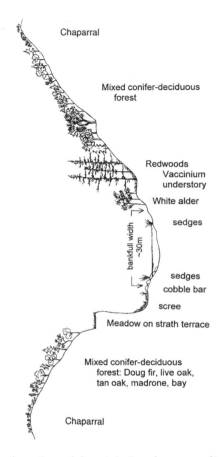

Chaparral

Mixed conifer-deciduous
forest

Redwoods
Vaccinium
understory

White alder

sedges

bankfull width ~30m

sedges
cobble bar

scree

Meadow on strath terrace

Mixed conifer-deciduous
forest: Doug fir, live oak,
tan oak, madrone, bay

Chaparral

Figure 15.2 Vegetational zonation and characterization of canopy profile from the South Fork Eel River to the ridge.

during the spring and summer season through September. Several small tributaries draining into the South Fork Eel are steeper (gradient 0.01), more extensively shaded by forest, and therefore lower in primary productivity and temperature. One tributary, however, has a drainage area slightly larger than that of the South Fork Eel at the point of confluence. Ten Mile Creek, which flows through what are now open, cattle-grazed grasslands, has a higher primary productivity and temperature than the South Fork Eel (Finlay et al. 1999).

Vegetation is elevationally zoned across the valley from the river to the ridge (fig. 15.2). Algae and aquatic mosses grow in the wetted summer

channel. Sedge tussocks (*Carex nudata*) and associated plants grow on rock
bars that are dry during summer and inundated during winter (see Levine
2000 for description of this vegetation). Riparian trees line the channel, dominated by white alder (*Alnus rhombifolia*), along with Oregon ash (*Fraxinus
latifolia*), big-leaf maple (*Acer macrophyllum*), and willow (*Salix* sp.). Farther
upslope in moist swales, large Coast redwood (*Sequoia sempervirens*) and understories of California huckleberry (*Vaccinium ovatum*) occur. Drier,
steeper, well-drained slopes support mixed conifer and deciduous forests,
including large Douglas fir (*Pseudotsuga menziesii*), live oaks (*Quercus
chrysolepis* and *Q. wislizenii,* canyon and interior live oak, are both common),
and tanbark oak (*Lithocarpus densiflora*), with patchy local stands of madrone
(*Arbutus menziesii*), bay (*Umbellularia californica*), and black oak (*Quercus
kelloggii*). In understories of upland forests, poison oak (*Toxicodendron diver-
silobum*) is common. Meadows occur within the Angelo Reserve on higher
river terraces. Chaparral grows along the ridgetops (generally only several
hundred meters from the channel margins), dominated by manzanita
(*Arctostaphylos* spp.), ceanothus (*Ceanothus* spp.), and chamise (*Adenostoma
fasciculatum*). Canopy height is taxon-, slope-, and soil-dependent and so is
irregular, with numerous gaps, but exceeds 50 m in redwood stands near the
channel and decreases upslope in drier oak stands or chaparral to only a few
meters near the ridge (fig. 15.2).

VARIATION IN SUBSIDIES

Emerging aquatic insects vary in their availability to consumers over space
and time. Factors driving this variation act on both the rate of emergence
of insects from the river and the lateral penetration of these insects into the
watershed.

Variation in Emergence

Rates of emergence of insects are higher from channels with higher algal
primary productivity. Channel productivity varies among tributaries and
longitudinally with insolation, sediment texture, and nutrient availability,
which in turn are determined by geology, climate, and land use in the watershed. A smaller-scale, but particularly intense, source of spatial variation
in our system is the distribution of floating mats of algae (primarily
Cladophora glomerata, with associated epiphytes). Floating algal mats accumulate in slack-water areas or lodge on emergent rocks in riffles (fig. 15.3).
The rate of insect emergence is 3 to 6 times greater from floating mats

Figure 15.3 Floating algal mats are hotspots of insect emergence in the South Fork Eel River.

than from submerged algae and many times greater than from bare gravel substrates (Power 1990b).

Floating mats could increase insect emergence for at least four reasons. First, they are convenient oviposition sites that minimize the exposure of female insects to aquatic predators. Second, they are rich feeding arenas, with copious growths of epiphytic diatoms and organic seston filtered from the river. Third, the mats trap sun-warmed water and, on a daily basis, become up to 8°C warmer than the surrounding water column (M. E. Power, unpublished data). For larvae of insect taxa that can tolerate the diel fluctuations of temperature and oxygen in these habitats, floating mats are food-rich incubators that accelerate their growth and development.

A fourth reason that floating mats may increase insect emergence is that they serve as partial refuges from aquatic predators. Larger fish in our system rarely forage from algal mats at the water surface (Power 1990b). Therefore, floating algal mats not only increase the secondary production of certain aquatic insects, but also route more of this production away from aquatic consumers, into the watershed.

Variation in insect emergence caused by the presence or absence of floating algal mats, the smallest spatial scale we are currently studying, is driven by the longest temporal scale of variation that we investigate: year-to-year variation in algal accrual that occurs because of hydrologically mediated trophic interactions. When northern California experiences its "normal" Mediterranean climate regime, scouring winter floods eliminate most algal and benthic biomass from the riverbed. During the following spring (April–June), the river subsides, clears, and warms. During this window of time, before animals become dense, the river food web can be thought of as having only one functional trophic level (sensu Fretwell 1977). By late spring or early summer, *Cladophora* blooms. As the summer drought progresses, consumer density increases as fish and invertebrates recruit, grow, and concentrate in the seasonally contracting river habitat. During this period, interactions mediated through three or four trophic levels influence the persistence of the algal standing crops that develop in the spring.

Food chains are shorter in northern California rivers when channels do not experience flood scour, following drought years or in artificially regulated channels (Power 1992a). Under prolonged stable flow conditions, benthic insect assemblages become dominated by slow-growing, heavily armored or sessile grazers: late successional taxa that, while slow to recover from physical scour disturbance, are relatively invulnerable to fish and other predators once they do. When they dominate the food web, food chains with only two functional trophic levels develop, and algae are suppressed (Power 1992a, 1995; Power et al. 1996; Wootton et al. 1996).

Year-to-year variation in nutrient flushing into the channel with variable rainfall could also contribute to annual variation in algal blooms, as it does in desert streams (Grimm et al. 1981; Grimm 1987). Experiments, however, suggest that top-down control by herbivores plays a dominant role in driving annual variation in Eel River algae. For example, removing a caddisfly, *Dicosmoecus gilvipes,* from experimental enclosures during drought years released algal standing crops that had been previously suppressed by this important predator-resistant but flood-vulnerable grazer (Wootton et al. 1996).

In addition to variation in insect emergence between high-algae years and low-algae years, we have observed strong within-year seasonal variation. Peak insect emergences follow algal mat formation in July, and emergences decline after algae senesce and disappear in August (Power 1990b). In addition, striking diel and day-to-day variation occurs because of the pulsed, often synchronized emergence patterns of common mayfly, caddisfly, stonefly, aquatic lepidopteran, and megalopteran taxa in our systems. Various degrees of synchrony have been documented for aquatic insect emergences in other fresh waters (Butler 1984; Sweeney 1984; Flecker et al. 1988). The post-emergence activity of aquatic insects may vary as well; for example, with temperature (Waringer 1991; Ward et al. 1996). Such variation would affect the lateral spread of these insects into the watershed and their availability to consumers.

Lateral Penetration into the Watershed

Quantifying the flux of emerging insects from rivers into watersheds is a daunting task, not only because of the considerable spatial and temporal variation in the subsidy source (emergence), but also because of the varied three-dimensional movements of insects after emergence. Jackson and Resh (1989) found that taxa caught in their canopy-level sticky traps dispersed a greater distance laterally into the forest than those trapped at stream height. The aggregation and dispersal of aquatic insects above forest canopies is largely unstudied, but in open settings over large African rivers, ground-based radar regularly detected discrete clouds of insects at heights of 30–130 m, which drifted over the shores in response to wind variations (Reynolds and Riley 1979). We have not yet appraised vertical variation in insect fluxes in our watershed, although we hope to initiate canopy-level sampling programs in the near future. The patterns reported here derive only from sampling using pitfall traps at ground level, sticky traps positioned 1.5 m above the ground, and light traps hung about 2 m above the ground. We used these sampling techniques on river-to-ridge transects, which extended 150–250 m upslope (perpendicular from the river) along the mainstem of the South Fork Eel and 10 m away from two less productive tributaries: Elder Creek and Fox Creek (see fig. 15.1C).

The pitfall and sticky trap data show a strong concentration of both aquatic and terrestrial arthropods near the river. Insect fluxes declined exponentially (both in numbers and in biomass) away from the river, with reductions of 50% within 10 m of the river's edge (fig. 15.4). Comparisons of the South Fork Eel with its darker tributaries showed, unsurprisingly, that lateral fluxes of insects at comparable distances from the water margin declined with decreasing watershed productivity. The results of light trapping (fig. 15.5) were similar in midsummer, when massive emergences of caddisflies resulted in peak biomass near the river. During other seasons, however, light trapping revealed a different pattern, in which sampled insect biomass, dominated by noctuid moths, peaked in the forest up to 150 meters away from the river.

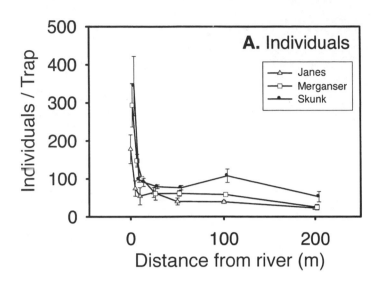

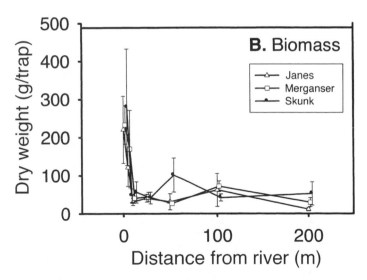

Figure 15.4 Numbers and biomass of invertebrates collected on sticky traps along transects perpendicular to the South Fork Eel River at three sites (averages of three replicate transects spaced 10 m apart per site) distributed over a ca. 5 km reach. Traps were 605 cm^2 acetate sheets, covered in Tanglefoot, rolled into a cylinder, and suspended from iron ("rebar") rods 1.5 m above the ground on 2–7 July 1998, a period of peak emergence. Biomass (dry weight) was estimated from length-weight regression (Rogers et al. 1976).

Figure 15.5 Light trap transect, showing river and riparian habitat along the South Fork Eel River.

RESPONSES OF WATERSHED CONSUMERS TO AQUATIC INSECT SUBSIDIES

The trophic and non-trophic requirements of consumer species determine how they are distributed over landscapes. Their local distributions and abundances, in turn, determine how closely consumers can track aquatic insect subsidies and what effects their subsidized populations may have on ecosystems. We will illustrate these classic ecological arguments with observations of consumers that forage over a range of spatial scales in the South Fork Eel watershed. These consumers include nearly stationary filmy dome spiders (*Neriene* sp., Linyphiidae), web-spinning, but potentially mobile, long-jawed spiders (*Tetragnatha versicolor*, Tetragnathidae), cursorial wolf spiders (Lycosidae), lizards (*Sceloporus* spp.) that defend territories along river rock bars and in upland meadows, and bats that forage by night over the river corridor and commute to roosts several kilometers away.

Filmy Dome Spiders

The least mobile consumers that we study are filmy dome spiders (*Neriene* sp.). Spider eggs overwinter in the forest floor litter. They hatch in the spring, and by late spring, the young spiders have spun distinctive webs with concave-down domes 30–60 cm above the ground in twiggy vegetation. As the spiders grow and mature, their webs become larger and denser. They shift their webs upward to positions 1–5 meters above the forest floor, still favoring sites where stiff, dense structure, such as the leafless, twiggy understory of oaks and tan oaks, is available for web support. Reproduction occurs in midsummer (July), with males moving into the webs of females. Aside from these local movements, filmy dome spiders appear to be relatively site-faithful. As their webs increase in size, they may become too massively invested in silk for the spiders to ingest and reposition them economically (M. S. Parker, personal observation).

The requirement for stiff web support may prevent the spiders from tracking river-to-ridge gradients in insect flux. Densities of filmy dome spiders censused from June through September in 1998 showed little change from the river to the ridge, 250 m away from the channel (fig. 15.6). Because of their low mobility and their failure to track river-subsidized insect fluxes spatially, individual performances of filmy dome spiders can serve as useful indicators of the importance of prey flux from rivers to the watershed. Michael Parker's observations during a year (1994) of high filmy dome spider abundance indicated that near the end of their growing season in September, spiders near the river had smaller webs, grew to larger sizes, and had longer times to starvation in laboratory tests than spiders of the same age 100–200 meters away from the river. Preliminary stable carbon isotope analysis suggests that filmy dome spiders may derive significant amounts of their carbon from algal-based river production, even when they are located several hundred meters from the river (J. C. Finlay, unpublished data).

Tetragnathid Spiders

Tetragnathid spiders (*Tetragnatha versicolor*), subjects of the dissertation research of Adrianna Smyth, are the second least mobile of the watershed consumers considered here. They spin orb webs, which they can ingest and relocate on a daily basis if foraging at a site proves unrewarding. Tetragnathids are extremely vulnerable to desiccation, and this limits their distributions worldwide to the margins of lakes and rivers. Along the South Fork Eel and its tributaries, tetragnathids are found only within about a meter of the water's edge during the summer drought. They commonly spin webs in riverside sedges (*Carex nudata*). Tetragnathid densities in the sedges that border much of the mainstem are much higher than are densities along a less productive, half-shaded tributary, Elder Creek (A. Smyth, unpublished data). Along the South Fork Eel and Elder Creek,

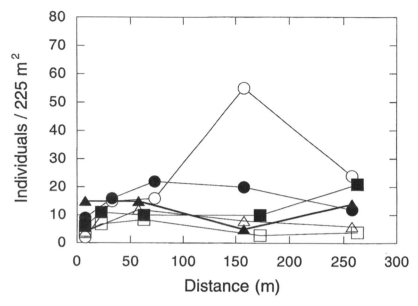

Figure 15.6 Abundances of filmy dome spiders in webs along six river-to-ridge transects at three sites along the South Fork Eel River, 26 June–8 July 1998. (From S. Khandwala, J. C. Finlay, and M. E. Power, unpublished data.)

Smyth experimentally introduced extra structure for web sites and supplemented food by importing floating algal mats. Her experiments have shown that in the mainstem of the South Fork Eel, tetragnathid densities increase only in response to the addition of structure, while in the less productive tributary, tetragnathid densities increase if food, structure, or both are supplemented. Corresponding to the difference in food limitation between the river and tributary spiders, she found a difference in their foraging patterns: tetragnathids along the mainstem foraged only by night, while those along the less productive tributary foraged 24 hours per day (A. Smyth, unpublished data). While release from food limitation is sufficient to explain the curtailed foraging by tetragnathids along the mainstem, the spiders there may also be more constrained by physical factors, such as dryness, heat, or wind, during the day (A. Smyth, personal observation). Whatever the cause of the shorter foraging period of mainstem tetragnathids, the difference has led Smyth to the counterintuitive prediction that in less productive, darker tributaries, these spiders may transfer more aquatic insect production to terrestrial predators than they do along sunny mainstem channels, where they are denser and probably more productive.

Smyth's recent discovery of two dipteran egg parasitoids on tetragnathid egg masses along the sunny mainstem may change this interpretation, however.

Lycosid Spiders

Cursorial wolf spiders (Lycosidae) do not spin webs, but stalk prey along the ground. Lycosids occurred in our watershed from the river margin up to the ridgetops, but were most abundant near the river (fig. 15.7A). In addition, lycosid densities were highest along the most productive mainstem habitat, intermediate along half-shaded Elder Creek, and lowest along a heavily shaded, less productive tributary, Fox Creek (K. Marsee, unpublished data; (fig. 15.7B). In addition to tracking differences in aquatic productivity among river channels, lycosids responded rapidly to algal mat manipulations that altered insect emergence over smaller scales. Where floating algal mats were naturally present, or where they were experimentally imported, insect emergence was higher. Within 24 hours, lycosids became more abundant along river shorelines adjacent to areas where mats had been imported or were naturally present than along shorelines where mats had been removed or were naturally absent (Parker and Power 1993). A positive correlation between lycosid spider densities and algal mat abundance was observed in five different river systems in northern California (Parker and Power 1993).

Lizards: *Sceloporus occidentalis* and *S. graciosus*

Sceloporus lizards in our watershed occur in upland meadow habitats and along river margins. Although these lizards have previously been studied only in upland habitats, John Sabo has found that during spring and summer in the South Fork Eel watershed, their densities are 7 times greater along river rock bars than in those habitats. In addition to the higher availability of insect prey in the river corridor, lizards find cover and favorable thermal environments in rock bar habitats. Sabo evaluated the importance of food subsidy relative to these other factors by experimentally reducing aquatic insect flux without influencing cover or thermal regimes. He constructed 2 m high fences of plastic and bird netting that enclosed about 90 m^2 of river margin rock bar habitat. Experimental treatments had a wall bordering the river (a "subsidy shield"). These enclosures reduced fluxes of aquatic insects by about 55% (numbers) and 70% (biomass) and those of terrestrial insects by a smaller amount (20% numbers and

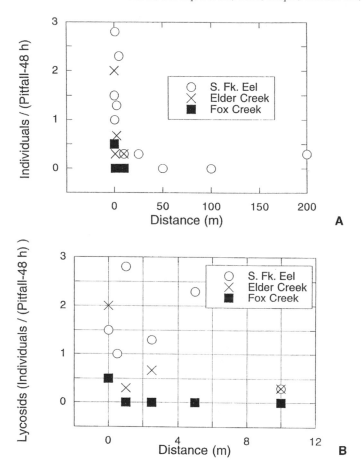

Figure 15.7 (A) Abundances of lycosid wolf spiders in pitfall traps (9.4 cm diameter), set for 24 h at various distances from the sunny, productive mainstem South Fork Eel (circles), a half-shaded tributary, Elder Creek (Xs), and a dark, unproductive tributary, Fox Creek (squares). (From K. Marsee, unpublished data.) (B) Same data truncated at 10 m from the river to show more clearly the abundances of riparian zone wolf spiders in channels of different productivities.

biomass). Control treatments were open to the river. In an experiment designed to examine the effects of subsidy reduction on lizard performance, the growth of lizards in enclosures with subsidy shields was reduced relative to the growth of lizards in control enclosures and free-ranging lizards (Sabo and Power 2002a). In a second experiment examining numerical responses by lizards to subsidies, the fences had open walls facing landward. More lizards emigrated from shielded areas than from unshielded controls (Sabo and Power 2002b).

Bats: Vespertilionidae and Molossidae

Bats are the most mobile of the small insectivores in our watersheds. Abundant species (*Myotis yumanensis, M. californicus, M. lucifugus, Eptesicus fuscus, Lasionycteris noctivagans, Tadarida brasiliensis*) forage commonly or primarily over still or slowly moving water and adjacent riparian vegetation. Most are aerial hunters, but some take insects from the surface of the water or glean them from vegetation.

William Rainey and Elizabeth Pierson have assessed bat foraging activity near ground level at our study site and along other California rivers. Their data, from multiple bat ultrasound detectors, are consistent with results elsewhere in demonstrating intensive foraging directly over the river channel by bats flying up to at least 20 m in height, but relatively little bat activity beneath or in the forest canopy up to similar heights (Thomas 1988; fig. 15.8). The data in figure 15.8 underestimate the steepness of the river-to-ridge gradient in bat foraging activity because this analysis is based on the number of 15-second intervals during which at least one bat was detected acoustically. In the forest, one bat is typically present in an acoustic sample, but we commonly observe groups of bats foraging simultaneously over the South Fork Eel. The steep decline in acoustic foraging activity away from the river indicates that bats, like lizards and lycosids, are tracking the exponential gradient in insect flux as measured by our sticky trap samples, and not the moths with the off-river density peak detected by

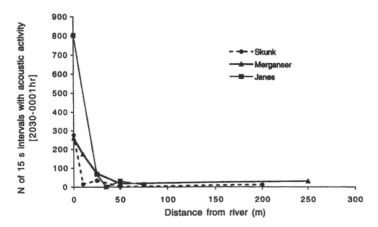

Figure 15.8 Distribution of bat foraging activity from the river to the ridge along three transects (see fig. 15.1C), as indexed by the number of 15-second intervals with at least one bat foraging call recorded by ultrasonic detectors.

our light trap samples (see fig. 15.4). Like our current insect trapping, however, ground-based bat detection does not permit the study of foraging above the canopy (Kalcounis et al. 1996).

CONSTRAINTS ON SUBSIDY TRACKING BY CONSUMERS AND THEIR ECOSYSTEM-LEVEL EFFECTS

Both trophic and non-trophic factors constrain the distributions and abundances of various consumers, determining both how closely they can track resource subsidies and the effects of their subsidized populations on local resources. We do not yet know enough to interpret the ecosystem-level effects of subsidies, but will offer some speculations based on our natural history observations to date.

The consequences of trophic subsidies to consumers in the receiving ecosystems depend on (1) the importance of the subsidy to the consumer, (2) the importance of the consumer in its ecosystem, and (3) the consumer's trophic position (i.e., where the subsidy enters the food web [Polis et al. 1998]). If consumers show strong numerical responses to a resource subsidy, closely tracking fluxes over space and time, we would suspect that the subsidy is important to them. If the resource subsidy determines environmental quality for a consumer species, close (nearly perfect) spatial tracking of the resource flux should result in an ideal free distribution of the consumer population (Fretwell and Lucas 1970), in which individuals in densely populated, heavily subsidized areas have the same fitness as individuals in sparsely populated, poorly subsidized areas.

It is when consumers are prevented from spatially tracking a subsidy that we have the opportunity to evaluate the effect of the subsidy on their individual fitnesses, which should ultimately affect their population dynamics. This is best illustrated in our system by filmy dome spiders, which appeared to be severely constrained by the requirement for stiff, dense structure to support their elaborate webs. These spiders had relatively flat river-to-ridge distributions, which did not match the exponentially declining flux of river-subsidized insects along this gradient. During 1995, a year of high population density for filmy dome spiders, their per capita growth and metabolic reserves decreased markedly with their distance from the river, while their foraging effort, as reflected by web diameter, increased (M. S. Parker, unpublished data). These trends were still evident, but weaker, in 1998, when spider densities declined sharply through the summer (S. Khandwala et al., unpublished data). Predation pressure on the spiders may have been more intense during 1998, as we measured 40–80%

declines over the summer in their censused densities, and we observed many webs without spiders that were nearly intact except for small (<2 cm long) tears. This type of damage could be done by wasps, which prey on similar spiders in other ecosystems (G. A. Polis and D. H. Wise, personal communication).

The year-to-year variation in spider densities and disappearance rates may suggest two alternative fates of the river insect subsidy to spider populations. During years of weaker predation, the subsidy may enter the watershed food web as detritus when the spiders die (during September and October) and their bodies and web contents are delivered to the forest litter. In years of higher predation, the subsidy may move up one or several trophic levels through predators (wasps, birds, other spiders) in the terrestrial food web. Because of their small per capita foraging rate and body size, both nutrient and food web transfers mediated through filmy dome spiders are probably not energetically important at the ecosystem level, except in years of extremely high spider abundance.

Tetragnathid spiders are restricted to water margins by their need for moisture. It therefore does not make sense to evaluate their river-to-ridge distributions in light of insect fluxes. Higher densities of tetragnathids along productive than unproductive channel margins (e.g., the South Fork Eel vs. Elder Creek) could reflect numerical resource subsidy tracking by these spiders over the spatial scales of hundreds to thousands of meters separating watersheds. We do not know enough about the mobility of this species over its annual lifetime to ascertain the degree to which individual movements versus demographic differences in population production among watersheds account for this pattern of tetragnathid density at the South Fork Eel. The demographic explanation seems more likely in light of the fact that egg masses are laid locally on sedges (A. Smyth and C. Anderson, personal observations). Spiders along dark tributaries may be both more food-limited and more subject to predation, because food limitation requires them to extend their foraging activity into dangerous daytime hours, when they are vulnerable to diurnal predators such as birds, wasps, and lizards. If Smyth's predictions are correct, the fate of ecosystem subsidies to tetragnathids should differ among streams of different productivity. More of their production should be transferred upward in terrestrial food webs along unproductive streams, while along more productive streams, more should be recycled directly back into the channel when winter floods inundate the spiders' bankside habitats.

Wolf spiders (lycosids) are cursorial hunters that do not use webs. They can be found along the entire river-to-ridge elevational gradient, so their moisture requirements do not appear to be limiting to their spatial distributions, although there may be differences among the species in our system in this regard. Lycosid densities closely correspond to the river-to-ridge exponential gradient in aquatic insect flux, although experimental work is required to establish whether this relationship is causal. Like tetragnathids, lycosids are more abundant along productive than unproductive streams, a pattern that could reflect either prey tracking or a response to some physical condition (wind, temperature, humidity) that also varies among watersheds. Lycosids numerically track aquatic insect fluxes over scales of meters to tens of meters along river channels, as demonstrated by their behavioral recruitment to emergence hotspots associated with floating algal mats (Parker and Power 1993). While we have much to learn about the factors limiting lycosids in the South Fork Eel watershed, their body size, tendency toward nocturnal activity, and life history characteristics suggest that they may be subject to heavy predation, as they are in other systems (Hering and Plachter 1997). Sabo and Smyth have hypothesized that the primarily nocturnal activity patterns of lycosids along the South Fork Eel may be a response to diurnal lizard predation. This hypothesis is supported by the greater diurnal abundances of lycosids observed along a small, but productive, tributary in the system where lizards are absent (J. C. Finlay and J. L. Sabo, personal observations). At present, our best guess for the fate of river subsidies routed through lycosids would be their transfer to higher trophic levels, such as lizards, birds, scorpions, carabid beetles, or other spiders, which abound in riparian corridors.

The sevenfold increase in the densities of western fence and sagebrush lizards, *Sceloporus occidentalis* and *S. graciosus*, from upland meadows to river margins can be attributed, at least in part, to tracking of river-subsidized insect production. Sabo's large-scale subsidy manipulations have shown that river insect subsidies enhance somatic growth and reduce emigration of lizards. In food webs, lizards are positioned at higher trophic levels than spiders (e.g., Spiller and Schoener 1988). Over small (behavioral) time scales, river subsidies lessen the effects of lizard predation on terrestrial arthropods (Sabo and Power 2002a). Over longer (population dynamic) time scales, however, numerical responses of lizards to river insect subsidies suggest the opposite effect (Sabo and Power 2002b), as has been found for aquatic subsidies to predators in other systems (Polis and Hurd 1995; J. Henschel, chap. 13 in this volume).

Flying predators have greater mobility, and hence more subsidy tracking potential, than ground-based spiders and lizards. Aerial vertebrates (birds and bats), with their high per capita feeding rates, have especially high potentials for subsidy-mediated effects. Swallows and black phoebes (*Sayornis nigricans*) forage along the South Fork Eel and its large, productive tributary, Ten Mile Creek. We are just beginning to observe these birds (J. Gutierrez, unpublished data) and can only speculate about how subsidies may influence their populations and ecosystem roles.

Mobility, vigilance, and parental care may restrict the amount of trophic transfer of subsidies up food chains by birds, particularly by colonially roosting bank and cliff swallows (*Riparia riparia* and *Hirundo pyrrhonota*), although trophic transfer to parasites may be important (Loye 1985a, 1985b; Polis, Anderson, and Holt 1997). The effects of birds as subsidized predators should depend on their degree of feeding specialization with respect to diet and habitat; black phoebes are more generalized in this regard, and so have more potential to exert subsidized predator effects on upland insects. Here, we will focus on a third effect, nutrient translocation from rivers to watersheds, as it is instructive to compare the potential effects of birds and bats.

The spatial scale of nutrient translocation by swallows may be limited by nest site selection that ties them to the river channel. Swallows in our system and along many other rivers nest in holes in the river bank (bank swallows) or construct mud nests on vertical rock faces or their anthropogenic equivalent, bridges (cliff swallows). They consume river insect production, sometimes in substantial amounts where colonies are large, but nutrient translocation away from the river is modest, as excretion is concentrated primarily within the winter-active channel. Black phoebes in our watershed forage over both rivers and meadows, and they nest in upland structures as well as along stream banks and under bridges. Therefore, they have slightly more potential than swallows to translocate river nutrients to higher positions in watersheds. In our study area, their numbers, activities, and movement scales are modest, however, compared with those of river-foraging bats.

Constraints on the landscape positions of both their foraging and their roosting sites may cause bats to translocate nutrients from river production to points higher in landscapes that may not receive comparable point inputs of nutrients from other sources (Rainey et al. 1992; Pierson 1998). In western North America, foraging by many bat species is concentrated, at least during the summer and fall, over open, quietly flowing streams, rivers, and ponds (Brigham et al. 1992; Herd and Fenton 1983; Brigham 1990). Bats foraging over water may choose pools and other quiet reaches rather than headwaters in part because turbulent water interferes with ultrasonic prey detection (von Frenckell and Barclay 1987). Foraging in narrow, higher-gradient tributaries is also constrained by the effects of vegetative clutter on prey detection and flight path (Brigham et al. 1997; Mackey and Barclay 1989). Lower insect densities in these typically shaded streams also may influence foraging site selection. Common aerially hunting bats are opportunistic foragers, recruiting rapidly to artificial (e.g., light-attracted) and natural insect aggregations and responding to foraging calls of conspecifics and insect flight sounds (Vaughan 1980; Fenton and Morris 1976; O'Farrell and Miller 1972).

In contrast, bat day roosts (usually located by radiotelemetry) are often high in forested watersheds. During warmer months, common species usually roost in cavities or crevices on large-diameter, canopy-emergent trees; some of these species also roost in caves, mines, and fractured vertical rock (Brigham and Barclay 1996; Pierson 1998). As small, long-lived heterotherms, bats select roosts offering thermal inertia, flight access, and isolation from diurnal predators. Such sites are often nearer drainage divides than lowland rivers for several reasons. Massive rock outcrops are often made of more resistant material than surrounding rocks, and after millennia of erosion tend to become high points in landscapes. Over historic time, logging may also have a role in the current landscape separation of bat roosting and foraging sites. Residual large trees and open-structured old-growth stands in many western U.S. forests are often on steep slopes far away from rivers. Commuting costs apparently do not play a dominant role in roost site selection, as one-way distances from day roosts to regularly occupied nocturnal foraging sites near and over water are usually several kilometers for 5–15 g bats (Pierson 1998).

Bats capture and ingest a significant fraction of their body weight in insects each night (estimated at 0.5–1 times body weight for the common aquatic forager *Myotis lucifugus* [Barclay et al. 1991]). Less nutritious parts of prey, such as wings of larger prey (e.g., megalopterans), may be culled and discarded in flight or at temporary feeding roosts, but typical soft-bodied aquatic insects are ingested whole. Gut passage times are rapid, so that foragers are redepositing comminuted prey and urine while they are foraging over the water. Ingestion usually exceeds excretion, however, and satiated bats of several species move to aggregated or solitary feeding roosts relatively near the river for intervals during the night. These night roosts may be tree exteriors and cavities or artificial structures with high thermal mass, such as concrete bridges. Both numbers of animals and individual fidelity over years at such sites can be high (Pierson et al. 1996). Annual guano deposition at such locations can range from grams to multiple kilograms dry weight. The temporal and spatial scale of river-to-watershed export of nutrients from bat night roosts depends on their location, which can be over the active channel (no export), in trees on rarely flooded terraces, or tens of meters above the flood zone. Day roosts are more consistent sites of nutrient export, as similar guano deposits accumulate at these sites, typically several kilometers from the river foraging site. The colony size and the persistence and character of the roost feature determine whether a day roost becomes a temporary or persistent hotspot of river-derived nutrients, which might locally affect upland decomposers, producers, or consumers.

We are planning to investigate the influence of translocations by bats of river subsidies at two different scales. Bat guano in tree hollows is consumed by some invertebrate detritivores, and culled insect parts are taken by ants and rodents. Guano in these settings is also frequently searched by vertebrate insectivores, and when exposed to rain or fog, it disappears rapidly. Nutrient translocation through detrital food webs to primary producers or decomposers has unknown but possibly interesting effects in N-limited old-growth conifer forests. Where bats roost in large numbers, input of river-derived nutrients to terrestrial landscapes has potential ecological significance. Several thousand Mexican free-tailed bats (*Tadarida brasiliensis*) roost on a cliff face at Rhyolite Dome, at Sutter Buttes, California. Their guano has elevated ^{15}N, consistent with their foraging over and around the agrochemically polluted Sacramento River, about 10 km away. Turkey mullein plants (*Eremocarpus setigerus*) growing downslope near the colony are ^{15}N enriched relative to conspecifics or other plant taxa growing lateral to or at a greater distance from the colony, offering the opportunity to trace river-derived nitrogen in an oak grassland food web in the upper watershed.

Nutrient translocation is a more likely consequence of river subsidies to bats than transfer up food webs to higher trophic levels. Data on bats in the diets of forest predators are limited, but their long mass-specific life spans and low reproductive rates (many bats have maximum longevity of more than 20 years, and only one offspring per year) suggest that they have not been subject to heavy predation over evolutionary time (Austad and Fischer 1991; Gillette and Kimbrough 1970).

In addition to translocating nutrients, bats may act as subsidized predators because there is a seasonal mismatch between the annual bat activity cycle and the flood and drought-mediated aquatic productivity cycle. Bats accommodate to low insect availability in colder seasons by episodic torpor, hibernation, or migration. When they first return or emerge with depleted energy reserves in the spring (April–May), food availability is low and uncertain. In wet years, river discharge may still be high, resulting in much lower aquatic insect emergence than will occur later. Terrestrial habitats around the Eel, though cool at this time, are also damp and green and are producing more insects than they will during the summer drought. We expect that open, productive terrestrial areas (meadows, forest edges, revegetating burns and landslide scars) experience intensified foraging by bats whose annual energy budgets are largely based on river production. Seasonal patterns of bat foraging habitat selection matched seasonal shifts in insect production among aquatic and terrestrial habitats in central Sweden (de Jong and Ahlen 1991). These seasonal shifts in foraging sites could be considered "subsidies in time," as described for pulse-and-release desert ecosystems and others reviewed by Polis and colleagues (Polis et al. 1996).

CONCLUSION

Investigations of river-to-watershed subsidies such as ours and those reported by Willson (chap. 19 in this volume) and others (e.g., Ben-David, Hanley, and Schell 1998) are beginning to reveal that rivers strongly influence watershed ecosystems far upslope from regions traditionally classified as riparian zones. These findings imply that our management (or mismanagement) of rivers may affect terrestrial ecosystems more extensively than was previously thought.

ACKNOWLEDGMENTS

We would like to acknowledge Stuart Fisher and John Jackson for our original inspiration for this research, the University of California Natural Reserve for protecting the Angelo Coast Range Reserve for university research and teaching, Bill Dietrich for supplying laser altimetry data for the South Fork Eel River watershed, the National Science Foundation and the California Water Resource Center for financial support, and Gary Polis and Shigeru Nakano for encouragement, friendship, and inspiration.

Chapter 16

Sources and Effects of Subsidies along the Stream-Estuary Continuum

Ralph H. Riley, Colin R. Townsend,
Dave A. Raffaelli, and Alex S. Flecker

Much has been made recently of the potential importance of allochthonous materials (subsidies) to the organization of food webs (Polis et al. 1996; Polis, Anderson, and Holt 1997; Polis and Hurd 1996b). The realization that most ecological systems are open to some degree and that allochthonous materials cross the boundaries between systems is not new; these ideas were a major feature of the findings in the 1970s of the International Biological Programme, which provides much of the backdrop for our current thinking on subsidies. However, new research on subsidies is not simply a reinventing of the ecosystem approach. The focus of research has moved away from mainly descriptive statements about the magnitude of allochthonous inputs to address their effects on food web dynamics, especially indirect effects.

Polis, Anderson, and Holt (1997) suggest that subsidy dynamics have several key features:

- Subsidy-consumer interactions are characterized by donor-controlled dynamics, in which the consumer has no effect on the production or supply rate of the subsidy.
- The subsidy crosses a boundary between two biologically distinct habitats or compartments.
- The interactions among the species in the recipient compartment have different dynamic features than in similar compartments not

receiving the subsidy. Thus, nutrients are most likely to enhance the primary productivity of basal species, detritivores will respond most to detrital material subsidies, and predator abundance will be enhanced by subsidies of their prey. Predators may even be enhanced by subsidies to a level at which they suppress their local prey and initiate a trophic cascade (Rosemond et al. 1993; Polis and Hurd 1996a).

Polis, Anderson, and Holt (1997) make a compelling case for the likely importance of subsidies for community organization, but much of the available evidence is circumstantial and supportive rather than confirmatory. In response to a questionnaire on subsidies, 44 responses from scientists (26 dealing with streams, 7 with floodplain rivers, and 9 with estuaries) included some evidence for the existence of subsidies, but provided largely inferential evidence for the origins of subsidies, and even less direct evidence of the effects of subsidies on aquatic systems. Results of rigorous experimental tests of the hypothesized effects of subsidies are only recently becoming available (as evidenced by this volume).

In the search for evidence of the key features listed above, there are some systems that are conceptually easier to investigate and offer more promise than others. For instance, defining the boundary between inshore and offshore marine systems is less straightforward than defining where two quite different systems abut. Any argument for the importance of subsidies to the organization of food webs is likely to be most convincing and defensible for systems where compartments are biologically distinct and the boundary between them is easily justified and conventionally accepted, such as the terrestrial-freshwater boundary, or where spatially separate habitats are characterized by strong unidirectional flows in the physical forcing factors that transfer the subsidized material, such as streams and rivers. Thus, the river continuum (Vannote et al. 1980) from headwater streams to the river estuary offers a promising system in which to evaluate the importance of subsidies in food web dynamics. There is evidence that downstream communities receive substantial subsidies of carbon and nutrients from upstream sections (e.g., Vannote et al. 1980; Naiman et al. 1987; Ward 1989) and, in the case of floodplain rivers, from the land (e.g., Junk et al. 1989; Webster and Meyer 1997a). By virtue of their location at the freshwater-marine interface, estuaries receive substantial subsidies of nonliving and living material from both upstream and downstream sources (e.g., Knox 1986).

In this chapter, we describe a series of investigations in two study systems in which subsidies seem likely to be major features of community organization: streams and rivers in temperate New Zealand and a single estuary in Scotland. Food webs in each of these systems have been intensively studied by the authors over many years. In the case of the New Zealand studies, investigations have been carried out on replicate water bodies within the Taieri River catchment and longitudinally along the main river. In contrast, in the Scottish estuary, the Ythan, the studies have been replicated within the system. Much of the work was not done with subsidy questions in mind, but the results are highly pertinent to the issue, and in several instances they provide for direct tests of the propositions described above.

The Taieri River

The Taieri River catchment (fig. 16.1) is an excellent setting in which to study the effects of subsidies. A large portion of the upper Taieri (> 100 km) has not been channelled or artificially constrained, the river maintains active floodplains with extensive riparian wetlands, and headwater tributaries are subject to a range of land use intensities and types. These factors offer a large scope to investigate two fundamental patterns of river subsidies: longitudinal (headwaters to estuary) and lateral (from adjacent wetlands and the riparian zone) patterns. Of particular note has been the extensive work done in the Taieri catchment on physicochemistry, disturbance, and the patterns and consequences of allochthonous inputs of nutrients and organic matter.

The Ythan Estuary

The Ythan estuary (fig. 16.2) in northeastern Scotland lies about 20 km north of Aberdeen. The Ythan River drains about 650 km^2 of agricultural land and enters the North Sea near the village of Newburgh. The tidal range is several meters, and there are extensive areas of mudflats and mussel beds. Here, too, we have detailed accounts of the estuary's physical and biological characteristics and one of the most detailed food webs in the literature. In addition, many controlled manipulative experiments have been carried out on the effects of enhancing particulate organic carbon, herbivores, detritivores, invertebrates, and invertebrate predators, and the effects of nutrients have been examined at the whole-estuary scale.

EFFECTS OF SUBSIDIES IN THE TAIERI RIVER SYSTEM

Effects of Nutrient Subsidies

Nutrient concentration data (Townsend, Arbuckle et al. 1997), nutrient addition data (R. H. Riley, unpublished data), and leaf decomposition data

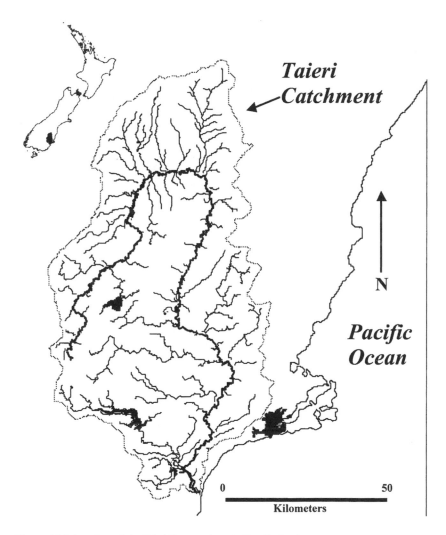

Figure 16.1 Location of the Taieri River catchment, New Zealand.

(Young et al. 1994) suggest that benthic algal growth and primary productivity in many headwater streams (predominantly draining native tussock grasslands) of the Taieri catchment are nutrient-limited (by N alone or N and P together). Management of "improved" pastures in the Taieri catchment, however, includes the application of superphosphate fertilizer and supplementation of clover seed to supply nitrogen through N fixation. Streams draining pasture catchments tend to be enriched in P or N or both relative to similar streams in tussock grasslands (Young et al. 1994).

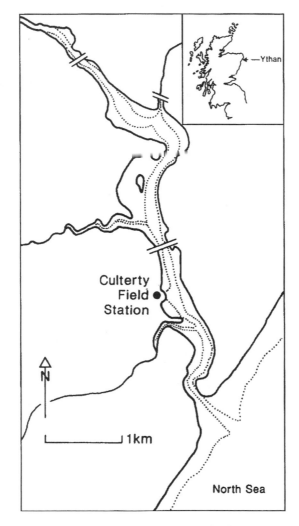

Figure 16.2 Location of the Ythan estuary, northeastern Scotland.

R. Young and A. D. Huryn (personal communication) studied community metabolism in pastoral and tussock catchments. They found that in spite of the higher nutrient concentrations in pastoral streams, their gross primary productivity (GPP) was comparable to tussock stream GPP in summer and autumn during periods when organic and inorganic seston concentrations were high. They speculate that light limitation associated with high turbidity reduced or negated any stimulatory effect of increased nutrient availability in the limited suite of streams in which they measured

Table 16.1   Measurements of selected water chemistry values, organic matter and seston, primary production to respiration ratio and food web statistics calculated for two headwater streams, Healy Creek and Dempsters Creek

Statistic	Healy (tussock)	Dempster's (pastoral)
Nutrients		
NO$_3$–N	7.28 µg l^{-1}	21.18 µg l^{-1}
DRP	4.61 µg l^{-1}	12.99 µg l^{-1}
Seston		
Organic	0.34(0.04) mg l^{-1}	0.52(0.13) mg l^{-1}
Inorganic	0.34(0.07) mg l^{-1}	0.06(0.02) mg l^{-1}
P:R	0.47	1.50
Invertebrate density	11,663 m^{-2}	15,101 m^{-2}
Number of links (maximum)	589	967
Number of chains (maximum)	2,506	14,534
Mean chain length (maximum)	3.02	4.67
Connectance	0.14	0.19

SOURCE: Jaarsma et al. 1998.

metabolism. We do not expect this to be true for all nutrient-enriched streams, since high nutrient concentrations and high turbidity do not always go hand in hand (Townsend, Arbuckle et al. 1997).

We have some information on the possible effects of nutrient subsidies on food web attributes. The food webs of two Taieri headwater tributaries have been characterized in detail: Healy Creek is in a tussock catchment, and an unnamed tributary of Dempsters Creek is in pastoral land (table 16.1; Jaarsma et al. 1998). These data reveal several notable patterns:

- NO$_3$, NH$_4$, DRP, total N, and total P concentrations in the pastoral stream were approximately 3 times higher than in the tussock stream.
- The productivity:respiration (P:R) ratio was 3 times higher in the pastoral stream.
- Invertebrate density was 1.3 times higher in the pastoral stream.
- The total number of food web links was about 1.5 times higher in the pastoral stream.
- The total number of food chains was 4 to 6 times higher in the pastoral stream.
- The maximum food chain length was about 2 times higher in the pastoral stream.
- The mean food chain length was 1.5 times higher in the pastoral stream.
- Connectance was higher in the pastoral stream.

This data set is consistent with bottom-up-driven increases in algal productivity that are reflected in substantial changes in food web characteristics. Not all pastoral streams can be expected to have high primary productivity, as we saw above. However, preliminary analysis of eight additional stream food webs has revealed a significant positive relationship between average food chain length and primary productivity (whether in pasture or tussock streams; Townsend et al. 1998). Thus, the influence of nutrient subsidies on instream primary productivity may well be reflected in changes in the structure and functioning of food webs.

A complication to this pattern is revealed by studies of the effects of exotic fish introductions. Several invertivorous fishes have been introduced to New Zealand since European colonization, including brown trout (*Salmo trutta*) in the Taieri. The introduced brown trout seem to be responsible for a trophic cascade in which changes at upper trophic levels are reflected by changes in nonadjacent trophic levels. Interactions between top-down fish effects and nutrient subsidies were studied using experimental channels containing trout, native galaxiid fishes, or no fish. At ambient nutrient levels, strong top-down effects (i.e., decreased invertebrate density and increased chlorophyll *a* density) of fish were detected, while no significant differences were observed between the galaxiid and no fish treatments (fig. 16.3).

We have further evidence that these trophic relationships may be altered by nutrient subsidies. In downstream channel sections that received nutrient subsidies, algal growth increased regardless of which fish were present, if any. Moreover, insect densities actually decreased in treatments with nutrient subsidies, presumably due to a shift in algal assemblage composition toward dominance by less preferred filamentous green algae. Thus, several mechanisms may act in concert to drive changes in food web structure in response to nutrient subsidies, including increased rates of primary productivity, algal assemblage compositional shifts, and resultant declines in grazer densities. Whatever the reason, it seems that a subsidy used by basal species in a food web (bottom-up) can alter the outcome of a top-down trophic cascade mediated by fish. These experiments were conducted over short time periods (approximately 1 month) and may therefore not reflect the dynamic changes in trophic structure that are possible in stable systems that are allowed to develop over long periods (Liebold et al. 1997). Few streams in the Taieri catchment can be considered stable, however, and disturbance regime (e.g., floods) has a strong influence on species composition (Townsend, Scarsbrook, and Dolédec 1997) and food web topology (Townsend and Riley 1999) in the

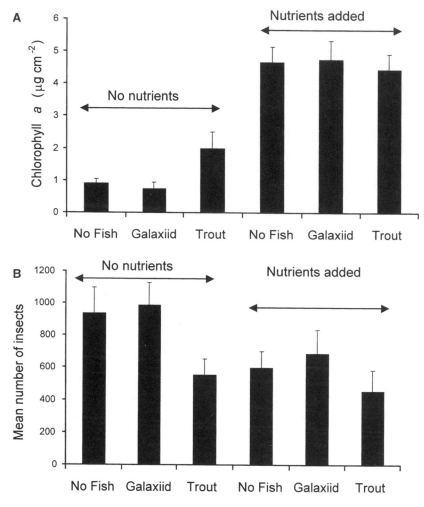

Figure 16.3 (A) Chlorophyll *a* density in artificial stream channels (in the Shag River, New Zealand) under three treatments in which trout, native galaxiid fishes, or no fish were present under ambient nutrient concentrations and under enhanced nutrient concentrations. Four channels were used per fish treatment (total = 12). Nutrients (N and P) were added to the lower half of each channel to achieve nutrient concentrations of approximately 3 times ambient concentrations. (B) Macroinvertebrate density under the same experimental conditions as in part A.

Taieri catchment. The short-term nutrient subsidy experiments thus may reflect trophic interactions relevant to Taieri stream communities for a significant portion of the year.

Nutrient subsidies also influence heterotrophic processes (Suberkropp and Chauvet 1995). Young et al. (1994) found higher decomposition rates of tussock grass leaves in pastoral streams than in tussock streams (a result, they suggest, of high NO_3^- concentrations in pastoral streams). Independently of their effects on GPP, nutrient subsidies may alter C availability to stream heterotrophs (see below).

In the mainstem Taieri, P and N concentrations increase from the relatively pristine headwaters to the estuary (Young and Huryn 1996). While benthic algae respond strongly to nutrient additions in the upper reaches, nutrient limitation of algal growth becomes less significant in the more turbid lower reaches. The longitudinal pattern of GPP reflects turbidity and nutrient availability, but the picture is not static. Maximum GPP and P:R levels are established within the first 100 km below the headwaters (Young and Huryn 1997). Consistent with the river continuum concept (Vannote et al. 1980), GPP declines in the lower river during the high-flow periods associated with higher turbidity and deeper water, which presumably lead to light limitation of primary productivity. During low-flow periods, however, high GPP levels may extend all the way to the lower reaches, concurrent with shallow water levels and high water clarity (Young and Huryn 1997).

Effects of Organic Carbon Subsidies

We find that some measures of C transport within streams are significantly related to land use, but we have only indirect information on the balance between C subsidies and allochthonous production. For instance, we find that organic seston transport is higher in pastoral than similar tussock streams (controlling for elevation, catchment area, and gradient) (Townsend and Riley 1999). This finding is consistent with the observation that C spiralling lengths are extremely short in Taieri mainstem reaches dominated by tussock grasslands (Young and Huryn 1997) and in tussock tributaries (R. Young and A. D. Huryn, personal communication), whereas spiralling lengths are particularly long relative to stream size in a pastoral catchment and in lower mainstem reaches. Conceptually, organic C spiralling length is the distance a C atom travels in a stream between the time it enters the stream (i.e., enters as allochthonous C or is fixed by stream autotrophs) and the time it is oxidized (Newbold et al. 1982). Organic C spiralling is a useful measure for comparing the efficiency of carbon use in different streams or reaches. The data of Young and Huryn (1997) suggest that biological oxidation of organic seston in native grassland and bush streams is rapid and limited by supply. The long spiralling lengths in pastoral streams may reflect both the abundance of C and the dominant effect of allochthonous material that may be metabolically recalcitrant (Wetzel et al. 1995).

While we have recorded longer food chains in streams with higher primary productivity (see above), there is no such relationship between any food web attribute and detritus availability (mainly allochthonous) (Townsend et al. 1998). It is tempting to speculate that higher rates of input of organic matter from the catchment will be reflected strongly in microorganism activity, while higher rates of production of good-quality algal biomass will be reflected more strongly in animal activity and longer food chains. Perhaps the microbial loop described by lake ecologists (Sherr et al. 1988) will also prove to be an important feature of heterotrophic stream communities.

Edwards and Huryn (1996) found that the movement of terrestrial invertebrates into adjacent stream reaches was related to land use: the biomass of terrestrial invertebrates entering streams was highest from native tussock grasslands and native forests and lowest from pastures composed of exotic grasses. They calculated that the biomass of terrestrial invertebrates consumed in a pastoral stream (Sutton Stream) was equal to 5% of the estimated food consumed by brown trout in that stream. We have no evidence yet that these terrestrial invertebrate subsidies enhance the resident predator populations or influence aquatic food web structure. However, Huryn (1998) found strong top-down effects of trout in Sutton Stream, where all invertebrate production was required to support trout production. Such effects were not found in a nearby trout-free stream, where only 18% of invertebrate production was consumed by the native *Galaxias eldoni*. Terrestrial invertebrate C subsidies may be especially important in trout-dominated streams and in streams that receive high inputs of terrestrial invertebrates. We mentioned above the probable existence of trophic cascades in Taieri tributaries involving trout. Subsidies of terrestrial invertebrate C may influence these trophic relationships, as already demonstrated for nutrient subsidies.

The data of Huryn (1998) also indicate that top-down suppression of grazing invertebrates, and the consequent reduction in consumption of annual primary productivity by invertebrates, has the potential to strongly alter subsidies of C (and presumably nutrients) from upstream to downstream communities. Huryn (1998) found the consumption of primary productivity in trout-dominated Sutton Stream to be 21%, while that in a galaxiid-dominated stream was 75% (primary productivity in the trout stream was 6 times higher than in the galaxiid stream). Mean surplus primary productivity in the galaxiid stream was not significantly different from zero, while that in the trout-dominated stream was 245 g AFDM m^{-2} yr^{-1}. Presumably more autochthonous (and allochthonous) carbon is washed downstream from the trout reaches than from the galaxiid reaches. Note that these upstream-downstream C subsidy patterns are likely to be strongly influenced by the existence of subsidies of limiting nutrients.

Adding to the set of variables that may confound predicted or simple effects of subsidies is streambed disturbance. Taieri tributaries whose beds are more frequently or intensely disturbed by discharge events tend to have less coarse particulate organic matter (C. R. Townsend, unpublished data) and probably have a lower biomass of algae on average (Biggs 1995). They also have a lower richness of invertebrate taxa (Townsend, Scarsbrook, and Dolédec 1997). Densities of all feeding groups in the most disturbed sites—at least soon after a storm event—decline, leaving a community dominated by browsing species and with a poor representation of shredders (fig. 16.4). Therefore, disturbance probably alters the total amount of allochthonous and autochthonous carbon in a stream, the relative proportions of the two, and the number and type of invertebrates available to consume particulate organic matter.

There is a distinct contrast between the patterns of nutrient subsidies and carbon subsidies in the mainstem Taieri. Whereas nutrient concentrations increase throughout the river continuum, consistently high concentrations of both DOC and POC are established within the first 70 km of the river below the headwaters as the Taieri passes through the extensive and unregulated upper floodplains (Young and Huryn 1997).

In the Taieri, we know that macroinvertebrate populations use temporally variable organic matter subsidies from the surrounding terrestrial systems. Using stable C and N isotope data, we determined that 30–90% of the diet of invertebrates in the mainstem Taieri might be derived from terrestrial organic matter (A. D. Huryn et al., unpublished data). The proportion depends upon river flow and geomorphological context: the terrestrial contribution to invertebrate diet is consistently higher in unconstrained (floodplain) reaches than in constrained reaches and is higher after prolonged flooding than after prolonged base flow.

We have no direct evidence of C subsidy effects on food web characteristics in the mainstem Taieri, but we do have information on organic C metabolism and organic C spiralling length. Young and Huryn (1997) found that organic C spiralling length increases significantly along the length of the Taieri, indicating that the degree to which C availability constrains river metabolism declines longitudinally in the Taieri. This finding suggests that the effects of subsidies will be greatest in headwater reaches, and that an

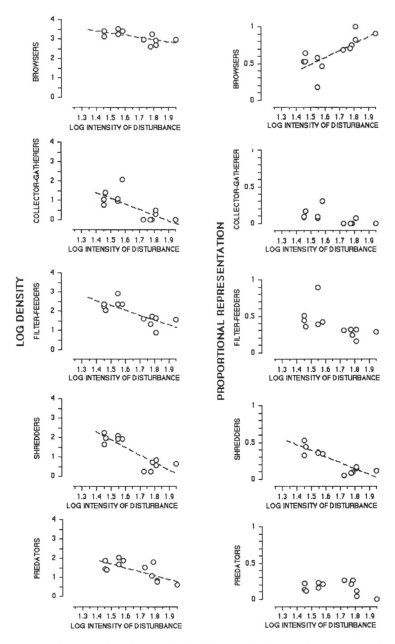

Figure 16.4 Relationships between the density (left column) and proportional representation (right column) of five trophic invertebrate groups and index of disturbance (IOD) in the Taieri catchment. IOD was measured as the proportion of painted rocks on streambeds that had moved immediately after a high-discharge event in 1994. Density is the number of individuals collected in four Surber samples. Regression lines are drawn where significant (*p* < .05). (From Townsend, Scarsbrook, and Dolédec 1997.)

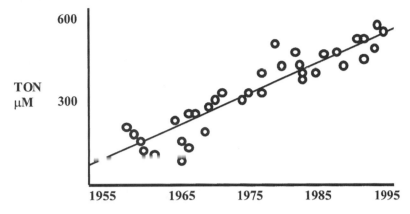

Figure 16.5 Changes in the concentration of total oxidized nitrogen (mainly nitrate) in the Ythan estuary over time.

excellent situation in which to investigate C subsidies is in contiguous constrained and unconstrained upper reaches, where we may expect significant natural differences in terrestrial C subsidies.

SUBSIDIES IN THE YTHAN ESTUARY SYSTEM

Effects of Nutrient Subsidies

Considerable information on nutrient subsidies to the Ythan estuary is available at the estuary scale through a long-term study of the effects of nutrient runoff from the catchment's intensively agriculturalized landscape (Raffaelli 1998). The concentration of N (almost entirely nitrate) in the river water entering the estuary has increased steadily from about 3 mg NO_3–N l^{-1} in the late 1950s to about 8 mg NO_3–N l^{-1} in the 1990s (Balls et al. 1995) (fig. 16.5). This increase has been accompanied by dramatic and large-scale increases in the biomass and coverage of green macroalgae (fig. 16.6). The only significant macrophytes in the estuary are green macroalgae in the genera *Enteromorpha*, *Ulva*, and *Chaetomorpha*. These macroalgae are now the main primary producers in this system, as phytoplankton production is relatively limited (Baird and Milne 1981). About 40% of the estuary's mudflats are now covered by algal mats with a wet weight biomass in the range 1–3 kg m^{-2} (Raffaelli et al. 1989; Raffaelli 1998). There is therefore evidence that nutrient subsidies from upstream and terrestrial sources enhance primary production in this system.

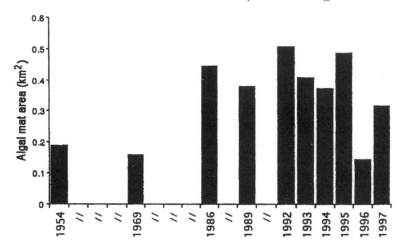

Figure 16.6 Area of intertidal flats in the Ythan estuary covered by macroalgal mats (> 1 kg m⁻²), from analysis of aerial photographs. The low values recorded in 1996 were associated with a major storm event.

The effects of enhanced primary production on other trophic levels in the Ythan estuary have been addressed through manipulative field experiments (Raffaelli et al. 1991, 1998). The effects are complex, species-specific, and often non-monotonic. With moderate enhancement of algae, the abundance of deposit feeders and predators increases (Raffaelli et al. 1991; Raffaelli 1999). We have been unable to demonstrate convincingly any increase in herbivores as a result of this level of enhancement, either for grazing invertebrates such as *Hydrobia ulvae* or for herbivorous wild-fowl (Raffaelli et al. 1999). At higher biomasses of macroalgae, only the polychaete *Capitella capitata* consistently increases in density (fig. 16.7A).

Since studies in the Ythan have been carried out within a single (un-replicated) system, we have had to follow the effects of nutrient subsidies on food web attributes over time, rather than making comparisons between enriched and non-enriched areas. Much of our information on the Ythan food webs' topological properties (Hall and Raffaelli 1991, 1996) is based on data collected in the 1960s and 1970s, when enrichment effects were slight (Raffaelli et al. 1989). Although the relative abundances of species have changed since then in response to eutrophication (Raffaelli et al. 1999), it is not immediately obvious how these changes might affect the Ythan's food web properties, such as chain length. However, given the results of the Taieri studies, a formal analysis of the highly enriched Ythan's food web would be interesting.

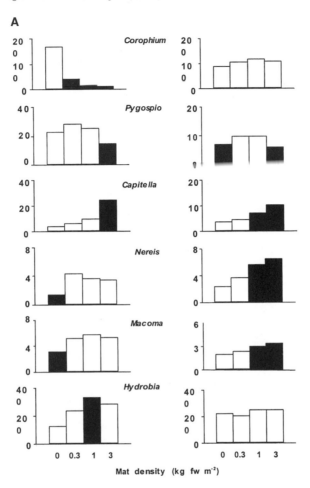

Figure 16.7 (A) Effects of small-scale experimental manipulation of macroalgal mat biomass on invertebrate densities (numbers per 45 cm² corer) 10 weeks into the experiment (left column) and after 22 weeks, when most algae had been removed from the sediment surface by wave action (right column). Solid bars indicate homogenous subsets. (B) Effects of small-scale experimental manipulations of organic carbon (O) and netting (N; this treatment was intended to mimic the physical as opposed to the organic enrichment effects of macroalgae) on invertebrate densities. C is a control treatment to which neither netting nor organic material was applied. OC is a second control treatment in which the sediment surface was disturbed in a similar way as when applying organic carbon in the O treatment. (From Raffaelli et al. 1998.)

Effects of Organic Carbon Subsidies

There is evidence that the abundances of deposit feeders have increased over the last 30 years in upstream sections of the Ythan estuary (Raffaelli et al. 1991; Raffaelli 1999). This increase has been attributed to enhanced

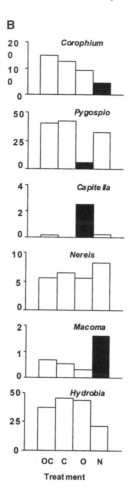

Figure 16.7 *Continued*

inputs of nutrients and organic material from freshwater inflow and from the high biomass of macroalgae farther downstream. Attempts to identify the source of the organic material using stable isotope (C, N, and S) analysis of putative sources, sediments, and invertebrates have been largely unsuccessful, mainly because of the high variation in the isotopic signatures of the macroalgae (Raffaelli 1999). While this enrichment effect on primary consumers is seen at the estuary scale, small-scale experimental manipulations in which organic matter was added directly to areas of mudflat did not result in significant increases in deposit-feeding species (fig. 16.7B), except for the opportunistic polychaete worm *Capitella capitata* (Raffaelli et al.

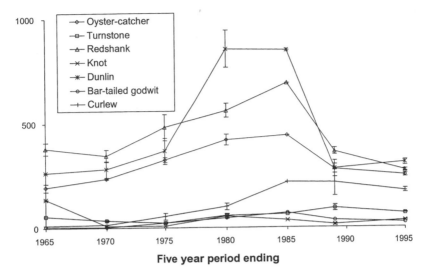

Figure 16.8 Number of shorebirds observed at the Ythan estuary during winter counts of shorebirds averaged over 5-year periods. Observed numbers of four of the six species increased until the early 1980s, in line with enhanced densities of deposit feeders. Recently there has been a decline in most species as macroalgal mats have continued to spread throughout the estuary.

1998). High inputs of organic matter, however, have pronounced negative effects as the sediment environment deteriorates (Raffaelli et al. 1991, 1998). The difficulty of detecting the effects observed at the estuary scale in small-scale experiments serves to underline the difficulties of interpreting the results of field experiments performed in open systems (Englund 1997).

High densities of primary consumers may exert both top-down and bottom-up effects. The potential top-down effects of enhanced densities of deposit feeders have been examined in a series of field and laboratory mesocosm experiments for several species, including the amphipod *Corophium volutator* (Limia and Raffaelli 1997), the mussel *Mytilus edulis* (Raffaelli et al. 1990), the mudsnail *Hydrobia ulvae* (Shearer 1997), and the polychaete worms *Nereis diversicolor* (Ventura 1997) and *Arenicola marina* (Ragnarsson 1996). The effects of these deposit feeders on sediment physicochemistry, meiofauna, and macrofauna were weak and inconsistent.

At the estuary scale, enhanced densities of primary consumers seem to have had a bottom-up effect on several species of shorebirds (Raffaelli et al. 1989) (fig. 16.8). Shorebirds increased on the Ythan between the early 1960s and the mid-1980s, probably in response to enhanced densities of invertebrates in areas not then affected by algal mats. However, since the 1980s, macroalgal mats have continued to spread, with a consequent decline in invertebrates, and hence shorebirds. These observed changes in shorebird abundance are compelling evidence of a bottom-up effect of nutrient subsidies on tertiary consumers. One shorebird, the shelduck (*Tadorna tadorna*), has declined in abundance over this period, probably due to physical interference with the birds' feeding behavior by the presence of algal filaments at the sediment-water interface (Raffaelli et al. 1998). However, this decline is unlikely to have led to a numerical response in other species. Long-term data sets comparable to those on shorebirds are not available for other consumers (fish and invertebrate predators) on the Ythan.

DISCUSSION

In river systems, we expect nutrient subsidies in upper reaches and tributaries to have a stimulatory effect on GPP, except when and where land use, hydrology, or geomorphology lead to concurrent conditions of high turbidity or deep water. Community respiration will generally be elevated under most nutrient subsidy conditions, assuming that periods of high turbidity are paralleled by elevated organic matter subsidies.

It is possible to discern both spatial and temporal patterns in the relative sizes of subsidies in the stream-river-estuary continuum. For example, while our results are consistent with the river continuum concept of Vannote et al. (1980), revealing a spatial pattern of nutrient and carbon subsidies in rivers that is strongly controlled by geomorphological context, these relationships are functions of seasonal patterns of river flow and are thus temporally variable. On a shorter time scale, spates and floods redistribute the stores of particulate organic matter on the riverbed and reset trophic organization. This kind of disturbance is probably a function of the size of a water body and its perimeter-to-volume ratio, and is therefore likely to be more important in small streams than in deep reaches of the lower river or estuary. Floods in floodplain rivers and the large-scale upstream-downstream movements of water masses in estuaries are not so much disturbances as predictable causes of organic and nutrient inputs (Junk et al. 1989). In both systems, there are marked temporal patterns. In the case of floodplains, these patterns are seasonal, while in estuaries there are highly predictable diurnal (tidal) and lunar (spring-neap) cycles that deliver subsidies to different points along the freshwater-marine gradient.

Strong spatial patterns in the nature and size of subsidies are related to land uses in the surrounding terrestrial catchment. For example, conversion of catchment vegetation from native tussock grassland to pasture, a predominant land use change in the Taieri region, can be expected to influence the role of subsidies, particularly through changes in (1) temperature and light regimes and thus rates of production and decomposition, (2) the chemistry of the stream water and thus rates of production and decomposition, (3) the input of dissolved and particulate terrestrial organic matter, (4) the riverbed disturbance regime and thus standing crops of autochthonous and allochthonous carbon, and (5) inputs of terrestrial invertebrates. Given this heterogeneity, the effects of subsidies on the receiving local stream food web are likely to be complex, but further along the river continuum such spatial effects will be more integrated. At the estuary, all heterogeneity in catchment subsidies may be fully integrated.

However, even if the subsidy itself is homogeneous in time and space, spatial heterogeneity in the receiving system may have a major influence on the way the subsidy affects food web dynamics. This is particularly true where there are discontinuities in the physical architecture of the receiving system. A good example is provided by Wallace et al.'s (1997) study of the role of CPOM subsidy from the riparian zone in a small forest stream. Over distances of only a few meters, patches of mixed substrate and of moss-covered bedrock responded quite differently to the experimental removal of the CPOM subsidy. The food web of the mixed substrate habitat was heavily dependent upon the lateral input of CPOM, whereas the moss-associated community depended on an input of a subsidy of FPOM from upstream. A similar physical discontinuity is generated by flow barriers in streams and rivers. In the Taieri system, trout occur downstream of many barriers to their upstream migration, while upstream of the barriers populations of native galaxiid fishes have a refuge from trout (Townsend and Crowl 1991). Benthic algae below the barrier can be expected to achieve a higher biomass than above the barrier because trout suppress grazing by invertebrates (Huryn 1998). However, if areas above and below the barrier receive a nutrient subsidy from the land, it is conceivable that the trout-grazer effect will be negated and algal biomass will be similar in both locations.

Increased primary production in Taieri streams is reflected in the general properties of the food web; for instance, by an increase in the average length of food chains. On the other hand, subsidies of organic matter are not reflected in the length of food chains. This difference is likely to be, at least in part, a reflection of the relative quality of algae and dead organic matter as foods for invertebrates and may partly explain the lack of obvious effects in the food web of the mainly decomposer-based Ythan estuary. Clearly, when dealing with subsidies, it is important to bear in mind not just their quantity, but also their quality in relation to autochthonous resources.

In summary, our analyses of well-documented and well-understood freshwater and estuarine systems confirm that subsidies of both nonliving and living materials can be significant, but that their effects on the receiving food web are likely to be variable in space and time. General enhancement of primary producers and their consumers by subsidies was apparent for both freshwater and estuarine sections of the river continuum. In the Taieri streams and rivers, there are tantalizing indications that the dynamics of the receiving food webs could be affected by locally enhanced densities of consumers. In the Ythan estuary, there were only weak or no effects of consumer enhancement (both primary and secondary) on the system's dynamics, underlining the need for experimental tests of hypotheses concerning the effects of subsides on food webs.

ACKNOWLEDGMENTS

We are grateful to the British Council for providing travel funds to facilitate collaboration between DAR and CRT and to the Royal Society for funding for DAR while in New Zealand. Thanks to Roger Young, Chris Arbuckle, and Ken Miller for their help. We are especially grateful to all those correspondents who took the time and trouble to respond to our original questionnaire. Although we were unable to use these replies quantitatively, they were immensely helpful in highlighting the gaps in our knowledge and our understanding of the dynamic role of subsidies in aquatic food webs.

PART III.

SUBSIDIES AT REGIONAL AND GLOBAL SCALES

Chapter 17

Integrating Food Web and Landscape Ecology: Subsidies at the Regional Scale

M. L. Cadenasso, K. C. Weathers, and S. T. A. Pickett

Gary Polis and his colleagues (Polis, Anderson, and Holt 1997) provided an organizing framework central to this book and, specifically, this section. Their seminal work synthesized food web ecology and landscape ecology by examining spatial flows among habitats. Furthermore, they showed how understanding the nature and dynamics of these spatial flows and the factors influencing them could contribute to a greater understanding of local food web structure and dynamics. Both of the areas of ecology that Polis and colleagues integrated are conceptually, theoretically, and empirically rich. Bringing them together requires the expansion of each field, beginning with expansion in the conceptual and theoretical realms and, eventually, empirical expansion. The consequent merging provides novel insights into both the structure and function of landscapes and the structure and dynamics of food webs. Here we will briefly discuss the main themes and concepts of food web and landscape ecology and how each has been expanded to facilitate their integration. We will then discuss how the authors of the chapters in this section have merged these two areas in their research, using highlights from each chapter.

Food webs are a central concern in community ecology because they describe the functional relationships among species in a community. The components of a food web that have been traditionally considered include the organisms that are interacting to make up the food web, nutrients, and detritus. Food web theory has focused on how life history characteristics, predator-prey interactions, and nutrient and detrital pools and processes contribute to the composition and dynamics of the web. This body of research addresses questions of community stability and its link to species diversity and web complexity, top-down versus bottom-up controls of community structure, web complexity and community response and resilience to disturbance, and the effects of invasions or extinctions on the stability of trophic interactions (Morin 1999).

Landscape ecology, on the other hand, has focused on spatial heterogeneity (Forman and Godron 1986; Turner 1989) and on interactions between and among patches across space (Pickett and Cadenasso 1995). This approach includes the concepts of patch dynamics (Pickett and White 1985), fluxes of organisms, materials, and energy among habitats (Wiens 1992; Forman 1995), and the permeability of boundaries between them, or "edge effects" (Stamps et al. 1987; Cadenasso and Pickett 2000, 2001). The emergence of landscape ecology has encouraged ecologists to consider ecological systems as open and exposed to influences arising outside their boundaries. These influences may interact with the structure or processes of the target system, and in some cases may dominate them. Exogenous influences may be direct, such as the input of organisms, materials, or energy from an adjacent system, or indirect, such as changes to an adjacent system that influence the structure or processes of the focal system. Common parameters to consider when evaluating the influence of exogenous factors include the perimeter-to-area ratio of the habitat as an index of habit openness, the identity of the habitats the focal habitat is adjacent to, and the identity and behavior of fluxes that may move between and among patches (Cadenasso and Pickett 2000).

In order to integrate food web ecology and landscape ecology, each must be expanded to ensure nodes of commonality between them. A critical expansion of food web theory is the recognition that food webs exist within the context of the local habitat they occupy and, consequently, are affected by interactions of that habitat with influences arising outside its boundaries. A greater understanding of food web structure and dynamics may be gained by incorporating the spatial context in which the food web resides and the spatial flows of organisms, materials, and energy to which the food web is exposed. Where in the landscape a food web resides may have critical implications for the identity of the organisms interacting in the web, consumer-resource dynamics, and the rates of nutrient and detrital production, storage, and movement. By integrating the traditional approach to studying food webs with the conceptual richness of landscape ecology, a better understanding of food web dynamics can be achieved through the investigation of novel research questions and approaches.

In addition, the merging of food web ecology and landscape ecology provides a functional basis for landscape ecology. Since its inception, landscape ecology has focused on elucidating pattern in the structure of landscape components (Forman and Godron 1986). Though this is a necessary preliminary focus, the field is poised to incorporate the explicit study of the function of landscape components (Cadenasso and Pickett 2000, 2001). Merging our understanding of landscape structure with our understanding of food web dynamics may provide a fruitful avenue toward this integration.

Subsidies to food webs have been defined by Polis, Anderson, and Holt (1997) as a transfer of resources from one habitat to another that results in a benefit to the recipient of that resource. Resources controlled by the donor habitat include prey, detritus, and nutrients. Resources from one habitat are transferred to a recipient—plant or consumer—from a second habitat. In order for the transfer to be considered a subsidy, there must be an increase in the population productivity of the recipient, which may, consequently, alter consumer-resource dynamics in the recipient system (Polis, Anderson, and Holt 1997).

All of the chapters in part 3 demonstrate a transfer of resources between habitats and suggest that these transfers influence the food web structure and dynamics of the recipient habitat. The movement of resources across landscape boundaries occurs by biological vectors such as migrating animals (Jefferies et al., chap. 18, and Willson et al., chap. 19) or by physical vectors such as water (Caraco and Cole, chap. 20). Subsidies may come from habitats with which the recipient habitat shares a boundary or from more distant donor habitats. Characteristics of the subsidy and the vector moving it influence the distance it can travel. Each of the chapters in part 3 contributes to the integration of landscape ecology and food web theory.

Jefferies et al. (chap. 18) address the flux of nutrients from one system to another through the movement of a biotic vector. They provide evidence that lesser snow geese feed in agricultural fields and acquire nutrients, particularly nitrogen, from these fields. Modern agricultural practices have increased crop size and fertilizer use, which has resulted in an increase in the quantity and quality of available food. Consequently, the geese are moving large amounts of nutrients from managed agricultural lands into wetland habitats where they roost. The feedbacks and interactions resulting from this cross-system transfer are numerous. The change in both quantity and quality of food has increased the population size of the geese and led to the expansion of their range. Some of the feedbacks of this subsidy include long-lasting changes in the vegetation structure of the wetlands due to fertilization through goose defecation and increased disturbance due to feeding by the geese.

Willson et al. (chap. 19) demonstrate that anadromous fishes function as ecological links between marine, freshwater, and terrestrial systems. Anadromous fishes facilitate the transfer of nutrients between marine and freshwater systems and between freshwater and terrestrial systems. The fish acquire biomass in the oceans and transport marine-derived nitrogen, phosphorus, and micronutrients into freshwater systems as they return to streams to spawn and die. Several vertebrate inhabitants of streamside forests, such as bears and eagles, feed on the fish. Their predation and consumption results in the movement of fish-derived nutrients and biomass to the adjacent terrestrial ecosystem. This subsidy transfer adds a new dimension to the paradigm of unidirectional flow of subsidies from terrestrial to aquatic systems. In addition, fish carcasses may be retained for variable amounts of time before being washed into downstream systems. How long the carcasses are retained is very difficult to measure, but may be a function of the heterogeneity of the stream and the possibility of the carcass getting caught in debris or topographic irregularities. This synthetic research combines terrestrial and aquatic food webs and requires that the characteristics of the habitats and organisms of both terrestrial and aquatic systems be explicitly linked.

Caraco and Cole (chap. 20) discuss the movement of organic carbon in large rivers and lakes. They conclude that inputs of organic carbon from outside the system—allochthonous inputs—contribute more to the respiration of large rivers or lakes than internally generated—autochthonous—sources. This has long been known for small streams, and it has been assumed that small streams receive relatively greater inputs from terrestrial systems than larger streams due to the greater difference in stream-to-drainage area ratios. Because smaller reaches are in closer contact with terrestrial inputs, it has also been assumed that large rivers are driven by the collection of organic material from their smaller upstream reaches. But, in fact, large rivers and lakes may receive even greater relative contributions of allochthonous organic carbon than small streams. Therefore, large rivers and lakes may be concentrators of organic carbon in the landscape, resulting in a heterogeneity of organic carbon pools.

All the chapters in part 3 address the subsidy of food webs in one system by nutrient flux from different, sometimes quite distant, systems in the landscape. These chapters show that the spatial scale at which food webs need to be considered is large and that the influences that must be incorporated in understanding web structure and dynamics are greater than previously considered. Accounting for these types of subsidies will enrich food web theory. In addition, a functional approach to landscape ecology is suggested. Investigating food web subsidies at regional scales provides a fruitful avenue to address how systems in landscapes interact. Research that increases the functional understanding of landscapes will improve landscape ecology. From this collection of chapters, it is apparent that subsidies occur between many types of habitats, across numerous scales, and move by physical or biotic vectors. The nature of the vector may determine the rate and extent of transfer as well as the spatial specificity of subsidies. The chapters in part 3 point the way toward greater emphasis on linkages between spatially distributed food webs.

ACKNOWLEDGMENTS

This chapter is a contribution to the program of the Institute of Ecosystem Studies. Support from the Andrew W. Mellon Foundation is greatly appreciated.

Chapter 18

Agricultural Nutrient Subsidies to Migratory Geese and Change in Arctic Coastal Habitats

Robert L. Jefferies, Hugh A. L. Henry, and Kenneth F. Abraham

Ecosystems both receive and donate allochthonous materials (Polis, Anderson, and Holt 1997). Allochthonous sources of nutrients and detrital material sustain net primary production and trophic dynamics and allow consumer populations to achieve greater numbers than those supported by in situ productivity (Power 1990b; Polis and Hurd 1995; Polis, Anderson, and Holt 1997). With the development of intensive agriculture in North America and elsewhere since World War II, there is mounting evidence of large-scale transfer of agricultural resources to semi-natural communities by both physical and biotic agencies. Birds, in particular, are both vectors and recipients of nutrient subsidies because of their mobility. Resultant changes in their population numbers have directly and indirectly affected food web dynamics in natural habitats. These effects have been accompanied by unforeseen, and possibly irreversible, modifications of natural ecosystems. This chapter provides evidence of changes in the trophic relationships of the midcontinental North American population of lesser snow geese (*Chen caerulescens caerulescens*) on their breeding grounds in coastal areas of Hudson and James Bays and on their wintering grounds in the Gulf states. The birds receive an agricultural food subsidy on their wintering grounds and along their migratory flyways that appears to have had a major influence on their population growth in recent decades.

268

DISTRIBUTIONS OF LESSER SNOW GEESE IN THE MISSISSIPPI AND CENTRAL FLYWAYS

The numbers and distributions of migrating and wintering lesser snow geese in different states have been determined annually in mid-December or mid-January on behalf of the U.S. Fish and Wildlife Service. Counts began in 1948 in the Central Flyway and in 1955 in the Mississippi Flyway; from 1969 to 1997, the official coordinated count was in mid December.

Before the 1920s, virtually all wintering lesser snow geese were found in coastal marsh habitats in Louisiana and Texas (Bent 1925; McIlhenny 1932; Lynch 1975). These marshes extended 800 km from Port Lavaca, Texas, to the Pearl River, Louisiana (Bateman et al. 1988). Historically, the geese arrived in large numbers in late October and remained until February. Significantly, in the last four decades, large numbers have not arrived until late November and early December.

From the 1930s onward, snow geese in Texas gradually expanded their winter range into rice prairies, where they utilized the green shoots and seeds of the crop as a food source (Alisauskas et al. 1988). The large increases in the numbers of birds that used the prairies in the 1940s were coincident with rapid changes in agriculture and industrialization on the Gulf coast (Robertson and Slack 1995). This range expansion continued and accelerated in the 1950s and 1960s (Stutzenbaker and Buller 1974), aided by rice irrigation and private landowners interested in conservation and commercial hunting (Hobaugh et al. 1989).

In Louisiana, snow geese wintered on coastal marshes until the late 1940s (Lynch et al. 1947; Lynch 1975), but thereafter expanded their winter range to include rice fields and other croplands. In addition, some snow geese began to winter in northeastern Louisiana in the late 1960s. Declines in numbers in major wintering areas of southeastern Louisiana were detected in the 1960s, and declines in numbers in southwestern Louisiana were noted in the 1950s, possibly because of delayed migration from the midwestern states (Lynch 1975).

The Missouri River valley in northwestern Missouri, southwestern Iowa, northeastern Kansas, and southeastern Nebraska historically hosted migrating snow geese, particularly in spring. Prior to 1940, few snow geese apparently used the area in autumn. However, autumn migration stopovers began in that year, and the number of birds using the area in autumn increased dramatically (20 times) from the early 1950s to 1971 (Burgess 1980). Snow geese now winter in the Missouri River Valley in most years.

Because it lies on the boundaries of both the Mississippi and Central Flyways, the area provides resources for populations of both flyways.

Snow geese began wintering in Arkansas in the 1970s, and their numbers rapidly increased throughout the 1980s (Widner and Yaich 1990). These birds are confined (>95%) to the eastern third of the state in the Mississippi alluvial valley adjacent to the Mississippi River; this habitat is additional to that in northeastern Louisiana.

CROP PRODUCTION, FERTILIZER USE, AND POPULATIONS OF LESSER SNOW GEESE

Boyd et al. (1982) suggested that these increases in the midcontinental population of lesser snow geese were the result of the birds' ability to benefit from changes in agricultural practices. Increased numbers and distributions of lesser snow geese along the Mississippi and Central Flyways and on wintering grounds in the United States appear to be linked to increases in crop production and to changes in agricultural practices. Figures 18.1–18.3 show the total area of cultivation of rice, wheat, corn (maize), and soybeans and the total yield of these crops for selected states. Amounts of nitrogen fertilizer applied to all these crops are also given, together with December or January counts of lesser snow geese in states along the Mississippi and Central Flyways (which are indices of abundance and not total counts). In midwinter (late December–January), over 70% of the birds are resident in the Gulf states. Geese actively forage in different croplands during their migration and on their wintering grounds. They consume spilt and wasted grain, sprouted seed, green stubble, and young seedlings; foraging preferences depend on season and type of crop (Stutzenbaker and Buller 1974; Hobaugh et al. 1989). Nutrient subsidies (fertilization) may increase not only the quantity but also the quality of food available in croplands, providing readily available sources of protein and carbohydrates in "an inadequate environment" (cf. White 1993). The states listed for each crop are those where production and area of cultivation are highest, and they are broadly coincident with the geographic areas of the flyway routes and wintering grounds of the birds.

A stepwise increase of about 30% in the area of cultivation of rice and wheat occurred between 1970 and 1975 (fig. 18.1). Although grain production has shown a nearly linear increase since 1950, a sharp rise in yield took place between 1970 and 1975. This rise was associated with selection for high-yielding crop varieties and increased use of nitrogen fertilizers. The rise in fertilizer application to these crops between 1950 and 1990 shows

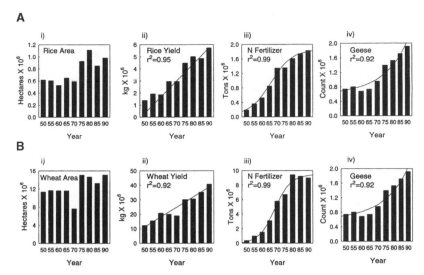

Figure 18.1 (A) Yearly data at 5-year intervals for (i) area of cultivation, (ii) yield, and (iii) total nitrogen fertilizer added to (A) rice in Arkansas, Louisiana, Missouri, Mississippi, and Texas and (B) wheat in the north central U.S. states of Iowa, Illinois, Indiana, Kansas, Michigan, Minnesota, Missouri, North Dakota, Nebraska, Ohio, South Dakota, and Wisconsin, along with (iv) annual 5-year means of winter Mississippi and Central Flyway lesser snow goose counts. Linear or curvilinear plots that gave the best fit in relation to r^2 values are shown. (Data from U.S.D.A. agricultural data, U.S. National Fertilizer and Environmental Research Center fertilizer summary data, and U.S. Fish and Wildlife Service winter waterfowl counts.)

a sigmoid relationship; the steepest increase in the rate of application occurred between 1965 and 1975. Counts of lesser snow goose numbers varied between 600,000 and 800,000 from 1950 to 1965. Thereafter, numbers progressively increased to about 2 million birds by 1990. Correlation coefficients (r^2) between total bird counts and the yield of rice and wheat are .91 and .90 respectively ($p = .0001$), and the corresponding correlations between yield and fertilizer use for the two crops are .91 and .80 respectively ($p = .0011$).

Overall, the same pattern of increases in yield and in application of nitrogen fertilizer is evident for corn yield (fig. 18.2A). The correlation coefficient between counts of lesser snow geese and corn yield is .89 ($p = .0001$), and that between corn yield and total fertilizer use is .84 ($p = .0005$). However, the total area of cultivation of corn has not increased, unlike that of rice and wheat.

The pattern of change in soybean production in the midwestern states (Ohio, Illinois, Indiana, and Iowa) is different from that of cereal crops (fig. 18.2B). The total area of cultivation has increased almost threefold

A

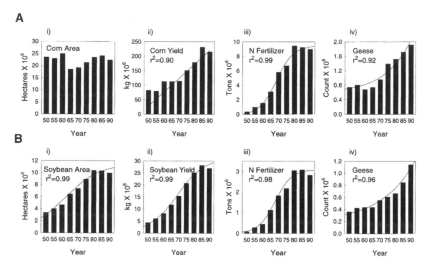

Figure 18.2 (A) Yearly data at 5-year intervals for (i) area of cultivation, (ii) yield, and (iii) total nitrogen fertilizer added to (A) corn in the north central U.S. states listed in figure 18.1B, and (iv) annual 5-year means of winter Mississippi and Central Flyway lesser snow goose counts. (B) Yearly data at 5-year intervals for (i) area of cultivation, (ii) yield, and (iii) total nitrogen fertilizer added for soybeans in Ohio, Illinois, Indiana, and Iowa, and (iv) annual 5-year means of winter lesser snow goose counts for the Mississippi flyway only.

since 1950, although there has been little change since 1980. Soybean production increased fivefold in this period, but in the last decade it has declined slightly. Applications of nitrogen fertilizer show a similar trend. The correlation between counts of lesser snow geese and soybean yield is weak ($r^2 = .75$, $p = .0026$), but there is a very strong correlation between fertilizer use and yield ($r^2 = .98$, $p < .0001$).

Bird counts for the Mississippi Flyway indicate a steep rise in the number of lesser snow geese since 1980 (fig. 18.3A) (comparable data for the Central Flyway are shown in fig. 18.3B). Much of the recent increase has occurred in Arkansas, Iowa, and Missouri, where the birds forage in rice fields and cornfields.

Approximately 39% of the nitrogen fertilizer used for crop production in the United States in 1990 was applied in the north-central states and in Texas (fig. 18.4) (Lanyon 1995). Much of the increase in crop production since 1950 was a consequence of the diversion of ammonia production for munitions to agricultural uses after World War II. However, just as important were changes in agricultural infrastructure associated with the ready availability of fertilizers (Lanyon 1995). Specialized crop production, which was no longer dependent on animal manure, but rather on nitrogen fertilizers

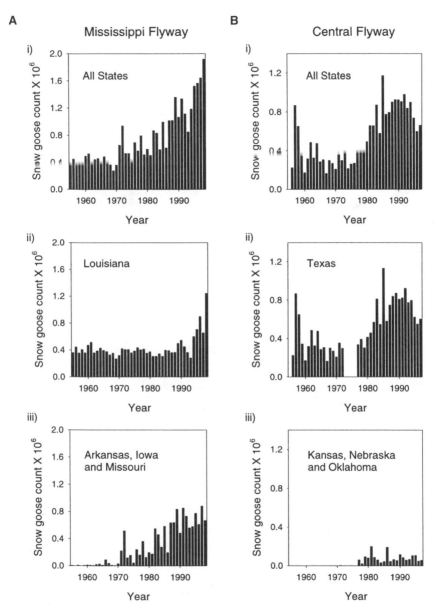

Figure 18.3 (A) Winter lesser snow goose counts for (i) all Mississippi flyway states (Alabama, Arkansas, Iowa, Illinois, Indiana, Kentucky, Louisiana, Michigan, Minnesota, Missouri, Mississippi, Ohio, Tennessee, and Wisconsin), (ii) only Louisiana and (iii) only Arkansas, Iowa, and Missouri. (B) Winter lesser snow goose counts for (i) all Central flyway states (Colorado, Kansas, Montana, North Dakota, Nebraska, New Mexico, Oklahoma, South Dakota, Texas, and Wyoming), (ii) only Texas, and (iii) only Kansas, Nebraska, and Oklahoma. Winter counts represent January counts from 1955 to 1968 and December counts from 1969 to the present. State-by-state counts are not available for the Central Flyway for 1965–1968. (Data from U.S. Fish and Wildlife Service.)

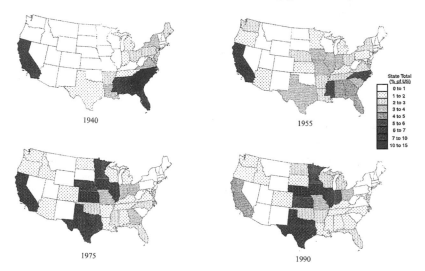

Figure 18.4 Proportional nitrogen fertilizer use by individual states of the United States in 1940, 1955, 1975, and 1990. (Adapted from Lanyon 1995.)

produced elsewhere, developed on a regional basis. As a result, single farm units were amalgamated into larger holdings. The rise in the lesser snow goose population appears to be inextricably linked to these overall changes in crop production, fertilizer use, and land use.

The dynamics of food availability and the diets of lesser snow geese vary greatly, both within and among marsh habitats, rice fields, and cornfields (Alisauskas et al. 1988). In rice fields, green shoots (87%), seeds (7%), and subterranean organs (4%) of rice are eaten, whereas in cornfields, geese tend to eat waste grain (88%) rather than green shoots (10%). Green shoots contain proportionally more protein, but less digestible carbohydrate, than waste grain (Alisauskas et al. 1988). Dry weather allows farmers to plow the land in winter (Stutzenbaker and Buller 1974), and geese may switch to eating ryegrass in pastures when this is done.

Similar changes in agricultural practice in the Netherlands have led to long-term shifts in the abundances of Anatidae (swans, geese, ducks): eleven out of seventeen herbivorous species have increased (Van Eerden 1990). In particular, improved quality of grasslands (crude protein, increased digestibility, longer season) has resulted in a higher carrying capacity for the true grazers among avian herbivores. There was a sixfold increase in nitrogen fertilizer (kg N ha^{-1} yr^{-1}) applied to permanent grasslands in the western section of the Netherlands between 1939 and 1992.

During the 1980s, the amount of added fertilizer was, on average, 300 kg N ha^{-1} yr^{-1}, which extended the growing season by 10 to 30 days (Van Steenbergen 1977). The timing of the start of the use of agricultural crops differs among bird species and appears to be related to body mass (Mattocks 1971; Owen 1971; Prop and Vulink 1992; Poorter 1981; Van Eerden 1984, 1990): larger avian herbivores preceded smaller species in the utilization of improved grasslands as a food source.

Dense aggregations of geese in agricultural lands have resulted in the birds being nutrient vectors (Post et al. 1998), and they have become important conduits of nutrients from fields to natural habitats at the local level. In New Mexico, daily feeding bouts by geese move large quantities of nutrients from farm fields to the managed wetlands where they roost (Post et al. 1998). Loading rates in a 50 ha wetland at Bosque del Apache National Wildlife Refuge peak at 300 kg N day^{-1} and 30 kg P day^{-1}. The birds supply 40% of the nitrogen and 75% of the phosphorus entering the wetland annually. Wetlands are generally considered to be N-limited or N-P co-limited, and N:P ratios below 29:1 promote dominance of cyanobacteria (Tilman et al. 1982; Smith 1983). This has occurred occasionally in winter in the Bosque del Apache wetland, but under existing flushing rates, such changes are not maintained. However, with ever-increasing densities of geese, long-term changes in water quality and species assemblages can be expected.

The principal cause of mortality of adult geese several decades ago was hunting (Owen 1980). There is no evidence of changes in the abundance of natural predators on geese. Harvest of birds, along with hunter numbers, has declined over the last 25 years in both the central United States and Canada (Abraham et al. 1996; Abraham and Jefferies 1997; B. Sullivan, personal communication, for the United States and K. Dickson, personal communication, for Canada; Cooke et al. 1999). Annual survival of adult lesser snow geese from La Pérouse Bay, Manitoba, on the Hudson Bay coast, increased from about 78% in 1970 to about 88% in 1987 (Francis et al. 1992). Although the decline in harvest numbers is not substantial, the increase in survival may have been due in part to a reduced overall harvest rate (measured as a proportion of the midwinter indexed population). A further possible cause of increased survival is the establishment of refugia along flyways where birds are protected from hunting and where supplementary food supplies are often provided (Abraham et al. 1996).

MacInnes et al. (1990) showed that lesser snow geese nested progressively earlier in the Hudson Bay region from 1951 to 1986. They suggested that this change was due in part to climatic warming. As reproductive success in Arctic-nesting geese is positively correlated with early spring melts (Owen 1980), climatic warming could have led to higher annual population growth rates for the midcontinental population of lesser snow geese.

The long-term effects of the increase of about 5% per annum in the midcontinental population of lesser snow geese are evident in coastal habitats around Hudson Bay, where the birds breed 5,000 km distant from the energy-nutrient subsidy provided to them in winter (Jefferies 1988a, 1988b, 1999; Abraham et al. 1996; Bazely and Jefferies 1997). Their use of agricultural crops as food during winter and the availability of food in refugia during migration appear to have released the midcontinental population from the density-dependent controls that existed when they fed only in coastal marshes of the Gulf of Mexico. We conclude that although other factors, such as shifts in decadal weather patterns, establishment of refugia, and changes in wildlife management may have contributed to population growth, this agricultural subsidy is the primary cause of increased goose numbers (cf. Abraham et al. 1996 and citations listed therein).

ARCTIC COASTAL HABITATS AND TROPHIC DYNAMICS

The biomass of primary producers may be severely reduced when a herbivorous species increases dramatically in the absence of control by a species at a higher trophic level. The herbivore may override top-down (and bottom-up) controls, and the rapid increase in its numbers can lead to a trophic cascade resulting in runaway consumption of the primary producers (Strong 1992). The net effect of a large energy-nutrient subsidy is equivalent to that of the removal of a predator in the upper trophic levels that controls herbivore numbers; hence, we have used the term "apparent trophic cascade" to describe this effect (Jefferies and Rockwell 2002). Theoretical studies of food web dynamics predict that high allochthonous inputs of nutrients to primary producers may destabilize systems (Rosenzweig 1971; DeAngelis 1992). Similar studies of the direct effects of large additions of allochthonous materials at the consumer level also indicate that food chains may become decoupled as feeding preferences of consumers are changed in favor of the external sources, which can lead to decreases or extinction for local prey (Huxel and McCann 1998; McCann et al. 1998).

As we describe below, widespread foraging by migratory lesser snow geese on fertilized agricultural crops in one season (winter) and location (the southern and midcontinental regions of the United States) has resulted in a sharp decrease in in situ plant productivity, as well as local extinctions of forage species, in another season (summer) and location (the Arctic breeding grounds). As such, the population changes brought about by the agricultural subsidy represent biomanipulation of Arctic-breeding goose populations on a massive scale (Abraham et al. 1996; Bazely and Jefferies 1997). This apparent trophic cascade is driven by bottom-up processes (agricultural crops and fertilization) on the wintering grounds and the strong top-down effects of geese at their summer coastal breeding sites around Hudson Bay and elsewhere.

The species richness of the Arctic salt marsh communities around Hudson Bay is low, and there are no major herbivores besides geese. Hence, as an approximation, the simplified food web of these communities can be treated as a trophic ladder. The changes in Arctic coastal vegetation (salt marsh and freshwater plant assemblages) caused by the foraging activities of increasingly larger numbers of geese are ongoing and cumulative. The effects of colonially nesting populations of lesser snow geese on vegetation are density-dependent, and as numbers of geese increase, the spatial and temporal scales of their effects on vegetation and other organisms change in a discontinuous manner that results from a switch between two positive feedback processes (fig. 18.5).

At low goose densities (characteristic of the late 1960s and early 1970s at most breeding sites on the south coast of Hudson Bay), grazing by family groups of geese initiates a positive feedback loop whereby increased growth of intertidal swards of salt marsh graminoids (*Puccinellia phryganodes* and *Carex subspathacea*) is driven by increased nitrogen availability in a system deficient in this element (Cargill and Jefferies 1984a, 1984b) (fig. 18.5A). Within a growing season, the availability of nitrogen is increased by its rapid recycling from goose feces, which allows, in turn, rapid regrowth of these shallow-rooted, stoloniferous and rhizomatous, prostrate graminoids following their defoliation by geese (Bazely and Jefferies 1989). Geese defecate every 4–5 minutes, and fresh goose droppings contain high levels of soluble nitrogen. Also, intense grazing maintains open swards with bare microsites that are colonized by cyanobacteria that fix nitrogen, particularly early in the season (Bazely and Jefferies 1985). This nitrogen replaces that which is retained by the geese in their body tissues.

The second positive feedback loop operates at the high densities of geese that are characteristic of the 1980s and 1990s, and it initiates an apparent terrestrial trophic cascade that has led to the destruction of salt marsh swards and freshwater plant assemblages (Srivastava and Jefferies 1996; Kotanen and Jefferies 1997) (fig. 18.5B). The removal of surface vegetation by goose grubbing in spring results in the development of hypersaline soil

(A)

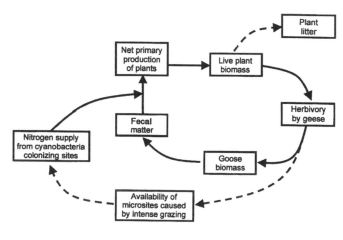

(B)

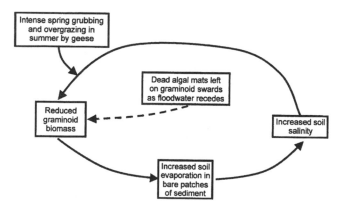

Figure 18.5 The interactions between lesser snow geese and the coastal salt marsh vegetation on which they forage at La Pérouse Bay, Manitoba, can be represented as two positive feedback loops. Solid lines represent direct effects; dashed lines, indirect effects. (A) In the first positive feedback loop, which occurred up to 1982, when goose numbers were low (<7,500 pairs), grazing by geese results in increased net aboveground primary production of forage species as a result of the availability of fecal nitrogen for plant growth and enhanced rates of nitrogen fixation by cyanobacteria growing on sediments where swards are heavily grazed. (B) In the second positive feedback loop, which has occurred since 1984 as goose numbers have increased to an estimated 44,500 pairs in 1996, removal of the vegetation cover by goose grubbing exposes the sediments, leading to increased evaporation and hypersalinity in the soil, which decrease plant growth. Summer grazing and algal mats that develop in spring because of fecal input from high goose densities both reduce available forage and accelerate the process. (Adapted from Bazely and Jefferies 1997.)

conditions, escalating the destruction of salt marsh graminoids (Iacobelli and Jefferies 1991; Srivastava and Jefferies 1995a, 1995b). Removal of vegetation by geese creates bare sediment in intertidal marshes, where only a few shoots of salt marsh graminoids remain. Increased evaporation from these patches leads to the development of hypersaline conditions in the upper layers, which are detrimental to the regrowth of the graminoids (Srivastava and Jefferies 1995a, 1995b, 1996). *Puccinellia phryganodes* and *Carex subspathacea* fail to recolonize patches of bare, hypersaline soil by clonal propagation, at least in the short term (Handa 1998). The former species is a sterile triploid, and seed set in the latter species is rare in swards grazed by geese.

In summer, families of lesser snow geese heavily graze the remaining salt marsh vegetation. This grazing exacerbates the effects of the degenerative positive feedback loop, with negative consequences for plant growth (Srivastava and Jefferies 1996) as well as gosling development and survival (Cooch et al. 1991; Francis et al. 1992) at the local level. Some experimental exclosures that were erected in denuded areas have been devoid of perennial plants for over 14 years. Weather can exacerbate the situation when late springs delay the northward migration of birds, so that staging birds, in addition to breeding populations, grub belowground vegetation in snow-free sites where the ground has thawed (Abraham et al. 1996; Skinner et al. 1998).

The degradation of these intertidal marshes is analogous to desertification (Graetz 1981) and has resulted in the loss of a "seral stage" in the development of vegetation in these coastal environments. Surveys conducted over 3 successive years (1993–1995) along the coast of the Hudson Bay lowlands (2,000 km distance) indicate that approximately a third of the intertidal salt marsh is now mudflats, a third of the swards are badly damaged but are capable of recovery in the absence of goose grazing, and a third remains as heavily grazed marsh (Abraham and Jefferies 1997). Losses of vegetation in coastal marshes during the last 20 years can be detected by LANDSAT imagery (Jano et al. 1998). There have been comparable changes in the vegetation of the freshwater sedge meadows adjacent to the intertidal marshes (Kerbes et al. 1990; Kotanen and Jefferies 1997).

EFFECTS OF GOOSE FORAGING ON REVEGETATION PROCESSES AND ON OTHER GROUPS OF ORGANISMS

Although natural revegetation and assisted revegetation can occur inside exclosures on bare intertidal mudflats, outside these exclosures, the continued foraging activities of geese preempt the reestablishment of plants.

Monitoring of vegetation over 12 years along transects (800 m) in an intertidal marsh indicates loss of species diversity and an increase in the area
devoid of vegetation cover (Jefferies and Rockwell 2002). Little or no natural
revegetation has been observed outside of exclosed areas, and soil degradation processes continue to accelerate. In assisted revegetation trials,
transplants of *Puccinellia phryganodes* (as plugs) became established in degraded sediments when soils had adequate moisture and low salinities
(<33 g of dissolved solids per liter of soil solution) (Handa and Jefferies
2000). The growth of plants increased significantly when single and combined additions of peat mulch and inorganic nitrogen and phosphorus
were made to the soil surface (Handa 1998). However, there was considerable spatial and temporal variation in the successful establishment of
tillers. This variation reflected fine-scale differences in edaphic conditions
in summer, particularly in soil moisture and salinity, that are difficult to detect at the beginning of the snow-free season (Handa and Jefferies 2000).

   The composition of the seed bank in the upper layers of the intertidal soil
changes dramatically at degraded sites where the veneer of organic material
has been removed (fig. 18.6; Chang et al. 2001). At these sites, short-lived
halophytic plants that are not eaten by geese, such as *Salicornia borealis* and
Spergularia marina, make up 82% of the seed bank, and the dicotyledonous
species characteristic of intertidal salt marsh swards are poorly represented
(3%). In addition, the density of all seeds is low in these degraded soils. In
soil where salt marsh swards are intact, seeds of species characteristic of
this habitat, such as *Potentilla egedii, Stellaria humifusa,* and *Ranunculus
cymbalaria,* constitute 54% of the total seed bank, and seed density is high
at 19 seeds per liter of soil. At both kinds of sites, "weedy" species that
colonize disturbed ground are common as long as moisture is adequate
and salinity low. Such species include *Senecio congestus* and *Matricaria
ambigua.* Overall, the soil seed bank becomes impoverished at sites where
the potential for revegetation from the depleted seed bank is low.

   The changes in soil properties associated with desertification also
adversely affect microbial processes. In particular, rates of soil nitrogen
mineralization are reduced where there has been loss of vegetation compared with rates in soils where swards are intact (Wilson and Jefferies 1996).

   Destruction of the plant community propagates up the food web to influence the dynamics of consumers. Terrestrial arthropod populations in
intertidal and supratidal areas have been affected by the habitat changes
where loss of vegetation has occurred (Milakovic 1999). Of the groups
examined, Heteroceridae, Saldidae, and Collembola increased in abundance, whereas Carabidae, Silphidae, Cicendeliae, Aranaea, and Acari

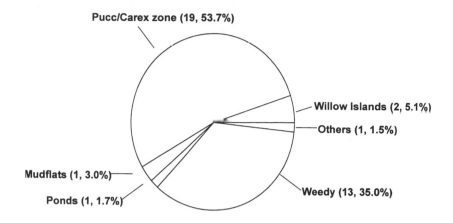

A

Pucc/Carex zone (19, 53.7%)

Willow Islands (2, 5.1%)

Others (1, 1.5%)

Mudflats (1, 3.0%)

Ponds (1, 1.7%)

Weedy (13, 35.0%)

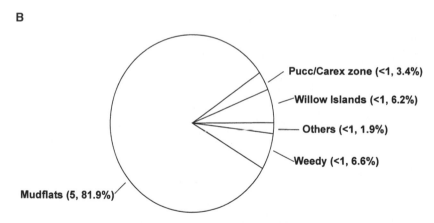

B

Pucc/Carex zone (<1, 3.4%)

Willow Islands (<1, 6.2%)

Others (<1, 1.9%)

Weedy (<1, 6.6%)

Mudflats (5, 81.9%)

Figure 18.6 Composition of the seed bank in the upper 5 cm of soil in (A) intact salt marsh swards and (B) degraded habitat where vegetation is absent at La Pérouse Bay, Manitoba. Seed species are grouped into different ecological classes, and the contribution of each class to the total seed bank is given as a percentage. The initial number in each set of parentheses refers to the average density of each seed class per liter of soil. "Pucc/Carex zone" refers to plant species characteristic of the intertidal graminoid swards of *Puccinellia phryganodes* and *Carex sub-spathacea;* "Willow Islands" refers to species common on islands in a braided estuary; "Weedy" refers to ruderal, short-lived opportunistic species found in disturbed habitats; and "Mudflats" indicates species found at degraded sites. (Data from Chang et al. 2001.)

decreased. These changes were associated with loss of vegetation and plant litter, increased soil salinity and aridity, and exposure. Early in the season, when soils were saturated from spring melt, collembolan abundance was high within the degraded salt marsh, and up to 20,000 individuals were captured, on average, per trap each day (in pan traps 15 cm in diameter). Collembolan numbers abruptly declined when soils dried out and salinity increased in late June.

The foraging activities of geese have also altered the hydrology of vernal ponds in these coastal marshes significantly. The removal of terrestrial vegetation surrounding the ponds by the grubbing activities of geese has led to increased rates of evaporation, high salinities, algal blooms in some ponds, and early drying out of ponds. Chironomids were the dominant invertebrates in ponds, but in ponds in denuded areas, only one species occurred, whereas in ponds with surrounding vegetation, up to eight species were present (Milakovic et al. 2001).

Some species of ducks, shorebirds, and passerines have experienced habitat loss and changes in the abundances of different foods through the destructive effects of foraging by geese (Gratto-Trevor 1994; Milakovic 1999; Rockwell et al. 2003). At an intensive study site, La Pérouse Bay, there have been local declines in breeding populations of American wigeon (*Anas Americana*), northern shoveller (*Anas clypeata*), semi-palmated sandpiper (*Calidris pusilla*), red-necked phalarope (*Phalaropus lobatus*), and yellow rail (*Coturnicops noveboracensis*) that are coincident with depletion of invertebrates and loss of habitat. Semi-palmated plovers (*Charadrius semipalmatus*), however, nest in degraded areas, and Ross's geese (*Anser rossii*) forage in intertidal marshes where the standing crop of aboveground biomass is very low (<15 g m^{-2}). Both of these species have increased in recent years.

CONCLUSION

In the Arctic, breeding and staging populations of geese act as a concentrating mechanism to produce the apparent trophic cascade we have described at the local scale (<100 km^2). The cumulative effects of this trophic cascade have led to biological impoverishment and the modification of trophic interactions in these coastal habitats. Because of the coalescence of damaged, fragmented areas, these changes increasingly can be seen at larger spatial scales. Modifications of trophic dynamics in response to agricultural food subsidies and increasing numbers of geese are unpredictable and nonlinear. The increasing use of agricultural lands as feeding grounds in winter and the widespread availability of freshwater forage plants in summer in the Hudson Bay lowlands indicate that at present, classic density-dependent controls on the midcontinental population of lesser snow geese are weak or absent on both wintering grounds and summer breeding grounds (except at local sites such as La Pérouse Bay). In addition, there is little evidence as yet of increased rates of predation or frequent outbreaks of disease that may limit the overall population growth of the geese. The midcontinental lesser snow goose population has "escaped" from these controls, and continues to increase. Ultimately, food resources in the northern wetlands will become limiting and the reproductive output of the population will start to fall. The decline, however, will be slow because of the life expectancy of adult birds (average 7 years) and their lifetime fecundity. Important research priorities include a coordinated monitoring program to record changes in the population size and distribution of the lesser snow goose, as well as studies of the direct and indirect effects of these birds on Arctic coastal communities.

POSTSCRIPT

Since this manuscript was prepared, new hunting regulations have come into force in both the United States and Canada that allow a restricted spring hunt of snow geese. It is too early to predict the effect of these changes on numbers of snow geese, but winter counts of the midcontinental population have fallen in the last 3 successive years.

ACKNOWLEDGMENTS

We wish to thank Michael Johnson and David Sharp for providing us with midwinter snow goose counts for the Central Flyway. Dave Ankney, Tom Nudds, and two anonymous referees made valuable suggestions that considerably improved the manuscript. Two of us (RLJ and HALH) gratefully acknowledge financial support from the Natural Sciences and Engineering Research Council of Canada. Mrs. Catherine Siu kindly typed the manuscript.

Chapter 19

Anadromous Fishes as Ecological Links between Ocean, Fresh Water, and Land

Mary F. Willson, Scott M. Gende, and Peter A. Bisson

Oceans, freshwater streams and lakes, and terrestrial "ecosystems" are usually thought of as separate entities, studied by different sets of scientists who seldom interact. Interactions among these systems are commonly seen as unidirectional: a flow of materials from land to fresh water to sea. For instance, stream ecologists often emphasize the importance of terrestrial inputs such as large woody debris for stream structure and allochthonous litter and insects as food for freshwater organisms. Marine ecologists note that streams carry silt and sediment to estuaries (e.g., Jickells 1998). The possibility of reverse flow patterns has received relatively little attention, although it has long been recognized (Leopold 1941, cited in Likens and Bormann 1974). This volume, and the relatively recent investigations that stimulated it, are proof that the long-lived classic paradigm of unidirectional flow is being overturned thoroughly.

Our thesis is that anadromous fishes form ecological links from ocean to fresh water to land. Anadromous fishes, returning from the ocean to freshwater streams to spawn, enrich the freshwater food chain. Terrestrial carnivores and scavengers, foraging on these fishes, transport marine-deived nutrients in the form of fish carcasses and digesta from fresh water to land, enriching the riparian food chain and influencing the biology of major terrestrial consumers.

Table 19.1 Taxonomic distribution of anadromy in fish families

Family	Anadromous species	Total species	% anadromous	Common names
Petromyzontidae	9	37	24	Lamprey
Acipenseridae	8	27	30	Sturgeon
Salmonidae	28	68	41	Salmon, trout, whitefish
Osmeridae	6	12	50	Smelt
Salangidae	13	14	93	Noodlefish
Retropinnidae	3	4	75	New Zealand smelt
Aplochitonidae	3	3	100	Whitebait
Clupeidae	<16	180	<9	Herring
Ariidae	1	120	<1	Sea catfish
Syngnathidae	1	175	<1	Pipefish, seahorse
Gasterosteridae	2	8	25	Stickleback
Gadidae	1	55	2	Cod
Percichthyidae	<3	40	8	Temperate perch
Cottidae	<3	300	<1	Sculpin
Gobiidae	<14?	800	<2	Goby

SOURCE: McDowall 1987; taxonomy mostly according to Nelson 1994.

That anadromous fishes contribute to the productivity of fresh waters has been known for decades (Gilbert and Rich 1927; Juday et al. 1932; Nelson and Edmondson 1955). Occasional papers on this subject appeared throughout the ensuing decades, but concerted approaches to understanding the cascading interactions based on anadromous fishes have appeared only recently. Popular articles have even appeared in newspapers and popular journals about the increasing evidence that "it takes a salmon to make a salmon," not only genetically but ecologically (e.g., Hunt 1997; Levy 1997a, 1997b). Nevertheless, we have a long way to go before we comprehend the detailed workings of these interactive systems, including their numerous and important variations in time and space, and the relationships between aquatic and terrestrial habitats in systems enriched by nutrient subsidies from anadromous fishes.

ANADROMOUS FISHES: DISTRIBUTION AND LIFE HISTORIES

"Anadromous" means "to run up," thus including any organism that regularly migrates upstream. In general usage, the term is restricted to organisms that migrate from ocean to fresh water to spawn. We use the second meaning of the term here, although many of the principles we set out could be applied to the first meaning.

About fifteen fish families have anadromous representatives (table 19.1; McDowall 1987, 1988). Of these, the lamprey, sturgeon, salmon and

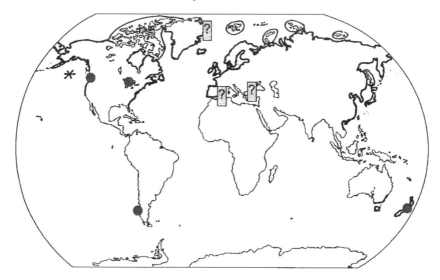

Figure 19.1 Global distribution of anadromous fish species. Heavy continental outlines indicate historical distributions; solid circles, introduced populations; asterisks, regions still retaining large populations of native anadromous fishes; ?, no information available. (From McDowall 1987, 1988.)

whitefish, smelt, noodlefish, and herring families contribute the most speies, but the habit is most widely expressed (as a percentage of species in species-rich families) in the sturgeon, salmon, smelt, and noodlefish families. Anadromy is well developed in some ancient lineages but also occurs sporadically in more recently evolved groups. In short, it has evolved independently many times but remains a relatively uncommon life history (< 1% of fish species).

Anadromy is overwhelmingly most common in the north-temperate and boreal regions and occurs at low frequency in the tropics and the Southern Hemisphere (McDowall 1987, 1988). The extensive coastline along the large northern land masses may have contributed to its high frequency there. In North America, the frequency of anadromy changes little with latitude but constitutes an increasingly dominant proportion of the life histories at higher latitudes. Similar latitudinal patterns of frequency and proportion usually occur within single species (McDowall 1987, 1988). Thus, on a global scale (fig. 19.1), most of what we say about the ecological importance of anadromous fishes pertains to north-temperate and boreal zones, and perhaps especially to the far north, because soils and freshwater streams are more often nutrient-poor there (e.g., Ulrich and Gersper 1978; Matthews 1992; Oswood 1997).

Life histories of anadromous fishes vary from fully iteroparous (sturgeon) to fully semelparous (five species of Pacific salmon, *Oncorhynchus*). Presumably the evolutionary choice between these life history patterns was set by the relative mortality rates of adults and juveniles: if adults have a low probability of living to reproduce again, semelparity may be favored (reviewed in Stearns 1992). A low probability of surviving to reproduce again is thought to be related to the cost of the spawning migration and the risks of freshwater reproduction—including the risk of predation by terrestrial carnivores (Willson 1997).

Most of the biomass of anadromous fishes accumulates during their growth at sea and therefore is composed of marine-derived materials. At maturity, these fishes return to fresh water to spawn. When semelparous fishes senesce and die, their carcasses often accumulate on rocks, woody debris, and gravel bars in streams (e.g., Cederholm and Peterson 1985; Cederholm et al. 1989). There they rapidly dwindle away (Minshall et al. 1991; Piorkowski 1995; M. F. Willson, S. M. Gende, and P. A. Bisson, personal observations), fed upon by scavenging vertebrates (eagles, corvids, gulls, carnivorous mammals), macroinvertebrates, and microorganisms. Both semelparous and iteroparous spawners may be preyed upon by terrestrial carnivores and scavengers, including bears, wolves, coyotes, and foxes (and many others of lesser size) (e.g., Willson and Halupka 1995; Marston et al. 2002). These consumers often carry fish from fresh water to land, where they cache or consume part of the prey, leaving the remnants and excreta to decompose in terrestrial, chiefly riparian, habitats. The input from semelparous species is potentially greater than that from iteroparous species because all adults die and decompose in or near fresh water and, within the salmon family, because the semelparous *Oncorhynchus* apparently maintain larger population sizes than their more iteroparous relatives (e.g., *Salmo*; T. P. Quinn, personal communication).

Thus, anadromous fishes bring marine-derived nutrients to fresh water, and terrestrial consumers carry some of those nutrients to land. Some proportion of those nutrients is recycled from land to fresh water and from fresh water back to the sea, but we focus here on the flow of materials from ocean to land.

NUTRIENT TRANSFER BETWEEN HABITATS

Nutrient Content of Anadromous Fishes

Little has been published on the nutrient content of anadromous fishes. Just as most studies of the effects of artificial fertilization on fresh waters (e.g.,

Table 19.2   Nutrient content of some anadromous fishes (g/fish)

	Unspawned	Spawned
Sockeye salmon (*Oncorhynchus nerka*; male + female)		
C	392	162
N	73	50
P	13	9.5
S	3.5	2.5
Atlantic salmon (*Salmo salar*; female)		
N	67	60
P	8	11.5
Alewife (*Alosa pseudoharengus*)		
C	33	21
N	6	5
P	1	0.9

SOURCES: Sockeye salmon, Mathisen et al. 1988; Atlantic salmon, Talbot et al. 1986; alewife, Durbin et al. 1979.

NOTE: Values are rounded.

Elser et al. 1990; Hart and Robinson 1990; Jorgenson et al. 1992; Peterson et al. 1993; Stockner and MacIsaac 1996; Perrin and Richardson 1997) and many studies of soil nutrients (e.g., Tarrant and Miller 1963; Russell 1966; Wollum and Davey 1975; several papers in Tieszen 1978) emphasize N and P, so do most of the available data on the nutrient content of fish or fish-fed vertebrates on land (table 19.2, Lyle and Elliott 1998; but see also Sugai and Burrell 1984; Talbot et al. 1986; Williams et al. 1978; Shearer et al. 1994). In general, the N content of whole salmon ranges between 2% and 4%, and the P content is less than 1% (Haywood-Farmer 1996; S. M. Gende et al., unpublished data). Many studies have documented the ecological effects of N and P enrichment (see below).

Some micronutrients in anadromous fish bodies may also have important ecological effects. For example, calcium (< 1% of whole-fish biomass; S. M. Gende et al., unpublished data) reduces the acidity of water, and streams flowing over limestone are often relatively productive (Egglishaw and Shackley 1985; Bryant et al. 1997). Ca in solution may directly benefit developing juvenile salmonids (McCay et al. 1936). Soil acidity and availability of base cations influences soil fertility (Vitousek, Aber et al. 1997), the productivity and composition of boreal forests (Giesler et al. 1998), and tree mortality (e.g., Likens et al. 1998). The effects of soil Ca from anadromous fishes on higher trophic levels are unknown but could include eggshell thickness in birds and bone density (Barclay 1994; Eeva and Lehikoinen 1995; Graveland and Drent 1997).

Sizes and Numbers of Anadromous Fishes

The size of the individual package of marine-derived nutrients represented by an anadromous fish varies enormously: anadromous fishes range in size from less than 40 g (smelt, alewife) to many kilograms (some chinook salmon, *Oncorhynchus tshawytscha;* sturgeon). The number of such nutrient packages that arrive from the ocean also varies greatly. Run sizes of wild salmon range from just a few fish in a tiny stream to over 40 million (and perhaps as many as 100 million) in the Fraser River system, often with considerable annual variation within a system (e.g., Ricker 1987; Northcote and Larkin 1989; Roos 1991; National Research Council 1996; Larkin and Slaney 1997; Halupka et al. 2000), although many runs are now seriously depleted, especially south of Alaska (Roos 1991; National Research Council 1996; Nehlsen 1997; Regier 1997). Less is known about the density of fish in spawning reaches, but estimates range from fewer than 300 coho salmon (*O. kisutch*) per kilometer in Washington (Cederholm et al. 1989; Washington Department of Fisheries, Washington Department of Wildlife, and Western Washington Treaty Indian Tribes 1993) and Oregon (Nickelson et al. 1992) to about 170,000 sockeye salmon (*O. nerka*) per kilometer in the Adams River in a peak year (Lewis 1994) to perhaps almost a million sockeye salmon per kilometer in another part of the Fraser system (Ricker 1987).

Delivery of Nutrients from One Habitat to Another via Anadromous Fish Bodies

Ocean to Fresh Water (and Back)

British Columbia and the Yukon Territory once had over 9,660 anadromous salmon stocks, of which over 100 are recently extinct (Slaney et al. 1996). Southern British Columbia alone has over 900 salmon streams with 2,300–2,400 distinct runs (Larkin and Slaney 1997). Larkin and Slaney estimate that salmon can increase regional levels of P by 200 times and levels of N by 18 times over the levels that would prevail without salmon. In southeast Alaska, over 5,200 anadromous fish streams, with a total length of over 40,000 km (Halupka et al., unpublished data), support several anadromous species: seven salmon and trout (*Oncorhynchus*), a char (*Salvelinus*), and three anadromous smelts (including the eulachon, *Thaleichthys pacificus*). The levels of fertilization (per unit area) provided by these fishes to fresh waters often exceed those of most agricultural ecosystems. Perhaps 100 million adult salmon return to southeast Alaska streams to spawn each year (Alaska Department of Fish and Game 2003), bringing an estimated 100 million kg of carbon, 10 million kg of N, and 2 million kg of P to freshwater streams. An additional 150–200 million fish, which would have returned to spawn, are harvested, thus reducing the nutrient input to about a third of what it would be without commercial harvest; most are wild fish, as only some (about 7% over all species; Alaska Department of Fish and Game 2003) of the harvested fish are hatchery-bred in southeast Alaska. On a local scale, a small run of salmon (e.g., 1,000 fish km^{-1}) would provide an estimated 9.4 kg P km^{-1} to its natal stream, assuming that all fish died and decomposed in situ (Willson et al. 1998). Salmon-enriched fresh waters exhibit enhanced productivity of aquatic phytoplankton, invertebrates, and fish (table 19.3) and would be expected to exhibit more rapid breakdown of allochthonous litter (e.g., Webster and Benfield 1986). The magnitude of the fertilization effect varies enormously among fresh water bodies (e.g., Mathisen 1972; Stockner 1987; Stockner and MacIsaac 1996; Gross et al. 1998) and among potentially responding organisms (e.g., Hershey 1992).

When the fish die, some are carried downstream to the estuary (Brickell and Goering 1970; Sugai and Burrell 1984; Reimchen 1994). The quantity of fish carcasses and dissolved nutrients thus returned to the sea is seldom known for any system but undoubtedly varies from almost 100% in short streams with few obstacles to retain carcasses (e.g., Brickell and Goering 1970) to a very small proportion in inland streams with ample woody debris and boulders (or lakes) to hold carcasses (Cederholm and Peterson 1985). Estuarine scavengers such as halibut, crabs, and snails gather to take advantage of the influx of salmonid material (e.g., Reimchen 1994). Decomposing carcasses increase the growth of estuarine algae, a principal food of harpacticoid copepods, which in turn are fed upon by juvenile salmonids (Fujiwara and Highsmith 1997). Some carcasses may decompose in intertidal meadows, fertilizing the growth of sedges and other major foods of herbivorous migratory waterfowl on the west coast of North America (Hutchinson et al. 1989; Verbeek and Butler 1989; R. H. Armstrong, personal communication).

Fresh Water to Land (and Back)

The Tongass National Forest occupies 16.9 million acres (about 6.8 million ha), covering most of southeast Alaska; about 59% of this area is forested (> 10% tree cover; Everest et al. 1997). Of the forested land, about 47% is within 0.5 km of an anadromous fish stream (G. Fisher, personal communication, from GIS maps); we estimate that over 90% of the forested land may lie within 5 km of an anadromous fish stream. Plainly, most of the anadromous fish streams in this region are within easy reach of terrestrial vertebrate predators and scavengers.

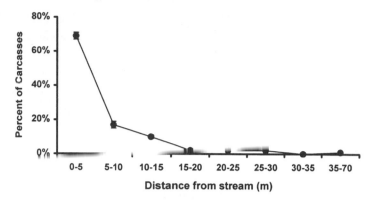

Figure 19.2 Distribution of salmon carcasses near spawning streams in southeast Alaska. Data points are the average proportions of carcasses (*N* = 900) at given distances from several streams. (From S. M. Gende and M. F. Willson, unpublished data.)

Many vertebrates forage on living, dying, or dead fish (Willson and Halupka 1995; Willson et al. 1998; Marston et al. 2002). These foragers include bears, canids, eagles, and corvids, which often carry whole fish, fish parts, or digested fish to land. Both undigested and digested fish are deposited, sometimes at high densities, on the soil, where this nutrient-rich material usually is quickly decomposed.

Estimates of the proportion of spawners captured by terrestrial foragers range up to 60–70% or even 100% in small streams (Shuman 1950; Gard 1971; T. P. Quinn, personal communication), although it commonly must be much less (e.g., Frame 1974). The amount of salmon eaten daily by a black bear (*Ursus americanus*) may be about 21 kg (Reimchen 1994). Even more may be taken from the stream to land, especially by large brown bears (*U. arctos*) when they are selectively foraging on particular portions of the carcass (Gende et al. 2001).

In southeast Alaska, bears usually deposit the partially eaten bodies of salmon near the stream, although some carcasses are carried over 75 m away (fig. 19.2). Elsewhere, carcasses may be carried up to 150 m into the forest (Reimchen 1994), or "several hundred yards" from the stream (Shuman 1950). Females may be more likely than males to be carried to land by black bears (Reimchen 1994), especially if they are still gravid (Frame 1974). Both black and brown bears are likely to carry large salmon farther into the forest than small ones (Reimchen 1994, 2000; S. M. Gende and M. F. Willson, unpublished data), and fresh kills are more likely to be carried into the forest than are carcasses of senesced fish

Table 19.3 Effects of anadromous fish carcasses on aquatic systems: some examples of "biogenic eutrophication" from the literature

Water body	Location	Fish biomass or abundance	Species	Nutrients		Effect on			Reference
				N	P	Microorganisms/ plankton	Invertebrates	Fish	
Karluk Lake	South-western Alaska	2 ×10^6 kg	Sockeye salmon		5,000 kg	Increase		Increase	Juday et al. 1932; Nelson and Edmondson 1955; Koenings and Burkett 1987
Iliamna Lake	South-western Alaska	340,000 to 24 million	Sockeye salmon		3.6–170 metric tons	Uptake of MDN[a]		Uptake of MDN	Donaldson 1967; Hartman and Burgner 1972; Kline et al. 1993
Nushagak system	South-western Alaska	Up to 5.9 million	Sockeye salmon					Increase	Mathisen 1971, 1972
Dalnee Lake	Russia	Up to 15% of energy input	Sockeye salmon						Krohkin 1968, 1969, 1975; Sorokin and Paveljeva 1978
Pausacaco Pond	Rhode Island	4,530 kg C	Alewife	728 kg	115 kg	Increase	Increase		Durbin et al. 1979
Skagit River	Washington	Up to 0.7 million = 1.6 million kg	Pink salmon					Increase coho, decrease chum?	Michael 1995

Chulitna River tributaries	South-central Alaska	?	Various salmon				Increase of some taxa	Piorkowski 1995
Margaret Creek	South-eastern Alaska	75,000	Pink salmon			Increase	Increase of some taxa	Wipfli et al. 1998
Sashin Creek	South-eastern Alaska	30,000	Pink salmon	Increase		Uptake of MDN	Uptake of MDN	Uptake of MDN	Kline et al. 1990
Various streams	British Columbia	Various	Various salmon	Up to 18 times increase	Up to 200 times increase		Increase	Increase	Johnstone et al. 1997; Larkin and Slaney 1997
Snoqualmie River tributaries	Washington	?	Coho salmon					Increase	Bilby et al. 1996

[a]MDN = marine-derived nutrients

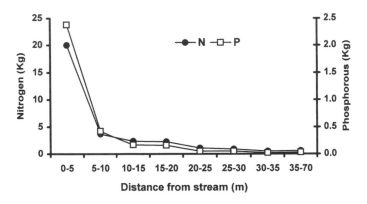

Figure 19.3 Distribution of nutrients (N and P) from salmon carcasses near spawning streams in southeast Alaska. Data points were calculated from figure 19.2.

(Reimchen 1994, 2000). The distribution of carcasses on the shore can be influenced by the abundance of spawning fish, the numbers of bears present, and social interactions among the bears (Luque and Stokes 1976; Egbert and Stokes 1976; Reimchen 1998; Gende 2002).

As a result of bear transport, salmon-derived nutrients are added to the terrestrial riparian system (Hilderbrand, Hanley et al. 1999) (fig. 19.3). Although nutrient deposition is clearly greatest near the stream, the long tail on the distribution indicates that some deposition occurs at some distance from the stream. In actuality, the tail of the distribution is more attenuated than illustrated in figure 19.3 because large carnivores range widely and some excreta will be deposited at greater distances. Furthermore, scavenging eagles (*Haliaeetus leucocephalus*), ravens (*Corvus corax*), and crows (*C. caurinus*) commonly carry salmon pieces (and smaller forage fishes) into the forest, sometimes over 100 m; some food items are stored in trees or meadows by corvids, dropped en route, or taken to nestlings.

Some marine-derived nutrients are recycled from the enriched land back to fresh waters via leaching and runoff, terrestrial insects that fall into the stream (Hunt 1975; Wipfli 1997; Nakano and Murakami 2001), riparian zone vegetation litter, which is often a major nutrient source for streams (e.g., Durbin et al. 1979; Sidle 1986), and vertebrate excreta deposited in streams.

Our emphasis on biological agents of nutrient transfer should not obscure the fact that some transfer, of variable magnitude, also occurs by abiotic means, such as hyporheic flow (Gende et al. 2002).

EFFECTS OF NUTRIENT TRANSFER

In Fresh Water

Increased nutrients commonly increase the productivity of freshwater food chains, especially in oligotrophic waters. Most information on this subject comes from studies in which inorganic supplements were added (e.g., Nelson and Edmondson 1955; Ashley and Slaney 1997; Kyle et al. 1997; Perrin and Richardson 1997; Budy et al. 1998). Some studies, however, focused specifically on nutrients arriving via anadromous fishes (see table 19.3; Lyle and Elliott 1998; Wipfli et al. 1999) or their analogs in strictly freshwater systems (Richey et al. 1975; Rand et al. 1992; Kraft 1993; Schuldt and Hershey 1995). In general, salmon subsidies are incorporated into aquatic organisms (Chaloner et al. 2002; Bilby et al. 1996) and can increase productivity at all trophic levels, including fish. Future work should disentangle the relative effects of different nutrients and nutrients following different pathways (Gende et al. 2002) as well as the interactions among nutrient subsidies (Treseder and Vitousek 2001).

On Land

Marine-derived nutrients (from decomposed salmon carcasses or carnivore excreta) can be taken up by streamside vegetation (Bilby et al. 1996; Reimchen et al., in press; Ben-David, Hanley, and Schell 1998; Ben-David, Bowyer et al. 1988), and some riparian plants exhibit enhanced growth (Helfield and Naiman 2001, 2002), but their possible effects on reproductive output or succession have not yet been assessed. Herbivorous and detritivorous invertebrates that forage on salmon-enriched vegetation and leaf litter accumulate salmon-derived nutrients (Reimchen et al., in press; Hocking and Reimchen 2002). Several species of invertebrates (flies, beetles) colonize salmon carcasses soon after deposition on land, rapidly turning such carcasses into a seething mass of maggots and beetle larvae (Johnson and Ringler 1979; M. F. Willson, S. M. Gende, and P. A. Bisson, personal observations). These invertebrates are not known to be eaten by terrestrial insectivores (Reimchen 1994) but, when washed into the stream, are fed upon by fish (Johnson and Ringler 1979; Wipfli 1997).

Dozens of species of vertebrates feed directly on anadromous fish bodies, carcasses, or eggs (Willson and Halupka 1995; Marston et al. 2002), sometimes gathering in large numbers near spawning streams (e.g., Luque and Stokes 1976; Marston et al. 2002). Vertebrates may also respond to salmon enrichment of terrestrial vegetation (and insects): in southeast Alaska, breeding bird density (in the spring) tended to be higher along salmon streams than along streams lacking salmon runs (Willson and Gende 2001), possibly because of increased arthropod abundances (Nakano and Murakami 2001). That breeding bird density might increase in response to salmon fertilization is plausible, given that bird density in the nearby Yukon Territory increased following aerial application of inorganic fertilizer (Folkard and Smith 1995).

The breeding biology and body size of consumers may also be affected by this rich food resource. The abundance of salmon in summer may increase survival of fledgling and juvenile bald eagles, as well as subsidizing the very high eagle densities in southeast Alaska. Spring foraging on eulachon probably helps pay the reproductive costs of Steller sea lions (*Eumetopias jubatus*) and harbor seals (*Phoca vitulina*) and the migration and prerepro-ductive costs of red-breasted mergansers (*Mergus serrator*) and several species of gulls (*Larus* spp.) (Marston et al. 2002). Mink (*Mustela vison*) in southeast Alaska give birth earlier than other high-latitude populations, perhaps because the lactation period then coincides with salmon availability (Ben-David 1997). Coastal bears with regular access to abundant fish runs grow bigger, mature earlier, have larger litters, and reproduce more often than interior bears (Herrero 1978; Spraker et al. 1981; Nowak and Paradiso 1983; reviewed by Welch et al. 1997; Willson et al. 1998), and they reach very high densities (over 200 bears 1,000 km^{-2}; Miller et al. 1997). Indeed, the large body size of coastal brown bears apparently cannot be supported by a diet of fruit (Welch et al. 1997), and the abundant food supply provided by salmon may have permitted their large size (Hilderbrand, Schwartz et al. 1999). However, inland bears with regular access to (formerly) large salmon runs on the Columbia, Fraser, and Yukon river systems (Hilderbrand et al. 1996) seem to have remained relatively small.

DISCUSSION

It may be a matter of semantics whether to discuss aquatic-terrestrial in-teractions as part of the ecological richness associated with ecotones (Naiman and Décamps 1990, 1998) or as integrated systems in and of themselves. Clearly, in any event, close integration of land and water ecology is potentially important wherever anadromous fish run. To illus-trate the relationships between land and water, we present a simplified version of the food web that links anadromous fishes to the terrestrial system (fig. 19.4). Although there are other interactors in this web (fleshy

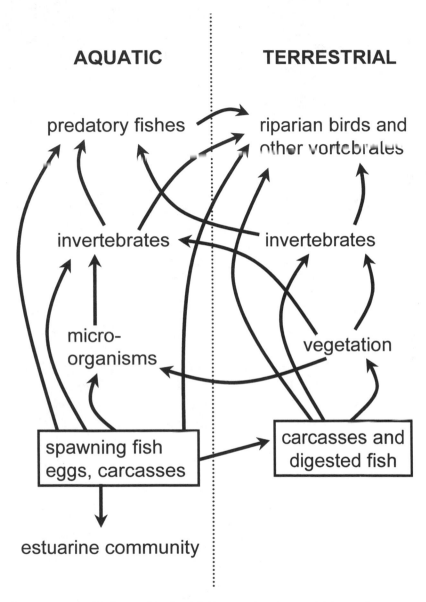

Figure 19.4 A simplified general food web linking aquatic systems containing anadromous fishes with terrestrial systems. Note the number of arrows that cross the border between land and water. Some locations may have additional aquatic-terrestrial interactions; for example, northern pike (*Esox lucius*) may feed on small mammals and birds. (From Willson et al. 1998.)

fruits and frugivores, for instance), we emphasize relationships centered on the fishes. Particularly noteworthy are the numbers of arrows that cross the border between the aquatic and terrestrial sides of the diagram and the numbers of types of organisms that in some way depend on anadromous fishes.

These integrated systems are still quite functional (though not without anthropogenic influences) in most of Alaska, but they are seriously impaired in most of the rest of the world. In most of the contiguous United States, southern Canada, and Eurasia, heavy fishing pressure, habitat destruction, and other factors have caused severe declines in many wild anadromous fish populations (e.g., Palmisano et al. 1993; National Research Council 1996). Historically, fish stocks were harvested very intensely—for example, over 80% of the returning Pacific salmon were commonly harvested, often wastefully (e.g., Ricker 1987). In addition, urbanization, other developments, and intense hunting have often led to the extirpation or near-extirpation of the carnivore populations that formerly served as nutrient transport agents between water and land (e.g., Storer and Tevis 1955; Kadosaki 1983; Miller 1989). Therefore, these highly modified systems will probably never be able to return to anything like their original ecological state.

On the other hand, anadromous salmonids have been introduced to several places in the world where they are not native, including Chile, New Zealand, and the North American Great Lakes, and shad have been introduced to the Pacific coast (see fig. 19.1). Neither Chile nor New Zealand has native populations of large carnivores, but the Great Lakes region still has some black bears and wolves (*Canis lupus*) that may take advantage of this introduced food resource.

The relative contribution of fish-borne nutrients may be greater in the upper reaches than in the mainstem of a river. Although fish abundance is higher in the mainstem of the river than in tributaries and headwaters, in many cases the mainstem is too deep or turbid to permit much fish capture by terrestrial carnivores (Frame 1974). Furthermore, nutrients derived from non-fish sources are often less abundant in small-order streams than in the mainstem (e.g., Vannote et al. 1980), and shallower waters permit heavier predation by terrestrial predators (Heggenes and Borgstrøm 1988; Lonzarich and Quinn 1995).

The principles described here for anadromous fishes sensu stricto clearly apply equally well to upstream fish migrations within fresh waters. Thus, the spawning migrations of suckers, inland trout (and introduced salmon), and other freshwater fishes also may support important components of the terrestrial community (e.g., Dombeck et al. 1984). These effects within fresh water may be more widespread in the world than the effects of strictly anadromous fishes.

Biotic aquatic subsidies of terrestrial communities have been described for islands in the Gulf of California (Polis and Hurd 1995, 1996b; Polis, Anderson, and Holt 1997; Polis et al., chap. 14 in this volume; Rose and Polis 1998), habitat and geographic islands in southern oceans (e.g., Atkinson 1964; Burger et al. 1978; Williams et al. 1978; V. R. Smith 1979; Siegfried 1981, Burger 1985; Myrcha et al. 1985; Panagis 1985; Ryan and Watkins 1989), cays in the Coral Sea (Heatwole 1971), and migratory birds in the Great Lakes (Ewert and Hamas 1995). In addition, biotic vectors of nutrients have been previously described (e.g., Bosman and Hockey 1986; Bildstein et al. 1992; Polis and Hurd 1996b) and the role of consumers in nutrient cycling emphasized (e.g., Kitchell et al. 1979; Schindler et al. 1993; Vanni and Headworth, chap. 4 in this volume). Therefore, the ideas presented here are not new (e.g., Likens and Bormann 1974). Our discussion suggests, however, that biotic subsidies and biotic transport agents moving across habitat borders may be more common and more influential than previously recognized.

The consequences of the ecological interactions described here are wide-ranging. These interactions obviously cannot be maintained without both predators and prey (and their habitats). Thus, maintenance of the fish populations is important to both the populations of wildlife consumers and the integrated aquatic-riparian ecosystem, and the predator populations are essential as nutrient transport agents maintaining the flow of nutrients to the terrestrial part of the system. In terms of conservation, clearly the functioning of the system depends on the presence of its parts. In terms of management, a wide perspective is needed if a functioning system is to be maintained in the face of exploitation or depletion of some parts of the system. Just as fisheries biologists have called for changes in forestry practices to preserve fish habitat, wildlife biologists may now call for changes in fisheries practices to preserve wildlife populations. Much research is needed to quantify the relative importance of these aquatic-terrestrial interactions in different situations.

The "ecological goods and services" rendered by natural communities are hard to quantify monetarily and are therefore typically undervalued, but true self-sustainability in aquatic and riparian ecosystems is undoubtedly an important ecological service to human societies (National Research Council 1992). If biological resources in ecological systems linked by anadromous fishes can be managed to allow them to sustain themselves without costly additions of fertilizer or artificially propagated fish, the long-term economic benefits in many cases could be great.

Naiman et al. (1995a, 1995b) discussed research priorities for North American freshwater ecosystems, urging an increased appreciation of the important trophic connections between water and land. We submit that reciprocal interactions between aquatic and terrestrial ecosystems mediated by the spawning migrations of anadromous fishes and their terrestrial consumers are important to all of the research priorities (ecological restoration, biodiversity, hydrological flow patterns, ecosystem services, predictive management, and future problem identification) recognized by Naiman et al. and should be considered in all of those contexts.

ACKNOWLEDGMENTS

We thank T. Reimchen and P. Slaney for sharing prepublication information, and T. P. Quinn, R. H. Riley, K. O. Winemiller, and the editors for comments on the manuscript.

Chapter 20

When Terrestrial Organic Matter Is Sent down the River: The Importance of Allochthonous Carbon Inputs to the Metabolism of Lakes and Rivers

Nina Caraco and Jonathan Cole

The fate of the majority of terrestrial primary production is to be respired in terrestrial environments. However, approximately 0.5 Gt yr^{-1} escapes respiration and is exported to aquatic systems (Meybeck 1982, 1993; Schlesinger and Melack 1981). Although this export is only about 1% of terrestrial primary production (Ludwig et al. 1997), it is concentrated from a large terrestrial area (e.g., the watershed) to the relatively small proportion of the watershed area that is occupied by drainage waters (Vander Leeden et al. 1990; Wetzel 1975). This fact, in addition to the relatively low primary production in many aquatic systems (del Giorgio et al. 1999; Meybeck et al. 1990), accounts for the potential importance of allochthonous inputs in these systems.

The importance of allochthonous inputs to small, first- and second-order streams has been extensively studied (e.g., Fisher and Likens 1973; Webster and Meyer 1997a). These streams are intimately connected with their terrestrial watersheds and receive organic materials directly as they leave the terrestrial environment. Some of these materials (e.g., leaf litter, leachates) are highly labile and are rapidly metabolized in streams (Meyer and Johnson 1983). Other materials are trapped in debris dams or in sediments, where they may be attacked by decomposers for some time (Bilby and Likens 1979). Because small streams have very short residence times

(Bencala et al. 1990), much of this terrestrial organic matter is transported farther downstream into larger aquatic systems (lakes and rivers), which have substantially longer residence times (Dillon and Molot 1997; Meybeck et al. 1990; Vollenweider and Kerekes 1980) and thus a greater potential to transform the organic inputs.

The fate of terrestrial organic matter in lakes and rivers is the subject of this review. We are particularly interested in the extent to which imported, allochthonous sources of organic matter affect the metabolism and organic carbon balances of lakes and rivers. We consider, using case studies and simple models (fig. 20.1), the magnitude of allochthonous inputs in comparison to the autochthonous primary production that occurs within these systems. Using the same approaches, we then consider the fates of autochthonous and allochthonous inputs within these systems. The fates of organic matter in these systems, and in rivers in particular, may largely determine patterns of organic matter transport and net heterotrophy at the regional and global scales (Ludwig et al. 1997; Hedges et al. 1997; Meybeck 1993).

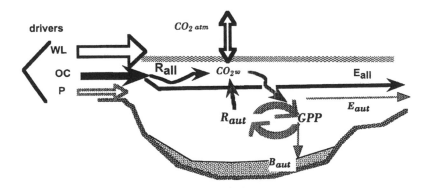

Figure 20.1 Calculation of the relative importance of allochthonous matter to organic matter loads in aquatic systems and to both total respiration and net ecosystem respiration from three parameters: the water load to the system (WL, in m yr^{-1}); the P input from the watershed (in g^{-2} m yr^{-1}), and the allochthonous organic matter input (OC$_{all}$, in g^{-2} m yr^{-1}).

Inputs

The inputs of organic matter are from allochthonous (OC$_{all}$) and autochthonous production (GPP). The allochthonous inputs are considered to be primarily from surface runoff and are calculated as

$$OC_{all} = [OC] \times WL$$

where WL is the water load (in m yr^{-1}) and [OC] is the average organic carbon concentration in the entering water. GPP was calculated according to Vollenweider and Kereke (1990) as

$$GPP = 6.99 \times \frac{X^{0.76}}{0.29 + 0.011X^{0.76}},$$

where $X = [P]/(1 + RT^{0.5})$, [P] is the average P concentration in inputs to the system, and RT is the water residence time, calculated as WL divided by the average depth of the system. In our calculations we considered the average depth to be 10 m.

Respiration

Both GPP and OC_{all} are respired within the system (R_{aut} and R_{all}, respectively). R_{all} was calculated as a deposition model according to Dillon and Molot (1997) as

$$R_{all} = OC_{all} \times \frac{DV}{DV + WL},$$

where DV is a deposition velocity (in m yr^{-1}) and faster velocities mean higher decomposition rates within the system. R_{aut} was calculated as

$$R_{aut} = GPP - B_{aut} - E_{aut},$$

where B_{aut} is burial of GPP and E_{aut} is the export of organic matter produced by GPP. E_{aut} was calculated as

$$E_{aut} = 50 \times [Chl] \times WL \times 0.001,$$

where [Chl] is the concentration of chlorophyll in μg l^{-1}, 50 is the C:Chl ratio of phytoplankton, and 0.001 transfers μg to mg (Caraco et al. 1997). [Chl] was calculated according to Vollenweider and Kereke (1990) as

$$\left[Chl\right] = 0.55 \times \left(\frac{[P]}{1 + RT^{0.5}}\right)^{0.76}.$$

B_{aut} was calculated as P burial in the system times the C:P ratio of sediments such that

$$B_{aut} = [P] \times WL \times \frac{DV_p}{DV_p + WL} \times C:P \text{ ratio}.$$

A DV_p of 10 m yr^{-1} was used in calculations (Vollenweider and Kereke 1990; Vollenweider 1968). The C:P ratio was assumed to vary with the trophy of the system from 50:1 in eutrophic systems to 200:1 in oligotrophic systems.

Note that the calculation of E_{aut} could be an underestimate in that some GPP being exported could be in forms in which chlorophyll has already degraded. On the other hand, the calculation of B_{aut} may be an overestimate of autotrophic burial in that some of the material being buried is of allochthonous origin.

The relative importance of allochthonous respiration to total respiration was calculated as

$$\% R_{all} = 100 \times \frac{R_{all}}{R_{aut} + R_{all}}.$$

The importance of allochthonous respiration to net ecosystem respiration (NER) was calculated as

$$NER = R_{all} - E_{aut} - B_{aut}.$$

Additionally, NER was calculated independently from CO_2 balance as

$$NER(CO_2) = (CO_2w - CO_2atm) \times k \times D,$$

where CO_2w and CO_2atm are the concentrations of CO_2 in the water and atmosphere, respectively; k is the air-water exchange coefficient (in m d^{-1}); and D is the exchange days (ice-free days) (in d yr^{-1}). For lakes and rivers, k values of 0.5 and 0.8 were used, respectively (Clark et al. 1994; Cole and Caraco 1998). D values of 250 for lakes and 330 for rivers were used.

MAGNITUDE OF ALLOCHTHONOUS ORGANIC
CARBON INPUTS

There are many ways that organic carbon produced in the watershed can enter aquatic systems. In very small lakes and narrow, tree-lined streams, falling litter and throughfall of precipitation from overhanging trees can be a significant route (Jordan and Likens 1975; Rau 1976). Wind can also carry materials into lakes and rivers, but is probably most important in arid regions where hydrological inputs are low. Wind typically transports small particles (Cole et al. 1990; Gasith and Hasler 1976), but can also move large debris, such as tumbleweeds, considerable distances into lakes (Galat 1986). Despite the wide range of mechanisms whereby organic carbon can enter lakes and rivers, generally the dominant route is in hydrological loads consisting of direct precipitation and runoff from watersheds (Jordan and Likens 1975); for most systems, runoff dominates inputs (Dillon and Molot 1997; Findlay et al. 1998).

The input from the watershed in runoff (OC_{all}, fig. 20.1) is equal to the product of the areal water load to the aquatic system and the organic carbon concentration ($[OC_{all}]$) in that water. Areal water load (WL), like precipitation, evapotranspiration, or watershed runoff, is expressed in m yr^{-1} and is the product of watershed runoff and the ratio of watershed area to the area of surface waters. Thus, high water loads occur either in wet regions with high runoff or in systems with large relative watershed areas (high watershed area:surface water area ratios). Lakes and reservoirs generally have water loads that vary between 3 and 30 m yr^{-1}, with reservoirs tending to have higher water loads than lakes (Meybeck et al. 1990). Rivers, with very large relative watershed areas, often have water loads of over 100 m yr^{-1} (Howarth, Billen et al. 1996; Limburg et al. 1986).

Although $[OC_{all}]$ in stream water and groundwater can range from less than 1 to over 100 mg l^{-1}, it generally varies between 3 and 30 mg l^{-1} (Wetzel 1975). Organic carbon consists of both dissolved (DOC) and particulate (POC) forms, with DOC generally dominating (Ludwig et al. 1997; Peterson, Hobbie, and Corliss 1986). The cause of variation in DOC and POC concentrations is not completely understood. DOC concentrations are negatively correlated with slope in the watershed and positively correlated with wetland area (Esser and Kohlmaier 1991; Kortelainen 1993). POC concentrations, in contrast, relate positively to the slope of the watershed (Ludwig et al. 1997). Humans can potentially affect POC and DOC concentrations in waters through agricultural and forestry activities (Esser and Kohlmaier 1991; Schlesinger and Melack

1981). Direct evidence for large-scale anthropogenic change, however, is weak.

Regardless of its cause, the variation in OC concentration ($[OC_{all}]$) contributes in part to variation in allochthonous loads to aquatic systems. Indeed, a tenfold range of $[OC_{all}]$, combined with a hundredfold range in WL, implies more than a thousandfold range in OC inputs to aquatic systems (fig. 20.2). For a lake with a moderate $[OC_{all}]$ of 10 mg l^{-1} (Ludwig et al. 1997) and a water load of 10 m yr^{-1}, OC_{all} loads would be 100 g C m^{-2} yr^{-1} (fig. 20.2). Mirror Lake, with lower WL and $[OC_{all}]$, has loads of about 40 g C m^{-2} yr^{-1} (Likens 1985; N. F. Caraco and J. J. Cole, unpublished data). Lakes with large WL or those surrounded by wetlands would tend to have loads greater than 500 g C m^{-2} yr^{-1} (del Giorgio and Peters 1994; Kortelainen 1993). Large rivers, with typical WL of about 100 m yr^{-1}, also have higher allochthonous C loads than most lakes. The lower Hudson River, located in a mesic area (New York State), has an OC load from the watershed of about 700 g C m^{-2} yr^{-1} (Howarth, Schneider, and Swaney 1996). The Ogeechee River in Georgia, U.S.A., with a slightly higher WL and $[OC_{all}]$, has an OC load of over 2,000 g C m^{-2} yr^{-1} (Meyer et al. 1997). For both of these

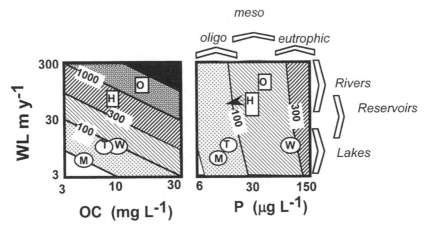

Figure 20.2 Organic matter inputs to aquatic systems from allochthonous (left panel) and autochthonous sources (GPP, right panel) as a function of entering water load (WL) and concentrations of organic matter ($[OC_{all}]$) and phosphorus ($[P]$). The calculation of these inputs is explained in figure 20.1. Allochthonous loads should vary from about 10 to 3,000 g C m^{-2} yr^{-1} with WL and $[OC_{all}]$. Autochthonous inputs, by comparison, should vary from about 30 to over 300 g C m^{-2} yr^{-1} with WL and $[P]$. Actual lake (circles) and river (squares) systems plotted on these graphs include Mirror Lake (M), Truite Lake (T), Waterloo Lake (W), the Hudson River (H), and the Ogeechee River (O). (Data from Caraco et al. 1996; del Giorgio et al. 1999; Howarth, Schneider, and Swaney 1996; and Meyer et al. 1997.)

riverine systems, the areal load of terrestrial organic carbon to the river is greater than the average areal carbon loads derived from primary production in most temperate and tropical watersheds (Ludwig et al. 1997), and may be large compared with primary production within the aquatic system (see below).

MAGNITUDE OF AUTOCHTHONOUS PRIMARY PRODUCTION

In most lakes, primary production is regulated by the supply of limiting plant nutrients, especially P and N, which tend to covary in inputs (Vollenweider 1968). As for OC_{all}, the P and N loaded to lakes and rivers can come from a variety of sources (Bildstein et al. 1992; Carlton and Goldman 1984; Caraco et al. 1992), but often the most important external source is runoff from watersheds (Carpenter, Caraco et al. 1998). The sources of the N and P include particles eroded from the terrestrial landscape and dissolved P lost by leaching (Carpenter, Caraco et al. 1998; Howarth, Billen et al. 1996). These inputs tend to vary naturally with soil type and hydrological regime (Dillon and Kircher 1975). Humans increase natural loads of N and P dramatically through changes in runoff patterns from land, fertilizer loads, increased erosional losses of soil, and in some cases, direct sewage inputs to surface waters (Caraco 1995; Caraco and Cole 1999; Carpenter, Caraco et al. 1998). These human activities can increase P concentrations entering lakes and streams from less than 10 μg l^{-1} to over 100 μg l^{-1} (Vollenweider 1968). Empirical studies in lakes suggest that this increase in nutrient loading will result in increases in primary production from about 50 to about 250 g C m^{-2} yr^{-1} (Vollenweider and Kerekes 1980).

Despite the strong human impact on nutrient load and primary production in aquatic systems in many areas of the world, surveys indicate that at present, a large proportion (48%) of lakes worldwide remain oligotrophic, with primary production less than 100 g C m^{-2} yr^{-1} (Meybeck et al. 1990). This distribution varies geographically, however. For example, in Germany, only 8% of surveyed lakes are oligotrophic, while in Canada over 73% are oligotrophic (Meybeck et al. 1990).

In rivers, P concentrations and loads are often greater than those in lakes (Caraco 1995; Meybeck et al. 1990), and this difference potentially could lead to greater autochthonous primary production in rivers. Actual measurements of primary production in rivers, however, are sometimes less than would be expected based on their nutrient loads. For example, from a knowledge of the P concentration entering the system (Lampman et al. 1999) and the water load, primary production in the Hudson River would be predicted to be about 200 g C m^{-2} yr^{-1}. Actual measured primary production in the Hudson (Cole et al. 1991), however, is less than 100 g C m^{-2} yr^{-1} in most locations, due to light limitation and the high abundance of benthic grazers there (Caraco et al. 1997) (see fig. 20.2). On the other hand, the Ogeechee River has a level of primary production close to that expected from its P loads. Overall, areal primary production in most rivers may be similar to that in lakes (Meyer et al. 1997; Peterson et al. 1985) despite the somewhat higher nutrient inputs to rivers.

RESPIRATION OF ALLOCHTHONOUS AND AUTOCHTHONOUS ORGANIC CARBON

Autochthonous Organic Carbon

Carbon produced within aquatic systems can be consumed and respired through grazer or microbial food chains, or it can be buried or exported. Budgeting of carbon balances in individual lakes and rivers suggests that for most aquatic systems, respiration is the major fate of autochthonous primary production. For example, in Mirror Lake, with a primary production of about 60 g C m^{-2} yr^{-1}, burial of autochthonous material is less than 12 g C m^{-2} yr^{-1} (Cole et al. 1989; Jordan and Likens 1975; Likens 1985), and export of phytoplankton carbon by surface runoff is less than 1 g C m^{-2} yr^{-1} (calculation shown in fig. 20.1). Thus, about 80% of primary production is respired within the system. In the Hudson River, with a primary production of about 100 g C m^{-2} yr^{-1}, burial is less than 30 g C m^{-2} yr^{-1} (Howarth, Schneider, and Swaney 1996), and the calculated export is about 10 g C m^{-2} yr^{-1}. Thus, over 60% of primary production is probably respired within the Hudson River by phytoplankton, microbes, and benthic and pelagic grazers (Findlay et al. 1991; Howarth, Schneider, and Swaney 1996; Strayer et al. 1996).

In addition to considering the organic carbon budgets of individual systems, we can model the fates of primary production by coupling simple predictive models that link water load and nutrient inputs to phytoplankton standing stock and nutrient burial (fig. 20.3). These models suggest that the importance of both burial and export will increase with nutrient input concentrations and water loads, while respiration of autochthonously produced material should decrease with these two factors. For the average oligotrophic-mesotrophic lake with a water load of about 10 m yr^{-1}, the combined loss to export and burial should equal about 15% of autochthonous primary production, while 85% should be respired (fig. 20.3). For

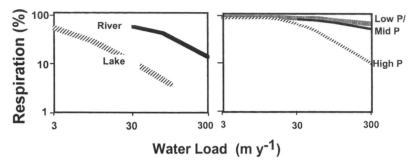

Figure 20.3 The respiration of allochthonous (left panel) and autochthonous (right panel) organic carbon inputs to aquatic systems. Both are expressed as a percentage of inputs. According to the model of Dillon and Molot (1997), described in fig. 20.1, respiration of allochthonous material should depend on water load and decomposition rate (expressed as a deposition velocity, DV). If this model is correct, DV in rivers (40 m yr^{-1}) is about 10 times that in lakes (4 m yr^{-1}). In general, autochthonous organic matter should decompose to a greater degree than allochthonous organic matter. Only in systems with high WL and high P loads is respiration expected to be low (export and burial high).

systems with high water loads (rivers) and moderate nutrient loads, about 70% of primary production should be respired within the system. Although this prediction for rivers is based on models developed for lakes, data for the Hudson suggest that the models may predict the metabolism of autochthonously produced organic carbon relatively well (fig. 20.3).

Allochthonous Organic Carbon

Despite the potentially refractory nature of allochthonous organic material, recent studies suggest substantial respiration of this material once it enters aquatic systems (del Giorgio and Peters 1994; Dillon and Molot 1997). Dillon and Molot's (1997) analysis of data for seven Canadian lakes suggests that this respiration can be described as

$$\% R = 100 \times \frac{DV}{DV + WL},$$

where %R is the percentage of allochthonous DOC respired and DV is a depositional velocity in m yr^{-1}.

For Dillon and Molot's (1997) lakes, DV varied from 2 to 5 m yr^{-1}. Using a DV of 4 m yr^{-1}, their model suggests that for a lake with a water load of 10 m yr^{-1}, about 30% of the allochthonous DOC load would be respired per year. These modeled values of allochthonous DOC decomposition are in general agreement with the respiration of allochthonous material measured in lakes of southern Quebec (del Giorgio et al. 1999) and may be widely applicable to lakes in general.

If the lake model of Dillon and Molot (1997) is extended to the higher water loads found in riverine systems and to both dissolved and particulate organic matter, the respiration of OC_{all}, as a percentage of input, should decline to quite low values (fig. 20.3). Due to this predicted lower percentage, respiration of allochthonous organic matter should be only slightly greater in rivers than in lakes, despite far higher allochthonous loads in rivers. For example, given a lake and a river both with $[OC_{all}]$ of 10 mg l^{-1}, but with water loads of 5 and 100 m yr^{-1}, the allochthonous loads would be 50 and 1,000 g C m^{-2} yr^{-1}, respectively, and the predicted respiration of the material would be 22 and 39 g C m^{-2} yr^{-1}. Thus, despite a twentyfold difference in OC loads, the respiration in the river and the lake would differ by less than twofold, based on predictions from the lake model.

Measurements of respiration in rivers suggest that allochthonous OC may be respired at rates substantially higher than those predicted by the lake model (fig. 20.3). For example, for the Hudson River, assuming that POC and DOC are respired at equal rates, the model of Dillon and Molot (1997) predicts that only about 20 g C m^{-2} yr^{-1} of OC inputs should be respired. Actual measurements, however, suggest values of about 200 g C m^{-2} yr^{-1} (Howarth, Schneider, and Swaney 1996; Raymond et al. 1997). Similarly, in the Amazon and the Ogeechee, respiration is equal to about 1,000 g C m^{-2} yr^{-1} (Meyer et al. 1997; Richey et al. 1990), while Dillon and Molot's model predicts values of about 100 g C m^{-2} yr^{-1}. Thus, rivers appear to decompose terrestrial organic material at rates 10 times greater than those found for lakes (fig. 20.3). The possible causes of this high respiration include a high rate of mixing, which results in more continuous exposure of DOC to light (Lindell et al. 1995), and the relatively high N and P content of rivers, which may increase decomposition of nutrient-poor materials (Caraco et al. 1998; Howarth and Fisher 1976; Meyer and Johnson 1983).

IMPORTANCE OF ALLOCHTHONOUS INPUTS TO TOTAL SYSTEM RESPIRATION

The total carbon remineralized is the sum of autochthonous and allochthonous carbon respiration. Our coupled models predict that in lakes, the importance of allochthonous respiration should vary substantially and should relate positively to OC inputs and negatively to P inputs (fig. 20.4).

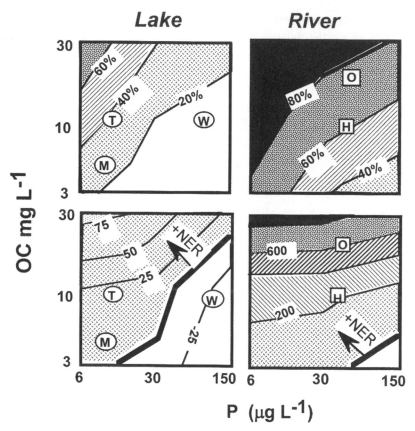

Figure 20.4 The effect of DOC and P inputs on the importance of respiration of OC_{all} to overall system respiration (top panels) and net ecosystem respiration (NER, bottom panels). Allochthonous respiration is expressed as a percentage of total respiration, and NER is given in g C m^{-2} yr^{-1}, with negative values indicating net autotrophy. Values are shown for both lakes (left panels, using a representative WL of 10 m yr^{-1} and a DV of 4 m yr^{-1}) and rivers (right panels, using a WL of 100 m yr^{-1} and a DV of 40 m yr^{-1}). For oligotrophic lakes such as Mirror and Truite, about 20–50% of the total respiration should be from allochthonous sources. For eutrophic lakes with high P inputs, this respiration becomes insignificant. Across almost all P loads, allochthonous respiration should be dominant in rivers. Similarly, NER should be greater in oligotrophic than in eutrophic lakes, and should be greatest in rivers.

For low-nutrient lakes with high [OC_{all}] in incoming water, respiration of allochthonous material should account for more than 50% of total system respiration, whereas in high-nutrient systems, respiration of allochthonous inputs is expected to be a small part of the total respiration. This prediction is supported by direct measurements of respiration and primary production in a series of lakes with a gradient from oligotrophic to eutrophic conditions (del Giorgio et al. 1999).

For rivers, as for lakes, both P inputs and OC inputs should play a role in controlling the relative importance of allochthonous loads to total system respiration. In rivers, however, allochthonous respiration generally appears to be of greater importance than in lakes. Our coupled models predict that for a river with moderate OC and P inputs, respiration should be dominated by respiration of allochthonous material (fig. 20.4). For example, for a system such as the Hudson River, with P input concentrations of about 40 μg l^{-1} and OC inputs of about 7 mg l^{-1} (Findlay et al. 1998; Lampman et al. 1999), total respiration should be about 350 g C m^{-1} yr^{-1}, 70% of which should be fueled by allochthonous inputs. Measured values of microbial and (at present) grazer respiration exceed measured primary production by about threefold (Cole et al. 1991; Caraco et al. 1997; Strayer et al. 1996), consistent with these predictions.

The significant respiration of allochthonous organic matter demonstrates that microorganisms assimilate this material and use some of it as an energy source. As a result of bacterivory and the microbial food web, some of this allochthonous organic matter may fuel components of the higher food web as well (Schell 1983). Alternatively, the fate of allochthonous organic matter assimilated by microbes may be simply to be respired by the microbes without any link to higher trophic levels (Cole and Caraco 1993). The relative importance of these two fates depends on microbial efficiency on allochthonous substrates (del Giorgio and Cole 1998) and the food web structure of the aquatic system (Meili et al. 1996).

THE ROLE OF ALLOCHTHONOUS INPUTS IN NER

The net balance of carbon between autochthonous primary production and respiration is very strongly affected by the respiration of allochthonous inputs from the watershed. Because we are interested here primarily in respiration, we use the term net ecosystem respiration (NER) to represent this balance. This term is analogous to net ecosystem production, but has the opposite sign:

$$NER = Total\ respiration - GPP.$$

When NER is less than 0, the system has a net autotrophic organic carbon balance. In such a system, burial plus export of OC is larger than the import of allochthonous OC from the watershed. When NER is greater than 0, the system has a net heterotrophic OC balance. A net heterotrophic system respires more OC than is produced by primary production within that system. Thus, the respiration of allochthonous OC must be greater than the sum of burial plus export (see fig. 20.1).

The NER of systems can be predicted from coupled models of autochthonous and allochthonous carbon processing (see fig. 20.1) and can be calculated from CO_2 or O_2 balance in aquatic systems (del Giorgio et al. 1999; see figs. 20.1, 20.5, and 20.6). The coupled models suggest that the overall OC balance (NER) for lakes should vary from about −25 to +100 across a P input range of 150 to 6 μg l^{-1} and a DOC input concentration of 3 to 30 mg l^{-1} (see fig. 20.4). As many lakes are oligotrophic, with P inputs less than 30 μg l^{-1}, the coupled models also suggest that NER for lakes should generally be positive (net heterotrophy), with values of about 0–50 g C m^{-2} yr^{-1} (see fig. 20.4).

NER calculated from the CO_2 balance in lakes is in general agreement with NER calculated from coupled metabolic models. NER calculated from CO_2 balance varies broadly among lakes, from about −20 to over 400 g C m^{-2} yr^{-1}, and most lakes appear to have an NER of 10–100 g C m^{-2} yr^{-1} (Cole et al. 1994) (fig. 20.5). Further, as predicted by the model, the overall variation in NER calculated from CO_2 balance is, at least in part, related to P concentrations in different lakes (del Giorgio et al. 1999). In a study of twenty lakes in southern Quebec, all but two of the lakes showed a positive

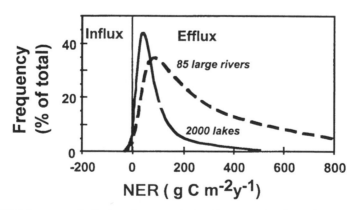

Figure 20.5 Frequency distribution of NER as calculated from CO_2 data for surface waters of lakes (solid curve) and large rivers (dashed curve). The equation relating CO_2 flux to NER is described in fig. 20.1. The exchange coefficients used were 0.5 and 0.8 m d^{-1} for lakes and rivers, respectively (based on Clark et al. 1994; Cole and Caraco 1998; Marino et al. 1988; and Raymond et al. 1997). Note that the generally positive NER in both lakes and rivers and the higher NER in rivers than in lakes are in agreement with model predictions (see fig. 20.4). (System CO_2 values are from Cole et al. 1994; Cole and Caraco 2001.)

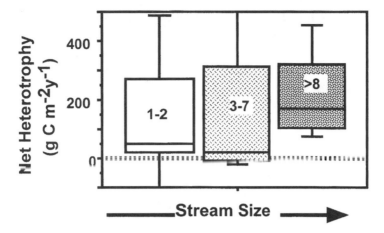

Figure 20.6 Distribution of NER in rivers of different sizes (as indicated by stream order). Systems are grouped into 13 small streams (order 1–2), 14 moderate-sized systems (order 3–7), and 43 large rivers (>8). The range of NER is shown by a box-whisker plot for each of these groups, with the 90%, 70% data inclusion being shown by the box and whiskers and the median being shown by the solid line in the box. (Data from Cole and Caraco 2001; Webster and Meyer 1997a.)

NER as calculated from CO_2 balance. The two lakes with a negative NER (including Waterloo Lake; see fig. 20.4) had relatively high P concentrations.

For rivers, like lakes, values of NER calculated from CO_2 balance suggest that systems vary substantially in NER, but generally show net heterotrophy (have positive NER; fig. 20.5). Values of NER vary from about -25 to over 800 g C m^{-2} yr^{-1}. On average, the NER of rivers appears to be substantially more positive than that of lakes, with the mean NER being over 200 g C m^{-2} yr^{-1} (figs. 20.5, 20.6). This high degree of net heterotrophy is also predicted from our organic carbon balance models if we use respiration coefficients (DV) for rivers 10 times higher than those found in lakes (figs. 20.1, 20.3, 20.4). Given a DV of 40 m yr^{-1}, we predict that an average river with P inputs of about 40 μg l^{-1} and DOC loads of about 10 mg l^{-1} (Meybeck 1982; Ludwig et al. 1997) should have NER of about 200 g C m^{-2} yr^{-1}.

DISCUSSION

A review of the existing data reveals a few striking points about the importance of allochthonous C inputs from terrestrial to aquatic systems. First, the magnitude of these inputs varies greatly with water load and the organic carbon concentration in the water. At typical water loads and OC concentrations, however, allochthonous inputs of carbon are often significant in comparison to autochthonous primary production. In most rivers and in oligotrophic lakes, allochthonous inputs are often much greater than autochthonous primary production, and may actually be greater than C inputs (from primary production) to terrestrial systems. These high loads reflect landscape concentrating mechanisms (fig. 20.7). For example, in the Hudson River, OC_{all} from an area of about 35,000 km^2 is concentrated in the lower river, which has an area of about 200 km^2 (Limburg et al. 1986), a nearly 200-fold concentration.

If terrestrial material entering aquatic systems were not respired, its influence on metabolism would be restricted to indirect effects brought about by the alteration of the physical and chemical environment of the system. For example, GPP might be lowered due to reduced light resulting from high allochthonous inputs of colored DOC (Carpenter, Cole et al. 1998), or it might be increased by DOC chelation of metals (Wetzel 1975). When allochthonous material is respired, however, not only are new indirect effects possible (e.g., lowering of pH by increased CO_2; Caraco and Miller 1998), but there are also clear direct effects on system metabolism. Although low rates of metabolism of the residual terrestrial material that is exported via hydrological flows might be expected, respiration of this material is apparently quite rapid in aquatic systems in general, and perhaps in rivers in particular (see fig. 20.3). The decomposition of terrestrial OC in aquatic systems overall is large enough to affect the total respiration in many lakes and most rivers (see fig. 20.4). Further, NER, which would be negative in the absence of allochthonous inputs, is positive in the majority of lakes and rivers due to decomposition of this material.

The effects of allochthonous inputs on small streams have been relatively well studied. Less is known, however, about the importance of these inputs to lakes or large rivers. The river continuum concept would suggest a lesser importance of allochthonous inputs in downstream locations, including large rivers, than in small streams (Vannote et al. 1980). This is, in part, because much of the material reaching these rivers has already been processed in small streams, suggesting delivery of somewhat lower loads of organic carbon and enrichment of refractory materials in rivers. Interestingly, however, NER in rivers appears to be as great as or greater than it is in small streams. While the median NER in small streams is about 50 g C m^{-2} yr^{-1}, the median value in large rivers is about 200 g C m^{-2} yr^{-1} (see fig. 20.6). This pattern is in part due to the low and sometimes negative NER in some streams (in particular, desert streams;

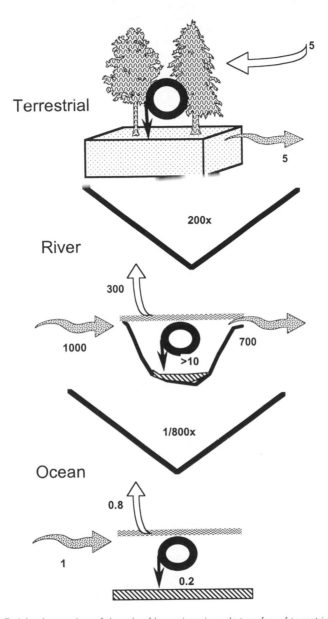

Figure 20.7 A landscape view of the role of large rivers in scale transfers of terrestrially derived organic carbon. The small percentage of terrestrial primary production that is lost hydrologically to runoff is concentrated in large rivers before being transferred to the ocean. The concentration factor is often over 200-fold. Thus, terrestrial loads to rivers are often greater than those in the surrounding terrestrial watershed. Because of this high load, the high decomposition of allochthonous organic materials in rivers, and the relatively low primary production in these systems, the metabolism of rivers is often dominated by allochthonous inputs. Further, due to high NER, driven by allochthonous loads, rivers can act as major "hotspots" of trace gases released to the atmosphere, including CO_2.

Peterson et al. 1985; Webster and Meyer 1997b). But it is also due to high NER in rivers, which is probably caused by both continuous new inputs of organic matter over the length of the river (e.g., lateral inputs; Richey et al. 1990) and the exceptionally high decomposition rate of even refractory material in rivers (Schell 1983; Cole and Caraco 2001).

If biological processes were equal over the entire planet, the earth would be covered by a thin, slightly autotrophic skin. Rather than being uniform, however, the earth is a patchwork of autotrophic and heterotrophic spaces, due in large part to aquatic transfers of organic carbon. Terrestrial environments tend, with some interesting exceptions (e.g., Polis, Anderson, and Holt 1997), to be autotrophic, while aquatic systems tend to be heterotrophic. It is well known that humans can influence the degree of heterotrophy in the world by fertilizing the environment with nutrients and CO_2 (Heiman 1997; Schindler and Bayley 1993). Discovering where these human alterations have occurred will be difficult, however, without understanding the background heterogeneity of heterotrophic and autotrophic processes. Heterotrophy in lakes and rivers affects not only the internal metabolism of these systems, but also the carbon balance of large regions of the earth's surface (Kling et al. 1991).

ACKNOWLEDGMENTS

This work was supported by grants from the National Science Foundation, the Hudson River Foundation, and the A. W. Mellon Foundation. The work is a contribution to the Institute of Ecosystem Studies.

Chapter 21

Trophic Flows and Spatial Heterogeneity in Agricultural Landscapes

Jacques Baudry and Françoise Burel

Ecologists increasingly recognize that ecological processes occur in landscapes that are heterogeneous in space and time and that this heterogeneity needs to be incorporated into developing ecological concepts and models (Forman 1995; Pickett and Cadenasso 1995). Heterogeneity refers to the fact that we perceive different types of structural elements (woods, crops, urban development, roads) when looking at landscapes. Trophic exchanges of energy and materials are major ecological processes that are strongly influenced by landscape heterogeneity (Polis, Anderson, and Holt 1997). Landscape ecology is a field of research that emerged from this recognition; heterogeneity, therefore, is at the core of the landscape ecology paradigm. Landscape ecology has two main objectives: to provide an understanding of the consequences of spatial heterogeneity for movements of materials, and to analyze the dynamics of landscape patterns (Burel and Baudry 1999). Risser (1987) stresses the importance of different scales of heterogeneity in space and time in understanding ecological processes. This chapter presents an overview of the ways in which species and nutrient flows are affected by the heterogeneity of agricultural landscapes.

One of the consequences of heterogeneity is a large number of ecotones (i.e., transition zones between adjacent landscape elements). This pattern is especially marked in agricultural landscapes, where field boundaries are numerous and are often associated with changes in vegetation structure (e.g., grassy strips, hedgerows, and ditches). Patterns and processes in agricultural landscapes, including those across ecotones, are worth studying from both basic and applied perspectives. The applied perspective aims at providing management guidelines for maintaining sustainable landscapes in which resources and biodiversity are conserved. In the European Union, agricultural land represents 44% of the total land area, whereas in North America, the corresponding value is only 12.6% (Food and Agriculture Organization 1998). Agricultural landscapes provide relatively simple, contrasting ecological systems for study. A major difficulty for ecologists, however, is that these contrasts are driven not by ecological processes per se, but by human decisions within a given socioeconomic and technical context. Farming systems research is the subdiscipline involved in the analysis of the overall processes in agricultural landscapes (Brossier et al. 1993; Fresco et al. 1994).

The goal of this chapter is to discuss trophic movements of materials in agricultural landscapes and their consequences for species richness and the distribution and availability of nutrients. We recognize two types of trophic movements: the first driven directly by human activity and the second the outcome of ecological processes that result from landscape changes, which, in turn, are driven by human activity.

Agriculture is based upon the manipulation of trophic flows. Additions of materials are particularly evident in intensive agricultural systems in the form of high inputs of inorganic nitrogen. Vitousek, Aber et al. (1997) have reported that human activity has doubled the rate of nitrogen input into the terrestrial nitrogen cycle. The current (but increasing) global N fixation rate for fertilizer use has now reached about 80 Tg yr $^{-1}$, and much of this N is no longer sequestered at the local level, but is moved across landscape boundaries by physical processes. Likewise, herbivores, predators, and plant propagules move across landscapes and transport nutrients such as nitrogen and phosphorus. We examine which components of the landscape promote or inhibit these flows of materials.

Because of the detrimental effects of habitat fragmentation by human activities on biodiversity, ecologists have stressed the role of remnant habitats at boundaries in agricultural ecosystems as corridors for the movement of materials and species across fragmented agricultural landscapes. However, these habitats also play a role as barriers to movement. For example, landscape features such as riparian zones act as barriers that trap sediment and as sinks for nitrogen, phosphorus, and pesticides (Haycock et al. 1996). Anthropogenic elements, such as roads, that prevent animal migration (Forman and Alexander 1998) also act as barriers.

The main thesis of this chapter is that the structural landscape elements mentioned above modulate ecological processes in ways that depend on the type of organism or material that is moving. Connectivity is a key concept that couples one type of landscape element to another through the flow of species or materials. We define connectivity as the intensity (rate) of exchange of species or materials between two landscape elements.

PATTERN GENERATION

One of the most studied landscape processes is deforestation and land abandonment. Deforestation occured even prior to agricultural development in temperate regions, and is now widespread in tropical and subtropical regions as well. Considerable land use changes may be expected to occur following deforestation and abandonment (Harris 1984; Baudry and Bunce 1991). Deforestation permits immigration of species, including deliberate introduction of crop species that require open conditions. Crops, with their associated weedy flora and insects, are moved between continents by human agencies following changes in land use. Abandonment of land following deforestation, or a cessation of agriculture, allows the invasion of uncultivated species, many of which are native species. When land is actively farmed, its use generates landscape patterns. For example, the planting of hedgerows and shelterbelts and the construction of other field boundary markers, including casual fences or ditches, result in landscape features that influence the movement of materials and species (Forman and Baudry 1984; Baudry et al. 2000). Farming operations themselves involve movements of allochthonous materials. Farmers transport seeds and fertilizers, remove nutrients (harvest), and move grazing livestock from site to site.

Traditional farming in Eurasia involved depletion of nutrients in some parts of the landscape in order to enrich other parts. Bazin et al. (1983) described the pastoral system in central France. During the day, sheep grazed on common land, and at night they stayed in a paddock on fallow arable land, where they deposited feces. The authors calculate that during the 150 days when sheep grazed on 100 ha of common land and stayed at night on 12 ha of arable land, they transported 92 kg of nitrogen to the latter site. They estimate that nitrogen fixation on the grazed land compensated for the removal of this element by sheep, until the number of sheep increased beyond the capacity of the nitrogen fixers to replace the nitrogen they removed. The increase in sheep, therefore, led to an unsustainable system on the grazed common land. In the Netherlands, Vink (1983) described the transport of sheep manure and sod to arable land that was obtained by stripping topsoil from grazed heathland. After hundreds of years of transfer, this practice resulted in the accumulation of a thick layer of humus on the arable land, which became elevated by as much as a meter above the surrounding fields. This type of agriculture is no longer practiced; the rich organic layer is disappearing, and the land is sinking.

Industrialized agriculture promotes nutrient movement across landscapes and continents, mainly as a consequence of cultivating food in one part of the world to feed domestic animals and humans in other parts. Water pollution in Western Europe is due, in large part, to the use of soybeans and manioc grown in Southeast Asia or the Americas as animal feed. For example, in 1996, the Netherlands had a net import of 3 millions tons of soybeans, and Europe imported over 13 million tons, while Brazil exported almost 4 million tons (Food and Agriculture Organization 1998). Given the production per hectare in Brazil, these figures are equivalent to an increase in the Netherlands' arable land by about 6%, and Europe's by 0.8%. Animals fed with these foodstuffs are usually concentrated in specific agricultural regions, and they are an important source of additional organic matter that is spread on fields as manure. In Brittany, France, the excess of nitrogen due to manure spreading is 100% of the amount required for the production of crops (about 100 kg ha^{-1} yr^{-1}). This example shows how nutrients are transferred between countries connected by trade. These additional sources of nitrogen and phosphorus may disrupt the functioning of non-agricultural ecosystems (Carpenter, Caraco et al. 1998) and augment forest clearance in the tropics (Kaimowitz et al. 1999).

LANDSCAPE PATTERNS

Within a landscape one can see three-dimensional elements, which are referred to as patches, and linear elements, called corridors or boundaries. A set of patches forms a mosaic, and a set of linear elements constitutes a network. Patches may be separated by distinct boundaries. The juxtaposition of two patches creates an ecotone. We use "ecotone" in a functional sense: an ecotone is a zone where within-patch processes are influenced by between-patch or neighboring patch processes. A riparian zone, for example, is flooded periodically due to variation in stream flow;

the biogeochemical functioning of the zone is related to the inflow of water and sediment, rich in nitrogen and phosphorus derived from other localities. Flooding and nutrient inputs interact, so that the changed conditions may lead to anaerobic conditions and denitrification, or the flooding may promote sediment buildup and the accumulation of nutrients and pesticides.

Numerous papers have been published on measuring landscape parameters. Often indices are calculated in order to compare landscapes at the global scale. The ecological significance of these indices is often dubious, however, as they aggregate spatial information as a single value (Haines-Young and Chopping 1996), and different patterns of landscape elements can produce similar values. We prefer to utilize Geographical Information Systems (GIS) to display results so that spatial dimensions are kept. For example, we can use the lengths of hedgerows (or their densities) to describe two landscapes, but this method provides only the average distribution of hedgerows across the two landscapes. Of more interest are the effects of the hedgerows on adjacent fields, which represent how environmental conditions change at various distances from the hedgerows. These conditions can be assessed with a map showing classes of distances from hedgerows. Microclimatic changes occur due to differences in the proximity of windbreaks and in inputs of solar radiation close to hedgerows. Another possible representation is the mapping of hedgerows as permeable and semi-permeable barriers (they never are a total barrier). This representation shows that materials and organisms are confronted by barriers and differences in microclimate (shelter) when moving across a landscape. The spatial distribution of these differences can provide a basis for sampling in heterogeneous environments, as we can combine patch type and the effects of neighboring elements.

We can also map landscape patterns resulting from farming practices. Such mapping can show the heterogeneity of levels of pesticide inputs or addition of fertilizers. The spatial heterogeneity of such inputs can be understood only with reference to individual farming practices. Hatfield et al. (1999) give an example of the spatial heterogeneity in inputs of nutrients in Iowa, U.S.A.

The grain of the landscape (field size, fragmentation of land cover types) also influences landscape processes. A consistent trend in intensive agriculture at the regional scale is a simplification of landscape structure and an increase in grain size associated with an increase in field size (Meeus 1993; Medley et al. 1995). Boundaries and uncultivated elements (habitats) are displaced farther away from the center of the crop patch.

SPECIES AND NUTRIENTS IN LANDSCAPES

Movement of organisms is usually constrained by vegetation. In agricultural landscapes, constraints on movement are high. Species dwelling in forests seek trees or hedgerows to traverse agricultural landscapes. Species of open habitats may thrive in fields if habitat conditions (e.g., level of pesticide use, fertilization, tillage) are suitable. The result is that species composition often varies over very short spatial ranges in agricultural landscapes. Landscape features (elements) controlling movements can act as corridors (flow enhancement) or barriers (flow inhibition). Boundaries can be physical or functional barriers. A physical barrier exists when a flow is arrested because a landscape element is not permeable; a functional barrier exists when a flow is reduced or stopped because the organisms or materials moving are preyed upon or otherwise transformed. The first barrier effect is exemplified by a hedgerow arresting windblown soil particles; the second by denitrification in riparian zones.

Organisms: Distribution and Movement

The distributions and movements of arthropods, birds, and small mammals across agricultural landscapes have been intensively studied (Bunce and Howard 1990; Merriam and Lanoue 1990; Ekbom et al. 2000). What are the landscape features (elements) associated with high species richness or a high density of individuals? What are the landscape features species use for different types of movements, and how do these features affect patterns of distribution and abundances of animals over time?

Arthropods are important as herbivores and predators, and they thus act as crop pests and as biological control agents; hence, the relative distributions of both functional groups are important. Small mammals are an important food source for birds of prey, but may act as crop pests as well. In many cases, boundary habitats harbor significantly more species than crops (Paoletti and Pimentel 1992). Table 21.1, based on our own study in Brittany, France, from 1993 to 1996, illustrates that the species richness of carabid beetles in grassy strips was particularly high, while beneath wooded hedgerows and in fields, species richness was low. Sown grassy strips were a mixture of species in the Poaceae and Leguminosae. A control boundary habitat with no sown strip was sampled on the opposite site of the field. There were fifteen replicates for each site. Table 21.2 shows that adjacent landscape elements contained different sets of species. Carabids were sampled during 4 consecutive years in nine fields

Table 21.1 Mean number of carabid species in different agricultural habitats from 1993 to 1996

Habitat type	1993	1994	1995	1996
Existing boundary: hedgerow		10.7	12.3	11
Sown grassy strip	16	15	15.3	18.8
Field	12.6	13	12	12.8
Opposite field boundary	15	15	20	

SOURCE: F. Burel, unpublished data.

Table 21.2 Percentage of sites where various carabid species were present in field boundaries (margins and hedgerows) and fields

Species	Under hedgerows	Margins with dense vegetation	Margins with grazed vegetation	Fields
Nebria brevicollis	78	93	74	100
Poecilus cupreus	50	68	69	100
Abax ater	21	17	34	0
Amara sp.	28	44	65	100
Diachromus germanus	7	17	8	66
Metallina lampros	64	65	43	100
Syntomus sp.	14	17	17	100
Anisodactylus binotatus	0	3	4	100
Chaetocarabus intricatus	35	10	17	0

SOURCE: F. Burel, unpublished data.

and associated hedgerows or grassy strips. *Rubus* sp. was the dominant plant species in field margins with dense vegetation. Some large, forest-dwelling species, such as *Chaetocarabus intricatus* or *Abax ater,* do not move into fields, while genera such as *Amara* are almost absent from wooded margins. Similar results from other studies have been reported by Kromp (1999). There is a close spatial correlation between the pattern of structural elements across the landscape and the distribution of species.

Three types of animal movement have been investigated: movements from boundaries to fields, movement along boundaries, which act as corridors, and movement across fields.

Movements between Boundaries and Fields
The movements of animals are frequently associated with foraging. Hedgerows are refuges for animals and sources of individuals of different species. Some of these are herbivores that will forage on crops when they

are present; others are predators that prey on potential pests. Animal movements may also be due to the multiple use of various habitats during the life cycles of animals.

Carabids moving from field margins exemplify predator movement from boundaries into fields. These predators overwinter in field margins. Dennis and Fry (1992) found that in June, aphid-feeding arthropods move more frequently into fields from field margins than the converse. Wratten and van Emden (1995) effectively demonstrated that wheat plants had fewer aphids in fields bordered by margins of flowering plants than in fields devoid of margins. Kromp (1999) challenges the assumption that carabids are effective pest control agents on the basis that most assays are carried out in laboratories, even though his review demonstrates the importance of carabid movements from boundaries into fields.

Ouin et al. (2000) conducted an experiment to study the fate of a small mammal, *Apodemus sylvaticus,* during summer. Individuals of this species were numerous in field boundaries during the spring, then declined. The trapping index in hedgerows and in crops reversed during the year. In crops, the index peaked in May and June, but was close to zero from September to January. *Apodemus* preferred wheat and peas to maize and carrots; hence, crop phenology influences the index. Overall, plant cover was the best predictor of abundance.

Movements along Corridors

The effect of corridors on the movement of organisms and materials is at the core of many studies of landscape ecology (for plants, see Baudry 1988; Marshall and Arnold 1995; for insects, see Burel 1989; Duelli et al. 1990; for birds, see Dmowski and Koziakiewicz 1990).

In her study of the distribution of carabids in hedgerows adjacent to a forest, Burel (1989) showed marked species differences. Some species were restricted to the forest, "peninsula" species were not found farther than a few hundred meters from the forest edge, and the "corridor" group was restricted to hedgerows irrespective of their distance from the forest. *Abax ater,* which feeds on earthworms, was the most common species in this latter group. Burel and her collaborators set up experiments to analyze rates of reproduction and movements of these beetles in hedgerows. Petit and Burel (1993) showed that *Abax* reproduce in the nodes of hedgerow networks. Charrier et al. (1997) used radio-tracking to follow movements of *Abax* in hedgerows and adjacent environments. They showed that hedgerows are more permeable to *Abax* than grassland or cropland. This finding permitted the research group (Petit and Burel 1998a, 1998b) to assess the influence of past and present landscape structural elements on the population dynamics of *Abax*. An important finding was that the distribution of *Abax ater* was related to both present and past landscape elements. Petit and Burel showed this by using maps of the landscape in 1950 and in 1994, at the time of their investigations. Numerous hedgerows had disappeared during that period. Hedgerows seemed to facilitate *Abax ater* colonization in the 1950s, but populations survived after hedgerow removal in 1994. Hence, the correlation of Abax abundance between two sampling points was higher when using 1950 maps than when using 1994 maps. At the time of high hedgerow density, *Abax* was able to colonize many hedgerows, where populations survived after the surrounding hedgerows were removed. In case of local extinction in a hedgerow no longer connected to another one, new colonization may not occur.

Boundaries as Barriers
A hedgerow provides a corridor for forest species to traverse an agricultural landscape; however, it may stop other types of species from moving across adjacent fields. For example, Mauremooto et al. (1994) conducted an experiment to test the ability of carabids (*Harpalus rifipes, Pterostichus melanarius,* and *Pterostichus madidus*) to move across hedgerows. They found that typical narrow English hedgerows slowed down carabid movement, but did not prevent their crossing. They also found that starvation before release increased the speed of movement.

Jepson (1994) reviewed the potential barrier effects of field boundaries and concluded that they may lead to local extinction. The barrier effect of field boundaries is dramatic for butterflies (Fry and Robson 1994). Dense or high hedgerows are not crossed by some butterfly species; even lines of shrubs 1 m high reduce movements between fields. Observations of movements of butterflies reveal that they tend to concentrate at and move along boundaries and cross when they meet a gap.

Delettre and Morvan (2000) have investigated the dispersal of Chironomidae in three landscapes, each with a different density of hedgerows. Adults emerge from streams and then fly onto adjacent land, but their flying capacity is generally weak. The average abundance of individuals decreases with increasing distance from a stream. In the landscape with the greatest density of hedgerows, some species are confined to sites close to a stream, as they cannot cross hedgerows. In the most open landscape, the species assemblage is homogeneous over a greater distance. Thus the movements of these Diptera from streams to terrestrial environments are constrained by landscape structural elements.

In their 6-year study of the population dynamics of the vole *Microtus arvalis* in eastern France, Delattre et al. (1999) found that in landscapes where hedgerows are present, as well as close to villages, vole populations do not fluctuate in numbers, as they do in open grassy landscapes. Hedgerows also appear to slow the diffusion of populations. The authors hypothesize that hedgerows harbor generalist predators that dampen population fluctuations of voles, whereas in open landscapes, specialist predators enhance population fluctuations. Thus, some types of landscape elements act as barriers at a regional scale as a result of the effects of predation.

Nutrients: Flows and Sinks

Nutrient loading of surface waters is a major environmental issue. Here we distinguish soluble nutrients dissolved in water, such as nitrate-nitrogen, from those attached to soil particles, such as phosphorus. These two major elements can cause trophic disturbances in surface waters and can also result in changes in species assemblages on land.

Nutrient Sources and Pathways

Natural processes as well as human practices result in nutrient flow from agricultural fields. The spatial distribution of use of nitrogen fertilizers by farmers across an agricultural landscape is often heterogeneous and is not correlated with soil or topographic gradients. Water and wind are carriers of nutrients, as in any type of landscape. In agricultural landscapes, heterogeneity in vegetation cover and linear features such as ditches and other types of field boundaries strongly influence water flow. Water not only flows across land surfaces, but may also flow in the subsurface zone or move down to the water table and emerge as a spring elsewhere.

The use of grassy or woodland strips to control erosion is found worldwide. These strips constitute buffer zones that prevent the movement of nutrients into adjacent fields. For example, they act as a physical barrier in stopping the dispersal of wind-blown soil particles. Where a strip is present along a contour, soil nutrients may be trapped and not moved downslope. We do not know of a study of the consequences of the accumulation of nutrients along field boundaries on biodiversity.

Buffers and Sinks

Lowrance et al. (1985) studied the nutrient budgets of four watersheds between 1,600 and 2,200 ha in area where farmland occupied 40–50% of the land area. They calculated nitrogen budgets based on inputs from precipitation and fertilization minus outputs from harvests and loss in stream flow. They found that inputs exceeded outputs by 34–45 kg ha^{-1} yr^{-1}; thus there must be sinks where nitrogen either accumulates or is lost in the watershed, or as a gas. The explanations most likely to account for the difference between inputs and outputs are uptake of nitrogen by vegetation or nitrogen loss by denitrification. Peterjohn and Correll (1984) found that in forest riparian zones, nitrate-nitrogen diminishes in groundwater.

These findings have promoted much research on riparian zones and their role as buffer systems, especially where denitrification occurs (Haycock et al. 1996). Nitrates are transformed into N$_2$, although the process may stop at N$_2$O, a greenhouse gas. Correll (1996) reports that in the coastal plain of the eastern United States, nitrogen budgets indicate that nitrogen retention in riparian zones can be as much as 74 kg ha^{-1} yr^{-1}, or 90% of inputs. Thus inputs of mineral nitrogen to stream water can be considerably less than nitrogen leached from fields. Denitrification is efficient in agricultural landscapes because the required nitrate-nitrogen is more abundant than in forested landscapes.

Hedgerows perpendicular to a slope also play an important hydrological role: they slow water flow and facilitate infiltration into the soil (Mérot 1999). This hydrological function may explain the nitrogen retention along hedgerows demonstrated by Caubel-Forget and Grimaldi (1999). As the water goes deeper into the soil, it moves more slowly, and is no longer pumped by trees. Ryszkowski et al. (1999) come to similar conclusions, as they emphasize that differential rates of evapotranspiration among different types of land cover (shelterbelts, grasslands, crops) strongly influence the hydrological cycle at the landscape scale. Their findings demonstrate the important role of shelterbelts as geochemical barriers. Riparian zones, hedgerows, and perennial vegetation also stop soil particles and pesticides (Papy and Souchere 1993). Thus, these barriers play an important role in disconnecting the flow of materials between terrestrial and aquatic ecosystems.

Nutrient Fluxes and Terrestrial Species Distributions

Heterogeneity in the distribution of nutrients at the landscape scale is a cause of heterogeneity in plant distributions, which affects animal distributions. For example, grassland vegetation has been intensively studied in relation to soil nutrient status (Balent 1986). Fertilization and high nitrogen loads are a major cause of decreased plant species diversity; competitive species in the Poaceae (*Lolium, Dactylis*) outcompete most

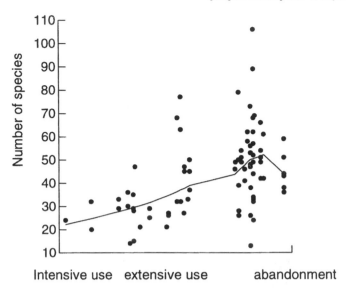

Figure 21.1 Relationship between herbaceous species richness and land use intensity in permanent grassland in Normandy, France. The x-axis is an ordination of fields according to farming practices.

dicotyledons under these conditions (Haggar and Peel 1994). Figure 21.1 shows the relationship between the intensity of usage of agricultural land and species richness. Intensity of usage is defined as increasing fertilization and higher stocking rates; the ordination of fields according to these variables provided the x-axis of the graph (Baudry, Alard et al. 1997). Species richness increases in less intensively utilized grassland, but abandonment of farmland reduces species diversity. A wide range in species richness occurs in less intensively managed areas because of variation in the natural nutrient content of soil. Thus intensification of land use leads to a more homogeneous distribution of nutrients and poorer species richness. A mosaic of nutrient availability in the landscape increases gamma diversity.

Effects of Intensive Agriculture on Food Webs

Studies of the effects of agricultural landscape patterns on food webs are scarce. We found little evidence of changes in food webs in relation to changes in land use or in landscape structural elements. Three factors appear to exert control over food webs: (1) plant communities, which in turn are related to soil nutrient status, use of herbicides, and crop type; (2) field boundaries; and (3) landscape geometry.

Butterflies are good candidates for the analysis of the first factor because they are important emblematic species, often associated with meadows rich in plant species. Erhardt and Thomas (1991) demonstrated a general correlation between plant richness and butterfly richness in grassland. Grazed grassland harbors the most butterfly species, and mowed, fertilized grassland the least. The number of butterfly species decreases faster than the number of plant species as land is increasingly fertilized and mowed. Erhardt and Thomas attribute this correlation to larval feeding preferences. The species most sensitive to changes in plant composition during either abandonment or intensification are those with narrow niches at the egg or larval stages.

Van Es et al. (1999) examined the butterfly communities of two riparian zones. They found that plant species diversity diminished as nitrogen availability increased, and that butterfly richness decreased significantly as well. The average number of butterfly species was seven in nutrient-rich areas and nine in nutrient-poor areas. In a comparison of species compositions in riparian and adjacent upland areas, they concluded that in oligotrophic riparian areas, butterflies stay in the riparian zone, whereas they move upslope in nutrient-rich riparian areas.

Wilson et al. (1999) stressed the importance of uncultivated grasslands, even strips of grass along field boundaries (the second factor affecting food webs), as habitats for arthropods, which are such an important part of the diet of birds. They analyzed the correlation between potential food source declines and declines in bird species. They found that declining groups of invertebrates, such as members of the Coleoptera, Diptera, Lepidoptera, Hymenoptera, Hemiptera, and Arachnida, accounted for a significant part of the diets of declining bird species. Birds also suffered from the disappearance of seeds of *Polygonum, Stellaria,* and *Chenopodium,* all weeds that vanished from arable fields treated with herbicides.

Kruess and Tscharntke (2000) have demonstrated that the geometry of the landscape is crucial: the increasing isolation and decreasing size of grassland fragments negatively affect species richness, as well as interactions among endophagous herbivores and their parasitoids. However, there appears to be a threshold at which these effects become evident, as Elliott et al. (1999) have shown that in South Dakota, U.S.A., (relatively) fine-grained landscapes harbor more aphid predator species than do coarse-grained landscapes.

All three of the factors affecting food webs (type of vegetation, presence of boundaries, and landscape geometry) vary with the intensification of agriculture, which brings in nitrogen, removes boundaries, and enlarges fields, and thus probably changes food webs to a large extent.

CONCLUSIONS

The ever-increasing consumption of primary production by humans is accelerating movements of organic and mineral nutrients at the global scale. This change is part of the industrialization of agriculture. In tandem with these increases are large changes in land use patterns, which have affected the distributions of species and generated new interactions among species at the landscape level, modifying trophic relations. Agricultural activities increasingly contribute to the fluxes of nutrients at the global and local scales. This duality of flows at different spatial scales has received little attention, and we are unaware of studies of nutrient feedbacks at these different spatial scales.

Farming activities are changing landscape heterogeneity. Whether they have led to an overall increase or a decrease compared with the landscapes that existed before intensive agriculture is difficult to assess. It depends partly on what we consider to be the different structural elements in the landscape today, and also on the spatial scale of measurement chosen to study this heterogeneity (Burel and Baudry 1999). This is one of the reasons why heterogeneity per se is not an operational concept. To understand ecological processes at the landscape scale, it is more important to look at the effects of spatial patterns on biodiversity and to address the question, "What are the important landscape features that influence movements of species and materials?" Some landscape elements and patterns that we perceive as generating heterogeneity also control movements of species and materials, demonstrating that "heterogeneity" is an important heuristic concept.

At a global scale, climatic and political differences induce flows of goods, and at a regional scale, climatic, historic, and cultural differences induce and maintain a diversity of farming systems (Baudry, Laurent, and Denis 1997). Within regions, the choices of production systems made by farmers and differences in farm sizes produce landscape patterns (Deffontaines et al. 1995). Landscape heterogeneity stems from the diversity of different types of land cover, linear structural elements across the landscape, and farming practices, including inputs of nutrients and pesticides. Thus, a hierarchical view of heterogeneities and of their consequences for ecological processes is required.

Within landscapes, species thrive or vanish; nutrients flow or are transformed or stored. Species richness in an agricultural field depends on the type of adjacent boundary, but the quantitative aspects of trophic flow and movements of species, which result from boundary margins that provide connectivity, are not clear. Furthermore, the grain of the landscape also influences trophic interactions. Large fields tend to receive fewer predators—there may be none at the center—and fragmented habitats do not allow full trophic webs to develop (Kruess and Tscharntke 2000). Species life history traits constitute another set of factors to consider. For example, some species in wooded areas do not move across the boundaries of woods into fields, but forage only at the boundaries. Some types of predators and biological trophic controls are maintained by the presence of perennial herbaceous vegetation at boundaries. Up to now, most studies of insect predators have been carried out in Europe, but it would be interesting to examine comparable situations in North America, where invertebrates have been subjected to intensive agriculture for fewer generations.

Temporal changes are also important in understanding current patterns of species distribution in relation to landscape patterns. Changes happen over the long and short term. Short-term changes (such as crop rotation) must be incorporated into analyses of landscapes. Kromp (1999) has found evidence that crop rotations affect carabid diversity; carabid communities are partly structured by the crop present the previous year. Long-term changes (such as shifts in patterns of land cover) bring about real changes that alter rates of ecological processes.

Flows of nutrients in landscapes, especially of nitrogen in its organic and mineral forms, are strongly influenced by the types of vegetation present. At a watershed scale, tons of nitrogen may be removed from terrestrial and aquatic systems into the atmosphere as a result of denitrification. The main sources and sinks of nitrogen at the landscape scale need to be established, together with the processes that control the flow of this element within and between landscapes. A strong connection exists between nutrient and biological flows, as areas enriched by fertilizers may experience losses of biodiversity. While amounts of arable land are a good predictor of nitrogen input into rivers at the coarse scale (Carpenter, Caraco et al. 1998), at a fine scale, landscape heterogeneity leads to differential levels of nitrogen enrichment in streams (Poiani 1996).

Because of the simplification imposed upon landscapes by most farming systems and the direct human influence on nutrient and species movements, agricultural landscapes possess structural elements and functional features that constrain the distribution and flow of resources. At a fine scale, high landscape heterogeneity may increase trophic interactions and the control of nutrient movement. Low heterogeneity, as is found in large fields with little perennial vegetation, facilitates herbivory by crop pests and nutrient leaching or runoff when land is fallow after a crop has been harvested. At a coarser scale, the globalization of the economy contributes to the globalization of land use changes; the coarse-scale heterogeneity resulting from different farming systems and political systems enhances long-distance movements of species and materials that may increase the risk of overfertilization and the introduction of pests.

ACKNOWLEDGMENTS

We thank the Ministry of the Environment and the Environment Program of the Centre National de la Recherche Scientifique and the European Commission for their financial support. Discussions with Gary Polis, along with comments and editing from Gary Huxel and reviewers, were essential for the writing of this chapter.

PART IV.

SYNTHESIS

The Influence of Physical Processes, Organisms, and Permeability on Cross-Ecosystem Fluxes

Jon D. Witman, Julie C. Ellis, and Wendy B. Anderson

Observations of feeding relationships are probably the oldest form of currency in ecology, dating back to ancient cave dwellers who painted predator-prey interactions on cave walls. When predator-prey observations were summarized as "food cycles" with trophic levels several millennia later, they included consumers, such as bears and foxes, that depended on food from another ecosystem, the ocean, in the form of carcasses of marine mammals cast ashore on Arctic beaches (Summerhayes and Elton 1923). Thus, the recognition that consumers are subsidized by food exported from other ecosystems (allochthonous inputs) has been a part of food webs since their inception. Fluxes of matter and nutrients between different habitats had been largely studied under the discipline of ecosystem ecology (Odum 1971). The possibility that cross-ecosystem links could actually regulate food webs was not widely appreciated until Gary Polis and his colleagues discovered that consumers in some terrestrial food webs were heavily dependent on allochthonous inputs of food from the ocean (Polis and Hurd 1995, 1996b; Polis, Anderson, and Holt 1997). The central thesis of this book is that such cross-ecosystem fluxes of food, nutrients, and consumers are widespread and that a complete understanding of population and community dynamics requires knowledge of these connections.

How, then, can we predict where material is likely to be exchanged across ecosystem or habitat boundaries? Recognizing that physical processes, organisms, and landscape elements all interact to determine how food, detritus, and organisms move across ecosystem boundaries, we synthesize the interaction of these components to predict where food webs are likely to be spatially subsidized. We use the term "spatial subsidy" to mean donor-controlled resources (prey, detritus, nutrients) moving from one habitat to another that increase the productivity of the recipient habitat (sensu Polis, Anderson, and Holt 1997). Habitats or organisms that are sources of material are referred to as donors, assuming that feedbacks to the originating habitat are minimal. Physical processes are key because they transport allochthonous materials to recipient habitats. On the largest scales, the rate and location of delivery depends on how wind and water flow are modified by the earth's topography (Atkinson 1981; Wolanski and Hamner 1988). The amount of allochthonous input to a habitat depends not only on the sizes of the pools of materials and organisms available for transport, but also on the distribution, size, and permeability of the habitat patches within the landscape that receives it (recipient habitats). Organisms greatly influence the flux of allochthonous material as bio-eroders, bioturbators (Levinton 1995), grazers, and predators, by dispersing in early life stages or as adults, and ultimately by dying and contributing to detrital pools. Biotic vectors of cross-ecosystem transfer are thus extremely important (Jefferies 2000). Since the myriad ways in which organisms and landscape elements influence allochthonous fluxes are specifically considered in other chapters of this volume, we restrict our discussion of organismal influences to those that create habitats, modify the physical environment, and subsequently affect the flux of allochthonous nutrients and other food resources. Much has been written recently about the ecological effects of habitat-forming species (Bell et al. 1991) and their roles as ecosystem engineers (Jones et al. 1994; Bruno and Bertness 2001). Here we highlight their potential to create or modify supplies of nutritive materials to other habitats or ecosystems.

Our approach begins with a consideration of physical vectors of transport across ecosystem boundaries and how they are modified by habitat-forming organisms. Recognizing that cross-ecosystem fluxes of energy, matter, and organisms are in large part determined by the permeability of habitat boundaries or ecotones (Gosz 1991), we focus next on the importance of permeability, distinguishing between topographic and habitat effects. We conclude the chapter by presenting a set of graphic models that predict the magnitude of cross-ecosystem flux using a range of independent variables.

It is our hope that this overview will motivate experiments to test hypotheses about the role of cross-ecosystem subsidies at the population, community, and ecosystem levels.

PHYSICAL PROCESSES

Physical processes and topography interact to regulate the flux and exchange of materials across ecosystems and habitats. The flux of materials is equivalent to the product of concentration and velocity, so it is useful to consider how the shape of landforms and habitats can alter both components of flux. The transfer of material suspended in moving air to the ground or from the water column to the bottom (benthic-pelagic) occurs through a laminar or turbulent layer of fluid closest to the solid surface. Turbulent flow is characterized by a strong mixing or eddying motion and thus can be an especially important mode of transfer from fluids to solids in a cross-ecosystem context.

Topographic Influences on Flow

This section is subdivided into discussions of cross-ecosystem effects related to cross-sectional area, elevation, and roughness. How organisms modify the area, elevation, and roughness of the environment they live in and the potential for these changes to create positive feedbacks by increasing allochthonous input is discussed in each section.

Cross-Sectional Area

Since fluids are incompressible and mass is conserved, the same amount of air or water that goes into a volume must come out. This relation is formalized as the equation of continuity, which states that velocity is inversely proportional to the cross-sectional area of the solid "container" of the fluid, whether it is a pipe, stream valley, or mountain pass. A constriction of topography will cause the velocity of moving air or water to increase, so that it will move through a narrow valley, river channel, or tidal estuary with a small cross-sectional area more rapidly than through a broad, less constricted feature with a large cross-sectional area (fig. 22.1A). Power et al. (1995) found that the cross-sectional area of a river channel strongly influences the dynamics of riparian food webs by regulating the velocity and discharge volume of the river. The implication of narrow topographic channels for the flux of materials across ecosystem boundaries is that the flux of passively transported particles (i.e., propagules, suspended food, detritus,

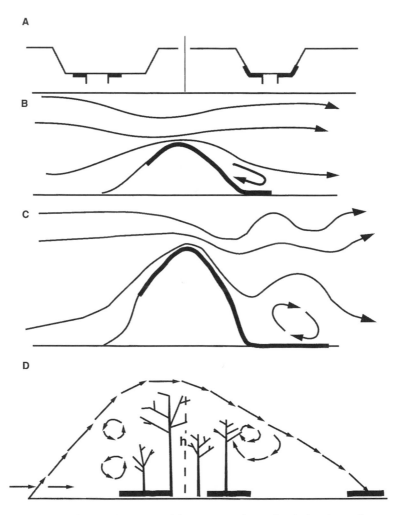

Figure 22.1 The influence of changes in (A) cross-sectional area, (B, C) elevation, and (D) roughness on the magnitude of allochthonous input. The area receiving allochthonous material is shaded in black. (A) A cross-sectional view of a valley or a stream or river with the channel in the center. Narrow valleys or channels have higher rates of input to the sides of the feature due to the increased velocity of air or water. (B, C) Airflow over an idealized hill or mountain, with the streamlines depicted as arrows. Velocity increases as the streamlines are compressed over the top of the elevation, resulting in a high supply of particulate material to the shaded areas. The boundary layer separates on the downstream side of the hill, possibly resulting in the deposition of allochthonous material in the recirculating flows. (C) Fully developed lee waves (a type of internal wave) behind the mountain. Here, rotary airflow may occur on the downstream side under the intense lee waves. Insects have been deposited in these flow environments. Allochthonous input would extend over a larger area in C than in B due to the vertical movement of the lee waves. (D) A dense windbreak of trees with a height h. The windbreak retards flow immediately in front, generating turbulent eddies. Eddies also develop on the downstream side of the windbreak. Generally, the area of reduced air velocity extends to a distance equal to 10–15 times the height of the vegetation in a dense windbreak. Allochthonous material transported by the wind would be deposited on the upstream and downstream sides of the windbreak. (B, C adapted from Stull 1988; D adapted from Atkinson 1981.)

periphyton, plankton, insects) will be highest through areas of constricted topography. This effect should apply not only to waterborne materials, but also to flying insects with weak locomotor abilities, since they effectively become passive particles when wind speeds exceed their flight speed (Drake and Farrow 1988).

As a consequence of flow constrictions, consumers foraging on the edge of narrow valleys, rivers, estuaries, or reef passes will have more food passively transported to them than those foraging in areas of wider topography with more slowly moving fluids. For example, more insects would be passively transported through narrow river valleys in this fashion than through broad ones. As one of the few groups of terrestrial suspension feeders, spiders (Ghiselin 1974) stand to benefit from increased delivery of insect prey. The large quantities of drift insects that float down streams and rivers (Brittain and Eikeland 1988) would similarly be transported at higher rates through constricted channels. A potential general consequence of this phenomenon is an enhanced food supply to be intercepted by consumers, including benthic suspension-feeding insects, spiders, fish, birds, and mammals, feeding on the sides and bottom of narrow channels or valleys (Hart et al. 1991; Merritt and Cummins 1996). The caveat that must be considered here for suspension-feeding organisms is a that higher velocity of food delivery is not always advantageous, since the feeding appendages or feeding structures (i.e., webs, nets) used by organisms to intercept food may deform at high velocities, resulting in decreased feeding success (Vogel 1981) at high flows. For other consumers, the reactions needed to capture food may be inadequate in high-flow regimes.

In estuaries, the effects of increased tidal current velocity at constrictions or "necks" can result in increased growth, recruitment, and abundance of intertidal suspension-feeding barnacles and mussels (Sanford et al. 1994; Leonard et al. 1998). Although the implications of high intertidal biomass at constricted areas of estuaries have not been explored in the context of cross-ecosystem subsidies, it is likely that land-based predators foraging in the intertidal zone, such as gulls and raccoons, would have higher energy return by foraging on the high biomass of mussels, barnacles, and their marine consumers (crabs, whelks, etc.) in such areas. For example, herring gulls were both more abundant and more actively foraging on mussels, crabs, and whelks at a tidal neck with high flow velocity than on an adjacent intertidal shore in Cobscook Bay, Maine (J. Witman and C. Marsh, unpublished data).

Organisms can modify the cross-sectional area of the flow environments that they live in by growing into and partially obstructing waterways. For instance, woody vegetation on islands in streams can create "root-defended" banks that create narrow channels with high velocities (McKenney et al. 1995). Vegetation on riverbanks and submerged aquatic vegetation can also play a large role in decreasing the cross-sectional area of streams and rivers (Kouwen and Li 1980; Watts and Watts 1990). Although it is likely that these specific habitat-modifying activities create a positive feedback by increasing the supply of allochthonous nutrients and other organic matter to the organisms modifying the cross-sectional area of waterways, we are not aware of any published studies demonstrating such feedbacks.

Changes in Elevation—Up

The slope of an obstruction to flowing air or water can predict the magnitude of allochthonous transport because the velocity of a fluid increases as it passes over the top of the obstruction (fig. 22.1B–C). The physics underlying the increase in flow speed over an obstruction depend on pressure-velocity relationships on the upstream, top, and downstream areas of the obstruction, and are summarized by the Bernoulli principle and the principle of continuity (Vogel 1981). Basically, the volume of fluid above the obstruction decreases as it becomes more elevated, and to satisfy the conservation of mass, the velocity of the fluid must increase as it passes over the elevation. Such elevated "ramps" are potential sites of high cross-ecosystem exchange because high particle or convective flux occurs in the high-velocity regime at the top of the ramp. On land, for instance, steeply sloping cliffs are ramps with enhanced wind speeds. Orb-weaving spiders were especially dense on cliff tops at seabird rookeries in the Galápagos Islands, where they tended to intercept large numbers of insects associated with the seabirds, which were blown into the webs (J. Witman, J. Ellis, and P. Wallem, personal observations). Crevices and depressions in the substrate of high intertidal rocky shores usually receive and accumulate organic matter derived from marine foam and wave splash, which attracts intertidal flies (Fariña 2000). These flies are the main prey of terrestrial lizards (*Microlophus atacamensis*) along the coast of the Atacama Desert, Chile (Fariña 2000). Increased airflow with increased elevation probably enhances allochthonous input of nutrients and prey on small landscape features such as cliffs and hillsides, but high winds on mountaintops are probably more of a stress.

In the ocean, rocky pinnacles, seamounts, and sills at the mouths of fjords are good examples of ramps where water flow can be higher at the top than at the lower sides (Farmer and Freeland 1983; Genin et al. 1986; Witman and Dayton 2001), resulting in high benthic-pelagic coupling. As an example, a 2.6-fold increase in average flow speed occurred at a pinnacle shoaling from 50 to 30 m depth over 100 m horizontal distance (Witman and Dayton 2001). On the smaller spatial scales of subtidal oyster reefs, a 2 m difference in height resulted in a 64% increase in flow velocity (Lenihan 1999). Oysters on tall reefs had higher growth rates than those on low reefs, which probably resulted from a higher flux of allochthonous food particles to these active suspension feeders at the tops of the reefs (Lenihan 1999). This finding suggests that oysters are ecosystem engineers that can indirectly enhance the supply of allochthonous food by increasing the height of their aggregations. A similar effect may obtain for mussels, which also create elevated banks that accrete vertically due to the biodeposition of mussel pseudofeces, the deposition of inorganic sediment, and the growth and settlement of mussels (Seed 1976).

In addition to increased flux associated with enhanced flow speed, changes in topography create another type of strong cross-ecosystem flux in the atmosphere and in the ocean: internal waves (Halpern 1971; Gossard and Hooke 1975; Christie et al. 1981; Witman et al., chap. 9 in this volume). These mesoscale waveforms are created when fluid is deflected by abrupt topography, causing vortex circulation (eddies) or internal waves on the downstream side (lee waves; fig. 22.1B–C). Lee waves in the ocean may be trapped or propagate forward when the tide changes. Due to the large amplitude of internal waves, they can be an important mechanism of vertical (from the water column to the ocean bottom; from the atmosphere to the land) and horizontal transport of particulate food and small invertebrates (Shanks 1983; Pineda 1991). Atmospheric lee waves are commonly trapped on the downstream side of hills and mountains (Forchgott 1949; Gossard and Hooke 1975; Atkinson 1981; fig. 22.1C). They have been linked to the deposition of massive numbers of potato beetles (Forchgott 1950) and to concentrations of moths in the downstream rotor circulation that may develop under lee waves (Pedgley et al. 1982). In Australia, radar revealed striking bands of migrating insects that were concentrated in trains of solitary internal waves (Drake 1985). Like sea breezes (see below), trapped lee waves over land have the potential to explain spatial patterns of herbivory on the ground if they retain herbivorous insects and transport them to the ground. Although little is known about this topic, it stands to reason that the ecological effect of internal wave circulation (e.g., herbivory, prey increase to consumers) will depend on the amplitude and duration of the internal wave feature and the length of contact with the ground. Clearly, slow-moving lee wave systems may have more of an ecological effect than transient features. The largest transfers of insects retained in internal waves from air to land may occur when the wave features relax and dissipate. In the sea, this flow/topography interaction can cause either downwelling or upwelling of nutrient-rich waters (Witman et al., chap. 9 in this volume; Leichter et al. 1998).

Changes in Elevation—Down

The velocity of water flowing down a mountain or hillside is directly proportional to the slope of the stream channel and inversely proportional to its cross-sectional area (Gore 1996). Other parameters, such as the area and geometry of the watershed, influence discharge from a drainage basin (Gregory and Walling 1973), but it is obvious that steep slopes produce a higher-velocity discharge than gentler slopes, with the potential for greater rates of allochthonous transport downstream. Like many of these simplified physical predictors, the consequences of slope of the landform for cross-ecosystem exchange remain to be explored. One effect would be similar to that predicted for constrictions in cross-sectional area (see above) in that consumers feeding on the sides of rivers or streams with steep inclines could expect a higher rate of allochthonous food delivery than those feeding along gentler slopes.

In stream and riparian landscapes, aspects of watershed geometry also influence the concentration component of flux—that is, the deposition of allochthonous material (see below). In swiftly moving streams or rivers, where the rate of allochthonous material transport is high, meanders, alluvial fans, and deltas would be especially rich in deposits of allochthonous material because it would accumulate rapidly there. The yield of sediment transported by water is inversely proportional to the area of the watershed because small watersheds tend to have steeper slopes than large ones, promoting greater erosion (Gregory and Walling 1973).

Roughness

Turbulent motion is largely determined by the roughness of the underlying solid surface, so there are probably feedbacks between roughness elements created by plants and sessile animals and the magnitude of mass transfer from fluids to solids (Rosenberg et al. 1983; Metais and Lesieur 1991). In air, turbulence is also influenced by the vertical gradient of temperature, but the frictional effects of surface roughness tend to be more important close to the ground up to a height of 50–100 m above the surface (Rosenberg et al. 1983). Turbulence invariably plays an important role in the delivery of allochthonous material to ecotones where large changes in the vertical structure and spatial complexity of plants and sessile animals occur.

The familiar logarithmic law characterizes the velocity profile near the substrate, where velocity increases as the log of elevation above the substrate (Robinson 1962). This log sublayer is part of the boundary layer, where flow is affected by friction with the underlying surface (boundary); at some distance above the boundary, fluid motions are no longer affected by the boundary and are governed by inertia. Laminar boundary layers are disrupted by objects protruding into them, which cause a reduction in flow speed and the generation of turbulent flow as eddies form and separate from the downstream side. Objects protruding into the boundary layer may be nonliving (topography) or living, such as plants and attached invertebrates.

A fluid dynamic predictor of the degree of turbulent fluid motion is u^*, or shear velocity, which is empirically determined as the slope of the regression of the log of elevation above the ground (z) on mean velocity, u, at a given elevation from a velocity profile (Nowell and Jumars 1984). As wind or water velocity increases, the shear velocity increases, and turbulent motion extends closer to the ground. Turbulent transport to the ground or sea floor occurs during random "sweeps"—downward motions that penetrate the boundary layer. This aspect of turbulent motion has been studied only in terms of sediment transport (Williams 1989). Future investigations are likely to shed light on how allochthonous materials are transferred from fluids to the substrate.

Whether fluid motion is laminar or turbulent is expressed by the magnitude of the Reynolds number (Re), where $Re = (\rho U L / \mu)$, which is equivalent to the product of the density of the fluid (ρ), the velocity of the fluid (U), and the characteristic length of the object in the flow (L) divided by the dynamic viscosity (μ) of the fluid (Vogel 1981). Flow switches from laminar to turbulent conditions when Re exceeds 2,000. The biological implication of the Reynolds number is that, if all other terms are unchanged, the degree of turbulence scales with the size of organisms or the structures they create (L). Thus, organisms may generate turbulence, and possibly enhance the deposition of allochthonous food and detritus, as they grow upward into the boundary layer. Similarly, flow around large biogenic structures is often turbulent.

On nearly all spatial scales, from large forested landscapes and coral reefs to small aggregations of worms, organisms modify the flow regime they live in by changing the roughness of the boundary layer (Eckman 1983; Nowell and Jumars 1984; De Bruin and Moore 1985; Thomas and Atkinson 1997). Some of the most dramatic flow effects of biogenic roughness occur on flat plains, where the roughness effects of tree stands (windbreaks or shelterbelts) greatly reduce wind velocity and create turbulence (fig. 22.1D). The effectiveness of windbreaks depends on their height, porosity, and length (Seginer 1975). Rosenberg (1974) observed that moderately penetrable windbreaks provide more effective shelter than dense ones by splitting the air into two streams—one bending over the tree canopy and the other passing between the trunks of the trees. When these two streams of air meet, turbulence is formed, slowing down airflow considerably on the leeward side. On plains in Poland, Ryszkowski (1992) found that the sharpest reduction in velocity on the leeward side of a shelterbelt corresponded to a horizontal distance 4 to 8 times the height of the shelterbelt (fig. 22.1D).

The biological ramifications of windbreaks are probably increased seed recruitment due to the trapping and retention of windborne seeds, increased soil moisture due to shading, and a greater retention of nutrients due to the trapping of detritus and reduction of soil erosion (Rosenberg 1974; Ryszkowski 1992). As Rosenberg et al. (1983) write, "Large contrasts of plant structure result in unique microclimates in ecotones." Similar reduction of wind speed occurs in the vegetated canopy of forests. De Bruin and Moore (1985) recorded a 10% reduction of wind speed in a pine forest 18 m tall. Another natural area where tall vegetation forms windbreaks is along the edges of rivers in arid regions (Malanson 1993). Riparian vegetation in these ecosystems stands higher than the surrounding vegetation due to increased groundwater and nutrients supplied by the river (Malanson 1993). Windbreaks may enhance seed deposition by biotic mechanisms in addition to passive deposition, as Harvey (2000) found that bird-dispersed seeds were more abundant in windbreaks than in pasture habitats in Costa Rica.

In aquatic and marine habitats, roughness can increase the turbulent mass transfer of food and nutrients from the water column to the aggregations of filter-feeding invertebrates, including soft corals (Patterson 1980), polychaetes (Eckman 1983), and mussels (Frechette et al. 1989; Wildish and Kristmanson 1997; Green et al. 1998), that create it. This transfer occurs when eddies formed by the protruding organisms increase the vertical mixing or residence time of fluids around them, increasing the supply of food suspended in the water column to these filter feeders (Patterson 1980; Butman et al. 1994; Bertness et al. 1998). For the filter feeders, there may be a trade-off between the benefits of living in an aggregation that passively increases allochthonous food supply from the water column and the potential disadvantages resulting from the depletion of food particles caused by the dense assemblage of filter feeders (Buss and

Jackson 1981). Aggregations of marine plants also increase the roughness of the bottom boundary layer, slowing water flow and increasing the uptake of nitrogen in kelp beds (Wickens and Field 1986) and of nutrients (ammonium) in sea grass beds (Thomas et al. 2000).

Depositional Environments

Predicting the location of cross-ecosystem effects requires an understanding of where materials carried by wind or water are likely to accumulate. What determines where small and large food particles are going to fall out of suspension from a moving fluid? Although many factors influence the deposition of food parcels in the form of detached algae and sea grasses, coarse woody debris, leaf litter, fine detritus, and individual or aggregated organic particles, two of the most influential are gravity and velocity. For example, the settling velocity of small passive particles depends on gravity, the velocity and the viscosity of the fluid, particle size, and buoyancy. These terms are components of Stokes' law, which determines the settling velocity of small particles in the range of 0.006 to 2.0 mm in diameter (phytoplankton to large sedimentary particles)—essentially, the conditions for small particle deposition (Jumars 1993). Generally, the sedimentation rate of passive particles is inversely related to the velocity of the moving fluid. This means that any habitats or geomorphological features that cause a sudden reduction in flow velocity will be a likely place for the deposition of food passively transported from another ecosystem.

Depositional environments caused by flow reduction are common in rivers, streams, and dunes and along beaches in the coastal zone. One of the places where flow reduction predictably occurs is the first widening of a river after a narrow constriction (Peterson 1986). In rivers with sedimentary shores, sediment (and presumably allochthonous nutrients) from upstream regions is deposited in these wide, low-velocity areas of the river (Peterson 1986). Abrupt elevations in topography can also cause the deposition of passively transported particles on the upstream side if the particle density exceeds that of the fluid medium, as the particles will not follow the streamline diverted over the top of the feature (Denny 1993). The buoyancy of some plant and animal material causes it to drift at the surface, where it is susceptible to wind-driven circulation that transports it past localized areas of low flow where it might have settled out.

Gravitational deposition can be predicted by a decrease in the angle of a slope, whether it is a mountainside or the slope of an undersea canyon (Gregory and Walling 1973). Alluvial fans form in these environments

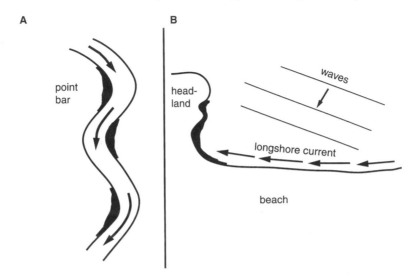

Figure 22.2 Depositional environments. (A) Deposition of material transported by flowing water occurs on the "inside" edges of river meanders, leading to the development of point bars. (B) The interface between beaches and headlands accumulates allochthonous material transported by longshore currents. These currents are generated when waves approach shorelines at an oblique angle. The area receiving allochthonous material is shaded in black.

and are sometimes enriched by nutrients or food parcels flowing downslope. For example, the transport of macroalgae and sea grass down submarine canyons is a major conduit of allochthonous transport to the deep sea (Josselyn et al. 1983; Suchanek et al. 1985; Harrold et al. 1998). It occurs in tropical and temperate ecosystems and may sustain deep populations of herbivores and omnivores (Stockton and DeLaca 1982; Suchanek et al. 1985; Vetter 1994). In some deep submarine canyons, downslope transport of macroalgae may account for 20–83% of the total particulate organic carbon deposited in the sediments (Harrold et al. 1998).

In riparian environments, which generally are locations of vigorous downstream nutrient transport (Riley et al., chap. 16 in this volume), deposition of allochthonous material occurs on the inside of meanders and on point bars, floodplains, and estuarine deltas (Malanson 1993; fig. 22.2A). Sediment deposition at river meanders leads to the development of point bars, which are readily colonized by riparian vegetation (Malanson 1993). Although the idea is untested, plant community dynamics in these depositional habitats are probably influenced by nutrient subsidizes from the river. Downstream deposition is one of the mechanisms behind the river continuum concept, in which plant communities downstream are subsidized by upstream nutrients (Vannote et al. 1980).

In a rare demonstration of a land-sea nutrient subsidy, Raffaelli et al. (1989) showed that the deposition of upstream nutrients on estuarine tidal flats enhanced the production of green macroalgae. Deposition of large allochthonous materials, such as woody tree debris transported downstream by rivers, can modify downstream habitats in addition to serving as an input into detrital food webs. Detached tree limbs, trunks, and root systems grounded in shallow water may trap sediment and other plant debris, leading to the accretion of riparian shorelines or the development of small islands (Malanson 1993; Naiman and Décamps 1997). Some of the woody tree material washing down rivers is deposited in the deep sea, where it is colonized and consumed by wood-boring bivalves (Turner 1973).

On coastal shores throughout many subtropical to cold temperate regions of the world, marine macroalgae are fragmented or detached from the substrate during storms, leading to massive deposits of plant material at terrestrial-marine ecotones (Branch and Branch 1981; Witman 1987). The resulting algal or sea grass "wrack" represents an important nutrient and food subsidy to consumers on the beach and farther inland (Branch and Griffiths 1988; Polis and Hurd 1996b). The onshore transport of this allochthonous plant material is governed by wave-induced nearshore currents. Strong longshore currents (moving parallel to shore) are generated when waves approach the coast at an oblique angle. Because of longshore transport and the reduction of flow where linear beaches abut headlands or promontories, the largest deposits of marine plant material often occur at the end of a beach abutting a headland (Komar 1998; fig. 22.2B). On the central coast of Chile, Rodriguez (1999) observed that the amount of drift algae (*Lessonia trabeculata* and *Macrocystis interglifolia*) deposited at the end of a rocky beach was twice that observed at open or exposed beaches. Consequently, the intertidal urchins (*Tetrapygus niger*) living on this enriched rocky beach were larger and more abundant than the individuals living on exposed or open rocky shores (Rodriguez and Fariña 2001).

Beavers (*Castor Canadensis*) are perhaps the most important animals modifying the flow and deposition of allochthonous carbon on land (Rudemann and Schoonmaker 1938; Naiman et al. 1986). Through their dam building and feeding activities, beavers alter all the geomorphological aspects of the watershed previously discussed, from the cross-sectional area of stream channels to the roughness of the land-water interface and the deposition of organic matter. Wood cut by beavers substantially increases the amount of allochthonous carbon in mid-sized streams (Naiman et al. 1986). The ponds created by beaver dams enhance deposition of benthic organic matter, which decays at a slower rate than in riffle streams, possibly as a result of reduced water velocity in the ponds (Naiman et al. 1986).

Thermally Driven Systems

Thermal convection is the driving force behind several processes transporting nutrients or organisms across ecosystem boundaries. Convection is simply the conduction of heat in a moving fluid. It occurs in the atmosphere when the temperature of the land or sea is warmer than the air above. Air that is warmed at the surface rises up to an altitude of 1–2 km. These upward convective currents can transport small insects higher than their normal altitude. Insects and birds use thermals, or convectively rising air masses, to soar to high altitudes to migrate or feed. In air, large fluctuations in velocity result from thermal turbulence, while smaller changes are governed by roughness effects (Stull 1988).

 Sea breezes represent an important thermal mechanism of cross-ecosystem transport from land to sea and from sea to land. They form along the coast during the day, since they are generated by temperature differences between the land and sea (Stull 1998). The circulation pattern consists of an onshore wind close to the ground ($\leq 1,000$ m) and airflow in the opposite direction (offshore) at higher altitudes (fig. 22.3A). Thus, depending on the altitude of flying insects, a sea breeze can either transport them far offshore or bring them from the sea to the land. Sea breezes are probably responsible for the large numbers of flying insects deposited in the ocean many kilometers off the coast (Cheng and Birch 1978). To our knowledge, no one has explored the consequences of wind-transported insects for marine food webs. Given the magnitude of input in some areas, it is a promising area of research. Flying insects may also be concentrated at a sea breeze front, where the onshore sea breeze meets the prevailing wind from land. These fronts have apparently retained swarms of desert locusts along the coast for days (Rainey 1963) and may also concentrate passively dispersed gypsy moth larvae in the coastal zone (Mason and McManus 1981). In addition to increasing the flux of insect prey to consumers such as spiders, the retention of herbivorous insects in sea breeze fronts could explain spatial gradients of herbivory perpendicular to the coast.

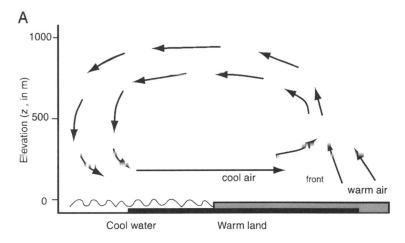

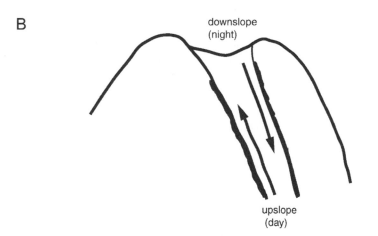

Figure 22.3 Effects of convection. (A) Formation of a sea breeze at the land-sea interface. (After Stull 1988.) (B) Diel airflow patterns in a mountain valley. Warm air moves up the valley during the day, followed by downslope movement of cold air at night. The area receiving allochthonous material is shaded in black.

Upslope and downslope patterns of airflow are produced by diel temperature changes that generate density gradients (fig. 22.3B). An upslope wind is created by warming of the land surface during the day. The reverse process occurs at night, with cool winds flowing downslope. These airflows are predictable features of mountain valleys (Atkinson 1981; Stull 1988). Upslope winds have been linked to large numbers of grasshoppers at mountain elevations higher than their normal altitudinal limits (Drake and Farrow 1998). This phenomenon may be significant for understanding spatially subsidized ecosystems, as consumers such as insectivorous birds at upper elevations may receive substantial inputs of passively transported prey.

Density gradients in aquatic and marine ecosystems are created by solar warming of surface waters, producing the well-known thermoclines and pycnoclines. Thermal stratification alone does not enhance the exchange of nutrients across ecosystem boundaries (i.e., from water column to benthic habitats) and may have the opposite effect of acting as a barrier to nutrient and oxygen exchange in lakes during summer. However, this barrier is eventually broken down by wind. Temperature/density stratification of the water column does contribute to the formation of internal waves, which together with tidal energy may result in vigorous vertical or horizontal transport of nutrients and particulate food to the bottom or coast (Shanks 1995).

Wind and thermally stratified currents contribute to large-scale upwelling in the ocean, which is just beginning to be explored as a bottom-up influence on marine communities that transcends traditional ecosystem boundaries (Bosman et al. 1987; Menge 1992; Bustamante et al. 1995b). Large upwelling systems, such as those located along the western margins of Africa and North and South America, are principally driven by wind blowing offshore. The surface water displaced offshore is replaced by deep, cold water that is usually rich in nutrients. These nutrients stimulate primary production (phytoplankton, algae), either at the site of upwelling or some distance away (Barber and Ryther 1969; Branch and Griffiths 1988). Upwelling is correlated with increased abundances of pelagic fishes (Barber and Smith 1981), intertidal algae, and filter-feeding invertebrates (Bustamante et al. 1995b; Menge et al. 1997a, 1997b, 1999), as well as kelp growth in the subtidal zone (Dayton et al. 1999). The trophic effects of upwelling may create greater linkage between bottom-up and top-down effects on food webs (Menge et al. 1999). Due to the advection of surface waters offshore, the supply of nutrients, particulate food, and larvae to the intertidal zone may actually be lower at sites of major upwelling than at non-upwelling sites (Roughgarden et al. 1988). The effects of upwelling may be more pronounced in subtidal than in intertidal communities, as they are in more direct contact with upwelled water. For example, striking changes in the growth and diversity of suspension-feeding invertebrates occurred at a rocky subtidal site in the Galápagos that regularly experiences upwelling currents with velocities of 3.0–32.0 cm s^{-1} (Witman and Smith 2003). Although upwelling ecosystems are exciting locations of cross-ecosystem transfer, much more work needs to be done before generalizations can be made about their influence on food web dynamics.

Hydrothermal vents are possibly the most vigorous source of allochthonous nutrients on the planet and represent one the most spectacular examples of thermally driven nutrient subsidies in the ocean. These "industrial smokestacks of the deep sea" are known from all major spreading centers on the sea floor (Van Dover 2000). Extremely hot hydrothermal fluid rich in hydrogen sulfide rises out of the vents into the cold nutrient-deficient sea water of the deep sea, causing turbulent mixing and immediate precipitation of sulfides. Free-living bacteria and symbiotic bacteria living in the tissues of tubeworms and bivalves metabolize the sulfides. Their metabolism drives a chemosynthetic food web, consisting of bacteria, mussels, clams, crabs, other filter-feeding invertebrates, and fish, that is entirely dependent on nutrient subsides from the earth to the sea.

PERMEABILITY

The flow of energy, matter, and organisms is in large part determined by the spatial configuration of the habitat elements in a landscape and the permeability of their boundaries (Wiens et al. 1985; Gosz 1991). Permeability can be quantified as a combination of the material flux (defined above as concentration times velocity) and the distance that the material travels into the recipient system. Permeability is determined by the topography of the boundary and by the density and spacing of structural elements in the recipient habitat. Permeability is probably more important in land-land material fluxes and water-land exchanges than in water-water exchanges, although aquatic thermoclines can restrict the exchange of nutrients between deep aphotic and euphotic surface waters (Eppley et al. 1983). We recognize two major categories of permeability: permeability controlled by topography and habitat permeability, which is controlled by biogenic factors. Abrupt discontinuities in the landscape created by features such as steep cliffs and mountains are good examples of topographic elements that may experience increased fluid flow (as above) and transport of fine particulates, but are less permeable to the horizontal transport of large organic debris by physical processes. Habitat permeability affects material flows across ecotones and is regulated by the structural complexity and size of physical and biological elements in the recipient habitat. Although we discuss the two categories separately, we recognize that allochthonous input is controlled by the interaction of topography and the biological complexity of habitat edges (see Cadenasso et al., chap. 10 in this volume).

A good example of topographically controlled permeability comes from the Gulf of California island system (see Anderson and Polis, chap. 6 in this volume). Algal wrack and carrion carried by ocean currents washes ashore only on gently sloping beaches or rocky shores, not on steep cliffs. In this case, the cliff tops are relatively impermeable to waterborne organic material, whereas the beach topography can enhance the concentration of such material (see fig. 22.2B). Marine material arriving on beaches travels farther inland via mobile consumers such as beetles or lizards than material arriving at the base of steep cliffs (W. Anderson, unpublished data). Conversely, cliffs attract seabirds for roosting and nesting. Seabirds are important vectors of marine nutrients and energy because they deposit large quantities of N- and P-rich guano, attract dipteran ectoparasites, and contribute fish carrion scraps and their own carcasses to the cliff habitat. The carbon and nitrogen isotopic signatures of these marine-derived seabird by-products can be found in plants, herbivores, detritivores, and higher consumers living in or moving through high-density seabird areas (Anderson and Polis 1998; Stapp et al. 1999). Thus, cliffs are highly permeable to materials imported by these biotic vectors, but are less permeable to materials transported by water.

Many types of organisms play important roles as mobile vectors for cross-ecosystem exchange, thus increasing habitat permeability. Habitat permeability can vary with the density of vegetation or other biogenic structure in the ecotone. For example, windborne nutrient and seed deposition patterns at meadow-forest edges vary with the density of edge vegetation. Intact edges limit the quantity and distance of seed dispersal and nutrient flux into the forest, but enhance seedling herbivory by meadow edge residents (see Cadenasso et al., chap. 10 in this volume). Emergence of aquatic insects from streams and lakes results in the transfer of a significant portion of aquatic insect biomass to terrestrial habitats, providing prey for insectivores in neighboring terrestrial ecosystems (Jackson and Fisher 1986; see Henschel, chap. 13 in this volume; Nakano and Murakami 2001). Vegetation density and consequent spider web structure can alter riparian zone permeability to these aquatic insects. Sanzone et al. (2000) found that web-building spiders were most dense, and contained ^{15}N tracers from aquatic insects, within the first 10 m of the riparian zone. The chances of an aquatic insect traveling farther inland were minimal within this structurally complex habitat. Another test of habitat permeability was conducted along a lakeshore in Missouri by clearing structurally complex shrubs from swaths 10 m wide by 40 m deep (W. Anderson, unpublished data). Emergent aquatic insects were trapped at 5 m intervals along cleared and control

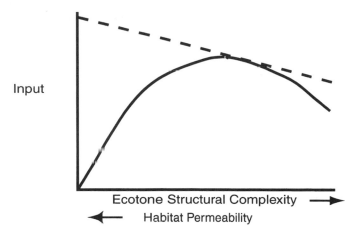

Figure 22.4 Relationship between the structural complexity of a recipient habitat (solid curve), the permeability of a recipient habitat (dashed line), and the magnitude or importance of allochthonous input. Input rises steeply with the structural complexity of the recipient habitat due to reduction of air or water flow and deposition of passively transported allochthonous material resulting from increased roughness, density, and size of the habitat elements. However, the permeability of the habitat decreases as it fills up with allochthonous material or as increased structural complexity results in the onset of skimming flow, where flow is diverted over, rather than through, the habitat. The maximum input should occur at intermediate structural complexity, at the intersection of the permeability line and the complexity curve.

shore-to-inland transects. In control transects, aquatic insects were found only within the first 5 m, but in cleared transects aquatic insects traveled up to 30 m inland (W. Anderson, unpublished data).

GRAPHIC MODELS

Since multiple physical and biotic processes influence the flux of allochthonous material, it is useful to summarize some of their effects in graphic models. Figure 22.4 posits a relationship between the structural complexity of a recipient habitat, its permeability, and the magnitude of allochthonous material crossing the ecotone to enter that habitat. Examples of ecotones that differ greatly in structural complexity include the forest-meadow edge (Cadenasso et al., chap. 10 in this volume), the interface between rivers and riparian vegetation (Malanson 1993), tree lines on mountain slopes, borders between dense subtidal algal or sea grass beds and rocky or sandy substrate habitats, and salt marsh-mudflat ecotones. Here we subscribe to Naiman and Décamps's (1997) view that the ecotone is analogous to a semipermeable membrane regulating the

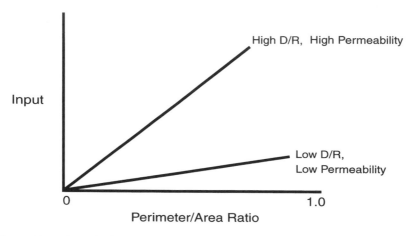

Figure 22.5 Relationship between the perimeter-to-area ratio of a habitat and the magnitude or importance of allochthonous input to that habitat. The relationship varies as a function of the relative productivity of donor (D) and recipient (R) habitats and the permeability of recipient habitats. (Adapted from Polis and Hurd 1996; Polis et al. 1997.)

flux of allochthonous materials across the ecotone boundary. Due to the effects of roughness in reducing flow speeds and generating turbulence (including recirculating eddies), the amount of allochthonous material trapped rises steeply with the structural complexity of the elements constituting the habitat edge. Input trapped by the recipient habitat will level off, however, after some threshold complexity because the permeability of the habitat declines as the concentration of structural elements become so dense that the air or water flow transporting allochthonous materials passes over, rather than through, the habitat. This phenomenon, known as skimming flow (Morris 1955), has been documented from dense wind-breaks (Rosenberg 1974), stands of submerged aquatic vegetation (Madsen and Warnke 1983), and beds of bivalves (Green et al. 1998). The magnitude of allochthonous input should also level off at high structural complexity as the edge "fills up" with organic material. The maximum input should occur at intermediate structural complexity, at the intersection of the permeability line and the complexity curve.

The relationship between perimeter-to-area ratio and permeability shown in figure 22.5 builds directly on the original ideas of Polis and Hurd (1996b) and Polis, Anderson, and Holt (1997) about how the magnitude or importance of allochthonous input varies with perimeter-to-area ratio and the difference in productivity of donor (D) and recipient (R) habitats. Overall, the magnitude of input increases with the perimeter-to-area ratio because the greater it is, the more edge is exposed to the donor ecosystem. However, the rates of increase of allochthonous input vary dramatically with the permeability of the edge (ecotone) and the ratio between the pro- ductivity of donor and recipient habitats. In systems in which the produc- tivity of the donor habitat is low and similar to that of the recipient habitat, there will be little increase in input with increasing perimeter relative to area. Similarly, low-permeability habitats will be little affected by in- creasing allochthonous input with increasing edge if the materials cannot got into the habitat. There will be a large input from other ecosystems (and presumably large ecological effects) in systems in which the ecotone is highly permeable and in which the ratio between the productivity of donor and recipient habitats is large. This situation occurs on beaches in the Baja California island system, which receive large amounts of marine-derived organic material (Polis, Anderson, and Holt 1997). It should be noted that the predictions of perimeter-to-area effects on cross-ecosystem subsidies have been restricted thus far to physical mechanisms of transport across ecosystem boundaries (Polis, Anderson, and Holt 1997). An important pri- ority for future research is to evaluate how the importance of cross- ecosystem exchange by biotic mechanisms (i.e., organism movement) relates to the perimeter-to-area ratios of islands. A recent study of this topic by Fariña et al. (in press) found that permeability, measured as elevation of the shoreline, was a better predictor of nutrient subsidy by sea lions in the Galápagos than the perimeter-to-area ratio of the islands.

Figure 22.6 shows the influence of runoff and frequency of flooding on allochthonous input. Rainfall creates runoff from the land, washing nutrients from the land into aquatic and marine habitats. Organisms may deposit nutritive materials in habitats, but they may remain unavailable to plants and other components of the ecosystem without rain. For ex- ample, terrestrial plant productivity was not stimulated in dry years on arid islands despite large inputs of seabird guano to the shoreline ecosystem (Anderson and Polis, chap. 6 in this volume). We postulate that allochthonous input increases linearly with rain, stream, or river runoff up to the highest levels of runoff, at which it decreases as it washes through the system as "throughput." The influence of flooding on allochthonous input is complex (Power et al. 1995; Wootton et al. 1996), but it generally increases the amount of allochthonous material released and delivered to ecosystems downstream. Therefore, we predict that the amount of input should increase with the frequency of flooding. As a form of physical disturbance, flooding kills large amounts of vegetation (Gregory and Walling 1973; Naiman and Décamps 1997) and distributes

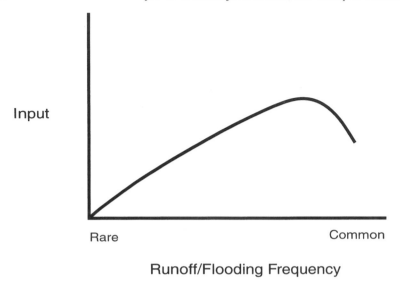

Figure 22.6 The relationship between runoff and or the frequency of flooding and the magnitude of allochthonous input to aquatic and marine ecosystems. Allochthonous input increases with rain, stream or river runoff and flooding frequency. We postulate that allochthonous input would decrease during extremely high runoff as nutrients are leached out of the watershed, as it washes through the system as throughput, and as chronic, high flooding removes all riparian vegetation.

it over the floodplain, where it enhances food resources for grazers (Power et al. 1995). However, low amounts of allochthonous material would be generated during chronic, high flooding because the amount of vegetation available for transport would decline due to insufficient time for recovery between floods. For this reason, we postulate that the relationship between allochthonous input and flooding frequency becomes hyperbolic at high flooding frequencies (fig. 22.6).

Without physical disturbance and grazing (bioerosion), the release of organic materials for transport across ecosystem boundaries would be dictated by natural mortality alone, resulting in rather low levels of cross-ecosystem exchange (fig. 22.7). Clearly, then, the input of allochthonous material to donor habitats should increase with physical disturbance and grazing activities. There are many examples of physical disturbance generating organic matter for grazer, detritivore, and omnivore components of food webs, from landslides and hurricanes affecting forested watersheds to macroalgae dislodged by storms and ice-scoured estuaries (Pickett and White 1985). Grazers are "sloppy" in the sense that they rarely consume all available plant material, except during episodic out-

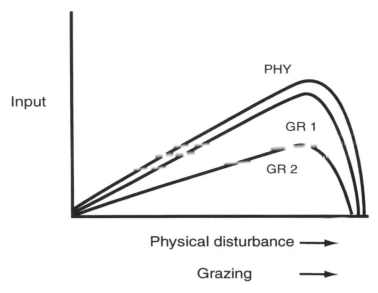

Figure 22.7 Variation in the magnitude or importance of allochthonous input with physical disturbance (PHY) and two levels of grazing (GR1, GR2). Input increases with physical disturbance and grazing, which remove standing stocks of plant biomass, making it available for transport to other habitats or ecosystems. Severe disturbance or grazing, however, removes nearly all plant standing stocks, so input declines steeply at high levels of disturbance and grazing, ultimately dropping to zero.

breaks (Myers 1993), and that they often shred organic material or make the remaining plant more susceptible to fragmentation by physical forces (Crawley 1983; Merrit and Cummins 1996). Similarly, forest canopy herbivores cause green leaves to fall from trees, and this "green-fall" supplies high-quality nitrogen for decomposers on the forest floor (Risley and Crossley 1993). We represent the variable effects of grazers as two curves on the model, GR1 and GR2. The GR1 curve represents the high levels of organic material contributed by canopy herbivores and shredders such as sea urchins in temperate marine habitats. Severe outbreaks of herbivory, such as gypsy moth and locust outbreaks, consume nearly all standing plant biomass (GR2), leaving little to enter other components of food webs. All three curves are hyperbolic, dropping to zero input at extremely high levels of disturbance or grazing because all plant biomass is removed in these conditions. Nonetheless, they suggest the important and rarely considered roles of disturbance and grazing in the supply of allochthonous food.

SUMMARY

With knowledge of physical transport processes and how they are modified by interaction with the topography of the landscape and the complexity of ecotones, we can predict, to a first approximation, the magnitude and location of allochthonous input. Identifying where communities and food webs are likely to be subsidized will foster a broader understanding of the interdependence of adjacent habitats and ecosystems and a better understanding of community ecology at larger spatial and temporal scales. As such, we suggest that significant advances in ecology will be made through further collaboration of ecologists with meteorologists, fluid dynamists, and oceanographers. Although physical processes often dictate the delivery of passive material, input is greatly influenced by organisms and by the permeability of the recipient habitat. A critical feature of permeability is the structural complexity of the ecotone. Permeability also modifies the relationship between island perimeter-to-area ratio and the magnitude or importance of allochthonous input. Finally, we point out that physical disturbance and grazing activities are major contributors of allochthonous material and predict that the magnitude or importance of allochthonous input is a parabolic function of disturbance or grazing.

ACKNOWLEDGMENTS

We are grateful to Gary Polis for the spark to write this chapter, for his boundless passion for community ecology, and for inspiring us to expand our horizons. The perseverance of Gary Huxel, Mary Power, and Bob Holt has enabled the book to go forward after the tragic loss of Gary. Support from the Andrew Mellon Foundation and the National Science Foundation contributed substantially to the development of ideas presented in this chapter. Thanks also to J. M. Fariña and M. R. Patterson for critiquing the manuscript.

Chapter 23

Feast and Famine In Food Webs:
The Effects of Pulsed Productivity

Anna L. W. Sears, Robert D. Holt, and Gary A. Polis

Food webs are a useful way of describing and organizing relationships among species, but for the most part descriptions of food webs have been closed in space and static in time. This volume emphasizes that food webs are part of dynamic landscapes. Wind, water, and animals move materials far from their point of origin, across habitat boundaries. This flow between habitats is an intrinsic part of most food webs. In this chapter we add the dimension of time, focusing in particular on temporal variation in productivity. Environmental conditions are in constant flux. Below the pulse of shifting seasons there is a low rumble of climate change. It is still impossible to predict the weather accurately more than a week in advance. Temporal variation, like spatial variation, is not well integrated into food web dynamics. However, ecologists are beginning to assemble the theoretical tools, the empirical evidence, and the computer power needed to analyze the temporal dimension of food webs.

Temporal variation in productivity is in many ways analogous to spatial variation. In each case, a more productive compartment (in space or time) can provide a subsidy to a less productive one. Rather than being transported from one place to another, resources move through time from one period to another. A temporal subsidy may consist of alternative prey species, available when primary prey are in low abundance, or fat stores and food caches preserving resources from a more productive season. Passive dispersal in spatially heterogeneous environments typically leads to diffusion from higher to lower densities, which in ecological systems usually means from more to less productive habitats. Spatial effects tend to be asymmetric, with flows having more pronounced effects in less productive habitats. In a similar way, seasons, years, or life stages experiencing resource abundance can lead to bottlenecks in other seasons, years, or life stages if, for example, high densities overexploit resource pools or there are strong time-lagged responses by predators or infectious disease agents.

The analogy between flows in space and time is not perfect, however. Time moves in a single direction, and there is no reciprocal flow between temporal compartments. However, if the temporal subsidy is a living resource, such as an emergence of seventeen-year cicadas, predation in a pulse year may limit the magnitude of the next pulse. In contrast, spatial subsidies are often donor-controlled. Donor control implies that recipients do not affect the abundance or renewal rate of the subsidy. Marine wrack washing onto beaches is an example of a donor-controlled spatial subsidy (Polis, Anderson, and Holt 1997). A temporal subsidy is donor-controlled if it involves a pulse of detritus or abiotic inputs such as rainfall.

There is also a fundamental difference in how variation in space and variation in time affect population dynamics. Consider a species with discrete populations: in spatial population models of well-mixed systems, the finite rate of increase (λ) is dependent on the arithmetic mean of conditions across habitats. By contrast, with temporal heterogeneity (in which all individuals in a population experience the same conditions in a given year), λ is dependent on the geometric mean of conditions across years. The effects of bad years tend to dominate geometric means. However, the effects of favorable patches—pulses of growth in space—have much stronger effects on λ than the effects of good years—pulses of growth in time (Chesson 1985, 2000a). In short, the effects of bad years are stronger, and the effects of good years weaker, than the equivalent variation in space. Thus, in space and in time, the quality of the variation determines how it quantitatively affects the magnitude of λ.

In many situations, spatial and temporal subsidies are closely bound together. Transport through space is often seasonal, and at the very least includes a time lag between production and utilization of resources. For example, there is a time lag in the spatial subsidies of riparian systems. Materials that enter small tributaries from leaf litter and runoff are slow to work their way into the food webs of larger rivers. Fish (Winemiller and Jepsen, chap. 8, and Willson et al., chap. 19 in this volume) and birds

(Jefferies, chap. 18 in this volume) transport nutrients between habitats, and their migratory patterns lead to pulses of productivity for their predators. Terrestrial-marine interchanges (Polis, Anderson, and Holt 1997) show tremendous temporal variation. On the Peruvian coast, El Niño leads to massive starvation of seabirds and marine mammals, whose carcasses litter the beaches. Ontogenetic shifts can also lead to time-lagged spatial subsidies, as when metamorphosing mayflies leave the aquatic environment for the terrestrial habitat. Terrestrial predators are thus subsidized in pulses by pond and stream communities (Nakano and Murakami 2001).

The purpose of this chapter is to review and synthesize current knowledge about the ways in which temporal heterogeneity in productivity may influence food web dynamics. There are three primary reasons for including this material in a book on spatial subsidies to food webs. First, as mentioned above, there are strong analogies between spatial and temporal variation in resources, and they can have similar dynamic effects. Second, because of these similarities, the theoretical work on temporal fluctuations is relevant to our development of spatial subsidy theory. Third, because of the pulsed nature of many spatial subsidies, researchers studying allochthonous inputs in food webs need to consider the possible influence of this temporal variation. At this time, there is little synthetic theory incorporating both temporal and spatial subsidies (but see Chesson 1985; Holt 2002; Holt and Barfield 2003; Holt et al., in press; Gonzalez and Holt 2002). Note that our focus on variation in production is only one slice through the broad topic of temporal variation in food web ecology, which includes disturbance regimes, fluctuations in interaction strengths, and colonization-extinction dynamics. Our empirical examples are biased toward terrestrial studies, but the theoretical insights are general to all systems.

As a prelude, we highlight the ubiquity of pulsed productivity in natural systems. We also note the importance of nonlinearities and how they affect population responses to environmental variation. As temporal fluctuations are rarely explicitly incorporated into formal food web theory, we start with what is known about the influence of environmental fluctuations on single-species population dynamics. We then examine predator-prey dynamics, first for specialist predators and then for generalist predators. A substantial amount of theory and empirical work has examined the effect of temporal variation on competition and diversity maintenance, especially in plant communities, and this work has suggestive messages for food web ecology. We end the chapter with a consideration of relevant empirical studies of food webs in fluctuating environments and draw some broad conclusions from our review.

PRELUDE

The Ubiquity of Temporal Variation in Productivity

Resources fluctuate for all organisms along the trophic spectrum. Plant productivity varies through time with rain, temperature, and sunlight. In aquatic systems, nutrient levels may be shifted by seasonal upwelling, turnover, or pulses of terrestrial runoff and dust. Variation in plant productivity translates into variable resource supplies for herbivores. Likewise, resources for detritivores vary with plant productivity, the timing of release, and seasonal variation in the microbial processes that break down plant material. These decomposition rates vary with temperature and moisture levels. Finally, resource availability for higher-level consumers is a function of secondary productivity and the accumulated biomass of herbivore and detritivore populations.

Resource variation may arise from a variety of abiotic forces acting on different time scales (fig. 23.1). On a short-term local scale, daily alterations in light availability (De Madariaga 1995) and water mixing by windstorms (Moline and Prezelin 1996) can cause high variation in photosynthesis and phytoplankton biomass. On longer time scales, but with regular periodicity, strong seasonal fluctuations in productivity reflect changes in light, precipitation, and temperature. Grasslands respond to annual rains (Sims and Singh 1978; Sala et al. 1988; Pandey and Singh 1992), litterfall in tropical forests occurs mainly in the dry season (Tanner 1980), and communities of fruit flies rely on winter frosts to create feasts of necrotic tissue in columnar cacti (Breitmeyer and Markow 1998). The arrhythmic pulse of El Niño Southern Oscillation (ENSO) events causes global-scale changes in precipitation patterns every 2 to 7 years (Enfield 1989). El Niño events bring heavy rains to the eastern Pacific and drought to India, Indonesia, Australia, and southern Africa, strongly affecting the productivity of these regions (Wright et al. 1999). For example, Cane et al. (1994) found that temporal variation in sea surface temperatures in the eastern equatorial Pacific could explain over 60% of variance in maize yield in Zimbabwe. Several longer-period oscillations in Pacific Ocean temperatures appear to modulate the intensity of El Niño events (Kerr 1999). On millennial time scales, warm and productive interglacial periods have alternated with Siberian conditions at temperate latitudes and aridity in tropical regions (Pons et al. 1995; Roy et al. 1996).

Environmental factors determine the basic characteristics of a productivity pulse: intensity, periodicity, the pattern of productivity decay, and

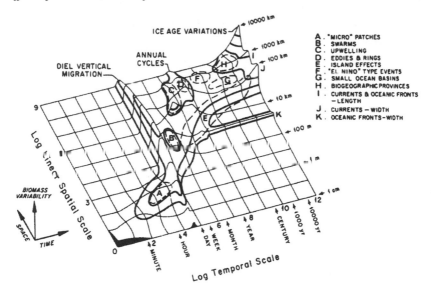

Figure 23.1 Stommel diagram of spatial and temporal scales of zooplankton biomass variability. (From Haury et al. 1978.)

pulse frequency. The intensity, or amplitude, of a pulse can be a function of the combined effects of temperature, rainfall, and organismal traits. Periodicity is also strongly influenced by environmental factors. Seasonal pulses are relatively regular. Other environmental factors, such as precipitation, may be highly erratic, increasing the overall unpredictability of a system. The periodicity of a pulse affects the amount of reserves that are stored. Many bird species experience a trade-off between fat storage and their ability to migrate and avoid predators. Where winters are milder and resources more predictable, birds tend to be leaner (Rogers and Smith 1993; Gosler 1996; Katti and Price 1999). Similarly, some rodents hoard less food when resources are more constant (Livoreil and Baudoin 1996). The shape of the productivity decay curve influences the ability of different ecosystem components to store productivity between good times and bad. Important factors here are the rate of nutrient decomposition, the rate of change in temperature, and water availability. The frequency of a pulse is felt differently by different organisms, depending on their life histories and intrinsic temporal scales. Short-term temporal cycles in resource productivity will be most important for organisms with short life spans. Larger, long-lived organisms may experience these oscillations as environmental noise.

It is crucial to examine the various temporal scales at which key organisms in the community respond to variation (Southwood 1977), particularly dominants, engineers (Jones et al. 1994), and keystone species (Paine 1966). Temporal rhythms in the lives of these species—including daily feeding patterns, behavioral responses to lunar cycles, and annual migrations, as well as generation times—may be particularly important in governing the overall response of the community to temporal variation in productivity. As Wiens et al. (1986) argued, it is important to remember that the dynamics of a long-lived species may be strongly affected by the dynamics of short-lived species with which it interacts (i.e., competitors or prey). In this way, environmental fluctuations may affect an organism both directly and indirectly through other members of the community. This is an essential reason for studying temporal heterogeneity in a food web context.

Note on Nonlinearity: How the World Really Works

Although linear relationships are easier to work with mathematically and statistically, the vast majority of biological and ecological processes involve nonlinearity. For example, a population's response to changes in resource availability is mediated by resource uptake, which is typically a nonlinear function of supply (i.e., a type II or III functional response, with saturation at high resource levels). The logistic model illustrates how density dependence implies a nonlinear relationship between population size and growth rate. Nonlinearities are also pervasive in the physiology of individual organisms. For instance, photosynthetic rate is a saturating function of light levels because reactions are limited by available enzymes.

Jensen's inequality provides a useful rule for predicting the qualitative effect of variation in resources (see Ruel and Ayres 1999). The basic principle here is that if the population size N_t, or some other variable, is a nonlinear function of resource levels, $N_t = f(R_t)$, then the average size of the population experiencing variable resource levels is not equal to the function evaluated at an average value of the resource levels: $\overline{f(R_t)} \neq f(\overline{R_t})$. Saturating or concave downward functions will have a lower value than expected in the absence of variation. Conversely, accelerating or convex functions will have a higher value than expected (fig. 23.2). The higher the variability of resource levels, the greater the deviation in the average population size from that expected in a constant environment.

Saturating responses arise for very basic biological reasons. Any organism, viewed as a machine, has a limited capacity to process material resources. Predators need time to capture, subdue, and digest prey; herbivores need

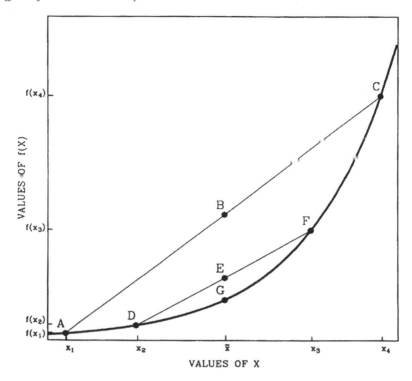

Figure 23.2 Illustration of Jensen's inequality, which is a consequence of nonlinear averaging. The variable X on the horizontal axis fluctuates between values x_1 and x_4 or the values x_2 and x_3. In both cases, the average value of X is $\bar{x}$. The function f, whose graph is given by the thick line, represents some property of a biological system dependent on the fluctuating variable X. Because the graph of f is nonlinear (not a straight line), the average value of $f(X)$ is not given by the point G (at $f(\bar{x})$), but by the point B or the point E, the midpoint of the straight line joining the points A and C or the points D and F. Note that the difference between $f(\bar{x})$ and the average of $f(X)$ increases with the variance of the fluctuations. (From Chesson and Huntly 1993.)

time to digest their cellulose-rich diet. Moreover, most animals are metabolically constrained in their ability to convert resources into fecundity. Even if intake rates are linear functions of resource availability, the net population response may reflect a saturating nonlinearity. Saturating responses in predator attack rates can lead to threshold effects in interspecific interactions, with systems pushed to high or low levels depending upon accidents of initial conditions (e.g., Ruesink 1998). A pulse in production can permit a species to "escape" a low equilibrium; this is a classic theoretical mechanism for insect outbreaks (Belovsky and Joern 1995). For these reasons, temporal variability tends to depress the average rate of resource consumption and realized population growth rates. However, accelerating responses can also arise in systems—for instance, because organisms adaptively shift between alternative activities (Abrams 2001). Nonlinearities also have profound implications for competitive coexistence. Species that would be competitively excluded under constant conditions can persist in a variable environment (see Chesson 2000b). To accelerate our understanding of the role of environmental variation in ecology and evolution, Ruel and Ayres (1999) recommend that ecologists become as familiar with Jensen's inequality as they are with the central limit theorem.

ENVIRONMENTAL FLUCTUATIONS IN SINGLE-SPECIES POPULATION MODELS

In complex communities in which consumers vary in their ability to respond to resource dynamics, it is difficult to separate out the relative importance of those processes that form the patterns we observe. Variation in productivity is one force that is likely to have strong effects on consumer dynamics. Rather than trying to average over exogenous variation in the environment in order to determine the "underlying" dynamics, we need to focus explicitly on the effect of variation on dynamics (Palmer et al. 1997). Ecological modeling is a good first step. We would like models to help us separate this complexity into comprehensible parts and to answer questions about both general processes and particular systems. Ideally, we would like to determine how the effects of temporal variation on populations translate through the food web to affect the dynamics of whole communities.

Population Dynamics as a "Transformation" of Environmental Fluctuations

Most theoretical models in ecology assume that populations have "autonomous" dynamics, in which the equation parameters are constant in time. In reality, the parameters of any model describing population or community dynamics can be expected to vary on different time scales (Holt and Barfield 2003). According to the jargon of systems theory, such systems are "driven" by external variation. Nisbet and Gurney (1982) describe a number of generalized examples of driven dynamics in their textbook on modeling fluctuating populations. If temporal variation is modest in scale, then nonlinear models of population and community dynamics reduce to well-understood linear approximations. For example, consider a population described by the usual logistic equation, but tracking temporal variation in productivity expressed as fluctuations in carrying capacity (K). If the intrinsic rate of increase (r) is very small, the population oscillates with a much lower amplitude than the oscillation of K, and out of phase with it by about one-fourth period. By contrast, if r is very large, the population closely tracks the carrying capacity, oscillating in phase with K.

Practically speaking, this model conforms to our expectations about what "should" happen to the density of a population in nature, given fluctuations over various relative time scales. We expect that organisms with long relative generation times and slow population growth (small r) will average over high-frequency short time scale environmental variations. Large-scale, low-frequency variations (the waxing and waning of ice ages is an extreme case) are much more likely to be tracked by populations (Chesson and Huntly 1993).

Integrating over a cycle in this linearized model, the time-averaged abundance of the population is $\bar{N} = K$. So temporal variation in productivity (as measured by K) does not alter the average population size. In general, however, this is not true, because with larger-scale fluctuations, nonlinearities (e.g., due to density dependence) become important. Moreover, additional dynamic phenomena may occur. In particular, an intrinsically stable but underdamped population (i.e., with an oscillatory approach to equilibrium) may oscillate at frequencies that are integer fractions (1/2, 1/3, etc.) of the frequency of the driving parameter. For instance, the environment may fluctuate annually, but the population may show multiannual cycles. This also implies that populations with intrinsic limit cycles in constant environments can become synchronized with the cycle of the driving parameter in variable environments.

We gain more insight by considering another population model with variable carrying capacity. Roughgarden (1975) analyzed a discrete-generation version of the Nisbet-Gurney model,

$$(N_{t+1} = r + 1 - \frac{r}{K_t}N_t)N_t.$$

He assumed that population fluctuations were relatively small and driven by changes in carrying capacity, with a low r so that, in a constant environment, the population was stable (i.e., $0 < r < 2$). Let $k_t = K_t - K'$ measure the deviation of current carrying capacity from the long-term mean, and let $n_t = N_t - K'$ measure the deviation of current population size from that same mean. The full model can be approximated by a linear model,

$$n_{t+1} = (1 - r)_{n_i} + rk_t.$$

Assume that the deviation in the current carrying capacity from a long-term average fits $k_t = qk_{t-1} + z_t$. This is a first-order autoregressive process, where z_t is a random normal deviate with mean zero and variance V_e (environmental variation in carrying capacity) and q measures environmental predictability. The variance in the population is then

$$V_N = V_e \frac{r}{2 - r} \times \frac{1 + (1 - r)q}{1 - (1 - r)q}.$$

The population dynamics thus transform variation in productivity into variation in abundance.

As before, this model predicts that populations with small r will respond sluggishly to variation in production, and so will have low variance in abundance. Populations with high r will respond more quickly to environmental change, and so will have greater variance in abundance for a given amount of environmental variation. Moreover, the expression for V_N reveals that greater environmental predictability (q) increases population variability. An increased q means that there are runs of good and bad years, reflecting sustained pulses of productivity. The variation in productivity experienced by higher-level consumers will then depend on how prey react to variation in basal productivity. Particular species can either magnify or buffer environmental variation for the remainder of their community. Thus, species with high r are potent transducers of variation into a community.

Ripa and Heino (1999) generalized Roughgarden's results and showed that the ratio between the variance of population size and the variance of the environment also depends on the nature of density dependence for that population. Assuming that environmental states are positively correlated, where, for example, good years tend to follow other good years (Halley 1996; this corresponds to $q \to 0$ in Roughgarden's model), populations that tend to overshoot their carrying capacity ("overcompensating" to density) can actually have a lower variance than populations that undercompensate (and "track" environmental states better). This is interesting, because in models of constant environments, overcompensation leads to unstable dynamics. The same population feature that increases variation in constant environments can thus decrease variation in fluctuating environments.

Interactions between environmental fluctuations and density dependence have other important implications for population stability and species coexistence, depending on the relative time scales at which the nonlinearities act. Long-lived species can be strongly affected by short-time-scale environmental variation if that variation affects their recruitment. The intrinsic rate of increase (*r*) may be very low in long-lived species, but if most of their density dependence occurs during infrequent bouts of recruitment, environmental variation may be very important for species coexistence, regardless of whether the species are "tracking" variation (Chesson 2000b).

In natural systems, there are many ways for species to have nonlinear responses to pulsed productivity (Ruesink 1998). Dixon et al. (1999) argue that nonlinear responses by larval fish to forcing variables in the physical environment can account for frequently observed erratic or episodic recruitment in marine populations. They reason that the sequential action of processes acting on different life stages can have a multiplicative effect. Although the authors focus on physical factors such as transport processes, it is easy to generalize this idea. Consider a species with two distinct larval stages, in which survival at each stage varies linearly with food abundance, which we assume remains fixed within a generation. If the slopes of these linearities differ, survival from the egg to adult stages will then be a nonlinear (quadratic) function of food availability. Nonlinear population responses can thus emerge from a sequence of stages, each of which itself has a linear response. This process is similar conceptually to scale transition dynamics, whereby spatial variation in density dependence (via spatial subsidies, for example) produces nonlinearities in net population-level growth rates, even when local-scale survival and fecundity are linear functions of resource availability (Chesson 1997, 2000a).

Sometimes, erratic population behavior can be the result of long-term transient dynamics (Hastings 1998). Over ecological time scales, populations with strong density-dependent recruitment may never (for all practical purposes) approach their projected "long-term" dynamics, especially in systems subject to frequent disturbance. Neubert and Caswell (1997) demonstrate that even simple ecological models with linear dynamics and stable equilibria can display surprisingly large deviations, moving temporarily away from equilibrium following a perturbation. These deviations are magnified by nonlinearities. Transient dynamics can make it difficult to predict how a population will "translate" environmental fluctuations into population fluctuations.

Effects of Temporal Variation on Average Abundance

How do large-amplitude fluctuations in productivity affect average population size? The outcome may depend upon which demographic variable is influenced by variation in productivity. Consider the logistic model,

$$\frac{dN}{dt} = N(r_t - \delta_t N_t).$$

Here r_t is the intrinsic growth rate at time t, and δ_t measures the strength of density dependence at t. In a constant environment, where $r_t = r'$ and $\delta_t = \delta'$, the equilibrial abundance, or carrying capacity, is $r'/\delta = K$. If temporal variation enters only in r_t, with a mean r equal to r', and δ is constant, then using time averaging (Levins 1979), it can be shown that

$$\overline{N} = \frac{\overline{r}}{\delta} = K.$$

In other words, temporal variation in the intrinsic growth rate with constant density dependence does not affect average population size (although if variation is too severe, the deterministic assumption of time averaging breaks down; Renshaw 1991). However, if r is constant, but the strength of density dependence (δ) fluctuates, it can be shown (using Jensen's inequality) that

$$\overline{N} > \frac{r}{\overline{\delta}} = K.$$

Therefore, if productivity mainly affects the strength of direct density dependence (e.g., because aggression or cannibalism is more prevalent when food is limited), then temporal variation in production should increase average abundance (but see Chesson 1991 regarding the assumptions producing these varied results). The underlying mechanism is that when density dependence is weak (e.g., because cannibalism is rare when food is abundant), a population can grow explosively to high numbers.

Lima et al. (1999) studied sporadic density dependence in populations of the leaf-eared mouse (*Phyllotis darwini*) of coastal Chile. This rodent species has dramatic outbreaks that are correlated with high rainfall during El Niño years, which increases primary production. Using an age-structured population model, Lima and colleagues argued convincingly that the fluctuations in population density reflected both this extrinsic factor (reproduction is correlated with rainfall) and strong, delayed density dependence in adult survival. This density dependence was most likely due to delayed responses by specialist predators, but also may have resulted from the rodents overexploiting vegetation. In any case, this model showed that an increase in population size during good years can lead to severe "crunches" in subsequent years. This finding also suggests that density dependence can vary greatly through time, associated with variation in production.

This scenario seems to apply to ungulates as well (Sæther 1997; Gaillard et al. 1998). For example, in the greater kudu, variation in calf survival through the nonbreeding season is correlated with annual rainfall, which determines food availability. The relationship between calf survival and rainfall is strongly magnified in years with high population density, presumably reflecting more intense exploitative competition for food. Populations are particularly sensitive to years with unfavorable abiotic conditions following years with pulses of production because this is when density will be high relative to food supply.

MULTISPECIES MODULATION OF TEMPORAL VARIATION IN PRODUCTION

Single-species models demonstrate that environmental variation and intraspecific density dependence interact to influence the dynamics of populations and the average population size. However, interspecific interactions can also modify species' responses to temporal variation in production. If one species in a food web is directly affected (numerically or behaviorally) by temporal variation in production, such variation will be transmitted to all other species with which that species interacts.

Specialist Predator-Prey Models

Nisbet and Gurney (1982) describe a Lotka-Volterra model with a time-dependent prey intrinsic growth rate,

$$r = r_0(1 + \phi_t).$$

They analyzed both the case in which ϕ_t is strictly "noise" and the case in which ϕ_t is periodic. If the Lotka-Volterra system approaches equilibrium monotonically, aperiodic fluctuations will produce noncyclic population fluctuations, while quasi-periodic oscillations in r will produce phase-remembering quasi-cycles. (The term "quasi-periodic" refers to complex patterns of fluctuations that are the sum of simpler periodic fluctuations or cycles.) If the Lotka-Volterra system has an oscillatory approach to equilibrium, aperiodic fluctuations produce quasi-cycles that eventually dampen out, and quasi-periodic fluctuations give complex quasi-cycles. Environmental noise alters the deterministic signal in either case, but intrinsic fluctuations have complex interactions with either random noise or periodic variation. The effects are particularly strong if the prey population has a large carrying capacity or the predator has a high attack rate and a low death rate.

Quasi-cycles may lead to the extinction of either predators or prey if their amplitudes become too great. With this same model, using time averaging (Levins 1979), it can be shown that variance in prey numbers is always less than variance in prey intrinsic growth rate. The predator responses absorb fluctuations in prey numbers. This finding illustrates that the transformation of environmental fluctuations into population fluctuations is influenced by feedbacks through other species in communities.

The above theoretical results focus on relatively rapid environmental change. If the change in production is sufficiently slow, the community may be viewed as being at a "moving equilibrium" (Holt and Barfield 2003). Simple food chain models (e.g., Oksanen et al. 1981; Oksanen and Oksanen 2000) predict that an increase in basal productivity should increase the abundance of the top trophic level, although the effect may be small if intermediate levels have saturating functional responses. However, this qualitative prediction may not hold if there are multiple species present at different levels (Abrams 1993) or if the system is unstable (but persistent) due to strong nonlinearities (Abrams and Roth 1994; Abrams et al. 1998). Increased production may indirectly depress the abundance of top consumers in communities with unstable dynamics (e.g., cycles), even to the point of extinction (Abrams and Roth 1994). The reason for this effect is that increasing production can increase the temporal variance of abundance in the system. If the top predator has a strongly saturating functional response, then (by way of Jensen's inequality), fluctuations will lower its overall feeding rate, and thus lower its average abundance. Short-term responses to a pulse of productivity can differ greatly from long-term responses (Ives 1995) because the latter include all the indirect effects acting within and among species.

In another theoretical study, Monger et al. (1997) explored how periodic fluctuations in carrying capacity influenced the dynamics of a three-link food chain with time-lagged recruitment in the intermediate trophic level. As in Nisbet and Gurney's results for intrinsically stable systems, high-frequency fluctuations had little effect on system dynamics (see also Holt and Barfield 2003). With low-frequency fluctuations and intrinsically stable population dynamics, abundances tracked the moving equilibrium. Intermediate frequencies could generate large-amplitude, erratic population fluctuations, or subharmonic oscillations. In systems that are marginally stable, top predators, which tolerate rapid fluctuations in production, could suffer extinctions in the presence of lower-frequency variation.

Another approach to analyzing environmental fluctuations is to perform spectral analysis on time series data. Using an analogy to the color of light, a power spectrum describes how total variability in a time series is partitioned among different wavelengths. White dynamics imply stochastic, unpredictable variation; red dynamics are slow, long-term fluctuations; and blue dynamics are high-frequency, short-term variations (Lawton 1988). Most population dynamics are white or red (Pimm and Redfearn 1988). Ripa et al. (1998) recently carried out a general analysis of the spectral character of noise in two-species food webs. They assumed that the species are interacting, and that in the absence of fluctuations, the community would exhibit stable dynamics. These assumptions permitted them to linearize the system. They also assumed that only one species directly experiences the environmental variation; for instance, it may be a prey species whose birth rate is correlated with stochastically varying rainfall. Their question was how, in general, the spectrum of fluctuations in the prey compares with that in the predator.

The results are somewhat different for species with continuous reproduction than for those with discrete, synchronized generations. In the former case, the species that experiences environmental variation will have a power spectrum that is dominated by higher frequencies (i.e., is "bluer") than that of the second species. Interestingly, this result does not depend upon the nature of the interaction (e.g., predation vs. competition), but is a general claim. In other words, the second species will show fluctuations over longer time scales than the directly affected species. The analytical results of Ripa et al. depend upon linearized analysis, but numerical studies suggest that this result is rather robust, even far from equilibrium. With discrete generations and undercompensation, the second species always has a redder spectrum than the directly affected species. In effect, time lags reduce the ability of populations to respond to rapid density changes in the species they interact with, buffering the effects of rapid environmental fluctuations.

A reddened spectrum has numerous implications for population and community dynamics. If birth rates closely track resource abundance and resources have a red spectrum, there will be "runs" of resource-poor years. In a closed population, this increases the risk of extinction, particularly for species with simple life cycles (e.g., annual insects). By contrast, in open populations maintained by immigration, a reddened spectrum tends to increase local population size (Holt et al., in press; Gonzalez and Holt 2002). Consider a population maintained by immigration in a habitat where, on average, it has a negative growth rate (i.e., a "sink" population). If there is a reddened spectrum in the local habitat, the local population may experience runs of good years with positive growth rates, even if, on average, it has a negative growth rate. Because it has persisted by virtue of immigration during poor years, it remains present and poised to exploit runs of good years and explode to high numbers. Averaging over the entire time series, one finds that the average abundance has been greatly increased by reddened variation, compared with a constant environment with the same average rate of decline in the local habitat.

This unexpected "inflationary" effect of temporal variation in open populations has many consequences. In general, reddened temporal variation magnifies the effect of system openness, making local processes even more sensitive to spatial fluxes of all sorts (Polis, Anderson, and Holt 1997; Holt and Barfield 2003). The results of Ripa et al. (1998) suggest that as variation percolates through a food web, its spectrum may redden. If so, the inflationary effect of temporal variation in spatially open communities may be important even in parts of the food web well removed from the direct effect of variation in production.

Generalist Predators and Temporal Subsidies

The most natural way to consider temporal subsidies in food webs is in the context of generalist consumers that use multiple resources. Generalist predators can respond both numerically and behaviorally to a resource pulse. Examples of this scenario were reviewed by Ostfeld and Keesing (2000). Such consumers exploit pulses of resources, such as masting fruit or insect outbreaks, and as the pulse declines, shift to another prey species or resource. In the short term, during the course of the pulse event, some species may experience a release from predation as predators become satiated by the pulsed resource. However, if predation limits total prey abundance overall, alternative prey should experience apparent competition (Holt 1977, 1984, and chapter 7 in this volume). These effects can destabilize predator-prey dynamics and may lead to the exclusion of one of the prey species.

Apparent competition via temporal subsidies is probably a frequent scenario in nature, with predators switching between food sources as they increase in response to abiotic environmental cues. At the onset of good periods, temporally subsidized consumers may have stronger top-down effects on their resources than would be expected in the absence of a subsidy, and can suppress resource eruptions. In apparent competition models, shared predation favors habitat partitioning among prey, similar to that resulting from resource competition (Holt 1984). In contrast, theory and empirical evidence suggest that shared predation can select for temporal clustering of prey species. This effect is particularly likely if predators have weak, short-term responses to upsurges in prey numbers and can become satiated, or if prey experience strong direct density dependence (Brown and Venable 1991; Holt 1997a). This effect has been suggested as the force responsible for synchronous, episodic masting of seed plants (see Janzen 1976; Ostfeld and Keesing 2000) and the concordant emergence of multiple species of the periodic cicada, *Magicicada* sp. (Williams and Simon 1995). Temporal clustering will decrease the predation risk to any particular prey population in the cluster, but can increase the intensity of predation on prey outside the cluster. Temporal "refugia" may also be important for reducing the impact of generalist predators. Seed banks, dormancy mechanisms, and long-lived, inedible adult or juvenile stages allow prey species to escape during pulses of intense predation (although these stages often have their own suite of predators).

Persistence in the community is facilitated by any factor that reduces the overall rate of population decline during "lean" periods. Often, predators have dormancy mechanisms, fat stores, or food caches that permit them to persist when prey are few and far between. In many cases, generalist predators would not be able to survive without some form of temporal subsidy to carry them through periods of low resources. Low levels of temporal subsidy may stabilize the community by permitting predator persistence. Predator populations will not die out during times of low resources, and thus prey populations will be less likely to overexploit their own resources.

In the case of spatial subsidies, Huxel and McCann (1998) found that simple food chains are stabilized by a low level of donor-controlled subsidy trickling in from another habitat. Low levels of allochthonous inputs stabilized food webs when species fed on autochthonous sources preferentially. Increasing the input level sufficiently or changing feeding preferences decoupled the food chain by permitting overexploitation of local prey, resulting in species losses (see also Holt 1984 and chap. 7 in this volume). Translating this finding to its temporal analog, low levels of alternative prey (particularly carrion or other donor-controlled food sources) may stabilize food webs. Predator populations that could otherwise fall to dangerously low levels when prey are dormant can subsist during lean periods on alternative prey and respond more rapidly to prey pulses during good periods (see also Ives and Settle 1997). McCann and Hastings (1997) examined models in which omnivorous predators could eat both intermediate consumers and basal resources. This type of omnivory stabilized the community by eliminating nonequilibrial dynamics (such as chaos), or bounded dynamics further from zero. Multi-trophic level omnivory may be an important way for higher-level consumers to use temporal subsidies.

Generalist consumers link spatial and temporal subsidies. If consumers have low mobility and tend to stay in the same habitat, seasonal prey switching may be considered a purely temporal subsidy. Consumers that move between habitats are likely to exploit both spatial and temporal variation in productivity. Ives and Settle (1997) used a metapopulation model to investigate the consequences of synchronous versus asynchronous crop planting for pest control by natural enemies. Synchronous planting provides a pulsed resource, and asynchronous planting provides a temporal subsidy for both prey and predator populations. Synchronous planting schemes are used in many parts of Asia because they are thought to reduce pest refuges and shorten the length of time that resources are available (Levins 1969). However, this planting system also makes it difficult for predator populations to persist, particularly if alternative prey are in low abundance. The model shows that synchronous planting gives the lowest pest (prey) densities in the absence of predators, but that asynchronous planting may give the lowest pest density in the presence of predators. Whether asynchronous planting will reduce pest densities depends on the population dynamics and relative migration abilities of predators and prey. For example, if prey are able to colonize new patches quickly and stay ahead of less mobile predators, asynchronous planting will be less effective. Predators with alternative prey species are predicted to be most effective in controlling pests because predator densities will not crash between cropping periods. This model successfully incorporates features of both temporal and spatial subsidies and demonstrates how they may be applied to agricultural systems.

COMPETITION IN A VARIABLE ENVIRONMENT

Competition is an important dynamic in food webs, as it represents a density-dependent interaction between two populations that are not necessarily linked as predators and prey, and that typically (though not always) share a trophic level. Most competition models, such as Tilman's R^* model (Tilman 1976, 1977), can be expanded to include predators as a source of density dependence (Holt et al. 1994). The best competitor in an environment will be that population that is able to persist using the lowest level of resource supply under the local regime of predation and abiotic mortality factors. Trade-offs may exist where species with greater resource requirements are better competitors when predators are present. If species do not use the same resources, but share the same predator, the species able to tolerate the highest level of predation will exclude its "apparent" competitors (the P^* rule; Holt et al. 1994). Keystone predators may allow competitor coexistence when they prefer the competitive dominant (Paine 1966).

These competitive interactions can be affected by temporal variation in productivity. Wiens (1977) proposed that competition will be most severe in years when resources are scarce, with relaxation of competition during more typical years. Tilman (1982) has suggested that competition will be important in most years, but that the limiting resource may vary. For example, Dayton et al. (1999) carried out an excellent experimental study of the effect of a large, dominant kelp, *Macrocystis pyrifera*, on a number of subordinate species. During nutrient-rich La Niña events, the dominant was able to grow rapidly and "shade out" the subordinate species. During El Niño events, growth rates were slower, and the dominant had a smaller effect on other species. Here, a relaxation in competition for nutrients led to increased competition for light. As most ecological studies are completed over relatively short time periods, it is important to keep in mind the potential importance of fluctuations in competitive interactions over longer temporal scales.

Hamback (1998) explored theoretically how seasonality affects plant species competing through resources and shared predation. This model includes two perennial shrubs that are grazed by voles from autumn through spring. In summer the voles switch to herbaceous plants, and the perennials are able to grow with little herbivory. Without selective herbivory, one perennial would competitively exclude the other. In the absence of seasonality, adding herbivory, either with fixed preferences or optimal diet choice (Schoener 1971), does not usually lead to coexistence. In the seasonal model, voles preferentially consume one perennial species in autumn, then by midwinter begin to include the other in their diet as the abundance of the first declines. Both perennials are able to recover during the summer. This pattern of seasonal diet switching allows plants to coexist in the presence of the herbivore. In seasonal environments, perennial densities regularly move across optimal foraging density thresholds of diet expansion or specialization. With fixed preferences, rather than optimal foraging, the herbivore is able to promote coexistence only within a narrow range of parameter values. Hamback argues that this seasonal optimal foraging scenario applies to a wide range of natural systems, such as desert ecosystems in which rodents intensely but selectively consume pulsed production of desert annuals.

Chesson and Huntly (1997) recently reviewed the long-term implications of temporal variation for competitive coexistence. While temporal fluctuations have little effect on coexistence in linear additive models of competition, they are an important factor promoting coexistence in models containing biologically reasonable nonlinearities. The primary role of temporal fluctuations in competition (or other sources of density dependence) is to provide opportunities for temporal niche partitioning, so that different species are better able to survive or acquire resources in different years. Niche partitioning means that intraspecific competition will always be more intense than interspecific competition because the greatest amount of resource overlap occurs between individuals of the same species. Temporal niche partitioning is an important component of the storage effect, which, along with nonlinear responses to fluctuating resources, can explain the puzzling coexistence of large numbers of similar species in plant communities.

The storage effect is easily observed in desert communities of winter annual plants. In the desert, rain falls sporadically among years, from the early fall through midwinter, at a range of temperatures. At each rain, a mixture of plant species germinate from the resting seed bank and compete for water and soil nutrients. These plants are short-lived as adults, often completing their entire aboveground life cycle in just a few weeks. Some species germinate disproportionately better following warm rains, others following cold rains. The seeds of each species enter the seed bank and germinate gradually through the coming years—a timed-release capsule of past recruitment. The "storage" of good recruitment years in a long-lived seed bank can facilitate competitor coexistence.

It is not variation alone that allows coexistence. Coexistence requires (1) niche partitioning, in which species are differentiated in their responses to the varying environment; (2) positive covariance between the environment and competition, such that individuals have a greater competitive effect when the environment is favorable and a lower competitive effect when the environment is unfavorable; and (3) buffered population growth, in which a seed bank or long-lived adult stage allows species to make low investments in reproduction during years when the environment is unfavorable. Storage at any trophic level allows a consumer (whether autotroph or heterotroph) to integrate pulses of resources metabolically over time. Organisms can persist through lean periods by using stored energy and exploiting resources as they become available. Chesson has recently shown that spatial variation can produce a closely analogous "spatial storage effect" (Chesson 2000a). Rather than good and bad periods being linked through time by seed dormancy, good and bad patches are linked through space by dispersal. This theory is very general, and is likely to be useful for the theoretical exploration of spatial subsidies in food webs.

Relative nonlinearity in competitive response also allows coexistence in temporally variable conditions, even in the absence of the storage effect. Assume that two consumer species exploit a single limiting resource and have saturating, nonlinear relationships between feeding rates and resource availability. Species A has a higher feeding rate at low resource levels, whereas species B has a higher feeding rate at high resource levels. In a constant environment the resources will equilibrate at a constant level, and the species with the lower R^* will win. However, if there is large-scale seasonal variation in resource supply rates, resources may fluctuate between high and low levels. Such variation can permit the coexistence of the two competitors on a single resource. This mechanism hinges on species-specific nonlinear responses to the shared limiting resource. If the feeding rate were to increase linearly with resource availability in both species, then even in the presence of high variation in resource availability, we would observe competitive exclusion.

Understanding the controls on competitive interactions and the maintenance of diversity within trophic levels is important for food web ecology because the existence of heterogeneities within levels influences how temporal variation in production is translated to higher trophic levels.

FEAST AND FAMINE IN FOOD WEBS

It is complicated to model temporal variation in multispecies communities. Early verbal models focused on the effects of temporal variation on life history patterns or considered temporal heterogeneity primarily in terms of disturbance. Recently, a growing number of empirical studies have forged ahead (see Ostfeld and Keesing 2000), looking for consistent patterns of response to pulsed productivity. These studies all emphasize the importance of some form of storage, as well as the importance of generalist predator species, which link the dynamics of their various resources.

Effects of Temporal Heterogeneity on Life History Patterns

In one of the first verbal models to explicitly connect temporal variation with community structure, Noy-Meir (1973, 1974) proposed the pulse-reserve hypothesis to explain population dynamics and life history strategies in deserts. He suggested that desert populations are buffered from their harsh and fluctuating environment because animals and plants store up reserves (fat, starch, etc.) during good (wet) periods, which they rely on to survive through bad (dry) periods. Similarly, detritus is "stored" by the ecosystem during good times and used by detritivore populations during bad times. These detritivore populations may later provide nutrients to plant populations and provide an alternative source of prey for consumers (Polis and Strong 1996). The gradual reinfusion of detrital material dampens the destabilizing effects of pulsed productivity. The community is stabilized because stored detritus provides a temporal subsidy for detritivores and allows this resource population to survive bad times. Consumer populations can be maintained through bad periods either by their own stored reserves or by reserves stored by other parts of the system: detritus, detritivores, roots, and so forth. These temporal subsidies, translocating organic matter and energy from good times to bad, are thought to be central to the dynamics of desert communities (Noy-Meir 1973, 1974; Polis 1991). Noy-Meir's pulse-reserve concept did not explore the importance of pulses for interspecific interactions (see Chesson 2000b), but brought attention to the importance of specialized life history characteristics (especially storage) in moderating the effects of harsh conditions.

Southwood (1977) also emphasized the importance of temporal and spatial variation in shaping life history characteristics. According to Southwood, organisms and their habitats are linked in a feedback system that determines when they will reproduce, migrate, or go dormant. His conceptual model involved a two-by-two "reproductive success" matrix, specifying whether an organism would breed here or elsewhere, and now or later. These choices were dependent on a "time heterogeneity index," based upon the length of time any particular location was suitable for breeding and the variation (heterogeneity) in the length of this period. During times of low productivity, resources are depleted, and fecundity is often lower for both plants and animals. The degree to which population size decreases with adversity depends on the life history characteristics of storage, longevity, and mobility. Long-lived sedentary species will average over fluctuations. This model was an ambitious effort to categorize habitat types and match them to forces of selection in complex communities. Like Noy-Meir, Southwood particularly emphasized the importance of dormancy mechanisms in unpredictable environments.

Effects of Temporal Heterogeneity on Diversity Gradients in Food Webs

Menge and Sutherland (1976) included temporal heterogeneity as a factor in their conceptual food web model. Although primarily concerned with temporal variation in disturbance, they argued that trophic complexity depends in part on the rate and stability of primary and secondary productivity. Resource population densities may fluctuate, and bad weather may reduce consumer foraging periods. In the long run, specialist consumers may be excluded due to unpredictable resources. Menge and Sutherland predicted that temporal heterogeneity could lead to trophic simplicity and increased competitive exclusion (as the effect of predation decreased), which would further reduce species diversity. The theory outlined earlier in this chapter does not give clear support to Menge and Sutherland's predictions. Temporal heterogeneity in resources can either increase or decrease average population size, depending on the openness or insularity of the focal populations, and moreover, can provide axes for temporal niche differentiation (Gonzales and Holt 2002; Holt et al., in press). Also, while eliminating keystone predators (Paine 1966) may increase competitive exclusion, increased exclusion is not a necessary outcome of reduced predation (Chesson 2000b, Abrams 2001). The outcome of reduced predation depends on whether predation is density-dependent or density-independent, the selectivity of the predators, and whether trade-offs exist between competitive ability and tolerance of predation.

Pulsed Resources and Community Dynamics

Ostfeld and Keesing (2000) reviewed a number of empirical food web studies that explicitly examined the effect of variation in resources. These studies provided excellent examples of temporal subsidies in terrestrial communities. Many of them followed the effects of mast seeding events, tracking the productivity pulse through primary consumers to their predators (fig. 23.3). Corresponding to the apparent competition model, consumers and predators responded numerically to the pulse of production and had unusually strong top-down effects on their alternative prey species. For example, mast seeding of oak and hornbeam trees in Poland led to high winter survival rates and strong numerical responses by rodents. The following year, high densities of mast-fed rodents had strong top-down effects on insects and ground-nesting songbirds. Rodent predators (owls and martens) also responded numerically and peaked a few months to a year after the rodents did (up to 2 years after the mast crop). These predators also preyed heavily on songbird populations (Jedrzejewska and Jedrzejewska 1998). Overall, Ostfeld and Keesing's review exposed a number of generalities: (1) consumers of pulsed resources are often generalists, which display a time-lagged numerical response; (2) as the pulsed resource diminishes, consumers tend to switch to alternative prey; (3) the generalist consumers

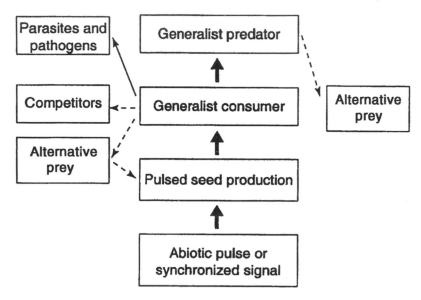

Figure 23.3 Conceptual model of the effects of pulsed resources permeating through a food web. The direction of the arrows represents the direction of causal change in abundance or biomass. Solid lines indicate a positive effect and dashed lines a negative effect of one trophic group on another. (Adapted from Ostfeld and Keesing 2000.)

act as a pulsed resource for their own predators; (4) the response of consumers varies as a function of their specialization, mobility, and rate of population response to the resource.

These findings fit well with our theoretical overview. Generalist predators are those most likely to exploit temporal subsidies because of their ability to switch between prey species. Mobility is an especially important trait for consumers of pulsed resources because it allows them to acquire spatial as well as temporal subsidies. The rate of numerical response to the pulsed resource depends on the intrinsic rate of increase (r) of consumer populations. Fast-reproducing (high r) rodents tracked resources fairly closely. As in the single-species models above, populations with low r tended to integrate across resource peaks, rather than tracking them. It is likely that resource pulses will slowly attenuate as they move through a food web. Through storage (seed banks and perennial tissue) and release (to herbivores or detritivores), producers buffer consumer populations from variation in resources (see also Levey and Stiles 1992). Similarly, storage by primary consumers buffers secondary consumers from fluctuations in primary productivity. Overall, storage mechanisms should dampen the effects of environmental stochasticity.

Time-lagged numerical and behavioral responses make it difficult to interpret ecological patterns. Wiens (1986) found that the dynamics of shrub-steppe bird populations in the Great Basin did not match the dynamics of their habitats. Variation in rainfall altered vegetation height and community composition, and plant species differed in their rate of response to these environmental changes. The plants were slow to respond to the changes in precipitation, the insect community (which the birds feed on) had a slow numerical response to the vegetation changes, and the birds were slow to make a behavioral adjustment to the changes in plant cover and insect densities. If this study had been conducted in a single season, it would have been impossible to determine the population dynamics from the observed patterns.

Productivity fluctuations may have complicated effects on species interactions. Hogstad (1995) found that fieldfare (*Turdus pilaris*) nesting behavior fluctuated in response to pulses of weasel predation. In years when there were few rodents, weasels attacked the fieldfares in a density-dependent manner, leading to a preponderance of solitary nesting. In high-rodent years, the weasels switched to rodents and ignored fieldfares; colonial nesting then became the norm, apparently to reduce predation by birds (the colonies protect themselves by mobbing the predator). In this case, the nesting behavior of birds was affected by fluctuations in the density of rodents, their apparent competitors.

Additional Complications

In open systems, and at the level of whole assemblages, it may be difficult to relate patterns of variation in resource availability to patterns of variation in consumer abundance. Herrera (1998) describes a 12-year study of frugivorous birds and fleshy fruits in a Mediterranean montane community that revealed decoupling at two levels. Total fruit abundance fluctuated temporally from 5 to 80 fruits m^{-2}. However, variation in each of the six most abundant fruit species was not related to annual variation in rainfall. It seems likely that primary production varies in response to rainfall, but this did not translate into predictable patterns of variation in fruit. This is another way in which temporal variation in one component of an ecological system is "filtered" through the dynamic responses of species (here, resource allocation by individual plants). Moreover, though there was substantial annual variation in frugivorous bird abundance, it was not correlated with variation in fruit availability. Herrera suggests that this decoupling of resources and consumers reflects the importance of abiotic factors (such as temperature) and, in particular, the wide range of the consumer populations. The abundance of migratory frugivorous birds at a particular site may be governed by processes operating over much of the Palearctic, obscuring any effect of local variation in resource availability.

If variation in productivity is driven by nutrients, then plant nutrient contents are likely to change. Huxel (1999) has argued that integrating food quality into standard consumer-resource models can be stabilizing. Turning his argument around, temporal variation in food quality driven by fluctuations in primary production could be an unsuspected source of variation in consumer numbers. Carpenter and Pace (1997) propose that nutrient fluctuations may be responsible for maintaining alternate stable states of dystrophy or eutrophy in whole-lake ecosystems. If so, then nutrient fluctuations could have huge ecosystem-level effects that determine which species can tolerate a particular habitat.

FOOD FOR THOUGHT

This chapter grew from an inspiration to extend the concept of spatial subsidies and landscape-level food web influences to the temporal scale. Time has always been included in population models, either implicitly or explicitly, but we have been slow to focus on temporal heterogeneity itself. The temporal subsidy concept reinforces our instinctive understanding that the natural world is constantly changing and that these changes happen over a variety of time scales and patterns. Jensen's inequality demonstrates that this variation may have large effects on the outcome of population dynamics and species interactions. Because of this, we cannot use population models to predict the trajectories of natural communities without taking into account the effects of temporal variation.

In the models and natural systems we have reviewed here, certain commonalities arise. General models demonstrate that the rate of population growth (r or λ) relative to the time scale of a productivity fluctuation will determine whether that population will track or average over the fluctuations. Storage is also extremely important, as it modifies how species respond to fluctuations in productivity. A population may have a rapid numerical response to favorable conditions, but through storage in long-lived adults or persistent stages, that population may not decline at the onset of unfavorable conditions. How well a population tracks environmental change will determine how productivity pulses are translated to the rest of the food web. It is difficult to predict how community dynamics will be altered by temporally varying productivity, and the outcome is contingent on the characteristics of the species making up the community (e.g., life history traits, the nature of density dependence) as well as on the web of species interactions. However, initial empirical work demonstrates that common patterns arise in diverse systems.

Because species and community characteristics strongly modify population dynamics, food web studies call for intensive natural history observations. Ecologists need long-term demographic data to quantify how communities react to variation in the environment. This requirement can be an obstacle to experimental testing because operational time scales may be very long, especially when predator populations have a time-lagged response to their prey. A traditional way to surmount this problem is to use organisms with very short life spans and fast reproductive rates. Mesocosms and microcosms have been useful for testing a wide variety of theoretical predictions, from Lotka-Volterra models (Gause 1934) to apparent competition (Lawler 1993; Bonsall and Hassell 1997). Although these systems may lack verisimilitude, they allow researchers (within a single lifetime) to follow a numerical response to system perturbations.

In light of recent successes in documenting the effects of pulsed productivity in field experiments, it may be reasonable to undertake some larger-scale experiments in agricultural systems. The rice planting scheme described by Ives and Settle (1997) is an excellent example. On a regional scale, fields are either planted synchronously, or planting is staggered throughout the year. Rice cultivation occurs year-round in many areas, and herbivorous insects and their predators have relatively rapid dynamics, so dynamics unfold swiftly. This type of study provides a wealth of information on both spatial and temporal food web subsidies and the factors that modify their effects (such as migration rates).

We are in a period of rapid global change—ironically, as a result of past productivity stored as petrochemicals that are now subsidizing human population growth. Global weather changes are altering local surface temperatures and precipitation patterns. Many resources are being depressed to low levels, while human waste products and toxins increase. One outcome may be an increase in environmental variability, both in resources such as water and nutrients and in disturbance forces such as storms and forest fires. Invasive species can provide sustained temporal pulses of novel resources, shifting resident predator-prey dynamics (Holt and Hochberg 2001). These global-scale changes are a critical reason for studying food webs in an explicit temporal context. Some outcomes may be predictable; others are likely to be complete surprises (Schneider and Root 1996). The experiment is on, but we may wish to modify some of the treatments.

ACKNOWLEDGMENTS

We would like to thank P. T. Stapp, F. Sánchez-Piñero, M. Rose, A. Boulton, and J. Vander Zanden for their intellectual insight and encouragement. We also thank P. Chesson, G. Huxel, E. Preisser, and L. Yang, as well as an anonymous reviewer, for their helpful comments on the manuscript. We are grateful to the National Science Foundation for support, including grants DEB9527888 and DEB9806657 to GAP and an NSF graduate research fellowship to ALWS, and to the University of Florida Foundation.

Chapter 24

Subsidy Effects on Managed Ecosystems: Implications for Sustainable Harvest, Conservation, and Control

Mary E. Power, Michael J. Vanni, Paul T. Stapp, and Gary A. Polis

THE USE OF SUBSIDY THEORY IN APPLIED CONTEXTS

It is conceivable that somewhere, in an isolated pocket of ancient groundwater deep in the earth's crust, there exists a hydrogen-based food web that has not yet felt the hand of man. Otherwise, it is hard to think of any ecosystem on earth that is not dominated, or strongly perturbed, by humans (Vitousek, Mooney et al. 1997; Kareiva et al. 1993; Crowder et al. 1996). Human effects have become so pervasive that most people in present and future generations will never experience "unmanaged" ecosystems. The effects of the human enterprise, intended or otherwise, have spread over regional or global scales (e.g., Riley and Jefferies, chap. 25 in this volume). To anticipate the consequences, we must better understand how our activities have changed the spatial and temporal scales of natural ecosystems. The study of "ecological subsidies," fluxes of organisms, energy, or materials across ecosystems boundaries, can make key contributions toward this understanding.

Ecological subsidy theory adds an explicit spatial context to the study of food web interactions. As explained by Polis, Anderson, and Holt (1997), it organizes the potentially overwhelming complexities of spatially registered food web ecology into a framework useful for exploring the community- or ecosystem-level consequences of fluxes between habitats. Just as economic subsidies (e.g., funding by the U.S. government of "below-cost" timber sales or interbasin water diversions) distort local economies and ecosystems, so do their ecological counterparts. Ecological subsidies of materials or organisms from distant sources, through arrays of direct and indirect effects, can change the structure and dynamics of local recipient food or interaction webs.

Applying the subsidy framework to food web ecology involves several steps: (1) characterizing the flux, including its variation in space and time; (2) identifying key members and linkages in recipient webs; (3) evaluating the population-level responses of recipient web members that directly intercept fluxes; and (4) analyzing (or predicting) the community- or ecosystem-level consequences emanating from the subsidy's direct and indirect effects on these and other web members. We begin to understand subsidy effects if we can answer the question, how would the web function if the subsidy changed or disappeared?

These steps can be tailored for application to management, in which explicit spatial information is crucial: (1) How did (would) the ecosystem function without human transfers of energy, materials, or organisms between previously isolated habitats? Where and when has land use or resource harvest reduced or stopped cross-habitat flows that were formerly important to ecosystems? Where and when have they distorted ecosystems by accelerating or concentrating flows, such as nutrient fluxes? (2) What are the direct and indirect effects (sometimes corresponding to the intended and unintended consequences) of human alterations of subsidies? (3) How will these consequences play out over years, decades, and centuries and over local, regional, and global scales? (4) How might system trajectories and feedbacks change under various management schemes or with changes in climatic or ecological conditions? Could failure to recognize food web linkages across larger spatial or temporal scales precipitate unpleasant surprises (management disasters)?

The global-scale consequences of human (postindustrial) effects on land cover, nutrient fluxes, and species distributions are reviewed by Riley and Jefferies (chap. 25 in this volume). In this chapter, we will focus on more local processes that mediate the effects of human subsidies on food webs we hope to manage. As Robert Holt (personal communication to MEP) has pointed out, for every subsidy, there is an "anti-subsidy," or "resource shadow": a zone with organisms from which resources of energy, materials, or organisms have been diverted. Human-induced resource shadows include the deserts that have followed water diversion, deforestation, and overgrazing (Reisner 1986, 1990; Southwick 1996; Sauer 1967; Perlin 1991) and the rivers that have suffered losses of huge fish migrations (salmonids, eels) following damming (e.g., National Research Council 1992). While resource diversion clearly influences the diversity and sustainability of both donor and recipient ecosystems, we will focus here on what happens to recipient ecosystems when subsidies from spatially extensive sources are discharged into them. We present several case histories that illustrate how subsidy theory can produce testable hypotheses about spatial food web and ecosystem processes that could inform adaptive management of ecosystems and species. Subsidy theory, by expanding the scope of ecological studies, can also aid in our recognition of how management for one target (e.g., agricultural production) may affect other societal goals (e.g., water quality or species conservation).

EFFECTS OF AGRICULTURAL SUBSIDIES

Agriculture is probably the oldest and most widespread human impact on the earth. Human agriculture appears to have begun around 8500 B.C. in the Fertile Crescent of southwestern Asia and less than a thousand years later in China. By 5500 B.C., it had spread to southwestern Europe, and it was independently initiated in the Americas about 2,000 years later (Diamond 1998). Over the subsequent millennia, agricultural production fueled increases in human populations, and humans, in turn, have intensified agricultural production by diverting and concentrating water, nutrients, and organisms (Matson et al. 1997). Subsidy analyses can help us recognize some of the interactions and controls that act over various spatiotemporal scales to influence the yields, as well as the impacts, of agriculture.

Agricultural food webs are less diverse than natural food webs, and they are often highly managed to maximize the yield of one or a very few species of crops. Nevertheless, they respond to the same direct and indirect processes that influence more natural communities. Cross-habitat fluxes of nutrients, plant competitors (weeds), herbivores, and predators from various spatial sources affect the dynamics of agricultural food webs and the productivity of target crops. Most of the important entities involved move across a variety of scales.

Nutrients

Terrestrial nutrients, particularly nitrogen and phosphorus, are distributed quite heterogeneously at almost all spatial scales, from within a watershed to the entire planet (Huston 1993). Before commercial fertilizers were available, our most fertile crop areas were enriched by allochthonous nutrients deposited by water. Rich floodplain soils were deposited annually by rivers (e.g., in the Nile delta and the "bottomlands" of the Mississippi) before human engineering isolated these rivers from their floodplains. Over a longer but more continuous time, soils from nutrient-rich areas are transported as atmospheric dusts at scales from meters to thousands of kilometers (Jackson 1971; Likens et al. 1990; Chadwick et al. 1999). These dusts transfer nutrients, especially phosphorus, over the entire world, from areas geologically rich in such elements to more depauperate areas. For example, pineapples and sugarcane in Hawaii are fertilized by phosphorus dusts from central Asia (Chadwick et al. 1999). Transfers of windblown iron dust from continents to oceanic phytoplankton may have been greater during glacial times, elevating production of iron-limited phytoplankton, which may have drawn down more CO_2, resulting in a cooler climate. Marine aerosols that contain rare micronutrients, macronutrients, and organic compounds travel, sometimes long distances, inland on continents (J. Noller, personal communication to GAP). The time-integrated contributions of marine aerosols to local soils may be very large, but their effects on plant productivity in agroecosystems and natural systems are poorly known.

The commercial fertilizers used today are imported from natural accumulations (e.g., seabird guano from oceanic nesting islands; phosphate mines), collected from concentrations of livestock, or derived by industrial nitrogen fixation using petrochemicals whose elements were assembled by ancient plant communities. Many problems and threats have arisen from the massive introduction of synthetic nitrogen fertilizers into groundwater, surface water, and terrestrial and marine ecosystems (Carpenter, Caraco et al. 1998; Paerl 1985; Howarth, Billen et al. 1996; Vitousek, Aber et al. 1997; Riley and Jefferies, chap. 25 in this volume). The dead zone in the Gulf of Mexico, apparently a result of nutrient export from U.S. agriculture in the Mississippi basin, is one notorious example (Rabalais et al. 1996). Scientists have also raised concern over the effects of atmospheric deposition of anthropogenic nitrogen on plant composition and productivity worldwide (e.g., Vitousek 1994; Vitousek, Aber et al. 1997; Paerl 1985; see Riley and Jefferies, chap. 25 in this volume).

Pests

Pests in agroecosystems also move among patches (of different plant species) and habitats. Many herbivores on crops are generalists that thrive on several species of host plants. Japanese beetles, Oriental and Mediterranean fruit flies, and the corn earworm (*Helicoverpa (Heliothis) zea*, economically the most harmful pest in North America) each infect many crop species. Microbial plant pathogens often move among hosts as spores blown over short and long distances from conspecific and heterospecific populations (Walker 1969; Roberts and Boothroyd 1972). Stages of some heteroecious rust fungi must move among different host species to develop; cedar-apple rust, for example, alternates between apples and eastern red cedar. Farmers decrease "take" by these agricultural competitors by manipulating the vegetation surrounding their fields to avoid combinations favorable to pests.

Some populations of agricultural pests move great distances. Probably most infamous are the migratory locusts (acridid grasshoppers) that destroy thousands of square kilometers of croplands annually. The corn earworm is another well-known migratory pest. Pollen tracer studies identified southern Florida, the Bahamas, Cuba, the Yucatán Peninsula, and northern Central America as potential source areas 1,515 km from capture sites in Oklahoma and Texas (Lingren et al. 1994). Biologists deduced that dispersing moths would have to fly over water for 72 hours or have very limited diurnal resting periods on seaweed (*Sargassum* spp.), ships, oil platforms, or the sea surface. Pair et al. (1991) used ecological and meteorological evidence and ground-based radar to identify irrigated corn grown in the Lower Rio Grande Valley of Texas and northeastern Mexico as the source of the migrant fall armyworm, *Spodoptera frugiperda*, which subsequently infested crops in Texas, Missouri, and Iowa, up to 1,900 km away. Individual corn earworms that infest the central to northern midsection of the United States come each year from a 200,000 ha area in northern Mexico (McCracken et al. 1996). Pupae do not overwinter north of mid-Texas, and populations hopscotch their way across America each year over several generations, wafting and flying up to 400 km in 9 hours. The corn earworm and fall armyworm do not overwinter in temperate areas. These and similar studies are revealing how weather systems, habitat modification, and biology underlie long-distance migrations of pests from defined source areas to crops in remote areas (Westbrook and Isard 1999). Vertebrate pests (starlings in North America, cockatoos and mice in Australia, and hippos, elephants, and quela and weaver birds in Africa) also move among habitats and sometimes cause crop damage.

Biological Enemies

Enemies of crop pests also move among patches (of plant species) and habitats. For example, swifts, swallows, martins, and bluebirds often either migrate varying distances or live in non-crop habitats adjacent to agricultural fields. Many arthropod predators have populations that live on prey from both crop and non-crop plants. In an apparent competition interaction, prey that are not crop pests may increase predator populations to levels capable of suppressing pests. For example, in a California vineyard (Napa), predatory two-spotted mites move from Johnson grass to relatively less productive grapevines; this steady influx allows higher populations of these mites to suppress an important in situ pest prey, the Willamette mite, to lower densities in the grapevines than if Johnson grass were absent (Flaherty 1969). Spiders of several varieties reside in ground detritus but move daily to row crops to feed on herbivores, significantly increasing crop yield (Riechert and Bishop 1990). Movements by predators and parasitoids are important to the population dynamics of crop pests (McCauley et al. 2000; Murdoch and Briggs 1996).

Parasitoid wasps are a key element in biological pest control. They require two distinct resources: host arthropods (insects or spiders) for their developing larvae and flowering plants to provide adult wasps with nectar and pollen for energy and egg production. Control of pest species by wasps appears to be more successful when the wasps have access to flowering plants that surround crops (Jervis et al. 1993). Polis et al. (1998) recognized the importance of adult resources during our work on spider dynamics on islands in the Gulf of California. Most years are exceptionally dry (< 20 mm of rain), and in those years, spider wasps (Pompilidae) are basically absent as a mortality factor, even though spider populations may be very dense. In wet (El Niño) years, however, the biomass of annual flowering plants increases by two orders of magnitude. Under these conditions, adult wasps had sufficient nectar and pollen resources to depress spider populations by an average of 90% or more on twenty-one different islands. These results suggest that resource subsidies provided to adults may increase the effect of larval parasitoids in agricultural habitats as well.

Knowledge of the spatial ecology of agricultural food webs has been used for centuries in traditional farming practices to enhance yields. For example, appropriate plant combinations in polycultures are used to nurture beneficial birds and arthropods. Traditional shade-grown coffee and cacao cultivation uses polycultures with overstory and understory plants that are important habitats for migratory and resident birds, which consume pests (Greenberg and Ortiz 1994). Hedgerows along field edges or other plants between crop rows (e.g., Johnson grass and grapes) are used to harbor natural enemies. Farmers also provide structures for predator habitats (e.g., ground detritus for spiders; bird boxes for martins). Trap crops (Scholte 2000; Barbercheck and Warrick 1997; Buntin 1998; Luther et al. 1996) or trap habitats adjacent to harvested crops are used to lure pests to sites where they can be easily destroyed (e.g., pest ant species can be concentrated and burned under hay bales, where they attempt to nest; P. Ward, personal communication). Manipulating habitat structure to influence organisms at trophic positions above and below pests can reduce crop damage without the toxic consequences of pesticides (Matson et al. 1997).

Agricultural Intensification

Traditional methods of pest control using local knowledge of the variation in weather, soils, plants, and animals are less and less practiced, however, as human population pressure drives the increasing industrialization of agriculture (Lal 1987). With agricultural intensification (Matson et al. 1997), the temporal scales of repeated extraction of crops or livestock from cultivated areas are shortening, while the spatial scales over which we are redirecting flows of nutrients and agricultural products are increasing.

Large-scale manipulation of the earth for agriculture is not a twentieth-century phenomenon. Massive irrigation projects have repeatedly subsidized crops and ultimately salinized soils, starting in Mesopotamia from 2400 to 1700 B.C. (Perlin 1991) and continuing to this day in the arid western United States and many other places (Reisner 1986; Southwick 1996). Globalization and intensification of agriculture on the modern scale, however, would not have been possible without industrial nitrogen fixation. The invention of the Haber-Bosch process for ammonia synthesis just before World War I, and its proliferation for global fertilizer production following World War II, released human agriculture from its previous nitrogen limitation and permitted the quadrupling of the human population during the twentieth century (Smil 1997). These changes have had obvious consequences.

In the next section, we discuss some of the more local effects of nutrient subsidies resulting from land use changes, including agriculture within a watershed, on the food web and water quality of a midwestern U.S. reservoir. In the following section, we discuss how regional nutrient subsidies from North American agriculture affect a wetland maintained for wildlife conservation in the arid Southwest. In both of these examples, enough has been learned about specific nutrient vectors and controls over the spatial dynamics of the subsidy to guide management responses.

WATERSHED MANAGEMENT: RESERVOIRS IN EASTERN NORTH AMERICA

Export of nutrients from terrestrial landscapes to aquatic ecosystems stimulates aquatic primary production and can modify food web structure in lakes, streams, and coastal environments (Carpenter, Caraco et al. 1998; Smith 1998). Watershed-scale transport of nutrients enhances the production and biomass of algae and vascular plants and sometimes alters the species composition of plant assemblages, which can then further alter food web structure. A well-known example is the shift toward cyanobacteria (blue-green algae) in fresh waters when anthropogenic increases in phosphorus inputs render nitrogen relatively more limiting (Smith 1998). Because cyanobacteria are less edible than other algae, shifts in the herbivore assemblage from generalist feeders (e.g., *Daphnia*) toward more specialized feeders (e.g., copepods and rotifers) may occur. Fish assemblages may also shift in response to increased nutrient inputs (e.g., Bachmann et al. 1996). Increased algal "blooms" and associated symptoms of eutrophication generally reduce water quality in a variety of ways, including the formation of surface scums of algae, depletion of deep-water oxygen and subsequent loss of fish habitat, and shifts toward fish species less desirable to humans. Eutrophication remains the most prevalent environmental problem facing fresh waters in terms of the number of lakes or the total length of rivers affected (Carpenter, Caraco et al. 1998).

Linkages among Watersheds and Reservoir Food Webs

Watersheds also export large quantities of nutrients in particulate form— for example, as nutrients attached to soil or sediment particles. These particulate-bound nutrients are generally much less available to primary producers than are dissolved nutrients. However, particulate inputs may subsidize aquatic food webs by providing a food source for certain key species. Reservoirs of eastern North America provide an excellent system for examining the consequences of these subsidies for several reasons (see Vanni and Headworth, chap. 4 in this volume). First, reservoirs have large watersheds (compared with natural lakes) because they are impounded rivers. (In contrast, most glacial lakes have small stream inflows or lack stream inflows altogether.) Therefore, reservoirs often receive large quantities of sediment and nutrients from their watersheds (Thornton 1990). Second, many reservoirs are constructed in agricultural areas; since these landscapes are often subject to high rates of soil erosion, reservoirs in agricultural areas are particularly likely to receive massive amounts of sediments and particulate nutrients (Renwick 1996; Vanni et al. 2001). Third, omnivorous gizzard shad (*Dorosoma cepedianum*) often dominate the fish assemblages of reservoirs of eastern North America (Stein et al. 1995; see Vanni and Headworth, chap. 4 in this volume). In reservoirs, postlarval gizzard shad often feed mainly on sediments (Mundahl and Wissing 1987; Yako et al. 1996; Schaus et al. 2002). By consuming sediment-bound nutrients and excreting nutrients in dissolved inorganic form into the water column, gizzard shad transport considerable quantities of dissolved inorganic nutrients (phosphate and ammonium), which increase phytoplankton biomass and nutrient standing stocks (Schaus et al. 1997; Schaus and Vanni 2000; Vanni and Headworth, chap. 4 in this volume).

Gizzard shad abundance is probably limited by subsidies from watersheds. These fish are more abundant in productive (eutrophic and hypereutrophic) lakes and reservoirs than in unproductive systems. Water quality in reservoirs is subject to positive feedback between watersheds and gizzard shad activities (Vanni and Headworth, chap. 4 in this volume). As land cover is converted from forest to agriculture, the export of both particulate and dissolved nutrients to reservoirs increases. Several "subsidy pathways" may then interact to affect reservoir food webs and degrade water quality (fig. 24.1). Suppression of large zooplankton species by high gizzard shad biomass (adult shad consume some zooplankton, and shad larvae are obligate phytoplanktivores that may exploitatively outcompete zooplankton) may also result in low rates of herbivory, further contributing to high algal biomass. In short, watershed degradation (increased nutrient and sediment export) leads to increased algal production and gizzard shad biomass; increased gizzard shad biomass further stimulates algal production.

At least three factors, however, can break this positive feedback loop and regulate shad biomass. First, gizzard shad growth rates are density-dependent and tend to be lower in highly productive habitats where shad are extremely abundant (e.g., DiCenzo et al. 1996). This factor could reduce reproductive output. Second, gizzard shad exhibit variable year-class strengths. Survival of young-of-the-year (YOY) shad varies considerably among years, in part due to events occurring at the larval stage, when gizzard shad are obligate planktivores. Third, relatively severe winters, when prolonged ice cover can lead to long periods of anoxia, depress shad populations, particularly in highly productive systems in which bacterial respiration rates are high. These factors probably interact to regulate gizzard shad abundance over interannual scales (Schaus et al. 2002).

Interactions of Cross-Habitat Subsidies and Water Quality Management

The potential interactions of nutrient subsidies have several implications for water quality management in ecosystems containing gizzard shad.

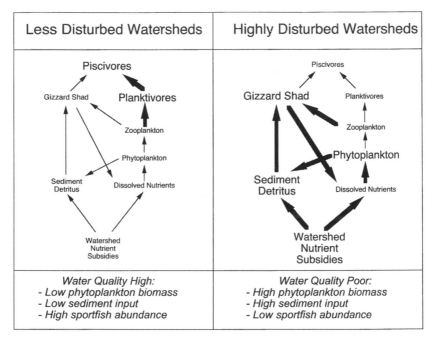

Less Disturbed Watersheds	Highly Disturbed Watersheds
Water Quality High: - *Low phytoplankton biomass* - *Low sediment input* - *High sportfish abundance*	*Water Quality Poor:* - *High phytoplankton biomass* - *High sediment input* - *Low sportfish abundance*

Figure 24.1 Proposed linkages between watersheds and reservoir food webs in reservoirs of eastern North America. The relative sizes of the arrows indicate the relative difference in a particular flux rate between the less disturbed and highly disturbed watersheds. Similarly, the relative font size labeling the system compartments represents the relative difference in biomass between the watershed types. In less disturbed watersheds, subsidies from the watershed are lower and gizzard shad are scarce. This allows planktivores (e.g., bluegill sunfish) to thrive; piscivores (many of which are sportfish) are abundant because they feed on these planktivores. In highly disturbed watersheds, subsidies of dissolved and particulate nutrients from watersheds are more substantial. These inputs stimulate phytoplankton productivity and also provide detrital resources for gizzard shad. Shad biomass therefore increases, leading to increased nutrient transport by shad. This transport leads to further increases in phytoplankton biomass. Gizzard shad larvae are obligate planktivores and may exploitatively outcompete zooplankton, leading to declines in planktivores. Shad also are not as vulnerable to piscivores as are other planktivores, and hence piscivores are less abundant in highly disturbed watersheds.

Improved watershed practices, particularly in agriculture, may yield relatively large water quality benefits—perhaps greater than those in systems lacking gizzard shad. Agriculturally derived nutrient inputs can be lowered by reduced fertilizer use, reduced soil erosion via improved tillage practices, and protection of riparian zones to reduce the movement of nutrients from land to water. Practices that lower inputs of both particulate and dissolved nutrients are likely to have the greatest effects, as they will reduce direct nutrient subsidies to phytoplankton (i.e., dissolved nutrient inputs) as well as food subsidies to gizzard shad. Furthermore, integrated management of fisheries and watersheds is likely to lead to the greatest water

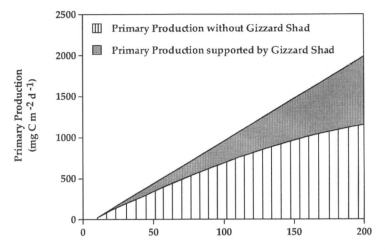

Figure 24.2 Predicted phytoplankton primary production in reservoirs with and without nutrients transported by gizzard shad. The open area represents predicted primary production without the nutrient transport process, and the shaded area represents additional primary production supported by nutrient transport by gizzard shad. The top line therefore represents total primary production supported by shad and all other sources. (Adapted from Vanni and Headworth, chap. 4 in this volume.)

quality benefits. For example, gizzard shad biomass might be lowered directly through enhancement of piscivorous fish populations in conjunction with reductions of watershed-derived nutrient subsidies. These two management practices should act synergistically to improve water quality.

If either gizzard shad populations or watershed-derived nutrient inputs are reduced, the model of Vanni and Headworth (chap. 4 in this volume) suggests that the greatest improvements in water quality may be achieved in the most highly productive reservoirs. This is because, according to this model, nutrient transport by gizzard shad sustains a greater proportion of phytoplankton primary productivity in highly productive systems than in unproductive systems (fig. 24.2). If shad biomass increases more than linearly with potential reservoir productivity, then reducing watershed-derived nutrient inputs should also have greater proportional effects on water quality in highly productive reservoirs (Vanni and Headworth, chap. 4 in this volume). Thus, relatively small improvements in watershed practices could lead to both direct improvement in water quality through a reduced supply of dissolved nutrients and indirect improvement via reduced gizzard shad biomass. Potentially synergistic measures such as these need to be better integrated into management efforts.

In another model, Carpenter, Ludwig, and Brock (1999) predicted that recycling of nutrients from sediments and lack of flushing could make lakes vulnerable to irreversible degradation. A more optimistic prediction for reservoirs may be in order because of the potential for flushing excess nutrients from reservoirs through outflow rivers. This mitigation, however, may simply transfer eutrophication problems to downstream water bodies (Rabalais et al. 1996; Riley and Jefferies, chap. 25 in this volume).

Potential Barriers to Water Quality Management

Even though the interactions of nutrient subsidies described above may facilitate management for water quality, there are a number of potential barriers to improving water quality in reservoirs dominated by gizzard shad. For example, reductions in gizzard shad populations via enhanced piscivory may not be feasible, as it appears that native piscivores (such as largemouth bass) are not likely to control gizzard shad populations (Stein et al. 1995). For example, in highly productive Ohio reservoirs, gizzard shad hatch earlier than other fish, and juveniles can switch to feeding on sediments once they develop a gizzard (which occurs when they are just a few months old). This developmental pattern results in relatively rapid growth of YOY gizzard shad, rendering them less vulnerable to YOY bass. Bass recruitment (and ultimately, population size) is greatly affected by success in the first year of life; thus, bass populations may not reach the densities necessary for regulation of gizzard shad populations (Stein et al. 1995, 1996). Introductions of exotic piscivores, such as striped bass or saugeye, may be more effective in controlling YOY gizzard shad, and hence overall gizzard shad biomass (Stein et al. 1996). Of course, introductions of exotic species are risky "experiments." Nevertheless, introductions of exotic piscivores are carried out all the time by state agencies, and these introductions may offer opportunities to test some predictions of food web theory (Stein et al. 1996).

Gizzard shad are virtually the only abundant fish in North American reservoirs that rely on sediment detritus. Therefore, reductions in watershed-derived sediment might effectively reduce gizzard shad populations. Efforts to control soil erosion in watersheds would also increase water clarity, an additional benefit. One effective means of reducing soil erosion is to reduce the extent to which croplands are tilled (conservation tillage) or eliminate tillage entirely (no-till methods). However, reduced tillage frequency may lead to increased export of dissolved phosphorus from croplands to water (Logan 1990; Gaynor and Findlay 1995) by at least two mechanisms (Logan 1990). First, P bound to sediments from conservation tillage fields tends to be more labile than P bound to sediments from conventional tillage fields because the former sediments tend to be higher in clay and organic matter. Second, the surface application of P fertilizer to the relatively undisturbed soils of conservation tillage fields can lead to a greater buildup of labile P at the soil surface than in the more frequently disturbed (tilled) fields of conventional agriculture. Because conservation tillage usually results in reduced export of particulate-bound P but sometimes results in increased export of dissolved P, the net effect of conservation tillage on total P export will depend on the relative contributions of dissolved versus particulate P export from a given watershed. Nevertheless, increased export of dissolved P under conservation tillage could increase algal biomass in recipient aquatic systems, particularly if light intensity penetration increases along with decreased sediment loading. Other strategies of reducing nutrient inputs from watersheds, such as reducing fertilizer use or protecting riparian habitats, therefore must be explored (Gaynor and Findlay 1995).

Carbon Sequestration in Reservoirs

Recent evidence suggests that exports of matter and nutrients from watersheds to aquatic systems and food web interactions within aquatic systems have implications for the global carbon cycle and hence global climate change. Based on recent data, Dean and Gorham (1998) concluded that 3 times as much organic carbon is buried annually in the sediments of freshwater ecosystems (lakes, reservoirs, and peatlands) as in the sediments of all of the world's oceans, even though freshwater ecosystems constitute less than 2% of the earth's surface. About half of all C buried in freshwater sediments is buried in reservoirs (Dean and Gorham 1998). Two mechanisms account for these high rates of burial in freshwater sediments. First, the rate at which materials are transported from land to water, expressed per unit surface area (of water), is much higher in fresh waters. This difference arises simply because freshwater systems are in closer contact with land than is the average ocean locale. This is analogous to the situation on oceanic islands, where inputs of ocean-derived detritus are greater (per unit island area) on small islands than on large islands (Polis and Hurd 1994). Second, because freshwater ecosystems are much shallower than the oceans, a much greater proportion of organic carbon becomes buried in sediments before being respired to CO_2. In the ocean, a much larger proportion of carbon is respired as organic matter sinks. Thus, freshwater systems may effectively trap considerable amounts of carbon. Reservoirs, having large watersheds, may play a critical role in storing carbon.

Food web structure and nutrient fluxes influence how much carbon is stored in lake sediments. Schindler et al. (1997) showed that the level of nutrient enrichment and food web structure affects whether lakes are net sources or sinks with respect to atmospheric CO_2. When lakes are fertilized or when herbivory is low (i.e., when planktivorous fish are abundant and large herbivores scarce), phytoplankton biomass is relatively high. High rates of photosynthesis draw CO_2 from the atmosphere into the lake water. Under these conditions, lakes are net sinks for carbon, presumably because much of the organic carbon is ultimately buried in sediments. When lakes have low nutrient inputs or when large grazers are abundant, phytoplankton biomass and photosynthetic rates are low. Under these conditions, respiration exceeds photosynthesis, and more CO_2 is released to the atmosphere than is drawn in.

Evidence strongly suggests that most lakes are net heterotrophic systems (i.e., respiration exceeds photosynthesis) because they are subsidized by organic carbon inputs from watersheds (Cole et al. 2000; Caraco and Cole, chap. 20 in this volume). This is almost certainly the case in most reservoirs, where allochthonous inputs of organic carbon are large. Indeed, it has been argued that reservoir fish production could not be sustained by in situ photosynthetic production (Adams et al. 1983). On the other hand, the high rates of organic carbon burial in reservoir sediments could cause reservoirs to be net carbon sinks. Food web interactions in reservoirs probably mediate the extent to which these ecosystems are net carbon sinks or sources. In Ohio reservoirs, phytoplankton abundance is much higher when sediment-feeding gizzard shad are abundant than when shad are scarce. This pattern has been demonstrated by enclosure experiments (Schaus and Vanni 2000) as well as whole-lake observations (M. J. Vanni et al., unpublished data). Thus gizzard shad abundance may be positively correlated with the net flux of carbon from the atmosphere to lake water, and ultimately with the rate at which autochthonously produced carbon is buried in sediments. On the other hand, sediment-feeding fish often resuspend sediments. Once resuspended, carbon may be exported via outflow streams or may be respired by bacteria. In other words, bioturbation of sediments by gizzard shad may decrease the rates at which organic carbon (including that from watershed-derived sources) is permanently buried in sediments. Thus it is clear that in reservoirs, watershed inputs and gizzard shad have the potential to affect net carbon flux, and that these factors may interact in complex ways. To determine whether reservoirs, or any other habitats, are net sources or sinks for carbon, we need to consider cross-habitat fluxes among their watersheds, the atmosphere, the water column, and the bed sediments and how these fluxes are mediated by food web processes.

WETLANDS AFFECTED BY REGIONAL AGRICULTURAL SUBSIDIES TO GEESE

Certain wetlands derive nutrients from spatial scales vastly greater than their local watershed through inputs mediated by the continental migrations of agriculturally subsidized geese. Along the marshes bordering Hudson Bay, tens of thousands of lesser snow geese arrive each summer to hatch and rear their goslings and graze on local graminoids (Jefferies et al., chap. 18 in this volume). Over recent years, populations of these geese have increased dramatically due to subsidies from grain fields in their wintering grounds in the United States and decreased human hunting. The increasingly crowded summering geese have had to switch from nondestructive clipping of aboveground graminoid leaves to destructive grubbing of the perennating roots and rhizomes. As a result, once productive marshes along Hudson Bay are being converted to mudflats, which are unlikely at these high latitudes to support much secondary invertebrate production. This conversion may be long-term, as it is stabilized by feedbacks such as salinization following graminoid extirpation (Jefferies et al., chap. 18 in this volume).

Wetlands in the southwestern United States that are wintering grounds for geese and other waterfowl are also being damaged by agricultural subsidies (Post et al. 1998). Wetlands in the arid Southwest have been largely lost to development and water diversion for human use. Flocks of lesser snow geese and Ross's geese winter on the small remnant wetlands that are managed for waterfowl production. Up to 40,000 geese winter in the Bosque del Apache National Wildlife Refuge along the middle Rio Grande Valley in central New Mexico, making up about 50% of the bird biomass on these wetlands, which also support sandhill cranes and ducks (Post et al. 1998). Using bioenergetics modeling calibrated with feeding, movement, and roosting observations, Post and colleagues estimated that geese translocated enough nutrients from local corn and alfalfa crops into these wetlands to account for 40% and 75% of the annual nitrogen and phosphorus inputs into their study area during one winter season.

These nutrient subsidies may have at least four consequences for the small areas of remaining wetlands. First, arriving goose populations (subsidized by agricultural production elsewhere before arriving on their wintering grounds) feed on marsh vegetation (such as bulrushes, *Scirpus pungens*), deplete it, and damage its regenerative potential. The damage may be direct, from overgrazing, as well as indirect, as goose-imported nutrients encourage the growth of periphyton and phytoplankton that shade submerged macrophyte leaves (Moss 1990). Second, after geese deplete the marsh vegetation, they expand their foraging area to graze local crops, which they can also damage (again, this effect may be intensified by subsidies from more remote agriculture). Third, geese translocate nutrients from croplands into the marsh, degrading wetland water quality and elevating nutrients to levels that support blooms of blue-green algae. Blue-greens may attain densities that are toxic to waterfowl and other animals. Finally, this eutrophication of wetlands enhances the probability of outbreaks of contagious avian cholera and type C botulism, which threaten other waterfowl of conservation concern, specifically the sandhill cranes (see Post et al. 1998 and references therein).

Post and colleagues studied these subsidy mechanisms and flow paths in sufficient detail to provide useful information to managers. First, they identified the vector organisms and the spatial and temporal scales of the fluxes. Geese (and not ducks, which fed within the wetlands, or cranes, which sometimes foraged outside wetlands for prolonged periods but farther afield) were largely responsible for importing nutrients from local agricultural fields into the wetlands. The amount of nutrients geese loaded into the wetlands was influenced by the weather. When wind speeds rose above a certain (temperature-dependent) threshold, geese tended to loaf at midday on a field, recycling some of their acquired nutrients locally. Under lower wind speeds and warmer temperatures, they returned midday to loaf and excrete on the wetlands. Loading also depended on whether geese fed on corn or alfalfa. Alfalfa had a higher gut passage rate, so a larger proportion of the nutrients in ingested biomass was excreted on agricultural fields. Alfalfa also, however, had a lower energy content than corn, so geese had to eat 8–9 times more biomass to meet their energy needs. Consequently, managers could reduce nutrient inputs to wetlands by planting only corn in areas foraged by geese. Of course, the most effective remedies for goose-induced eutrophication would be to flush more water through the wetland units where the geese rest, to reduce goose densities by hunting or harassment to overcome their colonial roosting habits (90% of the geese roost on 10% of the wetlands), or to expand available wetland habitat. Habitat expansion and increased flushing would require more water, an increasingly limiting resource in the arid western United States

(Reisner 1986, 1990; Gleick 1998; Power et al. 1997) and wherever human diversions and groundwater mining have altered food webs and ecosystems (Riley and Jefferies, chap. 25 in this volume; Kindler 1998).

SUBSIDIES AND CONSERVATION BIOLOGY

Food web connections across habitats affect focal species for conservation biology, whether these are endangered species themselves or key interactors that influence the fates of endangered species and their ecosystems.

Many key and endangered species forage in multiple adjacent habitats or migrate among distant habitats. Moose (Belovsky 1981), hippopotamuses (Naiman and Rogers 1997), eagles, and salmon (Spencer et al. 1991) use a mix of marine, freshwater, and terrestrial habitats. Consumers that depend on spatially and temporally variable resources (fruit, nectar, insects, forbs) must be highly mobile to track those resources, sometimes over intercontinental scales (Levey and Stiles 1992). Within a region, most birds and large herbivores and carnivores obtain foods from a mosaic of habitats. For example, Serengeti ungulates must migrate in order to acquire the proper mix of required elements (e.g., Na, P, Ca) from plants that grow in several different places (McNaughton 1990). Across regions, endangered migratory songbirds, water birds, and raptors, land and marine mammals, and some insects migrate long distances to feed in quite distinct habitats. Songbirds are vulnerable to changes in resource and habitat availability in either their northern (summer) or southern (winter) habitats.

Many key and endangered species use resources that arrive from more than one habitat. These include species living in habitat edges (such as riparian and coastal ecotones; see below) and species that use migratory foods. For example, approximately fifty species of terrestrial predatory and scavenging vertebrates in Alaska receive much of their annual energy budget by eating anadromous salmon (Willson et al. 1998 and chap. 19 in this volume). The orange roughy, an important commercial fish, lives on deep seamounts, areas of very low light and in situ productivity. These fish eat prey carried to them by currents and vertical migrations; their population biomass was once an order of magnitude higher than that of deep-sea fish populations that do not receive allochthonous prey. Nevertheless, their overexploited populations are now plummeting (Koslow 1997).

Enemies of key or endangered species are frequently subsidized by foods produced in other habitats. This problem worsens as habitats become more fragmented and edge-to-area ratios increase. For example, nest parasitism by cowbirds from nonforested habitats can greatly depress forest birds in fragmented forests because the smaller a forest fragment, the greater the proportion of that fragment that lies along a nonforested edge (Ambuell and Temple 1983; Wilcove et al. 1986; Andrén and Anglestam 1988). In a similar manner, subsidized domestic cats threaten English songbird populations and ground-dwelling birds in northern California.

Mobile animals can damage ecosystem structure and function. Concentrations of domestic herbivores (e.g., cattle, sheep, goats) denude vegetation worldwide around water holes, riparian areas, corrals, and feeding stations. These effects should be considered when aid agencies sink wells to ameliorate the effects of drought in developing countries. An unpleasant management surprise resulted from spatially rearranging water subsidies in Kruger National Park, South Africa (Starfield and Bleloch 1986). When game managers dug water holes in northern parts of the park to support the local endangered roan antelope, zebras were attracted to areas formerly too dry to support them. Lions followed the zebras, but wiped out the roan antelope instead. Wildlife modelers deduced this by noting that the roan antelope had good adult body weight but poor juvenile survival, indicating that its population declined because of apparent competition (sensu Holt 1977) rather than exploitative competition with zebras. Apparent competition may also affect a species' conservation status in natural circumstances. Nomadic herds of ungulates traveling through the Serengeti track rainfall to forage in relatively productive habitats (Sinclair and Norton-Griffiths 1979; Senft et al. 1987). Such migratory prey (e.g., wildebeest) are thought to allow resident lions to increase to the point at which they depress resident species (e.g., warthogs, impala; Schaller 1972). Subsidized lions are a key factor limiting endangered cheetah populations (Caro 1987). Heavy poaching of mobile Cape buffalo outside the Serengeti lowered buffalo numbers in the game park, causing lions to decrease substantially and resulting in increases in several alternative prey species (A. R. E. Sinclair, personal communication to GAP).

Trophic interconnections among habitats carry important implications for conservation efforts, which are usually directed at target species or circumscribed habitats. Most obviously, conservation may require the preservation or management of more than just focal species and their habitats. Allochthonous inputs may affect the success and abundance of a species directly (as food) or indirectly (via food web effects). Migrations to and inputs from external habitats can affect local community dynamics and ecosystem function. The loss of allochthonous resources could threaten species and whole communities. An excellent example of a non-trophic but crucial allochthonous resource is the shifting sands needed to preserve the

Coachella fringe-toed lizard and several endemic beetle species in the Coachella Valley of California (Turner et al. 1984; Beatley 1992; Barrows 1997). To preserve the unstable habitats that gave these rare species an edge over competitors that could exclude them in more stable habitats, The Nature Conservancy of California and other planners had to protect not only the lizard's immediate habitat, but also the source areas generating the sands and the landscape corridors that convey sands to the dunes actually inhabited by the lizard.

Overall, it is critical to understand that species success, community structure, and ecosystem function are often strongly connected to the dynamics of other habitats. Adequate conservation plans ideally should include all habitats influencing the dynamics of the target species and ecosystems.

CONSERVATION AT THE LAND-SEA INTERFACE

We now explore how the sea influences, both positively and negatively, conservation on islands and along the coastal ecotone. These habitats are home to many of the world's endangered and endemic species. Diverse terrestrial animals using marine resources attain high densities on coasts and small islands worldwide (Polis et al., chap. 14 in this volume). For example, many island endemics depend on foods associated with seabirds (e.g., the tuatara of New Zealand; Daugherty et al. 1990). Many large vertebrates flourish along coasts, including large birds of many types (e.g., eagles and large vultures) and mammalian carnivores (e.g., many canids and felids) (Rose and Polis 1998). These diverse species eat intertidal foods, marine carrion, and foods from colonies of marine birds and mammals. For example, photographs from the early 1900s show groups of the now endangered California condor foraging on whale carcasses stranded along southern California beaches. For all these creatures, input from the ocean is key to the success and numbers not only of coastal populations, but also possibly of inland populations via source-sink dynamics.

Humans have changed many aspects of marine ecosystems. Most notably, overfishing and pollution have decreased the productivity and species diversity of coastal and open ocean communities. Such changes must exert profound influences on the quality and quantity of input to islands and thus to island communities. For example, decreased fish populations depress the abundance of seabirds, a dominant group that structures entire communities on islands worldwide (Hutchinson 1950; Polis et al., chap. 14, and Anderson and Polis, chap. 6 in this volume). Decreases in the

abundance of marine fish and mammals also lower the numbers of carnivorous and scavenging vertebrates that forage on marine carrion along
the coast (Rose and Polis 1988) and on anadromous fish along river shorelines (e.g., Willson et al., chap. 19 in this volume, Ben-David, Bowyer et al.
1998; Wipfli 1997).

Humans have also damaged island communities by introducing exotic
animals (Atkinson 1994), either deliberately (e.g., grazing sheep, goats,
rabbits, and cattle; omnivorous pigs and boars; and predatory domestic
cats, mongooses, and foxes) or inadvertently (several species of rats). The
results have usually been catastrophic for native biota. It is not as well
recognized, however, that foods from the sea have magnified the harmful
effects of introduced species. Scattered reports document that most of
these invasive mammals use marine foods (either shore material or
seabirds). These subsidized exotics then depress populations of local
endemic species, sometimes to the point of extinction. In many cases, invasive species have largely extirpated native endemics but still occur at
high numbers, maintained by shore, intertidal, and seabird resources.
Such subsidized populations place continuous pressure on those few
natives that have escaped the initial depredation and thwart attempts to
reintroduce and restore native populations.

Domestic cats (*Felis catus*) and various rats are notorious for causing
extinctions on islands. Subsidized by seabirds, they also eat insular terrestrial species (Williams 1978; Burger 1985; Atkinson 1994). Cats on Ascen-
sion Island exterminated five species of seabirds, originally present in
thousands of breeding pairs. Norway rats (*Rattus norvegicus*) on South Georgia Island restrict the breeding areas and numbers of seven seabird species
by eating their eggs and chicks. Worldwide declines of small and medium-
sized seabirds are attributed to predation on adults, chicks, and eggs by
introduced rats and cats on nesting islands (Atkinson 1985; Stapp 2002;
McChesney and Tershy 1998; Hobson et al. 1999; and many others). Cats
have strong effects even on non-bird islands. For example, on islands on
the Pacific side of Baja California, cats have greatly depressed native reptiles and small mammals (D. Croll, personal communication). These cats
still occur in large numbers, but only along the shore, where they prey on
and scavenge marine foods.

Regurgitated scraps and corpses of seabirds may be important food
sources for mice and rats (e.g., Rowe-Rowe and Crafford 1992), supporting
high densities of both native and introduced rodents on some islands with
seabird colonies (e.g., Rowe-Rowe and Crafford 1992; Efford et al. 1988).
These omnivorous and opportunistic rodents subsequently eat and threaten

local populations of other native fauna (e.g., lizards [Cree et al. 1995; Towns 1991] and arthropods [Palmer and Pons 1996; Rowe-Rowe et al. 1989; Bremner et al. 1984]) and flora (Ryan et al. 1989) to an extent not possible if seabirds were absent. For example, kiore (a native New Zealand rat) accidentally introduced to offshore islands have greatly reduced populations of seabirds, endemic arthropods, lizards, and tuatara in New Zealand (Daugherty et al. 1990).

Even large herbivores such as cattle, sheep, and deer can maintain populations above the carrying capacity supported by terrestrial plant productivity by foraging on intertidal algae. Red deer feed on algae and then damage the terrestrial plant community by heavy grazing on Scottish islands (Clutton-Brock et al. 1983). On Auckland Island, introduced cattle survived on algae after removing almost all native land plants. Pigs on Auckland Island take eggs, young, and adults of burrow-nesting birds while also competing with insular herbivores (Atkinson 1994). The interplay among marine subsidies, introduced exotics, and declining native endemic and endangered species merits further investigation.

CONTROL OR LACK THEREOF: SUBSIDIES AND INFECTIOUS DISEASE

Humans have suppressed or eliminated most of the large species that threaten us. We may have accomplished this long ago: higher percentages of dangerous animals (mammoths, rhinos, bears) are prevalent in the earliest known cave paintings (e.g., in Chauvet, dated at 32,000 B.P.) than in more recent cave art painted 12,000–20,000 years B.P. (Jean Clottes, cited in Balter 1999). We have been less successful in controlling our smaller enemies, and may lose ground in this effort as the human population grows denser and more globally mixed. Zoonotic infections (yellow fever, typhus, Chagas' disease, hantavirus) increase with contact between humans and nonhuman hosts. Outbreaks can occur when and where human agriculture subsidizes rodent host populations and concentrates them near human dwellings (e.g., Lassa and Ebola fevers; Garrett 1994). In other cases, we rearrange the environment in ways that facilitate the contact. Felling tropical trees brings *Hemagoggus speggazini*, a canopy-dwelling mosquito that vectors yellow fever, into contact with humans (Southwick 1996). Air travel provides fossil fuel subsidies to pathogens, increasing contact among infected and susceptible human hosts. Infectious disease agents can now spread and explode in a world where the most distant major cities are only 16 hours apart.

Our subsidies to pigs, cattle, chickens, Atlantic salmon, and other live-stock, concentrated for industrial meat production, have polluted ecosystems over regional scales. Nitrogenous wastes from pig factories enter coastal rivers, estuaries, lagoons, and shallow coastal waters off the southeastern United States by two paths: continuously by ammonia volatilization and precipitation (Paerl 1985) and episodically, when sewage spills occur (Mallin et al. 1999). The resulting eutrophication of the habitats that once sustained valuable shellfish and finfish populations will be difficult to reverse. When phytoplankton blooms triggered by these nutrient additions sink, they create anaerobic conditions in bed sediments, which mobilize nutrients. These nutrients are then easily stirred by wind or currents back into the water column in shallow estuaries and offshore lagoons, stimulating subsequent phytoplankton blooms (Paerl 1985). These blooms lower the oxygen content of the water and often are dominated by harmful algae, including *Pfeisteria piscicida*. Increasingly frequent blooms and outbreaks off the southeastern United States have caused repeated, large-scale fish kills and human as well as environmental health problems (Burkholder et al. 1997).

Another threat to human health from industrial meat production is the use of antibiotics in highly crowded factory farms and their counterparts in aquaculture. About half of the antibiotics used annually in the United States are used "subtherapeutically" in animal feeds (American Society of Microbiology 1995). The profligate use of antibiotics for meat production selects for strains of drug-resistant bacteria (e.g., antibiotic-resistant *Salmonella* traced to pigs in Denmark; Hwang 2000), reducing our arsenal of antibiotics at a time when we are particularly likely to need them.

CONCLUSIONS

As the natural world falls ever more under human domination, we distort both local ecosystems and the flows among them at regional and global scales. The subsidy framework developed by Gary Polis and his colleagues can help organize our attempts to understand, predict, and manage the consequences. This framework requires first that the landscape positions of the sources and flow paths of fluxes that influence local ecosystems be identified. Failure to do this has precluded efforts to protect species and ecosystem services (e.g., by not distinguishing source from sink populations or by failing to recognize flow paths of enemies or crucial resources). The subsidy framework also requires expanded scales of study, as local short-term studies will not uncover causal linkages and feedbacks acting over large spatial or temporal scales. With models based on assumptions about key processes and interactions in both source and recipient habitats, we can ask how chains of consequences set in motion when we intensify or curtail ecosystem fluxes will play out over years, decades, and centuries and over local, regional, and global spatial scales. We can also ask how the system would change under various management schemes or under foreseeable changes in the environment.

These approaches could help us recognize when management for one target (e.g., agricultural production) affects other values or goals (e.g., species conservation, water quality, or human health). Understanding these interconnections and the large spatial scales over which the human enterprise distorts them is vital, not only for predicting and evaluating the consequences of our actions, but also for educating the public and marshaling the political will to change destructive practices and policies.

Chapter 25

Subsidy Dynamics and Global Change

Ralph H. Riley and Robert L. Jefferies

Global change is ever present. Change in the energy output of the sun, cyclic changes dictated by the rotation and tilt of the earth, and cyclic or random catastrophic events, including asteroid impacts, are just a few of the changes that have shaped the evolution of life and the biosphere. Biologically mediated changes, such as the development of an oxygen-rich atmosphere and nitrogen fixation, have also played an important role (Schlesinger 1997). In recent centuries, however, humankind has exerted a profound influence on the biosphere, and our influence grows as our population and technologies expand. How will current patterns of global change affect the dynamics of subsidies?

While many people associate the term "global change" with climate change, there are numerous environmental trends occurring on a global scale that are unrelated to climate (Vitousek 1994). Many of these are likely to have a more profound influence on the magnitude and effects of subsidies than climate change. While we recognize that any perturbation to a system could conceivably produce some effect on the components of that system, here we focus on those changes, perceptible at a global scale, that may have a substantial influence on the production, transfer, or use of subsidy materials that pass between donor and recipient systems (sensu Polis, Anderson, and Holt 1997). We include in this discussion changes in processes operating at global and regional scales (e.g., increase of green-house gases, changes in precipitation) as well as changes that occur on a very local scale (e.g., land use changes, alien species invasions), but whose cumulative effects on a global scale are substantial.

We also focus our discussion on global change trends that are rapid, as opposed to very long-term trends such as those in solar luminosity or plate tectonics. Specifically, in table 25.1, we highlight those trends that are occurring today and whose effects we are likely to see within the next generation. These trends are global in that that are likely to have a significant influence on a substantial proportion of the relevant systems on earth. While the causes of global change may include factors that are influenced by humans as well as some beyond human control, table 25.1 is dominated by those changes triggered by the accelerated pace of human-caused environmental change.

In table 25.2, we highlight some examples of the ways in which these global trends may affect subsidies. We have arranged this table to reflect the three spatial components of subsidies suggested by Polis, Anderson, and Holt (1997): donor systems, recipient systems, and the boundaries between them. Many global change trends will have variable effects on one or more of these components, and thus it is useful to consider them separately.

While it is impossible to anticipate all the consequences of global change, we have chosen to elaborate on three trends: land cover changes, exotic species invasions, and increased nutrient subsidies from human activities. These are not necessarily the most important global change trends, but they have fundamentally altered the rates and processes of subsidies around the globe. These examples also demonstrate the spatial and temporal linkages that extend subsidy effects through several systems of the environment.

LAND COVER CHANGES

Land cover changes affect both donor systems and the systems receiving subsidies from them. They lead to altered amounts of energy and nutrient subsidies moving across ecosystem boundaries. These changes affect both species assemblages and primary and secondary productivity. As Meyer and Turner (1992) indicate, land cover changes encompass both the conversion of one type of land cover to another and the modification of plant assemblages within an existing cover type, such as the conversion of forests to tree plantations or severe overgrazing in grasslands. Because

Table 25.1 Global change trends that are influencing, or may influence, the production, delivery, or utilization of subsidies

Global change trend	Comments on global rates and implications
Land cover changes	Large areas of the planet have been modified by humans. The area devoted to croplands has increased more than 5 times since 1700 on a global basis (Meyer and Turner 1992). Turner et al. (1990) estimate that half the surface of the earth (excluding permanent ice fields) has been transformed or exploited by humans.
Introduction of alien species	The distributions of species have been and continue to be dramatically transformed by human activities, both intentionally and unintentionally. In some countries, exotic plants constitute more than 25% of the flora (Heywood 1989).
Increased application of nutrients by humans	It is estimated that N fixation through human activities (fertilizer production, cultivation of N-fixing plant species, and industrial by-products) now exceeds N fixation from all combined natural sources (Vitousek 1994). Application of nitrogen fertilizer has more than doubled since 1970 to 80 Tg yr^{-1} in 1990, and it is expected to increase to at least 130 Tg yr^{-1} by 2050 (Matson et al. 1998).
Temperature changes	Computer models suggest substantial changes in surface temperatures around the globe, with most areas experiencing an increase in temperature of up to several degrees centigrade (Rind et al. 1990).
Changes in precipitation and evapotranspiration	Computer models suggest that the many effects of increasing greenhouse gas concentrations in the atmosphere may lead to substantial changes in the amount and timing of precipitation and to changes in water balance in different regions of the globe (Manabe and Wetherald 1986; Rind et al.1990; Loaiciga et al. 1995).
River channelization (e.g., dikes or deepening of channels)	Very few rivers near human settlement (as well as elsewhere) have not been channelized (e.g., 4,500 km of embankments on the Mississippi River) to alter their hydrology to be more favorable for certain human activities (e.g., flood control, drainage) (Ward 1978). Due to the prominent role water plays in transporting subsidies between systems, these changes in the relationship of a river to its basin are particularly important.
Human population growth	The explosion of the human population, apart from the many effects listed in this table, has also had a substantial effect on subsidies simply by virtue of our biomass and the diversion of subsidies needed to support that biomass. Even if each human were

Table 25.1 (Continued)

Global change trend	Comments on global rates and implications
	to consume a minimum amount of carbon and nutrients, the four- to fivefold increase of humans in this century alone (Smil 1997; Ojima, Galvin, and Turner 1994) represents a considerable rerouting of resources through human biomass. Our omnivorous habit and mobility lead us to be important vectors in moving carbon and nutrients across system boundaries (e.g., fishing reroutes marine and freshwater materials through terrestrial systems).
Dam construction	Dams substantially alter the shape and function of rivers. Few major rivers in developed countries do not have dams, and substantial dam building continues in developing countries. From 1950 to 1986, more than 36,000 large dams (> 15 m high) had been constructed or were under construction in the world (Goudie 1993).
Water diversion and irrigation	Irrigation reroutes the flow of water (and thus of subsidies), thereby establishing subsidy flows that previously did not exist (e.g., river to non-riparian terrestrial systems) and modifying existing ones (e.g., upstream to downstream or river to floodplain). The amount of irrigated cropland in the world has increased by 2,400% since 1700 (Meyer and Turner 1992). In 1985, approximately 2,710 km^3 of water per year was withdrawn from rivers for irrigation, of which the majority (approximately 2,340 $km^3 yr^{-1}$) was consumptive use (lost as evapotranspiration or seepage) (L'vovich and White 1990).
Harvesting (e.g., logging, cropping, fishing, etc.)	Although we apply nutrients and carbon to many systems, we also remove large fractions of these potential subsidies through harvest activities (e.g., 15–35 kg P ha^{-1} is removed in harvesting cereal crops; Smil 1997). Inasmuch as harvest and application of nutrients and carbon are never balanced spatially or temporally, these activities have the potential to dramatically shift pools of subsidies. Harvesting of organisms also alters the composition of communities, thereby altering food web structures.
Altered transport of subsidies by rivers (due to a number of factors including fertilizer use, land change, irrigation, etc.)	Transport of subsidies and sediment via river flow to the oceans has been substantially altered due to use a number of factors throughout the world (McIntyre 1992; Howarth, Billen et al. 1996). Of particular concern are the changes in subsidy delivery to coastal margins.

(Continued)

Table 25.1 (Continued)

Global change trend	Comments on global rates and implications
Loss of ozone and associated increase in UV-B exposure at the surface of the earth	As a result of atmospheric ozone depletion, solar UV-B reaching the surface of the earth is increasing (Rozema et al. 1997). This increase is not distributed evenly over the surface of the earth (Mathews and Keep 1993). UV-B has substantial effects on organisms, though the susceptibility of organisms to the deleterious effects of UV-B exposure varies considerably among species (Rozema et al. 1997).
Use of human-manufactured biocides in the environment and toxic waste	Pesticides and herbicides, by definition, alter the species composition of ecosystems. Although several classes of low-persistence biocides have been developed (after the recognition of the deleterious effects of DDT), contamination of nontarget ecosystems continues (Brown et al. 1990). Global consumption of pesticides exceeded 150,000 metric tons in 1980 (Brown et al. 1990). Discharge of many toxic materials into the environment is decreasing in most developed countries, but a considerable legacy of previous practices will persist (e.g., Schulz-Bull et al. 1995; Dahlgaard 1996; Holm 1996), and controls on toxic waste are less stringent in many developing countries.
Enhanced levels of carbon in the atmosphere	Carbon dioxide concentrations in the atmosphere have increased by more than 25% during the last century (Ojima, Galvin, and Turner 1994). This increase has substantial implications for plant physiology specifically and carbon metabolism in general, as well as numerous related effects listed in this table.
Acid deposition	Several important elements may be transported long distances and deposited as acidic precipitation (acid rain). This deposition has many consequences, including deleterious effects. In some cases these elements act as subsidies. Although rates of acidic deposition from the atmosphere appear to be decreasing in some regions, recovery of aquatic systems is not uniform (Stoddard et al. 1999). There is also concern over acidic precipitation in rapidly developing countries.
Enhanced extinction rates	Current data suggest that extinction rates are increasing, and this increase is projected to continue (e.g., Lovejoy 1980; Simberloff 1986). We may

Table 25.1 (Continued)

Global change trend	Comments on global rates and implications
	expect alterations in many susceptible food webs resulting from the loss of constituent species alone.
Alteration of global circulation patterns	Some global climate model simulations support the suggestion that El Niño/Southern Oscillation events may become more frequent with global warming (e.g., Timmermann et al. 1999).
Genetic engineering of organisms	Genetic engineering has the potential to alter the effects of land use changes through changes in carbon metabolism and nutrient use efficiency.

NOTE: The first three listed trends are not necessarily the most important, but they are discussed in greater detail in the text.

Table 25.2 Examples of how global change trends may alter the mass or lability of subsidies across boundaries, the flux of subsidies from donor systems, the flux of subsidies across boundaries, or the utilization of subsidies in recipient systems

Global change trend	Effects on subsidy production in donor systems	Effects on subsidy transit through boundaries	Effects on subsidy use in recipient systems
Land cover changes	Nutrient flux to streams is often altered (Pimentel et al. 1995); conversion of forest to pasture; removal of riparian vegetation reduces direct detrital input (Gregory et al. 1991)	May alter processes that intercept/modify subsidies, such as riparian margin planting (Karr and Schlosser 1981); irrigation, drainage alters waterborne transport of subsidies (Ziemer and Lisle 1998)	Conversion of riparian forest to grassland or vice versa alters the energy flux to streams, thereby changing in-stream productivity and use of subsidies (Chamberlain et al. alter the timing or 1991); may magnitude of subsidies
Introduction of alien species	Enhanced terrestrial nitrogen pool due to invasion of N-fixer (Vitousek 1987) establishes an enhanced pool of subsidy available for export; top-down effects from trout introduction enhance organic matter production (Huryn 1998)	Introduction of anadromous salmon to Chilean streams (Willson, chap 19 in this volume) establishes a new vector and linkage between systems	Top-down effects from trout introduction alter consumer use of primary production (Huryn 1998) and potentially use of terrestrial carbon subsidies
Increased application of nutrients by humans	Alters primary and secondary productivity	Application techniques may enhance transport of subsidies by water or wind, depending on fertilizer type or whether applied on the ground or by air	Leads to a shift away from N limitation in coastal waters (Justic et al. 1995)
Temperature changes	May alter decomposition rates and chemical process rates, thereby altering the turnover of subsidy pools (Lükewille and Wright 1997; Trumbore 1997)	Hydrology may change in response to temperature changes (Loaiciga et al. 1995), thereby altering the configuration of boundaries or the capacity of water to transport subsidies between systems	May alter thermoclines in lakes, thereby altering nutrient utilization in lake compartments (Schindler et al. 1990)

Changes in precipitation and evapotranspiration	Alters the availability of organic matter subsidies through effects on production and decomposition (McNulty et al. 1996; Schimel et al. 1997); alters rates of chemical process such as denitrification	Change in water flux to rivers/lakes (Vitousek 1994; Jordan and Weller 1996; Peierls et al. 1991); may alter the size of the aquatic-terrestrial interface, such as lake margins, effective floodplain	The volume of aquatic systems (e.g., sea level rise; Paw and Chua 1991) may change, which could alter aquatic community structure, proportional loading of subsidies, or residence time of subsidies
River channelization	Alters the availability of subsidies through effects on primary and secondary productivity (Portt et al. 1986)	May interrupt two-way transport of subsidies (sensu Junk et al. 1989; Vannote et al. 1980; Ward and Stanford 1995)	May alter the local geomorphology and response to floods, reducing lateral connectivity and shifting the importance of upstream to downstream subsidies (sensu Ward and Stanford 1995)
Dam construction	Submerges a portion of the terrestrial environment, often eliminating highly productive floodplains as donor systems (Shuman 1995); reservoirs worldwide have drowned about 1 million km² of land (Roberts 1994); alters river environment	Alters the aquatic-terrestrial interface (Ward et al. 1999); alters lateral linkages through local terrestrial subsidies, subsidies in groundwater, or exchange with hyporheos (Ward et al. 1999; Shuman 1995)	Alters downstream hydrology, productivity, and community structure (Ligon et al. 1995; Puckridge et al. 1998)
Water diversion and irrigation	Alters water flow and subsidy mass transport from upstream systems to downstream systems or from land to water systems	Stream reaches become donor systems for terrestrial systems	May alter mass flux of subsidies to downstream systems or alter productivity in some recipient systems, such as terrestrial systems; such alterations may include a decrease in productivity as through salinization

(Continued)

Table 25.2 (Continued)

Global change trend	Effects on subsidy production in donor systems	Effects on subsidy transit through boundaries	Effects on subsidy use in recipient systems
Harvesting	Diverts materials out of the system that otherwise may have been subsidies to other systems	Elimination of salmon as a vector (Willson, chap. 19 in this volume)	May alter food web structure and energy flow
Altered transport of subsidies by rivers	Alters the primary and secondary productivity in estuaries or oceans (Rabalais et al. 1996), thereby affecting the availability of subsidies from these systems to other receiving systems		May alter food web structure and composition, as in oceanic dead zones (Rabalais et al. 1996; Andersson and Rydberg 1988; Sarmiento et al. 1988)
Loss of ozone and associated increase in UV-B exposure at the surface of the earth	May alter the primary or secondary productivity of UV-sensitive species (United Nations 1998; Sullivan 1997)	Change in populations of terrestrial-feeding invertebrates may lead to change in mass of terrestrial subsidy to streams (Williamson 1996); alters the transport of carbon	Increased exposure alters community dynamics (Bothwell et al. 1994) and consequently nutrient and carbon metabolism (Riley et al. 1997)
Use of human-manufactured biocides in the environment and toxic waste	Alters food web structure, may alter ecosystem processes of donor systems (Howarth 1991)	May eliminate some vectors, such as terrestrial insects, used by stream organisms; alters community composition and function	Alters the food web structure and composition (Howarth 1991; Levine 1989)
Enhanced levels of carbon in the atmosphere	May alter primary productivity of donor environments and decomposition rates (Schimel 1995; Luxmoore et al. 1993); alters nutrient cycling rates (Berntson and Bazzaz 1997)		May alter primary productivity or biogeochemistry in recipient system

Acid deposition	Alters primary productivity of donor environments (Galloway et al. 1994) or decomposition or biogeochemistry of soils, thereby shifting element ratios of leachate from soils (Heneghan and Bolger 1996)	Through changes in pH, alters the chemical composition of subsidies in solution	Alters species composition, nitrogen fixation, or consumer-resource relationships (Hermann et al. 1997; Findlay et al. 1999)
Enhanced extinction rates	May alter species composition and productivity in donor environments	Loss of anadromous fish reduces subsidies (Willson, chap. 19 in this volume)	May alter species composition and productivity in recipient environments
Alteration of global circulation patterns	May indirectly alter the productivity/decomposition regime of donor environments through the alteration of local climates	Interrupts deep-sea upwelling along the South American coast (Patterson et al. 1992); may alter salmon runs on Pacific coast of North America	Alters species composition and productivity (Patterson et al. 1992; Lima et al. 1999)
Genetic engineering of organisms	May alter the resource use efficiency of organisms, thereby affecting resource retention and flux within donor systems		May alter C:N ratios in organic matter, thereby influencing the utilization of subsidies

of the difficulty of classifying these changes accurately in different regions, the overall changes in land cover are uncertain. However, the global increase in cropland between 1700 and 1980 has been estimated at between 392% and 466%, and for irrigated cropland between 1800 and 1989, the estimated increase is 2,400% (Meyer and Turner 1992).

The major proximate agents of land conversion have been fire, clear-cut timbering, tillage of soils, and drainage of wetlands. When these changes occur, substantial alterations of movement of materials across boundaries can be expected to take place during the transition phase. Thereafter, the new plant assemblages and land management practices will impose further changes on the export of allochthonous materials. The classic study of Davis (1976) at Frains Lake, Michigan, indicates that at the time of woodland clearance between 1830 and 1860, soil erosion rates increased 30 to 80 times. When farmland was established after 1860, the erosion rate was only about 10 times that in the present undisturbed woodland. A number of other studies conducted elsewhere have confirmed the same general pattern of erosion rates. The overall erosional response, however, is often less than that at Frains Lake (cf. Foster et al. 1990). In another study, rates of erosion associated with various land use activities in Ohio were estimated at less than 0.01, 0.09, 25.54, 154.61, and 164.02 t ha^{-1} yr^{-1} for woodland, grassland, rotated fields, fallow fields, and cornfields, respectively (Bennett 1938). Overall, the United States has lost an estimated one-third of its topsoil in the last 200 years (Pimentel et al. 1995). Similar land cover changes occurred much earlier elsewhere, but data recording these changes are less readily available.

Collectively, these land cover changes represent biomanipulation on a continental scale and result in large changes in the rates of transfer of materials across land drainage system boundaries. Many of the effects of changes in land cover and similar phenomena are evident at the micro- and mesoscales (spatial scales of 10^0–10^{10} m^2 and temporal scales of 10^0–10^4 yr; Delcourt and Delcourt 1991). However, changes in the transfer of materials by water or air associated with land cover change and in their accumulation in a recipient system may be on a much larger spatial and temporal scale than local land cover changes.

Effects of Land Cover Changes on Riparian Systems

Riparian zone interfaces often are less modified by human activities than the surrounding cropland, and so frequently have a relatively high biodiversity and act as a refuge for pests and predators. Most of the changes in riparian zones are occurring at micro- and mesoscales. As such, these zones are sensitive indicators of changes in the movement of materials in drainage systems linked to alterations in land use (Naiman and Décamps 1997).

Riparian plant assemblages are composed of specialized and disturbance-adapted species that can be classified into functional groups. For example, early colonizers, such as species of *Salix, Alnus, Populus,* and *Sequoia,* invade new alluvial substrates. These trees produce adventitious roots where sediment deposition rates are rapid, and they are tolerant of flooding, soil anoxia, and unstable sediment conditions. Many of these species have high growth rates where nutrient loading and sedimentation rates are high. The discharge drainage system represents a mosaic of sites that may be aggrading, degrading, or maintaining a steady state associated with changes in sediment load. Rapid colonization of disturbed sites by riparian vegetation modifies the rate of organic and inorganic sediment movement in these aquatic systems.

A belt of riparian vegetation also acts as a nutrient filter, reducing runoff of nutrients and sediments from agricultural sources (Karr and Schlosser 1978), particularly nitrogen and phosphorus. In North Carolina, some riparian areas trap 80–90% of sediments eroded from agricultural fields (Cooper et al. 1987; Daniels and Gilliam 1997). Riparian grasses appear to be less efficient at collecting sediment than woody plants (Magette et al. 1989; Naiman and Décamps 1997). In addition to physical trapping, accumulation of nutrients by riparian vegetation, especially woody plants, contributes to the role of the riparian zone as a nutrient filter (Peterjohn and Correll 1984). Woody plants with high growth rates, such as species of *Populus,* have considerable potential to absorb nutrients (Cole 1981). Where soil nitrogen saturation occurs, phosphorus becomes limiting, and the vegetation acts as a temporary sink for this element.

The richest riparian communities also have the greatest proportion of exotics, possibly because of disturbance, nutrient availability, and the high amount of environmental heterogeneity. Percentages of exotic species in the total riverine flora range from 17% to 46% at different sites in North America and Europe. Some of these exotics, once established, can lead to a loss of native species, as in the case of *Impatiens glandulifera* and *Tamarix ramosissima* (Naiman and Décamps 1997).

In spite of the presence of riparian vegetation in some localities, agricultural and urban land use activities are still major sources of nonpoint pollution by nitrogen and phosphorus (Carpenter, Caraco et al. 1998). Although the magnitude of these inputs is difficult to measure, there is general agreement that in extreme cases they may result in toxic algal blooms, deoxygenation of water, fish deaths, loss of biodiversity, and loss of aquatic plants and coral reefs (Carpenter, Caraco et al. 1998). The food web structure of receiving systems may be altered by the input of non-nutritive materials through land cover changes. For example, Pacific salmon (*Oncorhynchus* spp.) have disappeared from 40% of their historic breeding ranges in the Pacific Northwest (Policansky 1998). Although the causes of the decline are varied, one influence is the effect of logging and associated activities (Ziemer and Lisle 1998), which result in an increase in sediment load in streams. The changes in the salmon population result in major shifts in aquatic food webs (Power 1990b; Polis, Anderson, and Holt 1997) and ultimately changes in subsidy supplies (see Willson et al., chap. 19 in this volume).

Land Use and Hydrological Changes

The hydrological effects (on surface water and groundwater) of land cover and land use changes include not only changes in water quality, but also changes in water flow. Irrigation is by far the largest component of global withdrawal of water from the hydrological cycle (accounting for about 75% of the total demand; Meyer and Turner 1992). In some cases, water for irrigation has been obtained by the diversion of surface waters (Bertoldi 1992; Margat 1996; Gleick 1998). In other cases, it has been "mined" from aquifers, thereby supplementing the regional surface water hydrology; this practice has sometimes been associated with the depletion or degradation of aquifers. Depending upon the source of water and on irrigation management practices, these hydrological changes have resulted in substantial changes in the flux of subsidies to the agricultural systems, even to the point at which the additional materials no longer act as subsidies, but are detrimental to the productivity of the agricultural system (e.g., salinization; El-Ashry and Gibbons 1988; Goudie 1993).

The diversion of water from rivers and lakes has also resulted in considerable changes to aquatic communities. In the Colorado River, for example, the construction of dams and regulation of water flow have led to an increased mineral load, algal growths of *Cladophora* spp., low diversity of the zoobenthos, and low population numbers of endemic fishes. These changes, coupled with the introduction of alien fish species and diseases, have contributed to major declines of indigenous fishes (Petts 1984; Stanford and Ward 1986) and substantial alteration of aquatic food webs. Similar changes have occurred elsewhere: the diversion of water from the Amu Darya and Syr Darya rivers for cotton irrigation during the last

35 years has led to the retreat of the Aral Sea shoreline by 120 km or more (Kindler 1998). The quantity of water diversion associated with land conversion to irrigated crops is related to substantial changes in subsidy supply, landscape structure (including boundaries between donor and recipient systems), and community structure in both donor and recipient systems.

There are also changes associated with the timing of water diversion. In natural drainage systems, periodic flooding at different temporal and spatial scales triggers a sequence of geomorphological and biological responses that may be maintained over years and decades. The flooding results in alterations to the structure of riverine and riparian habitats, as well as periodic increases in resource and food availability, that represent the outcome of changes in subsidy inputs (Michener and Haeuber 1998). These disturbances ensure the reproductive success of organisms by maintaining resources for growth and reproduction, and thus maintain the high species diversity found in these systems. Where disturbances and inputs of subsidy materials are reduced by artificial regulation of river flow, species diversity often declines. Since 1912, when water flow in the lower Missouri River floodplain began to be regulated, "16 species of fish, 7 of plants, 6 of insects, 2 of mussels, 4 of reptiles, 14 of birds, and 3 of mammals have been classified as endangered, threatened or rare" (Scientific Assessment and Strategy Team 1994; Galat et al. 1998). During the last 87 years, the river, which formerly consisted of braided channels, has been constrained by levees, river channelization, bank stabilization, and dam construction.

Although the long-term effects of the regulation the of lower Missouri River on the biota are poorly documented, extensive flooding in the 1990s has provided rare insights as to how the groups of species mentioned above became endangered or rare (Galat et al. 1998). In 1993 there was a catastrophic flood when peak flood discharges exceeded that expected in any 100-year period. Record floods occurred again along the lower Missouri River in 1995 and 1996. The connectivity of low-lying areas in the floodplain of the river at times of flooding led to major changes in limnological conditions that depended on the depth of remnant basins. Plants and animals immediately exploited the reconnection of the river to its floodplain. Perennial plant assemblages dominated by obligate emergent and floating-leaved species, which are adapted to flooding and undergo clonal reproduction, quickly recovered in flooded remnant basins where the water depth was suitable for growth (Mazourek 1998). Invasive species also colonized newly scoured sites. The frequency and timing of flooding affected seed dispersal, rates of germination, and transport of vegetative parts of plants. In contrast to the above, over the short term, the richness and density of insect assemblages in basins that were recurrently flooded declined due to scouring and a reduced food supply. However, large quantities of particulate organic matter in the flood water provided a long-term food base for insects; hence the decline is likely to be transitory.

The response of fish to flooding depended on the timing and degree of connectivity of river water and the presence of water temperatures appropriate for spawning. Temperatures from 15°C to 25°C, as well as access to floodplain aquatic vegetation, are required for successful reproduction of riverine fish. These conditions are met from April to late June when flooding of basins adjacent to the river occurs. Isolated basins, for example, contain only 50% of the fish species found in flooded sites.

Populations of bird species also change in relation to food resources and available habitat. Shorebirds require open habitats and exploit newly formed scours produced by flooding, whereas wetland birds, such as the American bittern (*Botaurus lentiginosus*), require emergent wetland vegetation characteristic of remnant, non-flooded basins. The absence of sand islands, which were present when the river consisted of a braided channel system, has led to a loss of nesting sites for the least tern (*Sternia antillarum*) and piping plover (*Charadrius melodus*). Likewise, the loss of deep, braided channels is thought to account for the disappearance of obligate large-river fishes, such as the pallid sturgeon (*Scaphirhynchus albus*) and sicklefin chub (*Macrhybosis meeki*) (Galat et al. 1998). These responses of individual species to periodic flooding indicate that the absence of these conditions is likely to lead to a rapid loss of species diversity in these remnant basins.

INVASIONS BY EXOTIC SPECIES

There are considerable differences in the proportions of alien species present in different countries. Values for plants range from a few percent in some countries to over 25% of the total flora in Australia, Canada, and New Zealand (Heywood 1989). Even higher percentages are recorded for specific local areas, especially for islands (Moore 1983). Biological invasions are thought to be the second largest cause of loss of biodiversity in the United States, after habitat destruction (Enserink 1999). In the state of Florida, one out of every three or four species is non-native, and in parts of San Francisco Bay, 99% of all biomass is produced by non-native species (Enserink 1999). Around the world, these invasions represent widely different chronological patterns and modes of entry and reflect the agricultural and economic history of a region, as they are caused directly or indirectly by human activities.

Invasion success is partly a function of community complexity (i.e., number of species) and the degree of trophic specialization of the invader (Drake 1983; Sugihara 1983). In addition, plant invaders are likely to become established, all other things being equal, where climatic and edaphic conditions are similar to those of their home range. From a broad perspective, invasions are less common in undisturbed intact systems. At the ecosystem level, it is often difficult to separate the effects of biological invasions per se from the secondary effects of disturbance caused by invaders, which may have a substantial effect on other invasions and on ecosystem processes. Sometimes these disturbances can be novel and large-scale. Exotic animals often interact with exotic plants, promoting the successful establishment of one or both of the invaders (Lawton and Jones 1995).

Most of the effects of invaders on ecosystem processes and the subsequent movements of allochthonous materials have been reported in systems undergoing secondary succession. This is partly because there are few sites where it is easy to observe primary succession and partly because rates of establishment of invaders are slower, and fewer species are involved, in early primary successional habitats. The latter factor reflects the low resource base, and especially the availability of nitrogen, in these habitats (Ramakrishnan and Vitousek 1989).

The effects of introduced species on ecosystem processes are varied. One obvious change is the acceleration of soil erosion rates brought about by grazing, browsing, and trampling by feral and semi-domesticated animals. The desertification of land as a result of depletion of vegetation by livestock in the Sahelian pastoral system during drought led to increased wind and water erosion of soil, including subsidy materials (Sinclair and Fryxell 1985; Graetz 1991). Marshall (1973) drew attention to the positive feedback between decreasing plant cover, increasing bare soil, and increasing rates of erosion in this system. Some of this windblown soil contributes to the long-distance movement of subsidies from this region to the Amazon basin (Polis, Anderson, and Holt 1997). Another example is the effect of the Himalayan tahr (*Hemitragus jemlahicus*) on the vegetation and soil dynamics at Table Mountain Nature Reserve in South Africa (Lloyd 1975). The overall effect of this alien species has led to a reduction in net primary production of 73% compared with that in exclosures. In addition, accelerated soil erosion has resulted in a loss from this system (and a subsidy to other systems) of 8 mm of soil and organic matter in 11 months in one area.

Alterations of other geomorphological processes by invaders are widespread, particularly at the interface between aquatic and terrestrial habitats.

A number of introduced plant species characteristic of habitats at this interface trap sediment and alter fluvial or marine geomorphology. Source-sink relationships governing sediment movement change at different spatial and temporal scales as these plants become established. Alien plant species that bring about these alterations include *Ammophila arenaria*, *Casuarina equisetifolia*, *Leymus mollis*, *Tamarix ramosissima*, *Spartina alterniflora*, and *S. anglica* (e.g., Hertling and Lubke 2000; Ndiaye et al. 1993).

Invasive introduced plants or fungi can trigger events that lead to the plant community utilizing more or less of the annual input of precipitation. The outcome of these changes can have a considerable effect on ecosystem processes. The invasion of nature reserves in the jarrah forests of south-western Australia by an introduced fungus, *Phytophora cinnamoni*, has led to 10% of all eucalyptus trees becoming severely diseased. A total area of 1.5×10^6 hectares is affected (Shea and Dell 1981). A consequence of the presence of large numbers of diseased trees is that the evapotranspiration demand of the forest has decreased. The underlying saline water table has risen to the surface, leading to runoff of saline water and presumably altered mass transport of subsidies from this system to aquatic systems.

In contrast, evapotranspiration by invasive *Tamarix* species in parks and reserves of the southwestern United States has lowered the water table, and springs have dried up (Stanford and Ward 1986; Vitousek 1986), reducing or eliminating a major mode of transport of subsidies. The lack of water flow has led to invasions of river courses by other introduced plants, such as *Datura innoxia* and *Prosopis* spp., that further lower the water table.

Introduced species have also altered fire regimes, leading to changes in ecosystem structure and function. In both North America and Australia, the invasion of introduced grasses has affected fire regimes (Parsons 1972; Christensen and Burrows 1986; Clark 1990; D'Antonio and Vitousek 1992). Generally, these changes have led to a higher fire frequency, partly because the introduced grasses increase the quantity and availability of fuel and, consequently, the extent and intensity of fires. This has occurred in Hawaii, where introduced species of *Andropogon* have become established (C. W. Smith 1985; Macdonald and Frame 1988). A similar situation exists in the rangelands of the western United States, where an introduced annual, cheatgrass (*Bromus tectorum*), is widespread. It matures early, unlike the native perennial grasses, and dries out quickly at the peak of the fire season, as it has a low moisture content (Klemmedson and Smith 1964). The invasion of cheatgrass has changed the fire regime so that spring and early summer fires have become more frequent. In contrast, in the fynbos of South Africa, the establishment of introduced trees such as *Hakea sericea* at the expense of the indigenous sclerophyll vegetation appears to have altered both the intensity and frequency of fires because of changes in fuel characteristics that have led to lack of recruitment of native species (Van Wilgen and Richardson 1985).

Where fires of a higher intensity occur as a consequence of invasive species producing more fuel, losses of essential elements such as nitrogen may be expected to rise (Chandler et al. 1983; Whelan 1995). In addition, losses of fine ash in the smoke column can lead to a loss of elements such as phosphorus. Fire mineralizes nutrients present in biomass and redistributes them over a wide area, depending upon its intensity. Such processes may have a significant effect on the supply of subsidies to other systems.

Within the semiarid and the humid, taller grasslands of the world, fire and herbivory are major consumers of aboveground plant biomass (Wedin 1995). Fire replaces grazing in the taller grasslands as the major "consumer" as high plant productivity leads to fuel accumulation. This shift is also a result of the high C:N and lignin:N ratios in the litter of the dominant grasses, which slows decomposition and leads to the accumulation of plant litter. Fire volatilizes most of the nitrogen in the litter and reinforces nitrogen limitation (Ojima, Schimel et al. 1994). At some sites within tallgrass prairies, C_3 species, such as *Agropyron repens* and *Poa pratensis*, are replacing the indigenous slower-growing C_4 species, such as *Schizachyrium (Andropogon) scoparium,* particularly where land is cultivated and fertilized. The higher-quality (lower C:N ratio) litter of these C_3 grasses decomposes rapidly and shows no net immobilization of nitrogen, unlike litter from the C_4 grasses (Wedin and Tilman 1990; Wedin and Pastor 1993). Where the quality of the litter changes (i.e., C:N ratio falls) as a result of invader species becoming established, there is a shift from fire consumption to greater detrital consumption of ungrazed vegetation, which reduces the export of nitrogen from the system. These changes illustrate the important effects plant invaders may have on the biogeochemical cycling of nitrogen.

The spread of exotic nitrogen-fixing plants in early primary successional habitats, such as the establishment of *Myrica faya* on young volcanic substrates in Hawaii, can alter nitrogen availability and increase the rate of soil development (Vitousek 1986, 1987). A number of introduced legume species appear to act as nitrogen "donor species," altering patterns of biogeochemical cycling and thus the supplies of subsidies. Pioneer native nitrogen-fixing species can also strongly alter community and ecosystem dynamics due to their ability to increase soil nitrogen and alter soil nitrogen dynamics. Along portions of the northern California coast, grasslands where bush lupine (*Lupinus arboreus*) populations grow have almost twice the amount of soil nitrogen compared with sites that lack lupine (Maron and Jefferies 1999). Lupines can grow at high densities, and while alive, they simultaneously enrich the soil and shade the ground beneath them (Maron and Connors 1966). Individuals and stands of lupine, however, frequently die from the effects of herbivory (Strong et al. 1995) and are replaced by several species of invasive C_3 annual grasses until a new cohort of lupine becomes established. In sites low in soil nitrogen where bush lupines historically have not become established, a more diverse coastal prairie plant assemblage is present, consisting of native perennial grasses and a mix of annual and perennial forbs, many of which are native. Establishment of this native nitrogen-fixing species, therefore, has led to soil nitrogen enrichment that promotes invasion of fast-growing exotic species at the expense of the indigenous flora. Removal of the lupine is insufficient to restore the native plant assemblage, however; reductions in soil nitrogen are required—a much more difficult and long-term project (J. L. Maron and R. L. Jefferies, unpublished data).

A case study reported by Ramakrishnan and Vitousek (1989) in northeastern India demonstrates the subtle interactions between cultivation, invasive plants, native vegetation, and biogeochemical cycling. In the traditional cultivation method, secondary tropical/montane forest is cut and the cut biomass dried for 1–2 months, after which it is burned and the land used for crop production for 1–2 years. Thereafter, the land remains fallow for up to 30 years, although recently the interval has decreased to as little as 5 years. The tropical soils of this area are highly leached. Forest clearance leads to a loss of species richness and the establishment of "weedy" communities that are replaced by bamboo and other shade-intolerant species before forest species are reestablished. Many of the weedy species are C_3 exotics. Cutting and burning of older fallows increases soil fertility to a greater extent because of increased fuel loads associated with the buildup of C_4 plant biomass, which replaces the C_3 species as secondary succession proceeds. Over a long period, however, exotic C_3 species, if they remain, also can accumulate large amounts of biomass. These species include *Eupatorium* spp. and *Mikania micrantha,* which predominate in the early years following the cessation of cultivation (Saxena and Ramakrishnan 1986). Continuous imposition of short agricultural cycles (5–6 years fallow) results in the successional communities being dominated by C_3 exotics. Under these conditions, nitrogen accumulation during the fallow period is inadequate to replace the amounts removed in harvest, volatilized in fires, or leached (Mishra and Ramakrishnan 1984). In addition, nutrients are lost by soil erosion. Furthermore, as the soils are in a disturbed state much of the time and continuous plant cover is absent, there is also likely to be a loss of soil nitrogen by leaching of nitrate that otherwise would be taken up by plants (cf. Addiscott et al. 1991). These shorter agricultural cycles thus promote successful invasions of C_3 exotics, losses of C_4 species, and a decline in the availability of nitrogen within the system.

HUMAN-DERIVED NUTRIENT SUBSIDIES

With the advent of fossil fuel-based transportation and industrial systems in the nineteenth century, the mass of nutrients that humans directly applied to agricultural systems increased dramatically. Between 1950 and 1995, about 600 million metric tons of P fertilizer were applied to the earth's surface, primarily on croplands (Food and Agriculture Organization 1996). Nitrogen fertilizer production rose from nil in 1900 to more than 70 million tons yr^{-1} in 1985 (Smil 1991). Nutrients are also added to ecosystems indirectly as a consequence of industrial activities and through wind and water erosion of fertilized regions (e.g., Paerl 1985). Today, some estimates suggest that N fixed through human activities (energy production, crop cultivation, and fertilizer production, for example) exceeds biotic N fixation (Vitousek 1994; Galloway et al. 1995).

These added nutrients have numerous consequences for fluxes of subsidies between systems. One obvious effect is an increase in the mass of nutrient subsidies in donor systems available for transport to other systems (primarily to aquatic systems, but wind transport to other terrestrial systems may be significant under some circumstances; Polis 1998). Mineral fertilizer application (and in some cases, nutrient input in acid precipitation) often exceeds the capacity of the fertilized system to take up the added nutrients (Greenwood 1990; Matson et al. 1998; Moffat 1998). In some cases, the application of fertilizers is out of phase with crop phenology, or fallow periods correspond to periods of peak nitrogen mineralization, resulting in losses of NO_3^- via leaching (Johnston 1994). Jordan and Weller (1996) estimate that up to a third of net anthropogenic nitrogen input to regions of the United States is discharged in rivers. Nitrogen concentrations in large rivers flowing into the North Atlantic have increased 2 to 20 times since preindustrial times (Justic et al. 1995), and these increases are correlated with rates of application of human-derived nitrogen in the river basins (Howarth, Billen et al. 1996; Howarth, Schneider, and Swaney 1996).

In some regions, additions of some human-derived nutrients have decreased through stricter controls on pollution sources (von Gunten and Lienert 1993), but the long-term trend for many of the world's rivers continues to be much higher nutrient loadings today than in preindustrial times.

Another effect is a change in the ratios of nutrients in subsidies. The ratios of nutrients in mineral fertilizers are unlikely to match existing nutrient ratios in soil solutions and will therefore alter the ratios of nutrients transported into other systems through runoff, erosion, and groundwater. Rabalais et al. (1996) note that N and P concentrations in the Mississippi River have doubled in recent decades, but silica concentrations have decreased by half (according to the authors, as a consequence of increased freshwater riverine diatom production stimulated by enhanced phosphorus inputs). Based on an analysis of current river nutrient loadings and historical records, Justic et al. (1995) suggest that prior to the increase in anthropogenic inputs of nutrients to the Mississippi (especially since the 1940s), concentrations of N and P were low relative to the availability of silica, and coastal productivity in the Gulf of Mexico was probably limited by N or P. Today the ratios of these three nutrients are much closer to the Redfield ratio (Si:N:P = 16:16:1; Redfield et al. 1963). This finding suggests not only that the mass delivery of subsidies to the Gulf of Mexico has changed, but that the utilization of N, P, and silica subsidies has changed substantially as well, with N, P, and Si now being supplied in a ratio more favorable to diatom growth and with none of these subsidies being substantially in excess of demand.

These changes in the Mississippi River, which contributes 90% of the freshwater input to the Gulf of Mexico, have had important effects on the functioning of the northern Gulf. These effects include changes in phytoplankton species composition (e.g., less heavily silicified diatom species have increased in number) and an associated increase in primary productivity since at least the 1950s (Eadie et al. 1994). Decomposition of enhanced organic matter has led to the creation and maintenance of a large zone of bottom-water hypoxia (Rabalais et al. 1996) that has caused high rates of mortality for benthic organisms (Dortch and Whitledge 1992). Similar "dead zone" phenomena are occurring in other estuaries and coastal seas around the world (Sarmiento et al. 1988; Justic et al. 1995; Jickells 1998).

Third, enhanced productivity in donor systems in response to fertilizers will directly alter carbon subsidies to recipient systems. All other things being equal, enhanced terrestrial biomass production in response to fertilization should eventually alter direct carbon inputs to streams, as well as carbon transport in overland flow of water and through dissolved organic carbon in groundwater. Another possible effect is shifts in aboveground versus belowground allocation of carbon by plants (Greenwood et al. 1982). Such shifts influence the production of soil organic matter (and therefore dissolved organic carbon) versus aboveground litter or woody biomass and thereby may alter the relative "quality" of organic matter subsidies to recipient systems (e.g., rivers; Wetzel et al. 1995).

Finally, enhanced productivity in terrestrial donor systems in response to fertilizers may alter subsidies indirectly through modification of water balance. Several factors that influence terrestrial water balance may respond to fertilization, including plant biomass (affecting evapotranspiration), plant physiology (affecting instantaneous water use efficiency), and plant community structural complexity (affecting overland water flow, fog interception, canopy throughfall, snowpack, etc.). The relative importance of these effects, and even whether they increase or decrease water yield from basins, will vary depending on the local vegetation, climate, and land use practices (Holdsworth and Mark 1990; Ziemer and Lisle 1998). In practice, fertilization is usually associated with other land management activities that have an equal or greater effect on landscape water balance. For instance, although pastoral systems in southern New Zealand receive substantial nutrient subsidies, primarily in the form of superphosphate fertilizers, these systems often have lower plant canopy heights than ungrazed grasslands due to heavier grazing and the common practice of substituting exotic pastoral grasses for the native grasses, which form upright tussocks. These differences in the structure of grassland plant canopies influence snow (and perhaps fog) interception. This appears to be the major reason why tussock grassland systems generally have higher water yield, and different timing of water yield, than tussock systems of lower canopy height or pastoral systems (Clearwater et al. 1999). Since water is a major vector for the transport of subsidies from donor to recipient systems, these changes in the mass and timing of water yield will influence the mass and utilization of subsidies.

THE GLOBAL PICTURE

Global change has dramatically altered the mass and ratios of subsidies in donor systems, the vectors of and barriers to transport of subsidies, and the community structure of recipient systems. Since many of the mechanisms by which subsidies are transported between systems operate over a long range, this means that these dramatic changes in one portion of the globe may have substantial implications for other regions. Thus, increased use of fertilizer in the headwater basins of the Mississippi valley or of Central Europe leads to substantial alterations in the mass and roles of subsidies in the Gulf of Mexico or the Baltic Sea, respectively. Alterations in land use or industry in northern China or North Africa have consequences for wind-driven transport of subsidies to the North Pacific and North America (Fanning 1989) or to South America (Polis 1998), respectively. Even the most remote regions of the earth are receiving enhanced nutrient subsidies (e.g., Mayewski et al. 1986).

Many of the changes to donor and recipient systems and their boundaries are essentially irreversible. For instance, it is practically impossible to eliminate many of the exotic species that have invaded ecosystems and altered community structure (e.g., trout; McIntosh and Townsend 1996) or ecosystem processes (e.g., nitrogen-fixing *Myrica faya;* Vitousek 1987). Nor is it likely that we will see a return to preindustrial nutrient subsidy patterns in many parts of the world within the next generation, even if fertilizer application were to be drastically curtailed, which is unlikely (Food and Agriculture Organization 1996). And with a burgeoning human population, it is unlikely that there will be significant restoration of pre-industrial hydrology in many of the world's rivers (Dudgeon 1995; Kawashima et al. 1997).

It is further very likely that we will continue to see changes in the rates of these trends. Different areas of the world are experiencing different rates of change. We can expect to see the dynamic effects of changes in subsidies for the foreseeable future in many, if not most, systems. For the near future, the global trends will be driven in large part by growth in the human population and industrialization in developing countries.

Many of these changes in subsidies are considered undesirable for economic, health, aesthetic, or other reasons, and consequently a great deal of effort is, or should be, expended to reduce or reverse these trends. Efforts in many regions, however, are severely hampered by the long-term nature or severity of the changes. For instance, the long-term eutrophication of the Baltic was concurrent with a staggering amount of toxic pollution in some areas (Schulz-Bull et al. 1995; Dahlgaard 1996; Holm 1996), suggesting that rehabilitation of the subsidy regime in the Baltic will be severely constrained by the long-term alteration of this receiving system. For those trends that operate on a larger, global scale (for example, CO_2 accumulation in the atmosphere), we have failed to substantially slow, let alone reverse, modifications to the environment. For many global change trends, the lag time between taking action to amend practices that affect the environment and significant reversal of the effects of those practices may be long. For example, even with good compliance with the Montreal Protocols (and amendments) to halt production of ozone-destroying chemicals, there is expected to be a delay before we see a substantial decrease in the rate of ozone destruction due to the longevity of some chemicals in the atmosphere.

Yet we have the ability to influence some of the global change trends affecting subsidies with coordinated action. In some cases, considerable modification to subsidy regimes has been achieved on regional scales (e.g., von Gunten and Lienert 1993; Stoddard et al. 1999), though most of these examples may be limited to highly constrained systems (e.g., point source pollution) in developed nations. On a global scale, several efforts can be fruitful in reducing the magnitude of changes to subsidy regimes, including slowing population growth, minimizing resource consumption, and substituting low-impact technologies during industrial development, education, and technology transfer. Of particular concern will be the need to minimize alterations of subsidy regimes in regions undergoing rapid development, though the burden of accomplishing this will fall on both developed and developing nations.

Chapter 26

At the Frontier of the Integration of Food Web Ecology and Landscape Ecology

Gary R. Huxel, Gary A. Polis, and Robert D. Holt

The dynamics of populations, communities, and ecosystems are strongly influenced by the flow of nutrients, food, and consumers across habitats. Trophic linkage across habitats is a common feature even when habitats differ moderately or greatly in structure and productivity. Because of asymmetries in transport processes (e.g., prevailing winds, ocean currents), the net flux of resources is typically in one direction. The mechanisms that transport resources across habitat boundaries and their penetration of the recipient habitat can vary. For example, leaves that fall into a headwater stream can affect the stream over its entire course (as acknowledged in the river continuum concept; Vannote et al. 1980). By contrast, forest-edge birds that forage in neighboring grasslands may move those resources only a few hundred meters into a forest. Because the permeability of habitats to different organisms can diverge significantly, the rate of movement of resources across boundaries will vary with those biotic and abiotic variables that influence permeability.

Resources can be moved by either physical or biotic vectors (see Polis et al. 1996 and references therein). Physical vectors include gravity, wind, and water; biotic vectors include mobile predators and prey. Wind-dispersed resources include nutrients and particulate material (either organic or inorganic—see Witman et al., chap. 22 in this volume). Water movement can transport large amounts of nutrients across small or great distances (Witman et al., chap. 9, Winemiller and Jepsen, chap. 8, Riley et al., chap. 16, and Menge, chap. 5 in this volume). Deposition of atmospheric nutrients can play a major role in maintaining high levels of productivity in humid regions. Chadwick et al. (1999) argue that humid environments can reach a state of severe nutrient depletion due to weathering and leaching of minerals in soil. However, nutrients derived from the atmosphere can sustain high levels of productivity in downwind terrestrial ecosystems. For example, Hawaiian forests are able to maintain high levels of productivity because of the phosphorous they receive from Asian dust (Chadwick et al. 1999). Ocean water masses can differ greatly in the amounts of biomass and nutrients they transport. Transport occurs both vertically, via up welling, downwelling, and detrital fallout to the benthos, and horizontally, via currents, tidal movement, and eddy diffusion. These transport mechanisms are important determinants of local marine productivity and food webs (Angel 1984; Barnes and Hughes 1988; Barry and Dayton 1991; Menge, chap. 5, and Witman, chap. 9 in this volume).

Large-scale migrations of birds, fish, and mammals can move nutrients and biomass across continents and oceans (McNaughton 1985; Willson and Halupka 1995; Jefferies 1988a; Willson et al., chap. 19, and Jefferies et al., chap. 18 in this volume). However, the effects of resource transport by mobile consumers can be significant even on relatively small scales (Kitchell et al. 1979). For example, Vanni and colleagues (1996; Vanni and Layne 1997; Vanni et al. 1997; Vanni and Headworth, chap. 4 in this volume) have demonstrated that gizzard shad maintain increased levels of nitrogen in the pelagic zone of lakes and reservoirs via excretion of nitrogen from consumed benthic prey.

Large-scale movements of resources also arise from anthropogenic processes and mechanisms (Baudry and Burel, chap. 21 in this volume). For example, in 1996 Europe imported 13 million tons of soybeans, while Brazil exported almost 4 million tons (Food and Agriculture Organization 1998). Increasing anthropogenic carbon, nitrogen, phosphorous, and sulfur emissions into the atmosphere contribute millions of tons to downwind habitats each year. Similarly, fertilizers applied to agricultural and other lands are a major factor in the eutrophication of streams, lakes, reservoirs, and coastal waters (Carpenter, Caraco et al. 1998; Carpenter, Ludwig, and Brock 1999). Thus, while allochthonous resources may play an important role in the determination of food web structure and dynamics across natural landscapes, large-scale anthropogenic inputs may play a role in the degradation of terrestrial, aquatic, and marine ecosystems.

The flux of allochthonous resources is governed by land use patterns and water movement within and across landscapes and bioregions (Baudry and Burel, chap. 21 in this volume). Watercourses may be constrained by dams, levees, riprapping, ditches, canals, channelization, loss of wetlands, and other human activities. The constraining of riverine systems worldwide has greatly altered the rate of movement of resources from terrestrial systems to aquatic to marine systems. For example, over 98% of the riparian habitat has been lost in the Sacramento River Valley of central California, significantly altering terrestrial inputs into rivers and finally the Pacific Ocean (Malanson 1993). Habitat connectivity has also been significantly reduced, limiting the movement of resources. It has been argued that this loss of connectivity is an important factor in the local extinctions of some species; however, the role of movement corridors remains controversial (Simberloff and Cox 1987; Simberloff et al. 1992). Changes in land cover from native vegetation to exotic pasturelands or agricultural fields can significantly change the rate of movement of allochthonous resources into adjacent habitats (Baudry and Burel, chap. 21, and Riley et al., chap. 16 in this volume). Exotic vegetation may support lower invertebrate biomass and thus reduce the rate of terrestrial invertebrate movement into adjacent streams (Riley et al., chap. 16 in this volume). Agricultural activities may increase the rate of nutrient inputs into adjacent aquatic systems, causing eutrophication and subsequent changes in plant communities and in the strength of trophic cascades (Carpenter, Brock, and Hanson 1999).

Allochthonous resources can greatly influence the energy, carbon, and nutrient budgets of recipient habitats. While demographic responses mediated by dispersal can be important in local food webs (Holt, chap. 7 in this volume), here we will focus mainly on the effects of resource inputs on local populations and communities. Allochthonous resources can contribute up to nearly 100% of organic matter in many types of habitats—including headwater streams, caves, mountaintops, snowfields, marine aphotic zones and central oceanic gyres, deep oceanic benthic zones, phytotelmata, deserts, and islands (see Polis et al. 1996 and references therein). The allochthonous resources that drive these specialized habitats are often detrital—either of plant or animal origin. Allochthonous resources can be relatively easily differentiated from in situ resources in these specialized habitats. However, mobile organisms moving between similar habitats (grassland to savanna or littoral to sublittoral) can also contribute significant amounts of resources.

Allochthonous resources vary greatly in kind and quality, limiting which species benefit from them. For example, soil microbes and plants can directly exploit nutrient inputs that are deposited in guano (Hutchinson 1950). Anderson and Polis (1999) showed that soils on islands with seabird colonies had 6 times more N than soils on non-bird islands. Plants on bird islands had 1.6–2.4 times greater N and P concentrations than did plants on non-bird islands. Allochthonous resources in the form of organic matter can range from particulate matter to whale carcasses. For example, particulate matter from oceanic upwelling or headwater streams can subsidize filter-feeding organisms (Witman, chap. 9, and Menge, chap. 5 in this volume). Species assemblages differ in their use of detritus depending on whether the material is plant or animal matter (G. R. Huxel, unpublished data). In model ecosystems, Huxel and McCann (1998; see also McCann et al. 1998) found that the trophic status of the recipient consumer(s) greatly influenced the stability of the food web.

Thus, allochthonous resources influence individual fitness, population dynamics, the outcome of interspecies interactions (Holt, chap. 7 in this volume), community structure and ecosystem function, and landscape patterns and processes. The effects of allochthonous resources on recipient species may be direct, indirect, or both. Theoretical and empirical studies often show that indirect effects may be greater than direct effects (Huxel and McCann 1998; Abrams et al. 1996; Menge, chap. 5 in this volume). In the rest of this chapter, we will address how allochthonous resources influence single species populations, pairs of interacting populations, communities, and ecosystems/landscapes. We hope to demonstrate that traditional lines of investigation by habitat—freshwater, marine, benthic, pelagic, terrestrial, wetland, and so forth—result in missing key processes and mechanisms that influence local community structure and dynamics.

EFFECTS ON SINGLE SPECIES

The benefits of allochthonous resources are realized through the capture of these resources (as with any resource) and resultant increases in individual survivorship, fecundity, body size, and condition. These resources have direct and indirect implications for populations. Their direct effects include numerical response by the recipient species via increased reproduction with addition of a limiting resource (e.g., C, proteins, fatty acids, N, or P) or via higher survivorship due to increased nutrient levels in tissues. Their indirect effects include enhanced density dependence as a result of numerical response; source-sink dynamics, in which members of the recipient population emigrate due to density-dependent factors (e.g., increased competition); and decreased competition for particular limiting resources (i.e., ecological release).

Recent studies have demonstrated that the flow of resources from marine ecosystems to terrestrial ecosystems can be important for coastal populations. For example, Rose and Polis (1998) analyzed how the distribution and abundance of the coyote (*Canis latrans*) in Baja California was affected by resources from the Gulf of California. Coyote populations were 2.4 to 13.7 times denser along the coast than at nearby inland sites. Scat analyses found that scat mass was more than doubled in coastal areas compared with inland areas, indicating greater resource availability in coastal regions. Furthermore, 47.8% of all items found in coastal scats came directly from the marine environment. Similar studies (see table 3 in Rose and Polis 1998 and references therein) revealed that allochthonous resources from the marine environment heavily subsidize many omnivorous terrestrial mammals in coastal areas.

Semelparous vertebrates such as salmonids or squid can strongly subsidize carrion-feeding species. For example, in watersheds where salmonids have disappeared, bald eagle (*Haliaeetus leucocephalus*) populations have declined dramatically. In Glacier National Park (Montana, U.S.A.), McClelland et al. (1994) observed that kokanee salmon (*Oncorhynchus nerka*) dropped from more than 100,000 spawners before 1987 to zero spawners in 1991. Bald eagle numbers in this region declined precipitously from a peak of 639 in 1981 to 25 in 1989. Similar declines probably have gone unnoticed as salmonid populations elsewhere in western North America have approached or reached extinction.

The effects of allochthonous resources can be particularly dramatic in ecosystems with low intrinsic productivity. Desert systems are typified by low, highly variable precipitation and therefore low productivity—usually one to three orders of magnitude below that of other habitats (Louw and Seely 1982; Ludwig 1986, 1987; Polis 1991; Anderson and Polis 1999). However, since most desert soils have low levels of available nutrients, this factor may also limit primary productivity (Peterjohn and Schlesinger 1990, 1991; Charley and Cowling 1968; Ettershank et al. 1978; West and Skujins 1978; Hadley and Szarek 1981; Schlesinger and Peterjohn 1991).

On islands in the Gulf of California, Sánchez-Piñero and Polis (2000) found that tenebrionid beetle abundance tracks inputs of either carcasses on islands used by seabirds for nesting or increased plant litter following El Niño events on islands used by seabirds for roosting. Tenebrionids are 5 times denser on roosting and nesting islands than on non-bird islands, and they are 6 times denser within than outside seabird nesting colonies.

These effects arise directly through feeding on bird carcasses or fish scraps, or indirectly through increased plant litter due to fertilization by bird guano (see also Anderson and Polis 1999).

Populations may also have a net negative response to increased allochthonous resources. For example, a numerical response to allochthonous resources could theoretically result in overconsumption of all resources and a subsequent crash in population density (as in the "paradox of enrichment"; Rosenzweig 1971). May (1974) showed that increasing the growth rate could cause a population to move from stable to chaotic dynamics. DeAngelis and others (DeAngelis 1992; Huston and DeAngelis 1994; Andersen 1997) have shown that increased nutrient supply rates also can destabilize population dynamics. Various mechanisms could be involved, including self-shading and overconsumption of resources (see Liebold 1997; DeAngelis and Mulholland, chap. 2 in this volume). If allochthonous resource input is seasonally pulsed (R. D. Holt, unpublished data; Sears et al., chap. 23 in this volume), there can be increased population densities and overconsumption of autochthonous resources in other seasons, resulting in severe population crashes.

EFFECTS ON PAIRS OF SPECIES

Predator-Prey Interactions

The quality of the resources available to a consumer can govern the consumer's growth rates. Urabe and Sterner (1996) demonstrated that under low light to nutrient ratios, algae and their herbivores are limited by energetic resources, whereas under high light to nutrient ratios, they are nutrient-limited. This finding may explain why experimental additions of nutrients to test trophic cascade hypotheses have had equivocal results (Brett and Goldman 1996, 1997; Strong 1992; Polis and Strong 1996; Polis et al. 2000).

The subsidization of prey, either directly (e.g., guano to plant) or indirectly through a resources species (e.g., guano to plant to herbivore), can result in an increase in the prey with or without a concomitant increase in predation. If predator numbers are limited by factors other than prey (e.g., nest sites, interference competition, intraguild predation), increased production makes it more likely that prey can escape predator regulation (Sinclair and Arcese 1995). With a sufficient increase in predation pressure, however, the prey species can become extinct. If the flow of allochthonous resources is great enough, then the prey can increase even if predation pressure increases.

However, the ability of the predator to respond numerically to increased prey density or quality (in terms of higher nutrient content) may allow for over-compensation by the predator, driving the prey to low numbers or even extinction (Huxel and McCann 1998; Huxel 1999).

A major source of nutrients for ecosystems is atmospheric deposition (Erelli et al. 1998; Chadwick et al. 1999; Rennenberg and Gessler 1999). Erelli et al. (1998) found that higher elevations receive greater atmospheric nitrogen inputs, and that birch (*Betula papyrifera*) at these elevations had a foliar N content 13% higher than at lower elevations. However, the growth of insect larvae can be either increased or decreased by changes in foliar N and phytochemistry. Erelli et al. (1998) found that growth of *Orgyia leucostigma* larvae decreased with decreasing leaf nitrogen, even though feeding increased. Yet larvae of *Lymantria dispar* maintained their growth rates by compensating for low N by increasing consumption. This difference may be due to the greater sensitivity of the first instar of *O. leucostigma* to tannins, which are negatively correlated with nitrogen levels. Similarly, Redak et al. (1997) found that simulated atmospheric nitrogen deposition produced chrysomelid beetles on treated plants that were 36% larger and had growth rates 31% greater than those on control plants.

Plant response to herbivory is a key factor in whether (and how) an increased nutrient supply rate influences herbivory. The ability of a plant to resist herbivory may be either constitutive (always present) or inducible (turned on by specific interactions). Plants with induced chemical defense pathways may use increased allochthonous nutrients either for increased growth and reproduction or for increased chemical defenses. Recent studies have suggested that induced defensive responses by plants may allow plants to respond in ecological time to increased herbivory, decreasing the need for constitutive defenses (Karban and Baldwin 1997). Plants need to somehow judge the potential for future herbivory in order to be able to partition resources between growth and reproduction and defense, and increased nutrient levels may influence their response. One major difficulty for researchers is in determining whether the benefits of increased defense outweigh the costs of the production of chemical defenses.

Menge and his colleagues (Menge et al. 1996; Menge, chap. 5 in this volume) observed differences between two wave-exposed sites along the Oregon coast. At one site—Strawberry Hill—Menge found that the abundance of macroalgae was low, while densities of filer feeders and invertebrate predators were high. In contrast, another site—Boiler Bay—exhibited a high abundance of macroalgae and low densities of filer feeders and invertebrate predators. Menge et al. (1996) experimentally demonstrated that at the Strawberry Hill site, rates of predation by the invertebrate predators and recruitment and growth rates of the filter feeders were higher than at Boiler Bay. Menge (2000b; Menge, chap. 5 in this volume) has argued that the stronger top-down effects at the Strawberry Hill site are due to higher levels of food input for the filter feeders there. The higher levels of allochthonous food resources result from greater entrainment of oceanic currents off the coast of Strawberry Hill than at Boiler Bay. Thus, the landscape contexts of these two biotically similar communities drive their food web dynamics (Menge, chap. 5 in this volume), highlighting the significant influence of the flux of resources across landscapes or seascapes in driving local community dynamics. The similarity between these sites in community composition, in contrast to their different dynamic properties, suggests that the processes that influence the structure of local communities may not be the same as those that drive their dynamics.

The direct subsidization of a predator can lead to one of two outcomes: either the predator will increase its predation pressure on in situ prey (in a process similar to apparent competition, sensu Holt 1977, 1984; Holt, chap. 7 in this volume), or it will feed mainly on the allochthonous resource, relaxing predation pressure on in situ prey (ecological release; Holt 1997; Huxel and McCann 1998). Abrams (1999) has shown that prey switching by predators can stabilize food webs. Similarly, Huxel and McCann (1998; Huxel et al. 2002) found that low to moderate levels of allochthonous input could stabilize food webs, while high levels would result in strong indirect interactions in which the predator would reach high densities and increase predation pressure, causing the in situ prey to become extinct.

The movement of consumers can greatly influence consumer-resource interactions. For example, since the 1930s, lesser snow geese have switched from traditional coastal winter foraging areas to inland agricultural areas (Robertson and Slack 1995; Stutzenbaker and Buller 1974). Across the Mississippi and Central Flyways, the geese have responded to additional resources provided by crop residue and changes in agricultural practices (Boyd et al. 1982), greatly increasing in numbers from 600,000–800,000 during 1950–1965 to about 2 million birds by 1990 (Jefferies et al., chap. 18 in this volume). In some river valleys the increase has been up to twentyfold (Burgess 1980). The geese migrate to arctic tundra for the summer breeding season, where their burgeoning numbers have resulted in great stresses on the local tundra habitat. Jefferies (2000; Jefferies et al., chap. 18 in this volume) has shown that the tundra vegetation has been stripped from large areas due to foraging by the geese.

The input of resources from terrestrial habitats into headwater streams has been shown to drive secondary productivity in downstream waters (Vannote et al. 1980; Rosemond et al. 1993). The effects of this increase in productivity on trophic interactions in stream food webs have been greatly understudied until recently. Nakano et al. (1999) recognized that terrestrial inputs into streams represented an important energy source for secondary production, but few studies had examined the role of non-plant terrestrial resources. Allen (1951) suggested that secondary production within streams may be insufficient to support the observed levels of fish production. Nakano et al. (1999) showed that terrestrial arthropods constitute a high-energy and high-nutrient resource in stream food webs that can exceed in situ productivity (Garman 1991; Cloe and Garman 1996). They covered sections of a stream to reduce terrestrial arthropod inputs and excluded or included a predatory fish (Dolly Varden) in some sections as well. Fish in the uncovered sections fed mainly on terrestrial arthropods, while those in the covered sections relied mainly on chironomids (aquatic herbivores). In the covered sections, the biomass of aquatic herbivores decreased and periphyton increased. The duration of the experiment precluded strong numerical responses by the predator; however, over short time scales, buffering of fish effects on in situ stream invertebrates by terrestrial arthropods was observed.

Changes in land cover can significantly alter the amount of invertebrate biomass moving into streams (Edwards and Huryn 1996; Kawaguchi and Nakano 2001; see also Riley et al., chap. 16 in this volume). Edwards and Huryn found that the biomass of terrestrial invertebrates entering adjacent streams in New Zealand was correlated with land use patterns, and that the highest rates of movement into the streams was from native vegetation, suggesting that native vegetation supports more invertebrate biomass. Huryn (1998) found that only the introduced brown trout was able to fully utilize these allochthonous resources, and that it actually required all of the invertebrate production to support its populations, whereas the native galaxiid consumed only 18% of the invertebrate production. Also, the subsidized brown trout caused a top-down cascade resulting in decreased aquatic herbivore and increased algal biomass. Increased nutrient loading from exotic grasses in pastoral habitats further complicates New Zealand stream food webs. High levels of nitrogen cause a shift in algal dominance to filamentous green algae, which are less preferred by aquatic herbivores. This shift in algal community composition acts together with increased predation by the exotic brown trout to decrease grazer densities (Riley et al., chap. 16 in this volume). These studies suggest that the connectivity of adjacent terrestrial and aquatic habitats plays an important role in maintaining fish populations, and that changes in land cover from native vegetation to managed systems will influence the conservation of fish and other aquatic organisms.

Biological control specialists have long argued that generalist predators may be effective in maintaining low densities of pest species if they are provided with additional resources (prey). Karban et al. (1994) demonstrated that subsidizing the predatory mite *Metaseiulus occidentalis* by releasing an alternative prey, the Willamette mite, results in greater control of the Pacific spider mite on grapevines. While competition alone between the two herbivorous mites results in lower abundances of the Pacific spider mite, the predator further suppresses the pest. With only the pest species present, the predator becomes extinct on some grapevines after driving the pest to low densities, and the pest recovers to higher densities after extinction of the predator.

Competitive Interactions

Allochthonous resources may allow competitive species to coexist in either of two ways. First, if the combination of autochthonous and allochthonous resources is great enough, and the competitors are both generalists and can utilize both resources, then competition is reduced. Second, if both resources are low, then specialization on the two different resource types allows for coexistence (Huxel et al. 2002). Competitive interactions may also be important in driving habitat selection whether resources are adequate or not (Brown 1990, 1996; Holt 1985, 1996a, 1997b). Specialization on allochthonous resources is seen in many communities. Some aquatic insects in headwater streams specialize on leaf matter; these "shredders" process the leaf material into finer particles that can be utilized by other aquatic insects. Kelp flies and carrion flies utilize different marine inputs onto desert islands (G. R. Huxel et al., unpublished data). Similarly, tenebrionids can specialize on plant detritus or carrion, or feed more generally on both. Competition among tenebrionid species changes with different levels of inputs of these resources (F.Sánchez-Piñero, unpublished data).

Bustamante et al. (1995a) found exceptionally high biomasses of two intertidal limpets in the mid- to low intertidal communities of rocky shores in South Africa. Previously these limpets had been regarded as generalist consumers; however, Bustamante and colleagues found that the two limpets had highly specialized feeding mechanisms. *Patella argenvillei* is found in low regions of semi-exposed shores in association with adjacent kelp beds, and *P. granatina* is found in sheltered boulder bays. *P. granatina* feeds on kelp and seaweed debris, and *P. argenvillei* feeds on kelp plants. Thus, the subtidal production of kelp fronds subsidizes both species; however, their coexistence is maintained by habitat selection and timing of feeding.

While allochthonous nutrients may have large effects on some systems, we do not suggest that all plant communities will exhibit strong responses to these additional resources, because many different resources may influence plant community structure. Competition among plants is often governed by the availability of nutrients, light, or space (Hutchinson 1967; Harper 1977). Recent theory on resource competition has been based on the importance of hypothesized trade-offs between minimum requirements for nutrient resources (Tilman 1982). This theory predicts that there should be negative correlations between the supply rate of major limiting nutrients and the availability of at least some secondary nutrients or among the availabilities of different limiting nutrients. However, Liebold's (1997) analysis of four data sets from large-scale surveys of lakes shows mostly positive correlations among the availabilities and supplies of nutrients. In contrast, he found that a fifth data set, obtained in an area of high acidification, did show several important negative correlations that are consistent with the resource competition models. Liebold suggested that negative correlations between nutrient levels and light levels indicate that an important trade-off among species may involve low light requirements versus low nutrient requirements. As mentioned above, Urabe and Sterner (1996) showed that algae populations were energy-limited at low light to nutrient ratios, while at high light to nutrient ratios they were nutrient-limited. Similarly, Goldberg and Miller (1990) found that nutrient additions had little effect on plant productivity in a dry year, and that changes in diversity in experimental plots depended upon which nutrients were added to the plot. They also found that plots with nitrogen additions had greater mortality due to increased light limitation. These findings suggest that systems with high rates of allochthonous nutrient input may not respond strongly to additional resources. Thus the role of allochthonous nutrients in light-limited systems needs to be more fully addressed.

EFFECTS ON MULTIPLE SPECIES INTERACTIONS

Allochthonous resources allow for the examination of one of the major areas of debate in community ecology: the relative importance of top-down versus bottom-up effects. Since Lindeman's (1942) study of food web dynamics in lake ecosystems, the debate over whether food webs are controlled by top-down or bottom-up effects has been ongoing. Lindeman suggested that the species and processes in communities formed cycles, with resources moving from detritus to plants (or directly to detritivores) to primary consumers to secondary consumers, and whatever resources are not utilized by these trophic levels back to detritus (fig. 26.1). Hairston et al. (1960) argued that control of food webs is top-down in terrestrial ecosystems, such that carnivores limit populations of herbivores, allowing plants to grow mostly unimpeded by herbivory and making the world green. This green world hypothesis was expanded into the exploitative ecosystem hypothesis (fig. 26.2) for most types of habitats by various food web ecologists (Fretwell 1977, 1987; Oksanen et al. 1981; Carpenter and Kitchell 1993). They suggested that the dynamics of food webs in general are dominated by top-down processes, such that in food chains with odd numbers of trophic levels, the odd-numbered trophic levels increase in biomass, while in even-numbered food chains, the even-numbered trophic levels increase. Other researchers, however, argued that bottom-up processes exert strong control on food web structure and dynamics (e.g., donor control, sensu DeAngelis 1980), in that food web dynamics are controlled by the rate of resource input (White 1978; McQueen et al. 1986; DeAngelis 1980, 1992; Strong 1992; Polis and Strong 1996; Huxel and McCann 1998). Here we argue that both top-down and bottom-up processes can influence food web structure and dynamics (Polis and Strong 1996; Persson et al. 1996; Rosemond et al. 1993; Vanni 1987; Lynch and Shapiro 1981). Thus, allochthonous resources can influence food web dynamics via two general pathways: as resources for predators, which then have cascading effects on prey, and as resources for primary producers or primary consumers, resulting in more resources for predators.

Thus the influence of resources on multiple species interactions within a food web or a food web module (sensu Holt 1997a) can either be direct or indirect. Polis and Strong (1996) suggested that most strong trophic cascades occur in systems that are significantly subsidized. In Huxel and McCann's (1998) model food web modules, allochthonous resources initiate strong indirect interactions from either bottom-up or top-down effects. Indirect interactions that are similar to apparent competition (Holt 1977, 1984) occur in these model systems when a consumer increases in density due to allochthonous resources, resulting in a top-down effect on its autochthonous resources. This effect can then cascade down to reduce predation pressure on the next lower trophic level. However, if the consumer is inefficient in converting the allochthonous resource into greater

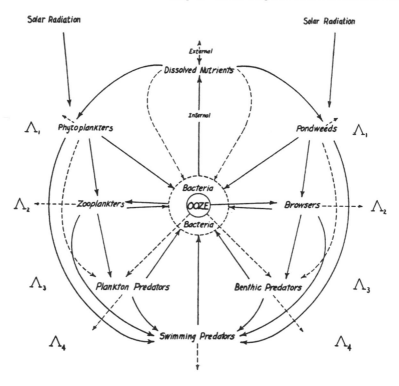

Figure 26.1 Generalized schematic of the food cycle envisioned by Lindeman. Notice that Lindeman stressed the linkages between the benthic and pelagic zones, as well as the movement of nutrients into and out of the system. (From Lindeman 1942.)

reproduction, then its prey may experience decreased predation pressure. Note also that by specializing on one prey, the predator may release another prey species (Abrams 1999) or, if its favored prey is a superior competitor, allow predator-mediated coexistence.

Again, the islands of the Gulf of California provide a good example of the effects of allochthonous resources on multiple species interactions. Following El Niño events, herbivore densities increase, providing increased resources for web-building spiders (Polis et al. 1998). Spider populations rise dramatically, but fall off sharply the next year. Polis (Polis et al. 1998; Polis et al., chap. 14 in this volume) found that in wet El Niño years, nectar resources are high enough to allow for increased feeding by pompilid wasps. This allows for greater egg production and increased parasitism on the spiders by the wasps. However, since herbivores are not very dense even when predators are at low abundances, no significant response by plants is detected.

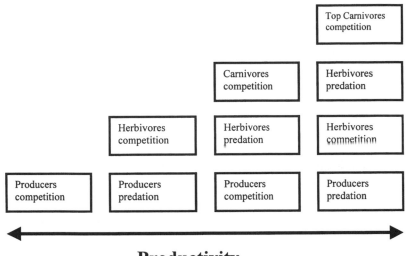

Productivity

Figure 26.2 The exploitative ecosystem hypothesis argued that productivity controlled the number of trophic levels (bottom-up control) while consumption by higher trophic levels controlled densities of trophic levels (top-down control). These controls could be experimentally manipulated by increasing nutrients or adding/removing trophic levels to demonstrate that "trophic cascades" occur. The role of spatial subsidies in influencing the strength of trophic cascades (and number of trophic links) needs to be more fully addressed. (From Fretwell 1977.)

Crooks and Soulé (1999) showed that strongly subsidized domestic cats prey heavily on native birds in small habitat fragments. However, in larger fragments, coyotes, which also can prey on birds (but are less efficient), actually cause an increase in bird densities due to their predation on domestic cats. This effect is typical of systems with generalized predators that are more efficient on mesopredators and less efficient on a shared prey (intraguild predators). A further example is the invasive brown tree snake, which has decimated native bird faunas on several Pacific islands. On Guam, the brown tree snake preys on domestic chickens that occur in high densities; this subsidy results in increased predation pressure on native birds (Savidge 1987).

ECOSYSTEM RESPONSES AND BIOLOGICAL DIVERSITY

Allochthonous resources can be the primary resource supplies for entire ecosystems. For example, over 10^8 adult salmon spawn in southeast Alaska streams each year, transporting an estimated 10^8 kg of carbon, 10^7 kg of nitrogen, and 2×10^6 million kg of phosphorus from marine to freshwater habitats (Willson et al., chap. 19 in this volume). Resources from these spawning salmonids boost productivity (1.5–6.8 times; Michael 1998) in freshwater streams and adjacent riparian zones (Lyle and Elliot 1998; Willson et al., chap. 19 in this volume).

The potential for overfertilization of terrestrial ecosystems is great. Fenn et al. (1998) have suggested that nitrogen deposition is leading to degraded conditions in many North American forests. The excess nitrogen can cause disruptions in plant-soil nutrient relations, increased soil acidification, increased runoff of nitrate into streams, decreasing water quality, and eutrophication of coastal waters. Increased nitrogen may cause a shift from slow N-cycling coniferous forest stands to fast-growing and fast N-cycling deciduous forests (Fenn et al. 1998).

Riverine, lake, estuarine, and coastal marine habitats are also greatly influenced by nutrient and organic matter loading (Sklar and Browder 1998; Carpenter, Caraco et al. 1998; Mallin et al. 1999). In these aquatic ecosystems, nutrient and organic matter input can result in toxic algal blooms, hypoxic and anoxic conditions, fish die-offs, reduced biological diversity, loss of aquatic bed plants and coral reefs, reduced water quality, and increased prevalence of disease. Carpenter, Caraco et al. (1998), in a literature review, found (1) that eutrophication is widespread in rivers, lakes, estuaries, and coastal ecosystems; (2) that agricultural practices and urban areas are the major sources of nonpoint pollution for these systems; (3) that inputs of N and P into agricultural land exceed outputs in produce; (4) that nutrient flows to aquatic systems are directly related to livestock densities, and that the manure outputs of these animals exceed the manure needs of crops to which manure is applied; (5) that P inputs to agricultural systems cause a surplus of P in soils, which often results in transport to aquatic systems; and (6) that excess fertilization and manure production on agricultural lands create surplus N and leaching of N into aquatic systems.

Large-scale disturbances can cause greatly increased inputs of nutrients and organic matter into aquatic systems from terrestrial systems. This is especially true in areas with high densities of livestock, as noted by Carpenter, Caraco et al. (1998). The Cape Fear watershed has high densities of humans, chickens, and swine; for example, the North Carolina swine population was 9.8×10^6 animals in 1997 (Mallin 1999). Hurricanes Dennis, Floyd, and Irene all struck this area in 1999, resulting in inputs to aquatic systems of over 265×10^6 l of human waste, and probably similar (if not greater) amounts of waste from chicken and swine operations.

Increased organic matter input into aquatic systems following such large disturbances can result in hypoxic or anoxic conditions, leading to mass die-offs of fish and shellfish. For example, following two hurricanes that struck the Cape Fear watershed, dissolved oxygen levels decreased to 2 mg l^{-1} in the mainstem Cape Fear River and to zero in the Northeast Cape Fear River for more than 3 weeks (Mallin et al. 1999). The Cape Fear estuary also experienced hypoxia for several weeks. The cause was heavy sediment loads from agricultural systems, including waste from large swine operations. Benthic organisms (both invertebrates and vertebrates) were especially hard hit, with densities of some organisms dropping to near zero or zero at some sampling sites.

Eutrophication in aquatic systems can lead to significant losses of species diversity and alter ecosystem processes. It can also result in turnover to "unfavorable" species such as *Pfiesteria* and noxious invaders such as *Caulerpa*. For example, Burkholder and Glasgow (1997) found that *Pfiesteria*, a causative agent of major fish kills, can be stimulated to produce toxic zoospores by inorganic and organic phosphates, which are major components of agricultural and urban runoff. *Pfiesteria* has become a major health threat in the Chesapeake Bay ecosystem, but is not the only species causing problems in eutrophic coastal waters. Blazer et al. (1999) suggest that *Aphanomyces* may be the cause of lesions of menhaden in the Chesapeake Bay. Outbreaks of nuisance algae cause extensive problems due to their production and release of toxins and result in anoxia as the dead algae are decomposed (Carpenter, Caraco et al. 1998). Seehausen et al. (1997) suggest that eutrophication is a major factor in the loss of fish diversity. In eutrophic freshwater systems, changes in planktonic communities from edible to toxic or inedible species occur frequently (Carpenter, Caraco et al. 1998). Blooms of cyanobacteria are common in many lakes that experience eutrophication. Blooms of these organisms can result in fish die-offs, decreased water quality, and anoxia.

While changes in community composition and subsequent changes in ecosystem processes are clear, the relationship between eutrophication and diversity is more problematic. The relationship between diversity and productivity is thought to be hump-shaped (Huston 1979, 1994; Rosenzweig 1995). Diversity should increase from low to moderate productivity levels, but then decrease from moderate to high levels. Rosenzweig (1971) termed this decrease in diversity with increasing productivity "the paradox of enrichment." It has been debated whether this paradox actually exists, with some suggesting that in systems with multiple limiting resources, the paradox will not hold, or if it does, it will do so only over a limited range of values (Huxel 1999; Liebold 1997). However, many experimental systems have shown that nutrient additions lead to higher productivity and a loss in diversity (Rosenzweig 1995). Anderson and Polis (1999) demonstrated that guano additions to desert islands in the Gulf of California resulted not only in increased productivity, but also in a change in species composition from perennials to annuals and a decrease in species richness.

Allochthonous resources can greatly benefit ecosystems when resources are added at low to moderate levels. As mentioned above, Chadwick et al. (1999) provided evidence that Hawaii's terrestrial ecosystems would be N-limited if not for the deposition of atmospheric N resources that originate in central Asia. Similarly, Polis and colleagues (Polis and Hurd 1995, 1996a, 1996b; Polis, Anderson, and Holt 1997; Polis et al. 1998; Anderson and Polis 1999) have demonstrated that islands in the Gulf of California, which have low terrestrial productivity due to low N and limited precipitation, have high levels of consumers due to inputs from the highly productive marine ecosystem of the Gulf.

CONCLUSIONS AND IMPLICATIONS

The flow of resources between habitats and across landscapes is an important factor in local communities. Two major factors are important determinants of the effects of these resources: the permeability of habitat boundaries to the resources, and the potential for utilization of the resources by the species of a local food web. Additionally, the strength of the environmental gradient across the habitats influences the movements of species (individuals) across the boundary. A further question is how species diversity changes across the boundary. One could imagine that diversity might be higher in the ecotonal area; however, this pattern varies among taxa and boundary types. Competitive and predatory interactions may greatly influence species habitat choices and therefore the potential for evolutionary change due to habitat selection (Brown 1990, 1996; Holt 1985, 1996a, 1997b).

Habitats of low productivity are typically net recipients of allochthonous resources. Thus, species that are inhabitants of these systems may rely greatly on the allochthonous resources. Population abundances in low-productivity sites will be dramatically reduced if these resources are missing. For example, Sánchez-Piñero and Polis (2000) found that abundances of tenebrionid beetles are significantly lower on desert islands in the Gulf of California that do not have allochthonous inputs from seabirds (either guano or carcasses and carrion). Thus, the conservation of species within low-productivity sites depends on the conservation of surrounding habitats with higher productivity because land use patterns can regulate the flux of resources and species (Baudry and Burel, chap. 21 in this volume, Carpenter, Caraco et al. 1998; Carpenter, Ludwig, and Brock 1999).

Finally, the linking of food web processes with landscape processes is an important step in understanding the structure and dynamics of communities. As pointed out in this volume by various authors, in order to understand food web interactions, we need to discern the fluxes of resources across landscapes and understand the implications of the contextual location of communities within a landscape. Species interactions can be significantly altered by this ecological context, making predictions and results from simple food web models difficult to interpret. Direct and indirect effects driven by allochthonous resources may alter the strength of competitive, predatory, and other interactions between and among species. The importance of allochthonous resources underscores the arguments of Polis and Strong (1996), who suggested that resources could enter a food web through many different channels (hence their term "multichannel omnivory"). Increased resources may cause significant changes in the structure and composition of communities and ecosystems. Therefore, in the conservation of sensitive and fragile habitats, the ecological and landscape milieus of these systems have to be taken into account.

ACKNOWLEDGMENTS

We thank the participants in the INTECOL Symposium on Food Webs at the Landscape Level and our co-authors of this book. Comments by various reviewers contributed greatly to the manuscript. Our research (past and present) is supported by the National Science Foundation Division of Environmental Biology, the University of California-Davis, the University of Kansas, the University of Florida, and the University of South Florida.

References

Abraham, K. F., and R. L. Jefferies. 1997. High goose populations: Causes, impacts and implications. Pp. 7–72 in D. J. Batt, ed., *Arctic ecosystems in peril: Report of the Arctic Goose Habitat Working Group*. Ducks Unlimited, Memphis, TN, USA.

Abraham, K. F., R. L. Jefferies, R. F. Rockwell, and C. D. MacInnes. 1996. Why are there so many white geese in North America? Pp. 79–92 in J. T. Ratti, ed., *Seventh International Waterfowl Symposium*. Ducks Unlimited, Memphis, TN, USA.

Abrams, P. A. 1993. Effect of increased productivity on the abundances of trophic levels. American Naturalist 141:351–371.

———. 1999. The adaptive dynamics of consumer choice. American Naturalist 153:83–97.

———. 2001. The effect of density-independent mortality on the coexistence of exploitative competitors for renewing resources. American Naturalist 158:459–470.

Abrams, P. A., R. D. Holt, and J. D. Roth. 1998. Apparent competition or apparent mutualism? Shared predation when populations cycle. Ecology 79:201–212.

Abrams, P. A., B. A. Menge, G. G. Mittelbach, D. A. Spiller, and P. Yodzis. 1996. The role of indirect effects on food webs. Pp. 371–395 in G. A. Polis and K. O. Winemiller, eds., *Food webs: Integration of patterns and dynamics*. Chapman and Hall, New York, USA.

Abrams, P. A., and J. D. Roth. 1994. The effects of enrichment of three-species food chains with nonlinear functional responses. Ecology 75:1118–1130.

Adams, S. M., B. L. Kimmel, and G. R. Ploskey. 1983. Sources of organic matter for reservoir fish production: A trophic dynamics analysis. Canadian Journal of Fisheries and Aquatic Sciences 40:1480–1495.

Addiscott, T. M., A. P. Whitmore, and D. S. Powlson. 1991. *Farming fertilizers and the nitrate problem*. CAB International, Wallingford, UK.

Agrawal, A. A., R. Karban, and R. G. Colfer. 2000. How leaf domatia and induced plant resistance affect herbivores, natural enemies and plant performance. Oikos 89:70–80.

Alaska Department of Fish and Game. 2003. Commercial Fisheries http://www.cf.adfg.state.ak.us.

Alisauskas, R. T., C. D. Ankney, and E. E. Klaus. 1988. Winter diets and nutrition of midcontinental Lesser Snow Geese. Journal of Wildlife Management 51:403–414.

Allen, K. R. 1951. The Horokiwi stream. Fisheries Bulletin no. 10. New Zealand Marine Department, Wellington, NZ.

Ambuel, B., and S. A. Temple. 1983. Area-dependent changes in the bird communities and vegetation of southern Wisconsin forests. Ecology 64:1057–1068.

American Society of Microbiology (ASM). 1995. Report of the ASM Task Force on Antibiotic Resistance, Washington, DC, USA.

Andersen, T. 1997. *Pelagic nutrient cycles: Herbivores as sources and sinks*. Springer-Verlag, Berlin, Germany.

Andersen, T., and D. O. Hessen. 1991. Carbon, nitrogen and phosphorus content of freshwater zooplankton. Limnology and Oceanography 36:807–813.

Anderson, D. T. 1994. *Barnacles: Structure, function, development and evolution*. Chapman and Hall, London, UK.

Anderson, W. B., and G. A. Polis. 1998. Marine subsidies of island communities in the Gulf of California: Evidence from stable carbon and nitrogen isotopes. Oikos 81:75–80.

————. 1999. Nutrient fluxes from water to land: Seabirds affect plant nutrient status on Gulf of California islands. Oecologia 118:324–332.

Andersson, L., and L. Rydberg. 1988. Trends in nutrient and oxygen conditions within the Kattegat: Effects on local nutrient supply. Estuarine, Coastal and Shelf Science 26:559–579.

Andrén, H. 1995. Effects of landscape composition on predation rates at habitat edges. Pp. 225–255 in L. Hansson, L. Fahrig, and G. Merriam, eds., *Mosaic landscapes and ecological processes*. Chapman and Hall, London, UK.

Andrén, H., and P. Angelstam. 1988. Elevated predation rates as an edge effect in habitat islands: Experimental evidence. Ecology 69:544–547.

Andrén, H., P. Angelstam, E. Lindstrom, and P. Widen. 1985. Differences in predation pressure in relation to habitat fragmentation: An experiment. Oikos 45:273–277.

Angel, M. V. 1984. Detrital organic fluxes through pelagic ecosystems. Pp. 475–516 in M. J. R. Fasham, ed., *Flows of energy and materials in marine ecosystems*. Plenum Press, New York, USA.

Angelstam, P. 1992. Conservation of communities—the importance of edges, surroundings, and landscape mosaic structure. Pp. 9–70 in L. Hansson, ed., *Ecological principles of nature conservation*. Elsevier Applied Science, New York, USA.

Angerbjörn, A., P. Hersteinsson, K. Liden, and E. Nelson. 1994. Dietary variation in arctic foxes (*Alopex lagopus*)—analysis of stable carbon isotopes. Oecologia 99:226–232.

Ankney, C. D. 1996. An embarrassment of riches: Too many geese. Journal of Wildlife Management 60:217–223.

Apel, J. R., M. H. Byrne, J. R. Proni, and R. L. Charnell. 1975. Observations of oceanic internal and surface waves from the earth resources technology satellite. Journal of Geophysical Research 80:865–881.

Apel, J. R., J. R. Holbrook, A. K. Liu, and J. J. Tsai. 1985. The Sulu Sea internal soliton experiment. Journal of Physical Oceanography 15:1625–1651.

Araujo-Lima, C., B. R. Forsberg, R. Victoria, and L. Martinelli. 1986. Energy sources for detritivorous fishes in the Amazon. Science 234:1256–1258.

Ashley, K. I., and P. A. Slaney. 1997. Accelerating recovery of stream and pond productivity by low-level nutrient replacement. Pp. 13-1–13-24 in *Fish habitat rehabilitation procedures for the watershed restoration program*. British Columbia Ministry of Environment, Lands and Parks, and Ministry of Forests, Watershed Restoration Technical Circular 9. Vancouver, BC, Canada.

Atkinson, B. W. 1981. *Meso-scale atmospheric circulations*. Academic Press, London, UK.

Atkinson, I. A. E. 1964. The flora, vegetation, and soils of Middle and Green Islands, Mercury Islands Group. New Zealand Journal of Botany 2:385–402.

————. 1985. The spread of commensal species of *Rattus* to oceanic islands and their effects on island avifaunas. Pp. 35–81 in P. J. Moors, ed., *Conservation of island birds*. ICBP Technical Publication no. 3. International Council for Bird Preservation, Cambridge, UK.

————. 1994. Introduced animals and extinctions. Pp. 54–69 in D. Weston and M. Pearls, eds., *Conservation for the twenty-first century*. Oxford University Press, New York, USA.

Austad, S. N., and K. E. Fischer. 1991. Mammalian aging, metabolism and ecology: Evidence from bats and marsupials. Journal of Gerontology 46:B47–53.

Avery, G., D. M. Avery, S. Braine, and R. Loutit. 1987. Prey of coastal black-backed jackal *Canis mesomelas* (Mammalia: Canidae) in the Skeleton Coast Park, Namibia. Journal of Zoology 213:81–94.

Bachmann, R. W., B. L. Jones, D. D. Fox, M. Hoyer, L. A. Bull, and D. E. Canfield. 1996. Relations between trophic state indicators and fish in Florida (USA) lakes. Canadian Journal of Fisheries and Aquatic Sciences 53:842–855.

Bagnall, R. G. 1979. A study of human impact on an urban forest remnant: Redwood Bush, Tawa, near Wellington, New Zealand. New Zealand Journal of Botany 17:117–126.

Baines, P. G. 1974. The generation of internal tides over steep continental slopes. Philosophical Transactions of the Royal Society of London A 277:27–58.

Baird, D., and H. Milne. 1981. Energy flow in the Ythan estuary, Aberdeenshire, Scotland. Estuarine and Coastal Shelf Science 13:455–472.

Balent, G. 1986. The influence of grazing on the evolution of botanical composition of previously cultivated fields: The example of the Pyrénées. Pp. 28–29 in P. J. Joss, P. W. Lynch, and O. B. Williams, eds., *Rangeland: A resource under siege.* Australian Academy of Science, Canberra, Australia.

Balls, P. W., A. Macdonald, K. Pugh, and A. C. Edwards. 1995. Long-term nutrient enrichment of an estuarine system: Ythan, Scotland (1958–1993). Environmental Pollution 90: 311–321.

Balter, M. 1999. Restorers reveal 28,000-year old artworks. Science 283:1835.

Barbarino, A., D. C. Taphorn, and K. O. Winemiller. 1998. Ecology of the coporo, *Prochilodus mariae* (Characiformes, Prochilodontidae) and status of annual migrations in western Venezuela. Environmental Biology of Fishes 53:33–46.

Barber, R. T., and J. H. Ryther. 1969. Organic chelators: Factors affecting primary production in the Cromwell current upwelling. Journal of Experimental Marine Biology and Ecology 3:191–199.

Barber, R. T., and W. O. Smith. 1981. The role of circulation, sinking and vertical migration in physical sorting of phytoplankton in the upwelling center at 15°S. Pp. 366–371 in F. A. Richards, ed., *Coastal upwelling.* American Geophysical Union, Washington, DC, USA.

Barbercheck, M. E., and W. C. Warrick. 1997. Evaluation of trap cropping and biological control against southern corn rootworm (Coleoptera: Chrysomelidae) in peanuts. Journal of Entomological Science 32:229–243.

Barclay, R. M. R. 1994. Constraints on reproduction by flying vertebrates: Energy and calcium. American Naturalist 144:1021–1031.

Barclay, R. M. R., M. A. Dolan, and A. Dyck. 1991. The digestive efficiency of insectivorous bats. Canadian Journal of Zoology 69:1853–1856.

Barkai, A., and C. D. McQuaid. 1988. Predator-prey role reversal in a marine benthic ecosystem. Science 242:62–64.

Barnes, H. 1959. Stomach contents and microfeeding of some common cirripedes. Canadian Journal of Zoology 37:231–236.

Barnes, R. S. K., and R. N. Hughes. 1988. *An introduction to marine ecology.* Blackwell Scientific, Oxford, UK.

Barrows, C. W. 1997. Habitat relationships of the Coachella Valley fringe-toed lizard (*Uma inornata*). Southwestern Naturalist 42:218–223.

Barry, J. P., and P. K. Dayton. 1991. Physical heterogeneity and the organization of marine communities. Pp. 270–320 in J. Kolasa and S. T. Pickett, eds., *Ecological heterogeneity.* Springer-Verlag, New York, USA.

Barth, J. A., D. Bogucki, S. D. Pierce, and P. M. Kosro. 1998. Secondary circulation associated with a shelfbreak front. Geophysical Research Letters 25:2761–2764.

Barth, J. A., and R. L. Smith. 1998. Separation of a coastal upwelling jet at Cape Blanco, Oregon, USA. South Africa Journal of Marine Science 19:5–14.

Barthem, R. B., M. C. L. B. Ribeiro, and M. Petrere, Jr. 1991. Life strategies of some long-distance migratory catfish in relation to hydroelectric dams in the Amazon basin. Biological Conservation 55:339–345.

Bateman, H. A., T. Joanen, and C. D. Stutzenbaker. 1988. History and status of midcontinental snow geese on their Gulf Coast winter range. Pp. 495–515 in M. W. Weller, ed., *Waterfowl in winter.* University of Minnesota Press, Minneapolis, USA.

Baudry, J. 1988. Structure et fonctionnement écologique des paysages: Cas des bocages. Bulletin d'Ecologie 19:523–530.

Baudry, J., D. Alard, C. Thenail, I. Poudevigne, D. Leconte, J.-F. Bourcier, and C. M. Girard. 1997. Gestion de la biodiversité dans les prairies dans une région d'élevage bovin: Le Pays d'Auge, France. Acta Botanica Gallica 143:367–381.

Baudry, J., and R. G. H. Bunce. 1991. Land abandonment and its role in conservation. Options Méditérranéennes A15:1–148.

Baudry, J., R. G. H. Bunce, and F. Burel. 2000. Hedgerow diversity: An international perspective on their origin, function, and management. Journal of Environmental
Management 60:7–22.

Baudry, J., C. Laurent, and D. Denis. 1997. The technical dimension of agriculture at a regional scale: Methodological considerations. Pp. 161–173 in C. Laurent and I. Bowler,
eds., *CAP and the regions: Building a multidisciplinary framework for the analysis of the
EU agricultural space*. INRA Editions, Paris, France.

Bazely, D. R., and R. L. Jefferies. 1985. Goose faeces: A source of nitrogen for plant growth in
a grazed salt-marsh. Journal of Applied Ecology 22:693–703.

———. 1989. Lesser snow geese and the nitrogen economy of a grazed salt marsh. Journal of
Ecology 77:24–34.

———. 1997. Trophic interactions in Arctic ecosystems and the occurrence of a terrestrial
trophic cascade. Pp. 183–208 in S. J. Woodin and M. Marquiss, eds., *Ecology of Arctic
environments*. Blackwell Science, Oxford, UK.

Bazin, G., G. R. Larrère, F. X. De Montard, M. Lafarge, and P. Loiseau. 1983. Système agraire
et pratiques paysannes dans les Monts Dômes. INRA, Paris, France.

Beatley, T. 1992. Balancing urban-development and endangered species—the Coachella-Valley
habitat conservation plan. Environmental Management 16:7–19.

Beckley, L. E., and G. M. Branch. 1992. A quantitative scuba diving survey of the sublittoral
macrobenthos at subantarctic Marion Island. Polar Biology 11:553–563.

Beeson, C. E., and P. F. Doyle. 1995. Comparison of bank erosion and vegetated and non-
vegetated channel bends. Bulletin of Water Research 31:983–990.

Beier, C. 1991. Atmospheric pollutants: Separation of gaseous and particulate dry deposition
of sulfur at a forest edge in Denmark. Journal of Environmental Quality 20:460–466.

Bell, S. S., E. D. McCoy, and H. R. Mushinsky, eds. 1991. *Habitat structure: The physical
arrangements of objects in space*. Chapman and Hall, London, UK.

Belovsky, G. E. 1981. Optimal activity times and habitat choice of moose. Oecologia 48:22–30.

Belovsky, G. E., and A. Joern. 1995. The dominance of different regulating factors for rangeland grasshoppers. Pp. 359–386 in N. Cappuccino and P. W. Price, eds., *Population dynamics: New approaches and synthesis*. Academic Press, San Diego, CA, USA.

Bencala, K. E. 1984. Interactions of solutes and streambed sediment. 2. A dynamic analysis of
coupled hydrologic and chemical processes that determine solute transport. Water
Resources Research 20:1804–1814.

Bencala, K. E., D. M. McKnight, and G. W. Zellweger. 1990. Characterization of transport in
an acidic and metal rich mountain stream based on a lithium tracer injection and
simulations of transient storage. Water Resources Research 26:989–1000.

Bencala, K. E., and R. A. Walters. 1983. Simulation of solute transport in a mountain pool-
and-riffle stream: A transient storage model. Water Resources Research 19:718–724.

Ben-David, M. 1997. Timing of reproduction in wild mink: The influence of spawning Pacific
salmon. Canadian Journal of Zoology 75:376–382.

Ben-David, M., R. T. Bowyer, L. K. Duffy, D. D. Roby, and D. M. Schell. 1998. River otter latrines
and nutrient dynamics of terrestrial vegetation. Ecology 79:2567–2571.

Ben-David, M., T. A. Hanley, and D. M. Schell. 1998. Fertilization of terrestrial vegetation by
spawning Pacific salmon: The role of flooding and predator activity. Oikos 83:47–55.

Bennett, H. H. 1938. *Soil conservation*. McGraw-Hill, New York, USA.

Bent, A. C. 1925. *Life histories of North American wildfowl*. Part II. Reprint 1962, Dover
Publications, New York, USA.

Berg, M. B., and R. A. Hellenthal. 1992. The role of Chironomidae in energy flow of a lotic
ecosystem. Netherlands Journal of Aquatic Ecology 26:471–476.

Berntson, G. M., and F. A. Bazzaz. 1997. Nitrogen cycling in microcosms of yellow birch exposed to elevated CO_2: Simultaneous positive and negative below-ground feedbacks.
Global Change Biology 3:247–258.

Bertness, M. D., and R. Callaway. 1994. Positive interactions in communities. Trends in
Ecology and Evolution 9:191–193.

Bertness, M. D., S. D. Gaines, and M. E. Hay, eds. 2001. *Marine community ecology.* Sinauer Associates, Sunderland, MA, USA.

Bertness, M. D., S. D. Gaines, and S. M. Yeh. 1998. Making mountains out of barnacles: The dynamics of acorn barnacle hummocking. Ecology 79:1382–1394.

Bertness, M. D., G. H. Leonard, J. M. Levine, P. R. Schmidt, and A. O. Ingraham. 1999. Testing the relative contribution of positive and negative interactions in rocky intertidal communities. Ecology 80:2711–2726.

Bertoldi, G. L. 1992. Subsidence and consolidation in alluvial aquifer systems. Pp. 62–74 in *Proceedings of the 18th biennial conference on ground water.* U.S. Geological Survey, Washington, DC, USA.

Bidleman, T. F., G. W. Patton, M. D. Walla, B. T. Hargrave, W. P. Vass, P. Erickson, B. Fowler, V. Scott, and D. J. Gregor. 1989. Toxaphene and other organochlorines in the Arctic Ocean fauna: Evidence for atmospheric delivery. Arctic 42:307–313.

Bierregaard, R. O., Jr., T. E. Lovejoy, V. Kapos, A. A. Dos Santos, and R. W. Hutchings. 1992. The biological dynamics of tropical rainforest fragments: A prospective comparison of fragments and continuous forest. BioScience 42:859–866.

Bigelow, H. B. 1926. Plankton of the offshore waters of the Gulf of Maine. Bulletin of the Bureau of Fisheries, 40, part 2.

Biggs, B. J. F. 1995. The contribution of flood disturbance, catchment geology and land use to the habitat template of periphyton in stream ecosystems. Freshwater Biology 33:419–438.

Bilby, R. E., B. R. Fransen, and P. A. Bisson. 1996. Incorporation of nitrogen and carbon from spawning coho salmon into the trophic system of small streams: Evidence from stable isotopes. Canadian Journal of Fisheries and Aquatic Sciences 53:164–173.

Bilby, R. E., and G. E. Likens. 1979. Effect of hydrologic fluctuations on the transport of fine particulate organic carbon in a small stream. Limnology and Oceanography 24:69–75.

Bildstein, K. L., E. Blood, and P. Frederick. 1992. The relative importance of biotic and abiotic vectors in nutrient transport. Estuaries 15:147–157.

Blais, J. M., D. W. Schindler, D. C. G. Muir, L. E. Kimpe, D. B. Donald, and B. Rosenberg. 1998. Accumulation of persistent organochlorine compounds in mountains of western Canada. Nature 395:585–588.

Blazer, V. S., W. K. Vogelbein, C. L. Densmore, E. B. May, J. H. Lilley, and D. E. Zwerner. 1999. *Aphanomyces* as a cause of ulcerative skin lesions of menhaden from Chesapeake Bay tributaries. Journal of Aquatic Animal Health 11:340–349.

Boisclair, D., and J. B. Rasmussen. 1996. Empirical analysis of the influence of environmental variables on perch growth, consumption, and activity rates. Annales Zoologica Fennici 33:507–515.

Bonsall, M. B., and M. P. Hassell. 1997. Apparent competition structures ecological assemblages. Nature 388:371–373.

Bootsma, H. A., R. E. Hecky, R. H. Hesslein, and G. F. Turner. 1996. Food partitioning among Lake Malawi nearshore fishes as revealed by stable isotope analyses. Ecology 77:1286–1290.

Bormann, F. H., and G. E. Likens. 1967. Nutrient cycling. Science 155:424–429.

———. 1979. *Pattern and process in a forested ecosystem.* Springer-Verlag, New York, USA.

Bosman, A. L., and P. A. R. Hockey. 1986. Seabird guano as a determinant of rocky intertidal community structure. Marine Ecology Progress Series 32:247–257.

Bosman, A. L., P. A. R. Hockey, and W. R. Siegfried. 1987. The influence of coastal upwelling on the functional structure of rocky intertidal communities. Oecologia 72:226–232.

Bothwell, M. L., D. M. Sherbot, and C. M. Pollock. 1994. Ecosystem response to solar ultraviolet-B radiation: Influence of trophic-level interactions. Science 265:97–100.

Botsford, L. W., J. C. Castilla, and C. H. Peterson. 1997. The management of fisheries and marine ecosystems. Science 277:509–515.

Bouchard, S. S., and K. A. Bjorndal. 1998. Sea turtles as biological transporters of nutrients and energy from marine to terrestrial ecosystems. 1998 Annual Meeting Abstracts, Ecological Society of America, Washington, DC, USA.

Boulinier, T., and E. Danchin. 1996. Population trends in kittiwake *Rissa tridactila* colonies in relation to tick infestation. Ibis 138:326–334.

Boutton, T. W. 1991. Stable carbon isotope ratios of natural materials: II. Atmospheric, terrestrial, marine, and freshwater environments. Pp. 173–185 in D. C. Coleman and B. Fry, eds., *Carbon isotope techniques*. Academic Press, San Diego, CA, USA.

Bowen, S. H. 1983. Detritivory in Neotropical fish communities. Environmental Biology of Fishes 9:137–144.

Boyd, C. E. 1973. Amino acid composition of freshwater algae. Archiv für Hydrobiologie 72:1–9.

Boyd, H., G. E. J. Smith, and F. G. Cooch. 1982. The lesser snow goose of the eastern Canadian Arctic: Their status during 1964–1979 and their management from 1982–1990. Occasional papers, no. 46. Canadian Wildlife Service, Ottawa, Canada.

Brabrand, A., B. A. Faafeng, and J. P. M. Nilssen. 1990. Relative importance of phosphorus supply to phytoplankton production: Fish excretion versus external loading. Canadian Journal of Fisheries and Aquatic Sciences 47:364–372.

Branch, G., and M. Branch. 1981. *The living shores of southern Africa*. Struik Publishers Ltd., Cape Town, SA.

Branch, G. M., and C. L Griffiths. 1988. The Benguela ecosystem. Part V. The coastal zone. Oceanography Marine Biology Annual Review 26:395–486.

Braunstrator, D. K., G. Cabana, A. Mazumder, and J. B. Rasmussen. 2000. Measuring life-history omnivory in the opossum shrimp, *Mysis relicta*, with stable nitrogen isotopes. Limnology and Oceanography 45:463–467.

Breitmeyer, C. M., and T. A. Markow. 1998. Resource availability and population size in cactophilic *Drosophila*. Functional Ecology 12:14–21.

Bremigan, M. T. 1997. Variable recruitment of gizzard shad, a strong interactor in reservoirs: Exploring causal mechanisms and implications for food webs. Ph.D. dissertation, Ohio State University, Columbus, OH, USA.

Bremigan, M. T., and R. A. Stein. 2001. Variable gizzard shad recruitment with reservoir productivity: Causes and implications for classifying systems. Ecological Applications 11:1425–1437.

Bremner, J. M., C. F. Butcher, and G. B. Patterson. 1984. The density of indigenous invertebrates on three islands in Breaksea Sound, Fiordland, in relation to the distribution of introduced mammals. Journal of the Royal Society of New Zealand 14:379–386.

Brett, M. T., and C. R. Goldman. 1996. A meta-analysis of the freshwater trophic cascade. Proceedings of the National Academy of Sciences, USA 93:7723–7726.

———. 1997. Consumer versus resource control in freshwater pelagic food webs. Science 275:384–386.

Brett, M. T., and D. C. Muller-Navarra. 1997. The role of essential fatty acids in aquatic food web processes. Freshwater Biology 38:483–499.

Brickell, D. C., and J. J. Goering. 1970. Chemical effects of salmon decomposition on aquatic ecosystems. Pp. 125–138 in R. S. Murphy, D. Nyquist, and P. W. Neff, eds., *International symposium on water pollution control in cold climates*. U.S. Government Printing Office, Washington, DC, USA.

Bridgeford, P. A. 1985. Unusual diet of the lion *Panthera leo* in the Skeleton Coast Park. Madoqua 14:187–188.

Brigham, R. M. 1990. Prey selection by big brown bats (*Eptesicus fuscus*) and common nighthawks (*Chordeiles minor*). American Midland Naturalist 124:73–80.

Brigham, R. M., H. D. J. N. Aldridge, and R. L. Mackey. 1992. Variation in habitat use and prey selection by yuma bats, *Myotis yumanensis*. Journal of Mammalogy 73:640–645.

Brigham, R. M., and R. M. R. Barclay. 1996. Bats and forests: Conference summary. Pp. xi–xiv in *Bats and forests symposium, October 19–21, 1995, Victoria, British Columbia, Canada*.

Working Paper 23/1996. British Columbia, Ministry of Forests Research Program, Victoria, BC, Canada.

Brigham, R. M., S. D. Grindal, M. C. Firman, and J. L. Morissette. 1997. The influence of structural clutter on activity patterns of insectivorous bats. Canadian Journal of Zoology 75:131–136.

Brinkhurst, R. O., and B. Walsh. 1967. Rostherne Mere, England, a further instance of guanotrophy. Journal of the Fisheries Research Board of Canada 24:1299–1309.

Brittain, J. E., and T. J. Eikeland. 1988. Invertebrate drift—a review. Hydrobiologia 166:77–93.

Broitman, B. R., S. A. Navarrete, F. Smith, and S. D. Gaines. 2001. Geographic variation of southeastern Pacific intertidal communities. Marine Ecology Progress Series 224: 21–34.

Brooks, D. A. 1985. Vernal circulation in the Gulf of Maine. Journal of Geophysical Research 90:4687–4705.

Brossier, J., L. de Bonneval, E. Landais, and J. Brossier. 1993. *System studies in agriculture and rural development.* Science Update. INRA Editions, Paris, France.

Brothers, T. S., and A. Spingarn. 1992. Forest fragmentation and alien plant invasion of Central Indiana old-growth forests. Conservation Biology 6:91–100.

Brown, A. C., and A. McLachlan. 1990. *Ecology of sandy shores.* Elsevier, Amsterdam, The Netherlands.

Brown, H. S., R. E. Kasperson, and S. S. Raymond. 1990. Use and transformation of terrestrial water systems. Pp. 437–454 in B. L. Turner II, W. C. Clark, R. W. Kates, J. F. Richards, J. T. Mathews, and W. B. Meyer, eds., *The earth as transformed by human action.* Cambridge University Press, Cambridge, UK.

Brown, J. S. 1990. Habitat selection as an evolutionary game. Evolution 44:732–746.

———. 1996. Coevolution and community organization in three habitats. Oikos 75:193–206.

Brown, J. S., and D. L. Venable. 1991. Life history evolution of seed-bank annuals in response to seed predation. Evolutionary Ecology 5:12–29.

Bruner, M. C. 1977. Vegetation of forest island edges. M.S. thesis, University of Wisconsin, Milwaukee.

Bruno, J. B., and M. D. Bertness. 2001. Habitat modification and facilitation in benthic marine communities. Pp. 201–218 in M. D. Bertness, S. D. Gaines, and M. E. Hay, eds., *Marine community ecology.* Sinauer Associates, Sunderland, MA, USA.

Brusca, R. C. 1980. *Common intertidal invertebrates of the Gulf of California.* University of Arizona Press, Tucson, USA.

Bryant, M. D., D. N. Swanston, R. C. Wissmar, and B. E. Wright. 1997. Coho salmon populations in the karst landscape of north Prince of Wales Island, southeast Alaska. Pacific Northwest Research Station, Forestry Sciences Laboratory, Juneau, AK. Unpublished report.

Budy, P., C. Luecke, and W. A. Wurtsbaugh. 1998. Adding nutrients to enhance the growth of endangered sockeye salmon: Trophic transfer in an oligotrophic lake. Transactions of the American Fisheries Society 127:19–34.

Bunce, R. G. H., and D. C. Howard, eds. 1990. *Species dispersal in agricultural habitats.* Belhaven Press, London, UK.

Bunn, S. E., and P. I. Boon. 1993. What sources of organic carbon drive food webs in billabongs?—a study based on stable isotope analysis. Oecologia 96:85–94.

Buntin, G. D. 1998. Cabbage seedpod weevil (*Ceutorhynchus assimilis,* Paykull) management by trap cropping and its effect on parasitism by *Trichomalus perfectus* (Walker) in oilseed rape. Crop Protection 17:299–305.

Burel, F. 1989. Landscape structure effects on carabid beetles: Spatial patterns in western France. Landscape Ecology 2:215–226.

Burel, F., and J. Baudry. 1999. *Ecologie du paysage: Concepts, méthodes et applications.* Lavoisier, Paris, France.

Burger, A. E. 1985. Terrestrial food webs in the Sub-Antarctic: Island effects. Pp. 582–591 in W. R. Siegfried, P. R. Condy, and R. M. Laws, eds., *Antarctic nutrient cycles and food webs.* Springer-Verlag, Berlin, Germany.

Burger, A. E., H. J. Lindeboom, and A. J. Williams. 1978. The mineral and energy contributions of guano of selected species of birds to the Marion Island terrestrial ecosystem. South African Journal of Antarctic Research 8:59–70.

Burgess, H. H. 1980. The Squaw Creek goose flocks with special attention to lesser snow geese. Preliminary report. U.S. Fish and Wildlife Service, North Kansas City, MO, USA.

Burkholder, J. M., and H. B. Glasgow. 1997. *Pfiesteria piscicida* and other *Pfiesteria*-like dinoflagellates: Behavior, impacts, and environment controls. Limnology and Oceanography 42:1052–1075.

Burkholder, J. M., M. A. Mallin, H. B. Glasgow, Jr., L. M. Larsen, M. R. McIver, G. C. Shank, N. Deamer-Melia, D. S. Briley, J. Springer, B. W. Touchette, and E. K. Hannon. 1997. Impacts of a coastal river and estuary from rupture of a large swine waste holding lagoon. Journal of Environmental Quality 26:1451–1466.

Burt, P. J. A., and D. E. Pedgley. 1997. Nocturnal insect migrations: Effects of local winds. Advances in Ecological Research 27:61–92.

Buss, L. W., and J. B. C. Jackson. 1981. Planktonic food availability and suspension feeder abundance: Evidence of in situ depletion. Journal of Experimental Marine Biology and Ecology 49:151–161.

Bustamante, R. H., and G. M. Branch. 1996a. The dependence of intertidal consumers on kelp-derived organic matter on the west coast of South Africa. Journal of Experimental Marine Biology and Ecology 196:1–28.

———. 1996b. Large scale patterns and trophic structure of southern African rocky shores: The roles of geographic variation and wave exposure. Journal of Biogeography 23:339–351.

Bustamante, R. H., G. M. Branch, and S. Eekhout. 1995a. Maintenance of an exceptional intertidal grazer biomass in South Africa: Subsidy by subtidal kelps. Ecology 76:2314–2329.

Bustamante, R. H., G. M. Branch, S. Eekhout, B. Robertson, P. Zoutendyk, M. Schleyer, A. Dye, N. Hanekom, D. Keats, M. Jurd, and C. McQuaid. 1995b. Gradients of intertidal primary productivity around the coast of South Africa and their relationships with consumer biomass. Oecologia 102:189–201.

Butler, M. G. 1984. Life histories of aquatic insects. Pp. 24–55 in V. H. Resh and D. M. Rosenberg, eds., *The ecology of aquatic insects.* Plenum, New York, USA.

Butman, C. A. 1987. Larval settlement of soft sediment invertebrates: The spatial scales of pattern explained by active habitat selection and the emerging role of hydrodynamical processes. Oceanography and Marine Biology Annual Review 25:113–165.

Butman, C. A., M. Frechette, W. Rockwell Geyer, and V. R. Starczak. 1994. Flume experiments on food supply to the blue mussel *Mytilus edulis* as a function of boundary layer flow. Limnology and Oceanography 39:1755–1768.

Cabana, G., and J. B. Rasmussen. 1994. Modelling food chain structure and contaminant bioaccumulation using stable N-isotopes. Nature 372:255–258.

———. 1996. Comparing the length of aquatic food chains using stable N isotopes. Proceedings of the National Academy of Sciences, USA 93:10844–10847.

Cabana, G., A. Tremblay, J. Kalff, and J. B. Rasmussen. 1994. Food chain length as a determinant of mercury levels in Ontario lake trout. Canadian Journal of Fisheries and Aquatic Sciences 51:381–389.

Cadenasso, M. L. 1998. Linking forest edge structure to edge function: An experimental and synthetic approach. Ph.D. dissertation, Rutgers University, New Brunswick, NJ, USA.

Cadenasso, M. L., and S. T. A. Pickett. 2000. Linking forest edge structure to edge function: Mediation of herbivore damage. Journal of Ecology 88:31–44.

———. 2001. Effects of edge structure on the flux of species into forest interiors. Conservation Biology 15:91–97.

Cadenasso, M. L., M. M. Traynor, and S. T. A. Pickett. 1997. Functional location of forest edges: Gradients of multiple physical factors. Canadian Journal of Forest Research 27:774–782.

Camargo, J. L. C., and V. Kapos. 1995. Complex edge effects on soil moisture and micro-climate in central Amazonian forest. Journal of Tropical Ecology 11:205–221.

Cane, M. A., G. Eshel, and R. W. Buckland. 1994. Forecasting Zimbabwean maize yield using eastern equatorial Pacific sea surface temperature. Nature 370:204–205.

Canfield, D. E., and A. Teske. 1996. Late Proterozoic rise in atmospheric oxygen concentration inferred from phylogenetic and sulfur-isotope studies. Nature 382:127–132.

Caraco, N. F. 1995. Influence of humans on phosphorus transfers to aquatic systems: A regional scale study using large rivers. Pp. 235–244 in H. Tiessen, ed., *Phosphorus in the global environment: Transfers, cycles and management.* John Wiley and Sons, Chichester, MA, USA.

Caraco, N. F., and J. J. Cole. 1999. Human impact on nitrate export: An analysis using major world rivers. Ambio 28:167–170.

Caraco, N. F., J. J. Cole, and G. E. Likens. 1992. New and recycled primary production in an oligotrophic lake: Insights for summer phosphorus dynamics. Limnology and Oceanography 37:590–602.

Caraco, N. F., J. J. Cole, P. A. Raymond, D. L. Strayer, M. L. Pace, S. E. G. Findlay, and D. T. Fischer. 1997. Zebra mussel invasion in a large, turbid river: Phytoplankton response to increased grazing. Ecology 78:588–602.

Caraco, N. F., G. Lampman, J. J. Cole, K. E. Limburg, M. L. Pace, and D. Fischer. 1998. Microbial assimilation of DIN in a nitrogen rich estuary: Implications for food quality and isotope studies. Marine Ecology Progress Series 18:59–71.

Caraco, N. F., and R. A. Miller. 1998. Effects of CO_2 on competition between a cyanobacterium and eukaryotic phytoplankton. Canadian Journal of Fisheries and Aquatic Sciences 55:54–62.

Cargill, S. M., and R. L. Jefferies. 1984a. The effects of grazing by lesser snow geese on the vegetation of a sub-arctic salt-marsh. Journal of Applied Ecology 21:669–686.

———. 1984b. Nutrient limitation of primary production in a sub-arctic salt-marsh. Journal of Applied Ecology 21:657–668.

Carlton, R. G., and C. R. Goldman. 1984. Effects of a massive swarm of ants on ammonium concentrations in a subalpine lake. Hydrobiologia 111:113–117.

Caro, T. 1987. Cheetah mothers' vigilance: Looking out for prey or for predators. Behavioral Ecology and Sociobiology 20:351–361.

Carpenter, S. R., W. Brock, and P. Hanson. 1999. Ecological and social dynamics in simple models of ecosystem management. Conservation Ecology 3(2):4.

Carpenter, S. R., N. F. Caraco, D. L. Carroll, R. W. Howarth, A. N. Sharpley, and V. H. Smith. 1998. Nonpoint pollution of surface waters with phosphorus and nitrogen. Ecological Applications 8:559–568.

Carpenter, S. R., D. L. Christensen, J. J. Cole, K. L. Cottingham, X. He, J. R. Hodgson, J. F. Kitchell, S. E. Knight, M. L. Pace, D. M. Post, D. E. Schindler, and N. Voichick. 1995. Biological control of eutrophication in lakes. Environmental Science and Technology 29:784–786.

Carpenter, S. R., J. J. Cole, J. F. Kitchell, and M. L. Pace. 1998. Variable productivity in whole-lake experiments: Roles of dissolved organic carbon phosphorus and grazing. Limnology and Oceanography 43:73–80.

Carpenter, S. R., K. L. Cottingham, and D. E. Schindler. 1992. Biotic feedbacks in lake phosphorus cycles. Trends in Ecology and Evolution 7:332–336.

Carpenter, S. R., and J. F. Kitchell, eds. 1993. *The trophic cascade in lakes.* Cambridge University Press, Cambridge, UK.

Carpenter, S. R., D. Ludwig, and W. A. Brock. 1999. Management of eutrophication for lakes subject to potentially irreversible change. Ecological Applications 9:751–771.

Carpenter, S. R., and M. L. Pace. 1997. Dystrophy and eutrophy in lake ecosystems: Implications of fluctuating inputs. Oikos 78:3–14.

Carter, J. C. H., M. J. Dadswell, J. C. Roff, and W. G. Sprules. 1980. Distribution and zoo-geography of planktonic crustaceans and dipterans in glaciated eastern North America. Canadian Journal of Zoology 58:1355–1387.

Case, T. J., M. E. Gilpin, and J. M. Diamond. 1979. Overexploitation, interference competition, and excess density compensation in insular faunas. American Naturalist 113: 843–854.

Caswell, H. 1989. *Matrix population models*. Sinauer Associates, Sunderland, MA, USA.

Caubel-Forget, V., and C. Grimaldi. 1999. Fonctionnement hydrique et géochimique du talus de ceinture de bas-fond: Conséquences sur le transfert et le devenir des nitrates. Pp. 169–189 in *Actes du colloque Bois et forêts des agriculteurs Paris*. Editions Cemagref, France.

Cederholm, C. J., D. B. Houston, D. L. Cole, and W. J. Scarlett. 1989. Fate of coho salmon (*Oncorhynchus kisutch*) carcasses in spawning streams. Canadian Journal of Fisheries and Aquatic Sciences 46:1347–1355.

Cederholm, C. J., M. D. Kunze, T. Murota, and A. Sibatani. 1999. Pacific salmon carcasses: Essential contributions of nutrients and energy for aquatic and terrestrial ecosystems. Fisheries 24:6–15.

Cederholm, C. J., and N. P. Peterson. 1985. The retention of coho salmon (*Oncorhynchus kisutch*) carcasses by organic debris in small streams. Canadian Journal of Fisheries and Aquatic Sciences 42:1222–1225.

Chadwick, O. A., L. A. Derry, P. M. Vitousek, B. J. Heubert, and L. O. Hedin. 1999. Changing sources of nutrients during four million years of ecosystem development. Nature 397:491–497.

Chaloner, D. T., K. M. Martin, M. S. Wipfli, P. H. Ostrom, and G. A. Lamberti. 2002. Marine carbon and nitrogen in southeastern Alaska stream food webs: Evidence from artificial and natural streams. Canadian Journal of Fisheries and Aquatic Sciences 59: 1257–1265.

Chamberlain, T. W., R. D. Harr, and F. H. Everest. 1991. Timber harvesting, silviculture, and watershed processes. Pp. 181–205 in W. R. Meehan, ed., *Influences of forest and rangeland management on salmonid fishes and their habitat*. American Fisheries Society, Bethesda, MD, USA.

Chandler, C., P. Cheney, P. Thomas, L. Trabaud, and D. Williams. 1983. *Fire in forestry*. 2 volumes. John Wiley and Sons, New York, USA.

Chang, E. R. 2000. Seed and vegetation dynamics in undamaged and degraded coastal habitats of the Hudson Bay lowlands. M.S. thesis, University of Toronto, Canada.

Chang, E. R., R. L. Jefferies, and T. J. Carleton. 200. Relationship between the vegetation and soil seed bank in a coastal arctic habitat. Journal of Ecology 89:367–384.

Charley, J. L., and S. L. Cowling. 1968. Changes in soil nutrient status resulting from overgrazing and their consequences in plant communities of semiarid areas. Proceedings of the Ecological Society of Australia 3:28–38.

Charrier, S., S. Petit, and F. Burel. 1997. Movements of *Abax parallelepipedus* (Coleoptera, Carabidae) in woody habitats of a hedgerow network landscape: A radio-tracing study. Agriculture, Ecosystems and Environment 61:133–144.

Chase, J. M. 2000. Are there real differences among aquatic and terrestrial food webs? Trends in Ecology and Evolution 15:408–412.

Checkley, D. M., and L. C. Entzeroth. 1985. Elemental and isotopic fractionation of carbon and nitrogen by marine copepods and implications to the marine nitrogen cycle. Journal of Plankton Research 7:553–568.

Chelazzi, G., and M. Vannini. 1988. Behavioural adaptation to intertidal life. NATO ASI series, Series A, Life Sciences, vol. 151. Plenum Press, New York, USA.

Chen, J., J. F. Franklin, and T. A. Spies. 1992. Vegetation responses to edge environments in old-growth Douglas-fir forests. Ecological Applications 2:387–396.

———. 1995. Growing-season microclimatic gradients from clearcut edges into old-growth Douglas-fir forests. Ecological Applications 5:74–86.

Cheng, L., ed. 1976. *Marine insects*. American Elsevier Publishing, New York, USA.

Cheng, L., and M. C. Birch. 1978. Insect flotsam: An unstudied marine resource. Ecological Entomology 3:87–97.

Chesson, P. L. 1985. Coexistence of competitors in spatially and temporally varying environments: A look at the combined effects of different sorts of variability. Theoretical Population Biology 28:263–287.

———. 1991. Stochastic population models. Pp. 123–143 in J. Kolasa and S. T. A. Pickett, eds., *Ecological heterogeneity*. Springer-Verlag, New York, USA.

———. 1997. Making sense of spatial models in ecology. Pp. 151–166 in J. Bascompte and R. V. Solé, eds., *Modeling spatiotemporal dynamics in ecology*. Landes Bioscience, Berlin, Germany.

———. 2000a. General theory of competitive coexistence in spatially-varying environments. Theoretical Population Biology 58:211–237.

———. 2000b. Mechanisms of maintenance of species diversity. Annual Reviews of Ecology and Systematics 31:343–366.

Chesson, P. L., and N. Huntly. 1993. Temporal hierarchies of variation and the maintenance of diversity. Plant Species Biology 8:195–206.

———. 1997. The roles of harsh and fluctuating conditions in the dynamics of ecological communities. American Naturalist 150:519–553.

Christensen, P. E., and N. D. Burrows. 1986. Fire: An old tool with new uses. Pp. 97–105 in R. H. Groves and J. J. Burdon, eds., *Ecology of biological invasions: An Australian perspective*. Australian Academy of Sciences, Canberra, Australia.

Christensen, V., and D. Pauly. 1992. Ecopath-II—A software for balancing steady-state ecosystem models and calculating network characteristics. Ecological Modeling 61: 169–185.

Christie, D. R., K. J. Muirhead, and R. H. Clarke. 1981. Solitary waves in the lower atmosphere. Nature 293:46–49.

Christie, W. J., K. A. Crossman, P. G. Sly, and R. H. Krus. 1987. Recent changes in the aquatic food web of eastern Lake Ontario. Canadian Journal of Fisheries and Aquatic Sciences 44 (Suppl. 2): 37–52.

Churcher, P. B., and J. H. Lawton. 1987. Predation by domestic cats in an English village. Journal of Zoology 212:439–455.

Clark, J. F., R. Waninkof, P. Schlosser, and H. J. Simpson. 1994. Gas exchange rates in the tidal Hudson River using a dual tracer technique. Tellus 46B:274–285.

Clark, J. S. 1990. Fire and climate change during the last 750 years in northwestern Minnesota. Ecological Monographs 60:135–159.

Clarke, R. D., and P. R. Grant. 1968. An experimental study of the role of spiders as predators in a forest litter community, Part 1. Ecology 49:1152–1154.

Clearwater, S., A. Mark, B. Fahey, R. Jackson, B. Fitzharris, and B. Thomas. 1999. Upland land use and water yield. Issues in Ecology, no. 1. Dunedin: Ecology Research Group, University of Otago, New Zealand.

Cloe, W. W., and G. C. Garman. 1996. The energetic importance of terrestrial inputs to three warm-water streams. Freshwater Biology 36:105–114.

Clutton-Brock, T. H., and S. D. Albon. 1989. *Red deer in the Highlands*. BSP Professional Books, Oxford, UK.

Clutton-Brock, T. H., F. E. Guinness, and S. D. Albon. 1983. *Red deer: Behaviour and ecology of two sexes*. University of Chicago Press, Chicago, IL, USA.

Coffin, R. B., B. Fry, B. J. Peterson, and R. T. Wright. 1989. Carbon isotopic compositions of estuarine bacteria. Limnology and Oceanography 34:1305–1310.

Cohen, J. E. 1978. *Food webs and niche space*. Princeton University Press, Princeton, NJ, USA.

Cole, D. W. 1981. Nitrogen uptake and translocation by forest ecosystems. Pp. 219–232 in F. E. Clark and T. Rosswall, eds., *Terrestrial nitrogen cycles*. Ecological Bulletin 33. Swedish Natural Sciences Research Council, Stockholm, Sweden.

Cole, J. J., and N. F. Caraco. 1993. The pelagic microbial food web of oligotrophic lakes. Pp. 101–112 in T. E. Ford, ed., *Aquatic microbiology: An ecological approach*. Blackwell, Cambridge, UK.

————. 1998. Atmospheric exchange of carbon dioxide in a low-wind oligotrophic lake measured by the addition of SF6. Limnology and Oceanography 43:647–656.

————. 2001. Carbon in catchments: Connecting terrestrial carbon losses with aquatic metabolism. Marine and Freshwater Research 52:101–110.

Cole, J. J., N. F. Caraco, G. W. Kling, and T. K. Kratz. 1994. Carbon dioxide supersaturation in the surface waters of lakes. Science 265:1568–1570.

Cole, J. J., N. F. Caraco, and G. E. Likens. 1990. Short-range atmospheric transport: A significant source of phosphorus to an oligotrophic lake. Limnology and Oceanography 35: 1230–1237.

Cole, J. J., N. F. Caraco, and B. Peierls. 1991. Phytoplankton primary production in the tidal, freshwater Hudson. Verhandlungen Internationale Vereinigung für Theoretische und Angewandte Limnologie 24:1715–1719.

Cole, J. J., N. F. Caraco, D. L. Strayer, C. Ochs, and S. Nolan. 1989. A detailed organic carbon budget as an ecosystem-level calibration of bacterial respiration in an oligotrophic lake during midsummer. Limnology and Oceanography 34:286–296.

Cole, J. J., M. L. Pace, S. R. Carpenter, and J. F. Kitchell. 2000. Persistence of net heterotrophy in lakes during nutrient addition and food web manipulations. Limnology and Oceanography 45:1718–1730.

Coll, M., and D. G. Bottrell. 1996. Movement of an insect parasitoid in simple and diverse plant assemblages. Ecological Entomology 21:141–149.

Connell, J. H. 1961a. Effects of competition, predation by *Thais lapillus*, and other factors on natural populations of the barnacle *Balanus balanoides*. Ecological Monographs 31:61–104.

————. 1961b. The influence of interspecific competition and other factors on the distribution of the barnacle *Chthamalus stellatus*. Ecology 42:710–723.

————. 1970. A predator-prey system in the marine intertidal region. I. *Balanus glandula* and several predatory species of *Thais*. Ecological Monographs 40:49–78.

————. 1975. Some mechanisms producing structure in natural communities: A model and evidence from field experiments. Pp. 460–490 in M. L. Cody and J. M. Diamond, eds., *Ecology and evolution of communities*. Belknap Press, Cambridge, MA, USA.

Connolly, S. R., and J. Roughgarden. 1998. A latitudinal gradient in northeast Pacific intertidal community structure: Evidence for an oceanographically based synthesis of marine community theory. American Naturalist 151:311–326.

————. 1999a. Increased recruitment of northeast Pacific barnacles during the 1997 El Niño. Limnology and Oceanography 44:466–469.

————. 1999b. Theory of marine communities: Competition, predation, and recruitment-dependent interaction strength. Ecological Monographs 69:277–296.

Conover, R. 1966. Assimilation of organic matter by zooplankton. Limnology and Oceanography 11:338–345.

Cooch, E. G., D. B. Lank, R. F. Rockwell, and F. Cooke. 1991. Long-term decline in body size in a snow goose population: Evidence of environmental degradation? Journal of Animal Ecology 60:483–496.

Cooke, F., C. M. Francis, E. G. Cooch, and R. Alisauskas. 1999. Impact of hunting on population growth of mid-continent lesser snow geese. Pp. 17–31 in H. Boyd, ed., *Population modelling and management of snow geese*. Occasional papers, no. 102. Canadian Wildlife Service, Ottawa, Canada.

Cooper, J. R., J. W. Gilliam, R. B. Daniels, and W. P. Robarge. 1987. Riparian areas as filters for agricultural sediment. Proceedings of the Soil Science Society of America 51: 416–420.

Corbett, A., and J. A. Rosenheim. 1996a. Impact of a natural enemy overwintering refuge and its interaction with the surrounding landscape. Ecological Entomology 21:155–164.

————. 1996b. Quantifying movement of a minute parasitoid, *Anagrus epos* (Hymenoptera: Mymaridae), using fluorescent dust marking and recapture. Biological Control 6:35–44.

Correll, D. L. 1996. Buffer zones and water quality protection: General principles. Pp. 7–20 in N. E. Haycock, T. P. Burt, K. W. T. Goulding, and G. Pinay, eds., *Buffer zones: Their processes and potential in water protection.* Quest Environmental, Harpenden, UK.

Crawford, G., and R. Brunato. 1978. An inventory of data on PCB and Mirex levels in Ontario sportsfish. Ontario Ministry of the Environment, Report 7801.

Crawford, S. S. 2001. *Salmonine introductions to the Laurentian Great Lakes: An historical review and evaluation of ecological effects.* Canadian Special Publication of Fisheries and Aquatic Sciences 132. NRC Research Press, Ottawa, Ontario, Canada.

Crawley, M. J. 1983. *Herbivory: The dynamics of animal-plant interactions.* University of California Press, Berkeley, USA.

Cree, A., C. H. Daugherty, and J. M. May. 1995. Reproduction of a rare New Zealand reptile, the tuatara, *Sphenodon punctatus,* on rat-free and rat-inhabited islands. Conservation Biology 9:373–383.

Crisp, D. J., and E. Bourget. 1985. Growth in barnacles. Advances in Marine Biology 22: 199–244.

Crisp, D. J., and A. J. Southward. 1961. Different types of cirral activities of barnacles. Philosophical Transactions of the Royal Society of London 243:271–308.

Crooks, K. R., and M. E. Soulé. 1999. Mesopredator release and avifaunal extinctions in a fragmented system. Nature 400:563–566.

Crowder, L. B., D. P. Reagan, and D. W. Freckman. 1996. Food web dynamics and applied problems. Pp. 327–336 in G. A. Polis and K. O. Winemiller, eds., *Food webs: Integration of patterns and dynamics.* Chapman and Hall, New York, USA.

Cubit, J. 1984. Herbivory and the seasonal abundance of algae on a high intertidal rocky shore. Ecology 65:1904–1917.

Cunningham, A., and R. M. Nisbet. 1983. Transients and oscillations in a continuous culture. Pp. 77–103 in M. J. Bazin, ed., *Mathematics in microbiology.* Academic Press, New York, USA.

Cyr, H., and M. L. Pace. 1993. Magnitude and patterns of herbivory in aquatic and terrestrial ecosystems. Nature 361:148–150.

Dadswell, M. J. 1974. Distribution, ecology and postglacial dispersal in certain crustaceans and fishes in eastern North America. National Museum of Canada, Ottawa, Ontario, Canada.

Dahl, E. 1952. Some aspects of the ecology and zonation of the fauna on sandy beaches. Oikos 4:1–27.

Dahlgaard, H. 1996. Polonium-210 in mussels and fish from the Baltic-North Sea estuary. Journal of Environmental Radioactivity 32:91–96.

Dahlhoff, E. P., and B. A. Menge. 1996. Influence of phytoplankton concentration and wave exposure on the ecophysiology of *Mytilus californianus.* Marine Ecology Progress Series 144:97–107.

Daniels, R. B., and J. W. Gilliam. 1997. Sediment and chemical load reduction by grass and riparian filters. Proceedings of the Soil Science Society of America 60:246–261.

D'Antonio, C. M., and P. M. Vitousek. 1992. Biological invasions by exotic grasses, the grass/fire cycle, and global change. Annual Review of Ecology and Systematics 23: 63–87.

Daugherty, C. H., D. R. Towns, I. A. E. Atkinson, and G. W. Gibbs. 1990. The significance of the biological resources of New Zealand islands for ecological restoration. Pp. 9–21 in D. R. Towns, C. H. Daugherty, and I. A. E. Atchinson, eds., *Ecological restoration of the New Zealand islands.* Conservation Sciences Publication No. 2. New Zealand Department of Conservation, Wellington.

Davis, B. N. K. 1983. *Insects on nettles.* Cambridge University Press, Cambridge, UK.

———. 1989. The European distribution of insects on stinging nettles, *Urtica dioica* L.: A field survey. Bollettino di Zoologia 56:321–326.

Davis, J. A., and C. E. Boyd. 1978. Concentration of selected elements and ash in bluegill (*Lepomis macrochirus*) and certain other freshwater fish. Transactions of the American Fisheries Society 107:862–867.

Davis, M. B. 1976. Erosion rates and land-use history in southern Michigan. Environmental Conservation 3:139–148.

Day, J. W., C. A. S. Hall, W. M. Kemp, and A. Yanez-Arancibia. 1989. *Estuarine ecology.* John Wiley and Sons, New York, USA.

Dayton, P. K. 1971. Competition, disturbance, and community organization: The provision and subsequent utilization of space in a rocky intertidal community. Ecological Monographs 41:351–389.

Dayton, P. K., M. J. Tegner, P. B. Edwards, and K. L. Riser. 1999. Temporal and spatial scales of kelp demography: The role of oceanographic climate. Ecological Monographs 69: 219–250.

Dayton, P. K., S. F. Thrush, M. T. Agardy, and R. J. Hofman. 1995. Environmental effects of marine fishing. Aquatic Conservation: Marine and Freshwater Ecosystems 5:205–232.

Dean, W. E., and E. Gorham. 1998. Magnitude and significance of carbon burial in lakes, reservoirs, and peatlands. Geology 26:535–538.

DeAngelis, D. L. 1980. Energy flow, nutrient cycling, and ecosystem resilience. Ecology 61: 764–771.

———. 1992. *Dynamics and nutrient cycling and food webs.* Chapman and Hall, New York, USA.

DeAngelis, D. L., M. Loreau, D. Neergaard, P. J. Mulholland, and E. R. Marzolf. 1995. Modelling nutrient-periphyton dynamics in streams: The importance of transient storage zones. Ecological Modelling 80:149–160.

De Bruin, H. A. R., and C. J. Moore. 1985. Zero-plane displacement and roughness length for tall vegetation, derived from a simple mass conservation hypothesis. Boundary-Layer Meteorology 31:39–49.

Deegan, L. A. 1993. Nutrients and energy transport between estuaries and coastal marine ecosystems by fish migrations. Canadian Journal of Fisheries and Aquatic Sciences 50: 74–79.

Deffontaines, J. P., C. Thenail, and J. Baudry. 1995. Agricultural systems and landscape patterns: How can we built a relationship? Landscape and Urban Planning 31:3–10.

de Jong, J., and I. Ahlen. 1991. Factors affecting the distribution pattern of bats in Uppland, central Sweden. Holarctic Ecology 14:92–96.

de Jong, J. J. M., and W. Klaassen. 1997. Simulated dry deposition of nitric acid near forest edges. Atmospheric Environment 31:3681–3691.

Delattre, P., B. De Sousa, E. Fichet-Calvet, J. P. Quéré, and P. Giraudoux. 1999. Vole outbreaks in a landscape context: Evidence from a six year study of *Microtus arvalis*. Landscape Ecology 14:401–412.

Delcourt, H. R., and P. A. Delcourt. 1991. *Quaternary ecology.* Chapman and Hall, London, UK.

Delettre, Y. R., and N. Morvan. 2000. Dispersal of adult aquatic Chironomidae (Diptera) in agricultural landscapes. Freshwater Biology 44:399–411.

del Giorgio, P. A., and J. J. Cole. 1998. Bacterial growth efficiency in natural aquatic systems. Annual Review of Ecology and Systematics 29:503–541.

del Giorgio, P. A., J. J. Cole, N. F. Caraco, and R. H. Peters. 1999. Linking planktonic biomass and metabolism to net gas fluxes in northern temperate lakes. Ecology 80:1422–1431.

del Giorgio, P. A., and R. H. Peters. 1994. Patterns in planktonic P:R ratios in lakes: Influence of lake trophy and dissolved organic carbon. Limnology and Oceanography 39: 772–787.

De Madariaga, I. 1995. Photosynthetic characteristics of phytoplankton during the development of a summer bloom in the Urdaibai Estuary, Bay of Biscay. Estuarine Coastal and Shelf Science 40:559–575.

DeNiro, M. J., and S. Epstein. 1978. Influence of diet on the distribution of carbon isotopes in animals. Geochimica et Cosmochimica Acta 42:495–506.

———. 1981. Influence of diet on the distribution of nitrogen isotopes in animals. Geochimica et Cosmochimica Acta 45:341–351.

Denman, K. L. 1977. Short term variability in vertical chlorophyll structure. Limnology and Oceanography 22:434–441.

Dennis, P., and G. L. Fry. 1992. Field margins: Can they enhance natural enemy population densities and general arthropod diversity on farmland? Agriculture, Ecosystems and Environment 40:95–116.

Denny, M. W. 1993. *Air and water.* Princeton University Press, Princeton, NJ, USA.

Dial, R., and J. Roughgarden. 1995. Experimental removal of insectivores from rain forest canopy: Direct and indirect effects. Ecology 76:1821–1834.

Diamond, J. 1998. *Guns, germs and steel.* Norton, New York, USA.

DiCenzo, V. J., M. J. Maceina, and M. R. Stimpert. 1996. Relations between reservoir trophic state and gizzard shad population characteristics in Alabama reservoirs. North American Journal of Fisheries Management 16:888–895.

Diehl, S., and M. Feissel. 2000. Effects of enrichment on three-level food chains with omnivory. American Naturalist 155:200–218.

Dillon, P. J., and W. B. Kircher. 1975. The effects of geology and land use on the export of phosphorus from watersheds. Water Research 9:135–148.

Dillon, P. J., and L. A. Molot. 1997. Dissolved organic and inorganic carbon mass balances in central Ontario lakes. Biogeochemistry 36:29–42.

Dixon, P. A., M. J. Milicich, and G. Sugihara. 1999. Episodic fluctuations in larval supply. Science 283:1528–1530.

Dmowski, K., and M. Koziakiewicz. 1990. Influence of a shrub corridor on movements of passerine birds to a lake littoral zone. Landscape Ecology 4:98–108.

Dombeck, M., J. Hammill, and W. Bullen. 1984. Fisheries management and fish dependent birds. Fisheries 9:2–4.

Donaldson, J. R. 1967. The phosphorus budget of Iliamna Lake, Alaska as related to the cyclic abundance of sockeye salmon. Ph.D. dissertation, University of Washington, Seattle, USA.

Dortch, Q., and T. E. Whitledge. 1992. Does nitrogen or silicon limit phytoplankton production in the Mississippi River plume and nearby regions? Continental Shelf Research 12:1293–1309.

Draaijers, G. P. J., W. P. M. F. Ivens, and W. Bleuten. 1988. Atmospheric deposition in forest edges measured by monitoring canopy throughfall. Water, Air, and Soil Pollution 42:129–136.

Drake, J. A. 1983. Invasibility in Lotka-Volterra interaction webs. Pp. 83–90 in D. DeAngelis, W. M. Post, and G. Sugihara, eds., *Current trends in food web theory.* Oak Ridge National Laboratory, Oak Ridge, TN, USA.

Drake, V. A. 1985. Solitary wave disturbances of the nocturnal boundary layer revealed by radar observations of migrating insects. Boundary Layer Meteorology 31:269–286.

Drake, V. A., and R. A. Farrow. 1998. The influence of atmospheric structure and motions on insect migration. Annual Review of Entomology 33:183–210.

Drenner, R. W., K. L. Gallo, R. M. Baca, and J. D. Smith. 1998. Synergistic effects of nutrient loading and omnivorous fish on phytoplankton biomass. Canadian Journal of Fisheries and Aquatic Sciences 55:2087–2096.

Drenner, R. W., J. D. Smith, and S. T. Threlkeld. 1996. Lake trophic state and the limnological effects of omnivorous fish. Hydrobiologia 319:213–223.

Drever, M. C., and A. S. Harestand. 1998. Diets of Norway rats, *Rattus norvegicus*, on Langara Island, Queen Charlotte Islands, British Columbia: Implications for conservation of breeding seabirds. Canadian Field Naturalist 112:676–683.

Dudgeon, D. 1995. River regulation in southern China: Ecological implications, conservation and environmental management. Regulated Rivers Research and Management 11:35–54.

Due, A. D., and G. A. Polis. 1985. The biology of *Vaejovis littoralis*, an intertidal scorpion from Baja California, Mexico. Journal of Zoology 207:563–580.

Duelli, P., M. Studer, I. Marchland, and S. Jakob. 1990. Population movements of arthropods between natural and cultivated areas. Biological Conservation 54:193–207.

Duffy, D. C. 1983. The ecology of tick parasitism on densely nesting Peruvian seabirds. Ecology 64:110–119.

————. 1991. Ants, ticks, and seabirds: Dynamic interactions? Pp. 242–257 in J. E. Loye and M. Zuk, eds., *Bird-parasite interactions: Ecology, evolution and behaviour.* Oxford University Press, Oxford, UK.

Duffy, D. C., and M. J. Campos de Duffy. 1986. Tick parasitism at nesting colonies of blue-footed boobies in Peru and Galapagos. Condor 88:242–244.

Dugdale, R. C., and J. J. Goering. 1967. Uptake of new and regenerated forms of nitrogen in primary productivity. Limnology and Oceanography 12:196–206.

Duggins, D. O., C. A. Simenstad, and J. A. Estes. 1989. Magnification of secondary production by kelp detritus in coastal marine ecosystems. Science 245:170–173.

Dunson, W. A., and J. Travis. 1991. The role of abiotic factors in community organization. American Naturalist 138:1067–1091.

Duran, L. R., and J. C. Castilla. 1989. Variation and persistence of the middle rocky intertidal community of central Chile, with and without human harvesting. Marine Biology 103: 555–562.

Durbin, A. G., S. W. Nixon, and C. A. Oviatt. 1979. Effects of the spawning migration of the alewife, *Alosa pseudoharengus,* on freshwater ecosystems. Ecology 60:8–17.

Eadie, B. J., B. A. McKee, M. B. Lansing, J. A. Robbins, S. Metz, and J. H. Trefry. 1994. Records of nutrient-enhanced coastal ocean productivity in sediments from the Louisiana continental shelf. Estuaries 17:754–765.

East, K. T., M. R. East, and C. H. Daugherty. 1995. Ecological restoration and habitat relationships of reptiles on Stephens Island, New Zealand. New Zealand Journal of Zoology 22:249–261.

Eckman, J. E. 1983. Hydrodynamic processes affecting benthic recruitment. Limnology and Oceanography 28:241–257.

Edwards, E. D., and A. D. Huryn. 1995. Annual contribution of terrestrial invertebrates to a New Zealand trout stream. New Zealand Journal of Marine and Freshwater Research 29:467–477.

————. 1996. Effect of riparian land use on contributions of terrestrial invertebrates to streams. Hydrobiologia 337:151–159.

Eeva, T., and E. Lehikoinen. 1995. Egg shell quality, clutch size and hatching success of the great tit (*Parus major*) and the pied flycatcher (*Ficedula hypoleuca*) in an air pollution gradient. Oecologia 102:312–323.

Efford, M. G., B. J. Karl, and H. Moller. 1988. Population ecology of *Mus musculus* on Mana Island, New Zealand. Journal of Zoology 216:539–563.

Egbert, A. L., and A. W. Stokes. 1976. The social behaviour of brown bears on an Alaskan salmon stream. Pp. 41–57 in M. R. Pelton, J. W. Lentfer, and G. E. Folk Jr., eds., *Bears: Their biology and management.* IUCN Publications, New Series 40. International Union for Conservation of Nature and Natural Resources, Morges, Switzerland.

Egger, M. 1979. Varied thrushes feeding on talitrid amphipods. Auk 96:805–806.

Egglishaw, H. J., and P. E. Shackley. 1985. Factors governing the production of juvenile Atlantic salmon in Scottish streams. Journal of Fish Biology 27 (Suppl. A):27–33.

Eggold, B. T., J. F. Amrheim, and M. A. Coshun. 1996. PCB accumulation by salmonine smolts and adults in Lake Michigan and its tributaries and its effect on stocking policies. Journal of Great Lakes Research 22:403–413.

Eickwort, K. R. 1977. Population dynamics of a relatively rare species of milkweed beetle (*Labidomera*). Ecology 58:527–538.

Eisenreich, S. J., B. B. Looney, and J. D. Thornton. 1981. Airborne organic contaminants in the Great Lakes ecosystem. Environmental Science and Technology 15:30–36.

Ekbom, B., M. E. Irwin, and Y. Robert, eds. 2000. *Interchanges of insects between agricultural and surrounding landscapes.* Kluwer Academic Publishers, Dordrecht, The Netherlands.

Ekerholm, P., L. Oksanen, and T. Oksanen. 2001. Long-term dynamics of voles and lemmings at the timberline and above the willow limit as a test of hypotheses on trophic interactions. Ecography 24:555–568.

El-Ashry, M. T., and D. C. Gibbons, eds. 1988. *Water and arid lands of the western United States.* Cambridge University Press, Cambridge, UK.

Elliott, N. C., R. W. Kieckhefer, J.-H. Lee, and B. W. French. 1999. Influence of within-field and landscape factors on aphid predator populations in wheat. Landscape Ecology 14:239–252.

Elser, J. J., and C. R. Goldman. 1991. Zooplankton effects on phytoplankton in lakes of contrasting trophic status. Limnology and Oceanography 36:64–90.

Elser, J. J., E. R. Marzolf, and C. R. Goldman. 1990. Phosphorus and nitrogen limitation of phytoplankton growth in the freshwaters of North America: A review and critique of experimental enrichments. Canadian Journal of Fisheries and Aquatic Sciences 47: 1468–1477.

Elton, C. 1927. *Animal ecology.* Macmillan, New York, USA.

Enfield, D. B. 1989. El-Niño, past and present. Reviews of Geophysics 27:159–187.

Englund, G. 1997. Importance of spatial scale and prey movements in predator caging experiments. Ecology 78:2316–2325.

Enserink, M. 1999. Biological invaders sweep in. Science 285:1834–1836.

Eppley, R. W., E. H. Renger, and P. R. Betzer. 1983. The residence time of particle organic carbon in the surface layer of the ocean. Deep Sea Research 30:311–323.

Erelli, M. C., M. P. Ayres, and G. K. Eaton. 1998. Altitudinal patterns in host suitability for forest insects. Oecologia 117:133–142.

Erhardt, A., and J. A. Thomas. 1991. Lepidoptera as indicators of change in the semi-natural grasslands of lowland and upland Europe. Pp. 213–236 in N. M. Collins and J. A. Thomas, eds., *The conservation of insects and their habitats.* Academic Press, London, UK.

Erisman, J. W., and G. P. J. Draaijers. 1995. *Atmospheric deposition in relation to acidification and eutrophication.* Elsevier, Amsterdam, The Netherlands.

Esser, G., and G. H. Kohlmaier. 1991. Modelling terrestrial sources of nitrogen, phosphorus, sulphur and organic carbon to rivers. Pp. 297–322 in E. T. Degens, S. Kempe, and J. E. Richey, eds., *Biogeochemistry of major world rivers.* John Wiley and Sons, New York, USA.

Estes, J. A., and D. O. Duggins. 1995. Sea otters and kelp forests in Alaska: Generality and variation in a community ecological paradigm. Ecological Monographs 65:75–100.

Estes, J. A., N. S. Smith, and J. F. Palmisano. 1978. Sea otter predation and community organization in the western Aleutian Islands, Alaska. Ecology 59:822–823.

Estes, J. A., M. T. Tinker, T. M. Williams, and D. F. Doak. 1998. Killer whale predation on sea otters linking oceanic and nearshore ecosystems. Science 282:473–476.

Ettershank, G., J. A. Ettershank, M. Bryant, and W. G. Whitford. 1978. Effects of nitrogen fertilization on primary production in a Chihuahuan desert ecosystem. Journal of Arid Environments 1:135–139.

Evans, D. O., and D. H. Loftus. 1987. Colonization of lakes in the inland region of the Great Lakes by rainbow smelt (*Osmerus mordax*); their freshwater niche and effects on indigenous fishes. Canadian Journal of Fisheries and Aquatic Sciences 44 (Suppl. 2): 249–266.

Evarts, J. L. 1997. Spatial and temporal variation in nitrogen and phosphorus release from sediments in a eutrophic reservoir. M.S. thesis, Miami University, Oxford, OH, USA.

Everest, F. H., D. N. Swanston, C. G. Shaw III, W. P. Smith, K. R. Julin, and S. D. Allen. 1997. Evaluation of the use of scientific information in developing the 1997 forest plan for the Tongass National Forest. General Technical Report PNW-GTR-415. USDA Forest Service, Portland, OR, USA.

Eviner, V. T., and F. S. Chapin. 1997. Plant-microbial interactions. Nature 385:26–27.

Ewert, D. N., and M. J. Hamas. 1995. Ecology of migratory landbirds during migration in the Midwest. Pp. 200–208 in F. R. Thompson III, ed., *Management of Midwestern landscapes for the conservation of Neotropical migratory birds.* General Technical Report NC-187. USDA Forest Service, St. Paul, MN, USA.

Fagan, W. F., R. S. Cantrell, and C. Cosner. 1999. How habitat edges change species interactions. American Naturalist 153:165–182.

Fanning, K. A. 1989. Influence of atmospheric pollution on nutrient limitation in the ocean. Science 239:460–463.

Fariña, J. M. 2000. Structure and organization of intertidal communities affected by copper mine tailings: Changes on the relative importance of bottom-up and top-down process. Ph.D. dissertation, Pontificia Universidad Catolica de Chile, Chile.

Fariña, J. M., S. Salazar, P. Wallem, J. D. Witman, and J. C. Ellis. In press. Nutrient exchanges between marine and terrestrial ecosystems: The case of the Galapagos Sea Lion (*Zalophus californianus wollebaecki*). Journal of Animal Ecology.

Farmer, D. M., and H. J. Freeland. 1983. The physical oceanography of fjords. Progress in Oceanography 12:147–219.

Farrell, T. M., D. Bracher, and J. Roughgarden. 1991. Cross-shelf transport causes recruitment to intertidal populations in central California. Limnology and Oceanography 36: 279–288.

Fausch, K. D. 2000. Shigeru Nakano—an uncommon Japanese fish ecologist. Environmental Biology of Fishes 59:359–364.

Fausch, K. D., M. E. Power, and M. Murakami. 2002. Linkages between stream and forest food webs: Shigeru Nakano's legacy for ecology in Japan. Trends in Ecology and Evolution 17:429–434.

Fenn, M. E., M. A. Poth, J. D. Aber, J. S. Baron, B. T. Bormann, D. W. Johnson, A. D. Lemly, S. G. McNulty, D. F. Ryan, and R. Stottlemeyer. 1998. Nitrogen excess in North American ecosystems: Predisposing factors, ecosystems responses, and management strategies. Ecological Applications 8:706–733.

Fenton, M. B., and G. K. Morris. 1976. Opportunistic feeding by desert bats (*Myotis* spp.). Canadian Journal of Zoology 54:526–530.

Fernandez, J. M. 1993. Fontes autotroficas de energia em juvenis de Jaraqui, *Semaprochilodus insignis* (Schomburgk, 1841) e curimata, *Prochilodus nigricans* Agassiz, 1829 (Pisces: Prochilodontidae) de Amazonia central. M.S. thesis, INPA/Fed. Univ. de Amazonia, Manaus, Brazil.

Findlay, D. L., R. E. Hecky, S. E. M. Kasian, M. P. Stainton, L. L. Hendzel, and E. U. Schindler. 1999. Effects on phytoplankton of nutrients added in conjunction with acidification. Freshwater Biology 41:131–145.

Findlay, S., M. L. Pace, D. Lints, J. J. Cole, and N. F. Caraco. 1991. Weak coupling of bacterial and algal production in a heterotrophic ecosystem: The Hudson River estuary. Limnology and Oceanography 36:268–278.

Findlay, S., R. L. Sinsabaugh, D. T. Fischer, and P. Franchini. 1998. Sources of dissolved organic carbon supporting planktonic bacterial production in the tidal freshwater Hudson River. Ecosystems 1:227–239.

Finlay, J. C. 2000. Stable isotope analysis of river food webs and carbon cycling. Ph.D. dissertation, University of California, Berkeley, USA.

———. 2001. Stable carbon isotope ratios of river biota: Implications for energy flow in lotic food webs. Ecology 82:1052–1064.

Finlay, J. C., S. Khandwala, and M. E. Power. 2002. Spatial scales of carbon flow in a river food web. Ecology 83:1845–1859.

Finlay, J. C., M. E. Power, and G. Cabana. 1999. Effects of water velocity on algal carbon isotope ratios: Implications for river food web studies. Limnology and Oceanography 44:1198–1203.

Finney, B. P., I. Gregory-Eaves, J. Sweetman, M. S. V. Douglas, and J. P. Smol. 2000. Impacts of climatic change and fishing on Pacific salmon abundance over the past 300 years. Science 290:795–799.

Fisher, S. G., and G. E. Likens. 1973. Energy flow in Bear Brook, New Hampshire: An integrative approach to stream ecosystem metabolism. Ecological Monographs 43: 421–439.

Fittkau, E. J. 1973. Crocodiles and the nutrient metabolism of Amazonian waters. Amazoniana 4:103–133.

Flaherty, D. L. 1969. Ecosystem trophic complexity with Willamette mite, *Eotetranychus willamettei* Ewing (Acarina: Tetranychidae), densities. Ecology 50:911–915.

Flanagan, P. W., and K. Van Cleve. 1983. Nutrient cycling in relation to decomposition and organic matter quality in Taiga ecosystems. Canadian Journal of Forest Research 13: 795–817.

Flecker, A. S. 1996. Ecosystem engineering by a dominant detritivore in a diverse tropical stream. Ecology 77:1845–1854.

Flecker, A. S., J. D. Allan, and N. L. McClintock. 1988. Swarming and sexual selection in a Rocky Mountain mayfly. Holarctic Ecology 11:280–285.

Fogel, M. L., E. K. Sprague, A. P. Gize, and R. W. Frey. 1989. Diagenesis of organic matter in the Georgia salt marshes. Estuarine and Coastal Shelf Science 28:211–230.

Folkard, N. F. G., and J. N. M. Smith. 1995. Evidence for bottom-up effects in the boreal forest: Do passerine birds respond to large-scale experimental fertilization? Canadian Journal of Zoology 73:2231–2237.

Food and Agriculture Organization (FAO). 1986. *1985 Fertilizer Handbook.* Food and Agriculture Organization of the United Nations, Rome, Italy.

———. 1996. *Fertilizer Yearbook 1995.* FAO statistics series, no. 45. Food and Agriculture Organization of the United Nations, Rome, Italy.

———. 1998. *FAO Yearbook: Trade.* FAO statistics series, no. 138. Food and Agriculture Organization of the United Nations, Rome, Italy.

Forchgott, J. 1949. Wave currents on the leeward side of mountain crests. Meterologicke Zpravy 3:49–51.

———. 1950. Transport of small particles or insects across Krusne Hory. Meterologicke Zpravy 1:14–16.

Forman, R. T. T. 1995. *Land mosaics: The ecology of landscapes and regions.* Cambridge University Press, New York, USA.

Forman, R. T. T., and L. E. Alexander. 1998. Roads and their major ecological effects. Annual Review of Ecology and Systematics 29:207–231.

Forman, R. T. T., and J. Baudry. 1984. Hedgerows and hedgerow networks in landscape ecology. Environmental Management 8:499–510.

Forman, R. T. T., and M. Godron. 1986. *Landscape ecology.* John Wiley and Sons, New York, USA.

Forman, R. T. T., and P. N. Moore. 1992. Theoretical foundations for understanding boundaries in landscape mosaics. Pp. 236–258 in A. J. Hansen and F. di Castri, eds., *Landscape boundaries: Consequences for biotic diversity and ecological flows.* Ecological studies, vol. 92. Springer-Verlag, New York, USA.

Forsberg, B. R., C. A. R. M. Araujo-Lima, L. A. Martinelli, R. L. Victoria, and J. A. Bonassi. 1993. Autotrophic carbon sources for fish of the central Amazon. Ecology 74: 643–652.

Forsberg, B. R., A. H. Devol, J. E. Richey, L. A. Martinelli, and H. dos Santos. 1988. Factors controlling nutrient concentrations in Amazon floodplain lakes. Limnology and Oceanography 33:41–56.

Foster, I. D. L., J. A. Dearing, R. Grew, and K. Orend. 1990. The sedimentary data base: An appraisal of lake and reservoir based studies of sediment yield. Pp. 119–143 in D. E. Walling, A. Yair, and S. Berkowicz, eds., *Erosion transport and deposition processes: Proceedings of a workshop held at Jerusalem, March–April 1987.* IAHS Publication 189. International Association of Hydrological Sciences, Wallingford, UK.

Fox, B. J., J. E. Taylor, M. D. Fox, and C. Williams. 1997. Vegetation changes across edges of rainforest remnants. Biological Conservation 82:1–13.

Frame, G. W. 1974. Black bear predation on salmon at Olsen Creek, Alaska. Zeitschrift für Tierpsychologie 35:23–38.

Francis, C. M., M. H. Richards, F. Cooke, and R. F. Rockwell. 1992. Long term changes in survival rates of lesser snow geese. Ecology 73:1346–1362.

Fraser, J. M. 1980. Survival, growth, and food habits of brook trout and F1 Splake planted in Precambrian lakes. Transactions of the American Fisheries Society 109:491–501.

Frechette, M., C. A. Butman, and W. R. Geyer. 1989. The importance of boundary layer flows in supplying phytoplankton to the benthic suspension feeder *Mytilus edulis* (L). Limnology and Oceanography 34:19–36.

Fredrickson, A. G., and G. Stephanopolous. 1981. Microbial competition. Science 109: 972–979.

Fresco, L. O., L. Stroosnijder, J. Bouma, and H. van Keulen, eds. 1994. *The future of the land: Mobilising and integrating knowledge for land use option.* John Wiley and Sons, New York, USA.

Fretwell, S. D. 1977. The regulation of plant communities by the food chains exploiting them. Perspectives in Biology and Medicine 20:169–185.

———. 1987. Food chain dynamics: The central theory of ecology? Oikos 50:291–301.

Fretwell, S. D., and H. L. Lucas. 1970. On territorial behavior and other factors influencing habitat distribution in birds. I. Theoretical development. Acta Biotheoretica 19:16–36.

Fry, B. 1988. Food web structure on Georges Bank from stable C, N and S isotopic compositions. Limnology and Oceanography 33:1182–1190.

Fry, B., and C. Arnold. 1982. Rapid C-13/C-12 turnover during growth of brown shrimp (*Penaeus aztecus*). Oecologia 54:200–204.

Fry, B., and E. B. Sherr. 1984. $\delta^{13}C$ measurements as indicators of carbon flow in marine and freshwater ecosystems. Contributions in Marine Science 27:13–47.

Fry, G. L. A., and W. J. Robson. 1994. The effects of field margins on butterfly movement. Pp. 111–116 in N. Boatman, ed., *Field margins: Integrating agriculture and conservation.* British Crop Protection Council, Farnham, UK.

Fujiwara, M., and R. C. Highsmith. 1997. Harpacticoid copepods: Potential link between inbound adult salmon and outbound juvenile salmon. Marine Ecology Progress Series 158:205–216.

Fuller, R. L., and R. J. Mackay. 1981. Effects of food quality on the growth of three hydropsyche species (Trichoptera, Hydropsychidae). Canadian Journal of Zoology 59:1133–1140.

Gaillard, J.-M., M. Festa-Bianchet, and N. G. Yoccoz. 1998. Population dynamics of large herbivores: Variable recruitment with constant adult survival. Trends in Ecology and Evolution 13:58–63.

Gaines, S. D. 1984. Herbivory and between-habitat diversity: The differential effectiveness of defenses in a marine plant. Ecology 66:473–485.

Gaines, S. D., and J. Roughgarden. 1985. Larval settlement rate: A leading determinant of structure in an ecological community of the marine intertidal zone. Proceedings of the National Academy of Sciences, USA 82:3707–3711.

Galat, D. L. 1986. Organic carbon flux to a large salt lake: Pyramid Lake, Nevada, USA. Internationale Revue der gesamten Hydrobiologie 71:621–65.

Galat, D. L., L. H. Fredrickson, D. D. Humburg, K. J. Bataille, J. R. Bodie, J. Dohrenwend, G. T. Gelwicks, J. E. Havel, D. L. Helmers, J. B. Hooker, J. R. Jones, M. F. Knowlton, J. Kubisiak, J. Mazourek, A. C. McColpin, R. B. Renken, and R. D. Semlitsch. 1998. Flooding to restore connectivity of regulated, large-river wetlands. Bioscience 48: 721–733.

Galloway, J. N., H. Levy II, and P. S. Kasibhatla. 1994. Year 2020: Consequences of population growth and development on deposition of oxidized nitrogen. Ambio 23:120–123.

Galloway, J. N., W. H. Schlesinger, H. Levy II, A. Michaels, and J. L. Schnoor. 1995. Nitrogen fixation: Anthropogenic enhancement-environmental response. Global Biogeochemical Cycles 9:235–252.

Gannes, L. Z., D. M. O'Brien, and C. Martinez del Rio. 1997. Stable isotopes in animal ecology: Assumptions, caveats, and a call for more laboratory experiments. Ecology 78: 1271–1276.

Gard, R. 1971. Brown bear predation on sockeye salmon at Karluk Lake, Alaska. Journal of Wildlife Management 35:193–204.

Garman, G. C. 1991. Use of terrestrial arthropod prey by a stream-dwelling cyprinid fish. Environmental Biology of Fisheries 30:325–331.

Garrett, L. 1994. *The coming plague: Newly emerging diseases in a world out of balance.* Farrar, Straus, and Giroux, New York, USA.

Gasith, A., and A. D. Hasler. 1976. Airborne litterfall as a source of organic matter in lakes. Limnology and Oceanography 21:253–258.

Gause, G. F. 1934. *The struggle for existence*. Williams & Wilkins, Baltimore, MD, USA.

Gaynor, J. D., and W. I. Findlay. 1995. Soil and phosphorus loss from conservation and conventional tillage in corn production. Journal of Environmental Quality 24:734–741.

Geiger, R. 1965. *The climate near the ground*. Harvard University Press, Cambridge, MA, USA.

Gende, S. M. 2002. Foraging behavior of bears at salmon streams: Intake, choice, and role of salmon energetics. Ph.D. dissertation, University of Washington, Seattle, USA.

Gende, S. M., R. T. Edwards, M. F. Willson, and M. S. Wipfli. 2002. Pacific salmon in aquatic and terrestrial ecosystems. BioScience 52:917–928.

Gende, S. M., T. P. Quinn, and M. F. Willson. 2001. Consumption choice by bears feeding on salmon. Oecologia 127:372–382.

Genin, A., P. K. Dayton, P. F. Lonsdale, and F. N. Spiess. 1986. Corals on seamount peaks provide evidence of current acceleration over deep-sea topography. Nature 322:59–61.

Genovese, S. J. 1996. Regional and temporal variation in the ecology of an encrusting bryozoan in the Gulf of Maine. Ph.D. dissertation, Northeastern University, Boston, MA, USA.

Genovese, S. J., and J. D. Witman. 1999. Interactive effects of flow speed and particle concentration on growth rates of an active suspension feeder. Limnology and Oceanography 44:1120–1131.

Ghiselin, M. T. 1974. *The economy of nature and the evolution of sex*. University of California Press, Berkeley, USA.

Giesler, R., M. Högberg, and P. Högberg. 1998. Soil chemistry and plants in Fenno-scandian boreal forest and exemplified by a local gradient. Ecology 79:119–137.

Gilbert, C. H., and W. H. Rich. 1927. Investigations concerning the red salmon runs to the Karluk River, Alaska. Bulletin of the U.S. Bureau of Fisheries 43:1–69.

Gill, J. M., and R. Coma. 1998. Benthic suspension feeders: Their paramount role in littoral marine food webs. Trends in Ecology and Evolution 13:316–321.

Gillette, D. D., and J. D. Kimbrough. 1970. Chiropteran mortality. Pp. 262–281 in B. H. Slaughter and D. W. Walton, eds., *About bats*. Southern Methodist University Press, Dallas, TX, USA.

Gleeson, J. P., and P. J. J. van Rensburg. 1982. Feeding ecology of the house mouse *Mus musculus* on Marion Island. South African Journal of Antarctic Research 12:34–39.

Gleick, P. H. 1998. Water in crisis: Paths to sustainable water use. Ecological Applications 8:571–579.

Glynn, P. W. 1976. Some physical and biological determinants of coral community structure in the eastern Pacific. Ecological Monographs 46:431–456.

———. 1988. El Niño-Southern Oscillation 1982–83; nearshore population, community, and ecosystem responses. Annual Review of Ecology and Systematics 19:309–345.

Goddard, C. I., D. H. Loftus, J. A. MacLean, C. H. Olver, and B. J. Shuter. 1987. Evaluation of the effects of fish community structure on observed yields of lake trout (*Salvelinus namaycush*). Canadian Journal of Fisheries and Aquatic Sciences 44 (Suppl. 2): 239–248.

Goering, J., V. Alexander, and N. Haubenstock. 1990. Seasonal variability of stable carbon and nitrogen isotope ratios of organisms in a North Pacific bay. Estuarine, Coastal, and Shelf Science 30:239–260.

Goldberg, D. E., and T. E. Miller. 1990. Effects of different resource additions on species diversity in an annual plant community. Ecology 71:213–225.

Gonzalez, A., and R. D. Holt. 2002. The inflationary effects of environmental fluctuations in source-sink systems. Proceedings of the National Academy of Sciences, USA 99: 14872–14877.

Gooday, A. J., and C. M. Turley. 1990. Responses by benthic organisms to inputs of organic matter to the ocean floor: A review. Philosophical Transactions of the Royal Society of London A 331:119–138.

Gore, J. A. 1996. Discharge measurements and streamflow analysis. Pp. 53–74 in F. R. Hauer and G. A. Lambert, eds., *Methods in stream ecology*. Academic Press, New York, USA.

Gosler, A. G. 1996. Environmental and social determinants of winter fat storage in the great tit *Parus major*. Journal of Animal Ecology 65:1–17.

Gossard, E. E., and W. H. Hooke. 1975. *Waves in the atmosphere*. Elsevier Press, Amsterdam, The Netherlands.

Gosz, J. R. 1991. Fundamental ecological characteristics of landscape boundaries. Pp. 8–30 in M. M. Holland, P. G. Risser, and R. G. Naiman, eds., *Ecotones: The role of landscape boundaries in the management and restoration of changing environments*. Routledge, Chapman and Hall, New York, USA.

Goudie, A. 1993. *The human impact on the natural environment*. 4th ed. Blackwell Publishers, Oxford.

Goulding, M. 1980. *The fishes and the forest*. University of California Press, Berkeley, USA.

Goulding, M., M. L. Carvalho, and E. G. Ferreira. 1988. *Río Negro: Rich life in poor water*. SPB Academic Publishing, The Hague, The Netherlands.

Graetz, R. D. 1991. Desertification: A tale of two feedbacks. Pp. 59–87 in H. A. Mooney, E. Medina, D. W. Schindler, E.-D. Schulze, and B. H. Walker, eds., *Ecosystem experiments*. SCOPE 45. John Wiley and Sons, New York, USA.

Graf, G. 1992. Benthic-pelagic coupling: A benthic view. Oceanography and Marine Biology Annual Review 30:149–190.

Grassle, J. F., and L. S. Morse-Porteous. 1987. Macrofaunal colonization of disturbed deep sea environments and the structure of deep-sea benthic communities. Deep Sea Research 34:1911–1950.

Gratto-Trevor, C. 1994. Monitoring shorebird populations in the Arctic. Bird Trends 3:10–12.

Graveland, J., and R. H. Drent. 1997. Calcium availability limits breeding success of passerines on poor soils. Journal of Animal Ecology 66:279–288.

Gray, L. J. 1989a. Correlations between insects and birds in tallgrass prairie riparian habitats. Pp. 263–265 in T. Bragg and J. Stubbendieck, eds., *Proceedings of the Eleventh North American Prairie Conference*. University of Nebraska, Lincoln, USA.

———. 1989b. Emergence production and export of aquatic insects from a tallgrass prairie stream. Southwestern Naturalist 34:313–318.

Green, M. O., J. E. Hewitt, and S. Thrush. 1998. Seabed drag coefficient over natural beds of horse mussels (*Atrina zelandica*). Journal of Marine Research 56(3):613–637.

Greenberg, R., and J. S. Ortiz. 1994. Interspecific defense of pasture trees by wintering yellow warblers. Auk 111:672–682.

Greenstone, M. H. 1978. The numerical response to prey availability of *Pardosa ramulosa* (McCook) (Araneae: Lycosidae) and its relationship to the role of spiders in the balance of nature. Symposium of the Zoological Society of London 42:183–193.

Greenwood, D. J. 1990. Production or productivity: The nitrate problem? Annals of Applied Biology 117:209–231.

Greenwood, D. J., A. Gerwitz, D. Stone, and A. Barnes. 1982. Root development of vegetable crops. Plant and Soil 68:75–96.

Gregory, K. J., and D. E. Walling. 1973. *Drainage basin form and process: A geomorphological approach*. John Wiley and Sons, New York, USA.

Gregory, S. V., F. J. Swanson, W. A. McKee, and K. W. Cummins. 1991. An ecosystem perspective of riparian zones. BioScience 41:540–551.

Gresh, T., J. Lichatowich, and P. Schoonmaker. 2000. An estimation of historic and current levels of salmon production in the Northwest Pacific ecosystem: Evidence of a nutrient deficit in the freshwater systems of the Pacific Northwest. Fisheries 25:15–21.

Griffiths, C. L., and R. J. Griffiths. 1983. Biology and distribution of the littoral rove beetle *Psamathobledius punctatissimus* (LeConte) (Coleoptera: Staphylinidae). Hydrobiologia 101:203–214.

Griffiths, C. L., and J. Stenton-Dozey. 1981. The fauna and rate of degradation of stranded kelp. Estuarine and Coastal Shelf Science 12:645–653.

Grimm, N. B. 1987. Nitrogen dynamics during succession in a desert stream. Ecology 68: 1157–1170.

Grimm, N. B., S. G. Fisher, and W. L. Minckley. 1981. Nitrogen and phosphorus dynamics in hot desert streams of southwestern USA. Hydrobiologia 83:303–312.

Grismer, L. L. 1994. Three new species of intertidal side-blotched lizards (genus *Uta*) from the Gulf of California, Mexico. Herpetologica 50:451–474.

Gross, H. P., W. A. Wurtsbaugh, and C. Luecke. 1998. The role of anadromous sockeye salmon in the nutrient loading and productivity of Redfish Lake, Idaho. Transactions of the American Fisheries Society 127:1–18.

Grover, J. P. 1988. Dynamics of competition in a variable environment—experiments with 2 diatom species. Ecology 69:408–417.

———. 1997. *Resource competition*. Chapman and Hall, London, UK.

Gu, B., C. L. Schelske, and M. V. Hoyer. 1997. Intrapopulation feeding diversity in blue tilapia: Evidence from stable-isotope analyses. Ecology 78:2263–2266.

Guichard, F., P. Halpin, G. W. Allison, J. Lubchenco, and B. A. Menge. In press. Mussel disturbance dynamics: Signatures of oceanographic forcing from local interactions. American Naturalist.

Gysel, L. W. 1951. Borders and openings in beech-maple woodlands in southern Michigan. Journal of Forestry 49:13–19.

Habicht, K. S., and D. E. Canfield. 1996. Sulphur isotope fractionation in modern microbial mats and the evolution of the sulphur cycle. Nature 382:342–343.

Hadley, N. F., and S. R. Szarek. 1981. Productivity of deserts. BioScience 31:747–753.

Haggar, R. J., and S. Peel, eds. 1994. Grassland management and nature conservation. BGS Occasional Symposium no. 28. British Grassland Society, Reading, UK.

Haines, E., and C. L. Montague. 1979. Food sources of estuarine invertebrates analyzed using $^{13}C/^{12}C$ ratios. Ecology 60:48–56.

Haines-Young, R., and M. Chopping. 1996. Quantifying landscape structure: A review of landscape indices and their application to forested landscapes. Progress in Physical Geography 20:418–445.

Hairston, N. G. S. 1989. *Ecological experiments*. Cambridge University Press, Cambridge, UK.

Hairston, N. G. S., F. E. Smith, and L. B. Slobodkin. 1960. Community structure, population control, and competition. American Naturalist 94:421–425.

Hall, C. A. S. 1972. Migration and metabolism in a temperate stream ecosystem. Ecology 53:585–604.

Hall, R. O., Jr. 1995. Use of a stable carbon isotope addition to trace bacterial carbon through a stream food web. Journal of the North American Benthological Society 14:269–277.

Hall, S. J., and D. G. Raffaelli. 1991. Food web patterns: Lessons from a species-rich web. Journal of Animal Ecology 60:823–824.

———. 1996. Food-web patterns: What do we really know? Pp. 395–417 in A. C. Gange and V. K. Brown, *Multitrophic interactions in terrestrial systems*. Blackwell, Oxford, UK.

Hall, S. J. G., and G. F. Moore. 1986. Feral cattle of Swona, Orkney Islands. Mammal Review 16:89–96.

Halley, J. M. 1996. Ecology, evolution, and 1/f noise. Trends in Ecology and Evolution 11:33–37.

Halpern, D. 1971. Observations on short period internal waves in Massachusetts Bay. Journal of Marine Research 29:116–132.

Halupka, K. C., M. D. Bryant, M. F. Willson, and F. H. Everest. 2000. Biological characteristics and population status of anadromous salmon in southeast Alaska. Pacific Northwest Research Station General Technical Report PNW-GTR-468:1–255.

Halweil, B. 2000. Where have all the farmers gone? World Watch 13:12–28.

Hamback, P. A. 1998. Seasonality, optimal foraging, and prey coexistence. American Naturalist 152:881–895.

Hamelink, H. L., R. C. Waybrant, and R. C. Ball. 1971. A proposal: Exchange equilibria control the degree chlorinated hydrocarbons are biologically magnified in lentic environments. Transactions of the American Fisheries Society 100:207–214.

Hamilton, S. K., W. M. Lewis, Jr., and S. J. Sippel. 1992. Energy sources for aquatic animals in the Orinoco river floodplain: Evidence from stable isotopes. Oecologia 89:324–330.

Handa, T. 1998. Revegetation trials in degraded coastal marshes of the Hudson Bay lowlands. M.S. thesis, University of Toronto, Toronto, Canada.

Handa, T., and R. L. Jefferies. 2000. Assisted revegetation trials in degraded salt marshes of the Hudson Bay lowlands. Journal of Applied Ecology 37:944–958.

Hansen, S. R., and S. P. Hubbell. 1980. Single nutrient microbial competition: Qualitative agreement between experimental and theoretically-forecast outcomes. Science 207: 1491–1493.

Hanson, B. J., K. W. Cummins, A. S. Cargill, and R. R. Lowry. 1985. Lipid content, fatty acid composition, and the effect of diet on fats of aquatic insects. Comparative Biochemistry and Physiology 80B:257–276.

Hansson, L.-A. 1992. The role of food-chain composition and nutrient availability in shaping algal biomass development. Ecology 73:241–247.

Hansson, S., J. E. Hobbie, R. Elmgren, U. Larsson, B. Fry, and S. Johansson. 1997. The stable nitrogen isotope ratio as a marker of food-web interactions and fish migration. Ecology 78:2249–2257.

Hargrave, B. T. 1973. Coupling carbon flow through some pelagic and benthic communities. Journal of the Fisheries Research Board of Canada 30:1317–1326.

———. 1978. Seasonal changes in oxygen uptake by settled particulate matter and sediments in a marine bay. Journal of the Fisheries Research Board of Canada 35:1621–1628.

Harper, J. L. 1977. *The population biology of plants.* Academic Press, London, UK.

Harrigan, P., J. C. Zieman, and S. A. Macko. 1989. The base of nutritional support for the gray snapper (*Lutjanus griseus*): An evaluation based on a combined stomach content and stable isotope approach. Bulletin of Marine Science 44:65–77.

Harris, L. D. 1984. *The fragmented forest: Island biogeography theory and the preservation of biotic diversity.* University of Chicago Press, Chicago, IL, USA.

Harris, L. G. 1976. Comparative ecological studies of the nudibranch *Aeolidia papillosa* and its anemone prey *Metridium senile* along the Atlantic and Pacific coasts of the United States. Journal of Molluscan Studies 42:301.

Harris, L. G., and N. R. Howe. 1979. An analysis of the defensive mechanisms observed in the anemone *Anthopleura elegantissima* in response to its nudibranch predator *Aeolidia papillosa.* Biological Bulletin 157:138–152.

Harrison, S., and A. D. Taylor. 1997. Empirical evidence for metapopulation dynamics. Pp. 27–42 in I. A. Hanski and M. E. Gilpin, eds., *Metapopulation biology.* Academic Press, San Diego, CA, USA.

Harrold, C., K. Light, and S. Lisin. 1998. Organic enrichment of submarine canyon and continental shelf benthic communities by macroalgal drift imported from nearshore kelp forests. Limnology and Oceanography 43:669–678.

Hart, D. D., R. A. Merz, S. J. Genovese, and B. D. Clark. 1991. Feeding postures of suspension-feeding larval black flies: The conflicting demands of drag and food acquisition. Oecologia 85:457–463.

Hart, D. D., and C. T. Robinson. 1990. Resource limitation in a stream community: Phosphorus enrichment effects on periphyton and grazers. Ecology 71:1494–1502.

Hartman, W. L., and R. L. Burgner. 1972. Limnology and fish ecology of sockeye salmon nursery lakes of the world. Journal of the Fisheries Research Board of Canada 29: 699–715.

Hartung, R. 1971. Effects of toxic substances. Pp. 325–335 in J. W. Davis et al. eds., *Infectious and parasitic diseases of wild birds.* Iowa State University Press, Ames, USA.

Harvey, C. A. 2000. Windbreaks enhance seed dispersal into agricultural landscapes in Monteverde, Costa Rica. Ecological Applications 10:155–173.

Hasler, A. D. 1975. *Coupling of land and water systems.* Springer-Verlag, New York, USA.

Hasselrot, B., and P. Grennfelt. 1987. Deposition of air pollutants in a wind-exposed forest edge. Water, Air, and Soil Pollution 34:135–143.

Hastings, A. 1996. What equilibrium behavior of Lotka-Volterra models does not tell us about food webs. Pp. 211–217 in G. A. Polis and K. O. Winemiller, *Food webs: Integration of patterns and dynamics.* Chapman and Hall, New York, USA.

————. 1998. Transients in spatial ecological models. Pp. 189–198 in J. Bascompte and R. V. Solé, eds., *Modeling spatiotemporal dynamics in ecology.* Springer-Verlag, New York, USA.

Hatfield, J. L., D. B. Jaynes, M. R. Burkart, C. A. Cambardella, T. B. Moorman, J. H. Prueger, and M. Λ. Smith. 1999. Water quality in Walnut Creek watershed: Setting and farming practices. Journal of Environmental Quality 28:11–24.

Haury, L. R., M. G. Briscoe, and M. H. Orr. 1979. Tidally-generated internal wave packets in Massachusetts Bay. Nature 278:312–317.

Haury, L. R., J. A. McGowan, and P. H. Wiebe. 1978. Patterns and processes in the time-space scales of plankton distributions. Pp. 277–327 in J. H. Steele, ed., *Spatial patterns in plankton communities.* Plenum Press, New York, USA.

Haycock, N. E., T. P. Burt, K. W. T. Goulding, and G. Pinay, eds. 1996. *Buffer zones: Their processes and potential in water protection.* Quest Environmental, Harpenden, U.K.

Hayes, W. B. 1974. Sand-beach energetics: Importance of the isopod *Tylos punctatus.* Ecology 55:838–847.

Haywood-Farmer, L. 1996. Nutrient content of salmonids: An annotated bibliography. Unpublished technical report, Ministry of Environment, Fisheries Research Section, University of British Columbia, Vancouver, Canada.

Heatwole, H. 1971. Marine-dependent terrestrial biotic communities on some cays in the Coral Sea. Ecology 52:363–366.

Heatwole, H., T. Done, and E. Cameron. 1981. *Community ecology of a coral cay.* Junk, The Hague, The Netherlands.

Hecky, R. E., P. Campbell, and L. L. Hendzel. 1993. The stoichiometry of carbon, nitrogen and phosphorus in particulate matter of lakes and oceans. Limnology and Oceanography 38:709–724.

Hecky, R. E., and R. H. Hesslein. 1995. Contributions of benthic algae to lake food webs as revealed by stable isotope analysis. Journal of the North American Benthological Society 14:631–653.

Hedges, J. I., R. Keil, R. Benner, K. Kvenvolden, and J. Curiale. 1997. What happens to terrestrial organic matter in the ocean. Organic Geochemistry 27:195–212.

Heggenes, J., and R. Borgstrøm. 1988. Effect of mink, *Mustela vison* Schreber, predation on cohorts of juvenile Atlantic salmon, *Salmo salar* L., and brown trout, *S. trutta* L., in three small streams. Journal of Fish Biology 33:885–894.

Heimann, M. 1997. A review of the contemporary global carbon cycle and as seen a century ago by Arrhenius and Hogbom. Ambio 26:17–23.

Helfield, J. M., and R. J. Naiman. 2001. Effects of salmon-derived nitrogen on riparian forest growth and implications for stream productivity. Ecology 82:2403–2409.

————. 2002. Salmon and alder as nitrogen sources to riparian forests in a boreal Alaskan watershed. Oecologia 133:573–582.

Heneghan, L., and T. Bolger. 1996. Effect of components of "acid rain" on the contribution of soil microarthropods to ecosystem function. Journal of Applied Ecology 33:1329–1344.

Henschel, J. R. 1986. The socio-ecology of a spotted hyaena *Crocuta crocuta* clan in the Kruger National Park. Ph.D. dissertation, University of Pretoria, Pretoria, South Africa.

————. 1995. Ein handliches Vakuumsammelgerät für die Erfassung von Spinnen und Insekten (A handy vacuum collector for catching spiders and insects). Arachnologische Mitteilungen 9:67–70.

Henschel, J. R., D. Mahsberg, and H. Stumpf. 1996. Mass-length relationships of spiders and harvestmen (Araneae and Opiliones). Revue Suisse de Zoologie, vol. hors série: 265–268.

————. 2001. Allochthonous aquatic insects increase predation and decrease herbivory in river shore food webs. Oikos 93:429–238.

Henschel, J. R., H. Stumpf, and D. Mahsberg. 1996. Increase of arachnid abundance and biomass at water shores. Revue Suisse de Zoologie, vol. hors série:269–278.

Hentschel, B. T. 1998. Intraspecific variations in $\delta^{13}C$ indicate ontogenetic diet changes in deposit-feeding polychaetes. Ecology 79:1357–1370.

Herd, R. M., and M. B. Fenton. 1983. An electrophoretic, morphological, and ecological investigation of a putative hybrid zone between *Myotis lucifugus* and *Myotis yumanensis* (Chiroptera: Vespertilionidae). Canadian Journal of Zoology 61:2029–2050.

Hering, D., and H. Plachter. 1997. Riparian ground beetles (Coleoptera, Carabidae) preying on aquatic invertebrates: A feeding strategy in alpine floodplains. Oecologia 111: 261–270.

Herman, R. L. 1992. Solitary waves. American Scientist 80:350–361.

Hermann, J., A. G. Degerman, C. Johansson, P. E. Lingdell, and I. P. Muniz. 1993. Acid-stress effects on stream biology. Ambio 5:298–307.

Herrera, C. M. 1998. Long-term dynamics of Mediterranean frugivorous birds and fleshy fruits: A 12-year study. Ecological Monographs 68:511–538.

Herrero, S. 1978. A comparison of some features of the evolution, ecology and behavior of black and grizzly/brown bears. Carnivore 1:7–17.

Hershey, A. E. 1992. Effects of experimental fertilization on the benthic macroinvertebrate community of an arctic lake. Journal of the North American Benthological Society 11: 204–217.

Hershey, A. E., G. M. Gettel, M. E. McDonald, M. C. Miller, H. Mooers, W. J. O'Brien, J. Pastor, C. Richards, and J. A. Schuldt. 1999. A geomorphic-trophic model for landscape control of Arctic lake food webs. BioScience 49:887–897.

Hershey, A. E., J. Pastor, B. J. Peterson, and G. W. Kling. 1993. Stable isotopes resolve the drift paradox for *Baetis* mayflies in an arctic river. Ecology 74:2315–2325.

Hertling, U. M., and R. A. Lubke. 2000. Assessing the potential for biological invasion—the case of *Ammophila arenaria* in South Africa. South African Journal of Science 96: 520–528.

Hesslein, R. H., M. J. Capel, D. E. Fox, and K. A. Hallard. 1991. Stable isotopes of sulfur, carbon, and nitrogen as indicators of trophic level and fish migration in the lower Mackenzie River basin, Canada. Canadian Journal of Fisheries and Aquatic Sciences 48:2258–2265.

Heywood, V. H. 1989. Patterns, extents, and nodes of invasions by terrestrial plants. Pp. 31–51 in J. A. Drake, H. A. Mooney, F. di Castri, R. H. Groves, F. J. Kruger, M. Rejmánek, and M. Williamson, eds., *Biological invasions, a global perspective*. John Wiley and Sons, New York, USA.

Hilborn, R., and M. Mangel. 1997. *The ecological detective*. Princeton University Press, Princeton, NJ, USA.

Hilderbrand, G. V., T. A. Hanley, C. T. Robbins, and C. C. Schwartz. 1999. Role of brown bears (*Ursus arctos*) in the flow of marine nitrogen into a terrestrial ecosystem. Oecologia 121:546–550.

Hilderbrand, G. V., C. C. Schwartz, C. T. Robbins, M E. Jacoby, T. A. Hanley, S. M. Arthur, and C. Servheen. 1999. Importance of meat, especially salmon, to body size, population productivity, and conservation of North American brown bears. Canadian Journal of Zoology 77:132–138.

Hilderbrand, G. V., S. D. Varley, C. T. Robbins, T. A. Hanley, K. Titus, and C. Servheen. 1996. Use of stable isotopes to determine diets of living and extinct bears. Canadian Journal of Zoology 74:2080–2088.

Hillier, F. S., and G. J. Lieberman. 1990. *Introduction to operations research*. McGraw-Hill, New York, USA.

Hinrichson, D. 1997. Humanity and the world's coasts: A status report. Amicus Journal 18:16–20.

Hobaugh, W. C., C. D. Stutzenbaker, and E. L. Flickinger. 1989. The rice prairies. Pp. 367–383 in L. M. Smith, R. L. Pederson, and R. M. Kaminski, eds., *Habitat management of migrating and wintering waterfowl in North America*. Texas Tech University Press, Lubbock, USA.

Hobson, K. A. 1993. Trophic relationships among High Arctic seabirds: Insights from tissue-dependent stable-isotope models. Marine Ecology Progress Series 95:7–18.

Hobson, K. A., M. C. Drever, and G. W. Kaiser. 1999. Norway rats as predators of burrow-nesting seabirds: Insights from stable isotope analyses. Journal of Wildlife Management 63:14–25.

Hobson, K. A., J. F. Piatt, and J. Pitocchelli. 1994. Using stable isotopes to determine seabird trophic relationships. Journal of Animal Ecology 63:786–798.

Hobson, K. A., and S. G. Sealy. 1974. Marine protein contributions to the diet of Northern Saw-Whet Owls on the Queen Charlotte Islands: A stable isotope approach. Auk 108: 437–440.

Hobson, K. A., and H. E. Welch. 1992. Determination of trophic relationships within a High Arctic marine food web using $\delta^{13}C$ and $\delta^{15}N$ analysis. Marine Ecology Progress Series 84:9–18.

Hocking, M. D., and T. E. Reimchen. 2002. Salmon-derived nitrogen in terrestrial invertebrates from coniferous forests of the Pacific Northwest. BMC Ecology 2:4. Available at http://www.biomedcentral.com/1472-6785/2/4.

Hogstad, O. 1995. Do avian and mammalian nest predators select for different nest dispersion patterns of fieldfares *Turdus pilaris*? A 15-year study. Ibis 137:484–489.

Holdsworth, D. K., and A. F. Mark. 1990. Water and nutrient input output budgets: Effects of plant cover at 7 sites in upland snow tussock grasslands of eastern and central Otago, New Zealand. Journal of the Royal Society of New Zealand 20:1–24.

Holland, J. M., and S. R. Thomas. 1997. Quantifying the impact of polyphagous invertebrate predators in controlling cereal aphids and in preventing wheat yield and quality reductions. Annals of Applied Biology 131:375–397.

Holligan, P. M., R. D. Pingree, and G. T. Mardell. 1985. Oceanic solitons, nutrient pulses and phytoplankton growth. Nature 314:348–350.

Holloway, P. E. 1987. Internal hydraulic jumps and solitons at a shelf break region on the Australian north west shelf. Journal of Geophysical Research 92:5405–5416.

Holm, E. 1996. Radioactivity in the Baltic Sea. Chemistry and Ecology 12:265–277.

Holt, R. D. 1977. Predation, apparent competition, and the structure of prey communities. Theoretical Population Biology 12:197–229.

———. 1984. Spatial heterogeneity, indirect interactions, and the coexistence of prey species. American Naturalist 124:377–406.

———. 1985. Population-dynamics in 2-patch environments—some anomalous consequences of an optimal habitat distribution. Theoretical Population Biology 28:181–208.

———. 1992. A neglected facet of island biogeography: The role of internal spatial dynamics in area effects. Theoretical Population Biology 41:354–371.

———. 1993. Ecology at the mesoscale: The influence of regional processes on local communities. Pp. 77–88 in R. Ricklefs and D. Schluter, eds., *Species diversity in ecological communities*. University of Chicago Press, Chicago, IL, USA.

———. 1996a. Adaptive evolution in source-sink environments: Direct and indirect effects of density-dependence on niche evolution. Oikos 75:182–192.

———. 1996b. Food webs in space: An island biogeographic perspective. Pp. 313–323 in G. A. Polis and K. O. Winemiller, eds., *Food webs: Integration of patterns and dynamics*. Chapman and Hall, London, UK.

———. 1997a. Community modules. Pp. 333–349 in A. C. Gange and V. K. Brown, eds., *Multitrophic interactions in terrestrial ecosystems: 36th Symposium of the British Ecological Society*. Blackwell, Oxford, UK.

———. 1997b. On the evolutionary stability of sink populations. Evolutionary Ecology 11: 723–731.

———. 2002. Food webs in space: On the interplay of dynamic instability and spatial processes. Ecological Research 17:261–273.

Holt, R. D., and M. Barfield. 2003. Impacts of temporal variation on apparent competition and coexistence in open ecosystems. Oikos 101:49–58.

Holt, R. D., M. Barfield, and A. Gonzalez. In press. Impacts of environmental variability in open populations and communities: Inflation in sink environments. Theoretical Population Biology.

Holt, R. D., J. P. Grover, and D. Tilman. 1994. Simple rules for interspecific dominance in systems with exploitative and apparent competition. American Naturalist 144:741–771.

Holt, R. D., and M. E. Hochberg. 2001. Indirect interactions, community modules, and biological control: A theoretical perspective. Pp. 13–37 in E. Waijnberg, J. K. Scott, and P. C. Quimby, eds., *Evaluation of indirect ecological effects of biological control.* CAB International, Wallingford, UK.

Holt, R. D., and J. H. Lawton. 1993. Apparent competition and enemy-free space in insect host-parasitoid communities. American Naturalist 142:623–645.

———. 1994. The ecological consequences of shared natural enemies. Annual Review of Ecology and Systematics 25:495–520.

Holt, R. D., J. Lawton, G. A. Polis, and N. Martinez. 1999. Trophic rank and the species-area relation. Ecology 80:1495–1504.

Holt, R. D., and M. Loreau. 2002. Biodiversity and ecosystem functioning: The role of trophic interactions and the importance of system openness. Pp. 246–262 in A. Kinzig and S. Pacala, eds., *Biodiversity and ecosystem function.* Princeton University Press, Princeton, NJ, USA.

Howard-Williams, C., and W. J. Junk. 1977. The chemical composition of Central Amazonian aquatic macrophytes with special reference to their role in the ecosystem. Archiv für Hydrobiologie 79:446–464.

Howarth, R. W. 1991. Comparative responses of aquatic ecosystems to toxic chemical stress. Pp. 169–195 in J. Cole, G. Lovett, and S. Findlay, eds., *Comparative analyses of ecosystems.* Springer-Verlag, New York, USA.

Howarth, R. W., G. Billen, D. Swaney, A. Townsend, N. Jarworski, K. Lajtha, J. A. Downing, R. Elmgren, N. F. Caraco, T. Jordan, F. Berendse, J. Freney, V. Kudeyarov, P. Murdoch, and Z. Zhao-liang. 1996. Regional nitrogen budgets and riverine inputs of N and P for the drainages to the North Atlantic Ocean: Natural and human influences. Biogeochemistry 35:75–139.

Howarth, R. W., and S. G. Fisher. 1976. Carbon, nitrogen, and phosphorus dynamics during leaf decay in nutrient-enriched stream microecosystems. Freshwater Biology 6:221–228.

Howarth, R. W., R. Schneider, and D. Swaney. 1996. Metabolism and organic carbon fluxes in the tidal freshwater Hudson River. Estuaries 19:848–865.

Hsu, S.-B., S. Hubbell, and P. Waltman. 1977. A mathematical theory for single-nutrient competition in continuous cultures of microorganisms. SIAM Journal on Applied Mathematics 32:366–383.

Hullar, M. B., B. J. Peterson, and R. T. Wright. 1996. Microbial utilization of estuarine dissolved organic carbon: A stable isotope tracer approach tested by mass balance. Applied and Environmental Microbiology 62:2489–2493.

Hunt, E. 1997. Salmon survival rates triple when carcasses are left. Salmon-Trout-Steelheader, Oct.–Nov., 18–19.

Hunt, G. L., R. W. Russell, K. O. Coyle, and T. Weingartner. 1998. Comparative foraging ecology of planktivorous auklets in relation to ocean physics and prey availability. Marine Ecology Progress Series 167:241–259.

Hunt, R. L. 1975. Food relations and behavior of salmonid fishes. Pp. 137–151 in A. D. Hasler, ed., *Coupling of land and water systems.* Springer-Verlag, Berlin, Germany.

Hunter, M. D., and P. W. Price. 1992. Playing chutes and ladders: Heterogeneity and relative roles of bottom up and top down forces in natural communities. Ecology 73:724–732.

Hureau, J. C. 1985. Interactions between antarctic and sub-antarctic marine, freshwater and terrestrial organisms. Pp. 626–629 in W. R. Siegfried, P. R. Condy, and R. M. Laws, eds., *Antarctic nutrient cycles and food webs.* Springer-Verlag, Berlin, Germany.

Hurlbert, S. H. 1984. Pseudoreplication and the design of ecological field experiments. Ecological Monographs 54:187–211.

Huryn, A. D. 1998. Ecosystem-level evidence for top-down and bottom-up control of production in a grassland stream system. Oecologia 115:173–183.

Huston, M. A. 1979. A general hypothesis of species diversity. American Naturalist 113: 81–101.

———. 1993. Biological diversity, soils and economics. Science 262:1676–1680.

———. 1994. *Biological diversity: The coexistence of species on changing landscapes.* Cambridge University Press, Cambridge, UK.

———. 1999. Local processes and regional patterns: Appropriate scales for understanding variation in the diversity of plants and animals. Oikos 86:3393–3401.

Huston, M. A., and D. L. DeAngelis. 1994. Competition and coexistence: The effects of resource transport and supply rates. American Naturalist 144:954–977.

Hutchinson, G. E. 1950. Survey of existing knowledge of biogeochemistry. Vol. 3. *The biogeochemistry of vertebrate excretion.* Bulletin of the American Museum of Natural History, no. 96. American Museum of Natural History, New York, USA.

———. 1961. The paradox of the plankton. American Naturalist 95:137–145.

———. 1967. *A treatise on limnology.* Volume 2. John Wiley and Sons, New York, USA.

Hutchinson, I., A. C. Prentice, and G. Bradfield. 1989. Aquatic plant resources of the Strait of Georgia. Pp. 50–60 in K. Vermeer and R. W. Butler, eds., *The ecology and status of marine and shoreline birds in the Strait of Georgia, British Columbia.* Special publication. Canadian Wildlife Service, Ottawa, Canada.

Huxel, G. R. 1999. On the influence of food quality in consumer-resource interactions. Ecology Letters 2:256–261.

Huxel, G. R., and K. McCann. 1998. Food web stability: The influence of trophic flows across habitats. American Naturalist 152:460–469.

Huxel, G. R., K. McCann, and G. A. Polis. 2002. Resource partitioning in food webs. Ecological Research 17:419–432.

Huyer, A. E. 1983. Coastal upwelling in the California Current system. Progress in Oceanography 12:259–284.

Huyer, A. E., R. Pillsbury, and R. Smith. 1975. Seasonal variation of the alongshore velocity field over the continental shelf off Oregon. Limnology and Oceanography 20:90–95.

Hwang, A. 2000. Tougher germs, at home and on the farm. World Watch 13:34–35.

Iacobelli, A., and R. L. Jefferies. 1991. Inverse salinity gradients in coastal marshes and the death of stands of *Salix:* The effects of grubbing by geese. Journal of Ecology 79:61–73.

Ingram, R. G., J. C. Osler, and L. Legendre. 1989. Influence of internal wave induced vertical mixing on ice algal production in a highly stratified sound. Estuarine, Coastal and Shelf Science 29:435–446.

Ives, A. R. 1995. Predicting the response of populations to environmental change. Ecology 76:926–941.

Ives, A. R., and W. H. Settle. 1997. Metapopulation dynamics and pest control in agricultural systems. American Naturalist 149:220–246.

Jaarsma, N. G., S. M. de Boer, C. R. Townsend, R. M. Thompson, and E. D. Edwards. 1998. Characterising food webs in two New Zealand streams. New Zealand Journal of Marine and Freshwater Research 32:271–286.

Jackson, J. K., and S. G. Fisher. 1986. Secondary production, emergence, and export of aquatic insects of a Sonoran Desert stream. Ecology 67:629–638.

Jackson, J. K., and V. H. Resh. 1989. Distribution and abundance of adult aquatic insects in the forest adjacent to a northern California stream. Environmental Entomology 18:278–283.

Jackson, M. L. 1971. Geomorphological relationships of tropospherically derived quartz in soils of the Hawaiian Islands. Soil Science Society of American Journal 35:515–525.

Jadranka, M. 1992. Spinnenfauna in Uferbereichen: Artengemeinschaften und ihre räumliche Einnischung. M.S. thesis, Ludwig Maximilians University, Munich, Germany.

Janetos, A. C. 1982. Foraging tactics of two guilds of web-spinning spiders. Behavioural Ecology and Sociobiology 10:19–27.

Jano, A. P., R. L. Jefferies, and R. F. Rockwell. 1998. The detection of vegetation change by multitemporal analysis of LANDSAT data: The effects of goose foraging. Journal of Ecology 86:93–99.

Jansson, B. O., ed. 1988. *Coastal-offshore ecosystem interactions.* Springer-Verlag, Berlin, Germany.

Janzen, D. H. 1976. Why bamboos wait so long to flower. Annual Review of Ecology and Systematics 7:347–391.

Jedrzejewska, B., and W. Jedrzejewska. 1998. *Predation in vertebrate communities: The Białowieża primeval forest as a case study.* Springer-Verlag, New York, USA.

Jefferies, R. L. 1988a. Pattern and process in arctic coastal vegetation in response to foraging by lesser snow geese. Page 281–300 in M. J. A. Wegner, P. J. M. van den Aart, H. J. During, and J. T. A. Verhoeven, eds., *Plant form and vegetation structure.* S. P. B. Academic Publishing, The Hague, The Netherlands.

———. 1988b. Vegetational mosaics, plant animal interactions and resources for plant growth. Pp. 341–369 in L. D. Gottlieb and S. K. Jain, eds., *Plant evolutionary biology.* Chapman and Hall, London, UK.

———. 1999. Herbivores, nutrients and trophic cascades in terrestrial environments. Pp. 301–330 in H. Olff, V. K. Brown, and R. H. Drent, eds., *Herbivores: Between plants and predators.* 38th Symposium of the British Ecological Society. Blackwell Scientific Publications, Oxford, UK.

———. 2000. Allochthonous inputs: Integrating population changes and food-web dynamics. Trends in Ecology and Evolution 15:19–22.

Jefferies, R. L., and R. F. Rockwell. 2002. Foraging geese, vegetation loss and soil degradation in an Arctic salt marsh. Applied Vegetation Science 5:7–16.

Jeffries, M. 1988. Individual vulnerability to predation: The effect of alternative prey types. Freshwater Biology 19:49–56.

Jenkins, R. M. 1982. The morphoedaphic index and reservoir fish production. Transactions of the American Fisheries Society 111:133–140.

Jepsen, D. B., and K. O. Winemiller. 2002. Structure of tropical river food webs revealed by stable isotope ratios. Oikos 96:46–55.

Jepsen, D. B., K. O. Winemiller, and D. C. Taphorn. 1997. Temporal patterns of resource partitioning among *Cichla* species in a Venezuelan blackwater river. Journal of Fish Biology 50:1–24.

Jepson, P. C. 1994. Field margins as habitats, refuges and barriers of variable permeability to Carabidae. Pp. 67–76 in N. Boatman, ed., *Field margins: Integrating agriculture and conservation.* British Crop Protection Council, Farnham, UK.

Jervis, M. A., N. A. C. Kidd, M. G. Fitton, T. Huddleston, and H. A. Dawah. 1993. Flower-visiting by hymenopteran parasitoids. Journal of Natural History 27:67–105.

Jickells, T. D. 1998. Nutrient biogeochemistry of the coastal zone. Science 281:217–222.

Jimenez, J. E., P. A. Marquet, R. G. Medel, and F. M. Jaksic. 1991. Comparative ecology of Darwin's fox (*Pseudalopex fulvipes*) in mainland and island settings of southern Chile. Revista Chilena de Historia Natural 63:177–186.

Jobling, M., E. H. Jorgensen, and S. I. Siikavuopio. 1993. The influence of previous feeding regime on the compensatory growth-response of maturing and immature Arctic char, *Salvelinus alpinus.* Journal of Fish Biology 43:409–419.

Johnson, J. H., and N. H. Ringler. 1979. The occurrence of blow fly larvae (Diptera: Calliphoridae) on salmon carcasses and their utilization as food by juvenile salmon and trout. Great Lakes Entomologist 12:137–140.

Johnston, A. E. 1994. The Rothamsted classical experiments. Pp. 9–37 in R. A. Leigh and A. E. Johnston, eds., *Long term experiments in agricultural and ecological sciences.* CAB International, Wallingford, UK.

Johnston, C. A., and R. J. Naiman. 1987. Boundary dynamics at the aquatic-terrestrial interface: The influence of beaver and geomorphology. Landscape Ecology 1:47–58.

Johnstone, N. T., J. S. MacDonald, K. J. Hall, and P. J. Tschaplinski. 1997. A preliminary study of the role of sockeye salmon (*Oncorhynchus nerka*) carcasses as carbon and nitrogen sources for benthic insects and fishes in the "early Stuart" stock spawning streams, 1050 km from the ocean. Fisheries Project Report, no. RD55. British Columbia Ministry of Environment, Lands and Parks, Victoria, BC, Canada.

Jones, C. G., and J. H. Lawton, eds. 1995. *Linking species and ecosystems*. Chapman and Hall, New York, USA.

Jones, C. G., J. H. Lawton, and M. Shachak. 1994. Organisms as ecosystem engineers. Oikos 69:373–386.

Jones, T. H., H. C. J. Godfray, and M. P. Hassell. 1996. Relative movement patterns of a tephritid fly and its parasitoid wasps. Oecologia 106:317–324.

Jonsson, P., and R. Carmen. 1994. Changes in deposition of organic matter and nutrients in the Baltic Sea during the Twentieth Century. Ambio 28:417–426.

Jordan, M. J., and G. E. Likens. 1975. An organic carbon budget for an oligotrophic lake in New Hampshire, U.S.A. Verhandlungen Internationale Vereinigung für Theoretische und Angewandte Limnologie 19:994–1003.

Jordan, T. E., and D. E. Weller. 1996. Human contributions to terrestrial nitrogen flux. BioScience 46:655–664.

Jorgenson, J. K., H. E. Welch, and M. F. Curtis. 1992. Response of Amphipoda and Trichoptera to lake fertilization in the Canadian Arctic. Canadian Journal of Fisheries and Aquatic Sciences 49:2354–2362.

Josselyn, M. N., G. Cailliet, T. Niesen, R. Cowen, A. Hurley, J. Conner, and S. Hawes. 1983. Composition, export and faunal utilization of drift vegetation in the Salt River Submarine Canyon. Estuarine and Coastal Shelf Science 17:447–465.

Jost, J. L., J. F. Drake, H. M. Tsuchiya, and A. G. Fredrickson. 1973. Interactions of *Tetrahymena pyriformis, Escherichia coli, Azotobacter vinelandii,* and glucose in a minimal medium. Journal of Bacteriology 113:834–840.

Juday, C., W. H. Rich, G. I. Kemmerer, and A. Mann. 1932. Limnological studies of Karluk Lake, Alaska, 1926–1930. Bulletin of the Bureau of Fisheries 47:407–434.

Jumars, P. A. 1993. *Concepts in biological oceanography*. Oxford University Press, Oxford, UK.

Junger, M., and D. Planas. 1994. Quantitative use of stable carbon isotope analysis to determine the trophic base of invertebrate communities in a boreal forest lotic ecosystem. Canadian Journal of Fisheries and Aquatic Sciences 51:52–61.

Junk, W. J. 1985. Temporary fat storage, an adaptation of some fish species to the water level fluctuations and related environmental changes of the Amazon river. Amazoniana 9:315–351.

Junk, W. J., P. B. Bayley, and R. E. Sparks. 1989. The flood pulse concept in river-floodplain ecosystems. Pp. 110–127 in D. P. Dodge, ed., *Proceedings of the International large river symposium*. Canadian Special Publication in Fisheries and Aquaculture Science 106. Department of Fisheries and Oceans, Ottawa, Ontario, Canada.

Justic, D., N. N. Rabalais, and R. E. Turner. 1995. Stoichiometric nutrient balance and origin of coastal eutrophication. Marine Pollution Bulletin 30:41–46.

Kadosaki, M. 1983. Food habits of the brown bear in Hokkaido. Journal of the Mammalogists' Society of Japan 9:116–127.

Kaimowitz, D., G. Thiele, and P. Pacheco. 1999. The effects of structural adjustment on deforestation and forest degradation in lowland Bolivia. World Development 27:505–520.

Kajak, A. 1978. Invertebrate predator subsystem. Pp. 539–589 in A. J. Breymeyer and G. M. van Dyne, eds., *Grasslands, systems analysis and man*. Cambridge University Press, Cambridge, UK.

Kalcounis, M. C., K. A. Hobson, and R. M. Brigham. 1996. Spatial and temporal habitat use by bats along a vertical gradient in temperate forest. Bat Research News 37:137.

Kapos, V. 1989. Effects of isolation on the water status of forest patches in the Brazilian Amazon. Journal of Tropical Ecology 5:173–185.

Kapos, V., E. Wandelli, J. L. Camargo, and G. Ganade. 1997. Edge-related changes in environment and plant responses due to forest fragmentation in central Amazonia. Pp. 33–44 in W. F. Laurance and R. O. Bierregaard, eds., *Tropical forest remnants*. University of Chicago Press, Chicago, IL, USA.

Karban, R., and I. T. Baldwin. 1997. *Induced responses to herbivory*. University of Chicago Press, Chicago, IL, USA.

Karban, R., D. Hougen-Eitzmann, and G. English-Loeb. 1994. Predator-mediated apparent competition between two herbivores that feed on grapevines. Oecologia 97:508–511.

Kareiva, P. M., J. G. Kingsolver, and R. B. Huey, eds. 1993. *Biotic interactions and global change*. Sinauer Associates, Sunderland, MA, USA.

Karr, J. R., and I. J. Schlosser. 1981. Water resources and the land-water interface. Science 201:229–234.

Katti, M., and T. Price. 1999. Annual variation in fat storage by a migrant warbler overwintering in the Indian tropics. Journal of Animal Ecology 68:815–823.

Kawaguchi, Y., and S. Nakano. 2001. Contribution of terrestrial invertebrates to the annual resource budget for salmonids in forest and grassland reaches of a headwater stream. Freshwater Biology 46:303–316.

Kawashima, H., M. J. Bazin, and J. M. Lynch. 1997. A modelling study of world protein supply and nitrogen fertilizer demand in the 21st century. Environmental Conservation 24:50–56.

Kerbes, R. H., P. M. Kotanen, and R. L. Jefferies. 1990. Destruction of wetland habitats by lesser snow geese: A keystone species on the west coast of Hudson Bay. Journal of Applied Ecology 27:242–258.

Kerr, R. A. 1999. Big El Niños ride the back of slower climate change. Science 283:1108–1109.

Kerr, S. R., and N. V. Martin. 1970. Tropho-dynamics of lake trout production systems. Pp. 365–376 in J. H. Steele, ed., *Marine food chains*. Oliver and Boyd, Edinburgh, UK.

Kidd, K. A., D. W. Schindler, R. H. Hesslein, and D. C. G. Muir. 1998. Effects of trophic position and lipid on organochlorine concentrations in fishes from subarctic lakes. Canadian Journal of Fisheries and Aquatic Sciences 55:869–881.

Kidd, K. A., D. W. Schindler, D. C. G. Muir, W. L. Lockhart, and R. H. Hesslein. 1995. High concentrations of toxaphene in fishes from a sub-arctic lake. Science 269:240–242.

Kim, B. K., A. P. Jackman, and F. J. Triska. 1992. Modeling biotic uptake by periphyton and transient hyporrheic storage of nitrate in a natural stream. Water Resources Research 28:2743–2752.

Kindler, J. 1998. Linking ecological and developmental objectives: Trade-offs and imperatives. Ecological Applications 8:591–600.

Kinzig, A., S. Pacala, and D. Tilman, eds. 2002. *The functional consequences of biodiversity: Empirical progress and theoretical extensions*. Princeton University Press, Princeton, NJ, USA.

Kiriluk, R., M. R. Servos, D. M. Whittle, G. Cabana, and J. B. Rasmussen. 1995. Using ratios of stable nitrogen and carbon isotopes to characterize the biomagnification of DDE, Mirex, and PCB in the Lake Ontario pelagic food web. Canadian Journal of Fisheries and Aquatic Sciences 52:2660–2674.

Kitchell, J. F., R. V. O'Neill, D. Webb, G. W. Gallepp, S. M. Bartell, J. F. Koonce, and B. S. Ausmus. 1979. Consumer regulation of nutrient cycling. BioScience 29:28–34.

Kitchell, J. F., D. E. Schindler, B. R. Herwig, D. M. Post, M. H. Olson, and M. Oldham. 1999. Nutrient cycling at the landscape scale: The role of diel foraging migrations by geese at the Bosque del Apache National Wildlife Refuge, New Mexico. Limnology and Oceanography 44:828–836.

Klemmedson, J. O., and J. G. Smith. 1964. Cheatgrass (*Bromus tectorum* L.). Botanical Reviews 30:226–261.

Kline, T. C., Jr., J. J. Goering, O. A. Mathisen, P. H. Poe, and P. L. Parker. 1990. Recycling of elements transported upstream by runs of Pacific salmon: I. $\delta^{15}N$ and $\delta^{13}C$ evidence in Sashin Creek, southeastern Alaska. Canadian Journal of Fisheries and Aquatic Sciences 47:136–144.

Kline, T. C., Jr., J. J. Goering, O. A. Mathisen, P. H. Poe, P. L. Parker, and R. S. Scalan. 1993. Recycling of elements transported upstream by runs of Pacific salmon: II. $\delta^{15}N$ and $\delta^{13}C$ evidence in the Kvichak River watershed, Bristol Bay, southwestern Alaska. Canadian Journal of Fisheries and Aquatic Sciences 50:2350–2365.

Kling, G. W. 1994. Ecosystem-scale experiments: The use of stable isotopes in fresh-waters. Pp. 91–120 in L. A. Baker, ed., *Environmental Chemistry of Lakes and Reservoirs*. Advances in Chemistry Series 237. American Chemical Society, Washington, DC, USA.

Kling, G. W., B. Fry, and W. J. O'Brien. 1992. Stable isotopes and planktonic structure in Arctic lakes. Ecology 73:561–566.

Kling, G. W., G. W. Kipphut, and M. C. Miller. 1991. Arctic lakes and streams as gas conduits to the atmosphere: Implications for tundra carbon budgets. Science 251: 298–301.

Knapp, A. K., J. M. Blair, J. M. Briggs, S. L. Collins, D. C. Hartnett, L. C. Johnson, and E. G. Towne. 1999. The keystone role of bison in North American tallgrass prairie. BioScience 49:39–50.

Knox, G. 1986. *Estuarine ecosystems: A systems approach*. CRC Press, Boca Raton, FL, USA.

Koenings, J. P., and R. D. Burkett. 1987. An aquatic Rubic's cube: Restoration of the Karluk Lake sockeye salmon (*Oncorhynchus nerka*). Pp. 419–434 in H. D. Smith, L. Margolis, and C. C. Wood, eds., *Sockeye salmon* (Oncorhynchus nerka) *population biology and future management*. Canadian Special Publication in Fisheries and Aquatic Sciences 96 Department of Fisheries and Oceans, Ottawa, Ontario, Canada.

Koepcke, H. W., and M. Koepcke. 1952. Sobre el proceso de transformacion de la materia organica en las playas arenosas marinas del Peru. Publicaciones del Museo de Historia Natural "Javier Prado" 8:49–125.

Komar, P. D. 1998. *Beach processes and sedimentation*. Prentice Hall, Saddle River, NJ, USA.

Konkle, B. R., and W. G. Sprules. 1986. Planktivory by stunted lake trout in an Ontario lake. Transactions of the American Fisheries Society 115:515–521.

Koop, K., and J. G. Field. 1980. The influence of food availability on the population dynamics of a supralittoral isopod, *Ligia dilatata* (Brandt). Journal of Experimental Marine Biology and Ecology 48:61–72.

Koop, K., and M. I. Lucas. 1983. Carbon flow and nutrient regeneration from the decomposition of macrophyte debris in a sandy beach microcosm. Pp. 249–262 in A. McLachlan and T. Erasmus, eds., *Sandy beaches as ecosystems*. Developments in hydrobiology 19. Junk, The Hague, The Netherlands.

Kortelainen, P. 1993. Content of total organic carbon in Finnish lakes and its relationship to catchment characteristics. Canadian Journal of Fisheries and Aquatic Sciences 50: 1477–1483.

Koslow, J. A. 1997. Seamounts and the ecology of deep-sea fisheries. American Scientist 85: 168–176.

Kosro, P. M., J. A. Barth, and P. T. Strub. 1997. The coastal jet: Observations of surface currents along the Oregon continental shelf from HF radar. Oceanography 10:53–56.

Kotanen, P. M., and R. L. Jefferies. 1997. Long-term destruction of sub-arctic wetland vegetation by lesser snow geese. Écoscience 4:179–182.

Kotler, B. P., and J. S. Brown. 1988. Environmental heterogeneity and the coexistence of desert rodents. Annual Review of Ecology and Systematics 19:281–308.

Kouwen, N., and R. M. Li. 1980. Biomechanics of vegetative channel linings. Journal of Hydraulics Division 106:1085–1100.

Kraft, C. E. 1992. Estimates of phosphorus cycling by fishes using a bioenergetics model. Canadian Journal of Fisheries and Aquatic Sciences 49:2596–2604.

———. 1993. Phosphorus regeneration by Lake Michigan alewives in the mid-1970s. Transactions of the American Fisheries Society 122:749–755.

Krohkin, E. M. 1967. Influence of the intensity of passage of sockeye salmon *Oncorhynchus nerka* (Wald.) on the phosphate concentration of spawning lakes. Izdanija "Nauka" 15:26–31. (Translated from Russian by Fisheries Research Board of Canada Translation Series 1273, 1968).

———. 1968. Effect of size of escapement of sockeye salmon spawners on the phosphate content of a nursery lake. Fisheries Research Board of Canada, Translation Series 1186: 1–45.

————. 1975. Transport of nutrients by salmon migrating from the sea into lakes. Pp. 153–156 in A. D. Hasler, ed., *Coupling of land and water systems*. Springer-Verlag, Berlin, Germany.

Kromp, B. 1999. Carabid beetles in sustainable agriculture: A review on pest control efficacy, cultivation impacts and enhancement. Agriculture, Ecosystems and Environment 74:187–228.

Kruess, A., and T. Tscharntke. 2000. Effects of habitat fragmentation on plant-insect communities. Pp. 53–70 in B. Ekbom, M. E. Irwin, and Y. Robert, eds., *Interchanges of insects between agricultural and surrounding landscapes*. Kluwer Academic Publishers, Dordrecht, The Netherlands.

Kyle, G. B., J. P. Koenings, and J. A. Edmundson. 1997. An overview of Alaska lake-rearing salmon enhancement strategy: Nutrient enrichment and juvenile stocking. Pp. 205–232 in A. M. Milner and M. W. Oswood, eds., *Freshwaters of Alaska: Ecological syntheses*. Ecological studies, 119. Springer-Verlag, New York, USA.

Lajtha, K., and J. D. Marshall. 1994. Sources of variation in the stable isotopic compositions of plants. Pp. 1–20 in K. Lajtha and R. H. Michener, eds., *Stable isotopes in ecology and environmental science*. Blackwell Scientific Publications, London, UK.

Lajtha, K., and R. H. Michener, eds. 1994. *Stable isotopes in ecology and environmental science*. Blackwell Scientific Publishers, Oxford, UK.

Lal, R. 1987. Managing the soils of sub-Saharan Africa. Science 236:1069–1076.

Lampman, G., N. Caraco, and J. Cole. 1999. Spatial and temporal patterns of nutrient concentration and export in the tidal Hudson River. Estuaries 22:285–296.

Lang, A., J. Filser, and J. R. Henschel. 1999. Predation by ground beetles and wolf spiders on herbivorous insects in a maize crop. Agriculture, Ecosystems and Environment 72:189–199.

Lanyon, L. E. 1995. Does nitrogen cycle? Changes in the spatial dynamics of nitrogen with industrial nitrogen fixation. Journal of Production Agriculture 8:70–78.

Larkin, G. A., and P. A. Slaney. 1997. Implications of trends in marine-derived nutrient influx to south coastal British Columbia salmonid production. Fisheries 22:16–24.

Larsson, U., R. Elmgren, and F. Wulff. 1985. Eutrophication and the Baltic Sea: Causes and consequences. Ambio 1:9–14.

LaViolette, P. E., D. R. Johnson, and D. A. Brooks. 1990. Sun glitter photographs of Georges Bank and the Gulf of Maine from the space shuttle. Oceanography 3:43–49.

Lawler, S. P. 1993. Direct and indirect effects in microcosm communities of protists. Oecologia 93:184–190.

Lawton, J. H. 1988. More time means more variation. Nature 334:563.

Lawton, J. H., and V. K. Brown. 1993. Redundancy in ecosystems. Pp. 255–270 in E. D. Shulze and H. A. Mooney, eds., *Biodiversity and ecosystem function*. Springer-Verlag, Berlin, Germany.

Lawton, J. H., M. P. Hassell, and J. R. Beddington. 1975. Prey death rates and rate of increase of arthropod predator populations. Nature 255:60–62.

Lawton, J. H., and C. G. Jones. 1995. Linking species and ecosystems: Organisms as ecosystem engineers. Pp. 141–150 in C. G. Jones and J. H. Lawton, eds., *Linking species and ecosystems*. Chapman and Hall, London, UK.

LeFevre, J. 1986. Aspects of the biology of frontal systems. Advances in Marine Biology 23:164–299.

Liebold, M. A. 1996. A graphical model of keystone predators in food webs: Trophic regulation of abundance, incidence and diversity patterns in communities. American Naturalist 147:784–812.

————. 1997. Do nutrient-competition models predict nutrient availabilities in limnetic ecosystems? Oecologia 110:132–142.

Liebold, M. A., J. M. Chase, J. B. Shurin, and A. L. Downing. 1997. Species turnover and the regulation of trophic structure. Annual Review of Ecology and Systematics 28:467–494.

Leichter, J. J., G. Shellenbarger, S. J. Genovese, and S. R. Wing. 1998. Breaking internal waves on a Florida (USA) coral reef: A plankton pump at work? Marine Ecology Progress Series 166:83–97.

Leichter, J. J., S. R. Wing, S. L. Miller, and M. W. Denny. 1996. Pulsed delivery of subthermocline water to Conch Reef (Florida Keys) by internal tidal bores. Limnology and Oceanography 41:1490–1501.

Leichter, J. J., and J. D. Witman. 1997. Water flow over subtidal rock walls: Effects on distribution and growth of suspension feeders. Journal of Experimental Marine Biology and Ecology 209:293–307.

Lenihan, H. S. 1999. Physical-biological coupling on oyster reefs: How habitat structure influences individual performance. Ecological Monographs 69:251–275.

Leonard, G. H., J. M. Levine, P. R. Schmidt, and M. D. Bertness. 1998. Flow-driven variation in intertidal community structure in a Maine estuary. Ecology 79:1395–1411.

Lesser, M. P., J. D. Witman, and K. P. Sebens. 1994. Effects of flow and seston availability on scope for growth of benthic suspension feeding invertebrates in the Gulf of Maine. Biology Bulletin 187:319–335.

Lester, D. C., and A. McIntosh. 1994. Accumulation of polychlorinated biphenyl congeners from Lake Champlain sediments by *Mysis relicta*. Environmental Toxicology and Chemistry 13:1825–1841.

Levey, D. J., and F. G. Stiles. 1992. Evolutionary precursors of long-distance migration: Resource availability and movement patterns in Neotropical landbirds. American Naturalist 140:447–476.

Levin, S. A., and R. T. Paine. 1974. Disturbance, patch formation and community structure. Proceedings of the National Academy of Sciences, USA 71:2744–2747.

Levine, J. M. 2000. Complex interactions in a streamside plant community. Ecology 81: 3431–3444.

Levine, S. 1980. Several measures of trophic structure applicable to complex food webs. Journal of Theoretical Biology 83:195–207.

Levine, S. N. 1989. Theoretical and methodological reasons for variability in the responses of aquatic ecosystems to chemical stress. Pp. 145–174 in S. A. Levin, M. A. Harwell, J. R. Kelly, and K. D. Kimball, eds., *Ecotoxicology: Problems and approaches*. Springer-Verlag, New York, USA.

Levins, R. 1969. Some demographic and genetic consequences of environmental heterogeneity for biological control. Bulletin of the Entomological Society of America 15:237–240.

———. 1979. Coexistence in a variable environment. American Naturalist 114:765–783.

Levinton, J. 1995. Bioturbators as ecosystem engineers: Control of the sediment fabric, inter-individual interactions, and material fluxes. Pp. 29–36 in C. G. Jones and J. H. Lawton, eds., *Linking species and ecosystems*. Chapman and Hall, New York, USA.

Levy, S. 1997a. Pacific salmon bring it all back home. BioScience 47:657–660.

———. 1997b. Ultimate sacrifice. New Scientist, 6 Sept. 1977:39–41.

Lewis, A. 1994. *Sockeye: The Adams River run*. Raincoast Books, Vancouver, BC, Canada.

Lewis, J. R. 1964. *The ecology of rocky shores*. The English Universities Press Ltd., London, UK.

Lewis, W. M., Jr. 1988. Primary production in the Orinoco River. Ecology 69:679–692.

Lewis, W. M., Jr., S. K. Hamilton, and J. F. Saunders, Jr. 1995. Rivers of northern South America. Pp. 219–256 in C. E. Cushing, K. W. Cummins, and G. W. Minshall, eds., *River and stream ecosystems*. Elsevier, Amsterdam, The Netherlands.

Lieth, H. 1978. Primary productivity in ecosystems: Comparative analysis of global patterns. Pp. 300–321 in H. Lieth, ed., *Patterns of primary production in the biosphere*. Benchmark papers in ecology 8. Dowden, Hutchinson and Ross, Stroudsburg, PA, USA.

Ligon, F. K., W. E. Dietrich, and W. J. Trush. 1995. Downstream ecological effects of dams: A geomorphic perspective. BioScience 45:183–192.

Likens, G. E. 1985. *An ecosystem approach to aquatic ecology: Mirror Lake and its environment*. Springer-Verlag, New York, USA.

Likens, G. E., and F. H. Bormann. 1974. Linkages between terrestrial and aquatic ecosystems. BioScience 24:447–456.

Likens, G. E., F. H. Bormann, L. O. Hedin, C. T. Driscoll, and J. S. Eaton. 1990. Dry deposition of sulfur: A 23-year record for the Hubbard Brook Forest Ecosystem. Bulletin of the Ecological Society of America 71:231.

Likens, G. E., F. H. Bormann, R. S. Pierce, J. S. Eaton, and N. M. Johnson. 1977. *Biogeochemistry of a forested ecosystem.* Springer-Verlag, New York, USA.

Likens, G. E., C. T. Driscoll, D. C. Buso, T. G. Siccama, C. E. Johnson, G. M. Lovett, T. J. Fahey, W. A. Reiners, D. F. Ryan, C. W. Martin, and S. W. Bailey. 1998. The biogeochemistry of calcium at Hubbard Brook. Biogeochemistry 41:89–173.

Lilyestrom, C. G. 1983. Aspectos de la biología del coporo (*Prochilodus mariae*). Revista de Ciencia y Tecnología UNELLEZ (Barinas) 1:5–11.

Lima, L., P. A. Marquet, and F. M. Jaksic. 1999. El Niño events, precipitation patterns, and rodent outbreaks are statistically associated in semiarid Chile. Ecography 22:213–218.

Lima, M., J. E. Keymer, and F. M. Jaksic. 1999. El Niño-southern oscillation-driven rainfall variability and delayed density dependence cause rodent outbreaks in western South America: Linking demography and population dynamics. American Naturalist 153: 476–491.

Limburg, K. E., M. A. Moran, and W. H. McDowell. 1986. *The Hudson River ecosystem.* Springer-Verlag, New York, USA.

Limia, J., and D. G. Raffaelli. 1997. Effects of burrowing by the amphipod *Corophium volutator* on the ecology of intertidal sediments. Journal of the Marine Biological Association of the United Kingdom 77:409–413.

Lindberg, S. E., and G. M. Lovett. 1992. Deposition and forest canopy interactions of airborne sulfur: Results from the Integrated Forest Study. Atmospheric Environment 26A: 1477–1492.

Lindberg, S. E., G. M. Lovett, D. D. Richter, and D. W. Johnson. 1986. Atmospheric deposition and canopy interactions of major ions in a forest. Science 231:141–145.

Lindeboom, H. J. 1984. The nitrogen pathway in a penguin rookery. Ecology 65:269–277.

Lindell, M. J., W. Graneli, and L. J. Transvik. 1995. Enhanced bacterial growth in response to photochemical transformation of dissolved organic matter. Limnology and Oceanography 40:195–199.

Lindeman, R. L. 1942. The trophic-dynamic aspect of ecology. Ecology 23:399–418.

Lingren, P. D., V. M. Bryant, J. R. Raulston, M. Pendleton, J. Westbrook, and G. D. Jones. 1993. Adult feeding host range and migratory activities of corn earworm, cabbage looper, and celery looper (Lepidoptera: Noctuidae) moths as evidenced by attached pollen. Journal of Economic Entomology 86:1429–1439.

Lingren, P. D., J. K. Westbrook, V. M. Bryant, J. R. Raulston, J. F. Esquivel, and G. D. Jones. 1994. Origin of corn earworm (Lepidoptera: Noctuidae) migrants as determined by Citrus pollen markers and synoptic weather systems. Environmental Entomology 23: 562–570.

Littler, M. M., D. S. Littler, and E. A. Titlyanov. 1991. Comparisons of N and P limited productivity between high granitic islands versus low carbonate atolls in the Seychelles Archipelago: A test of the relative dominance paradigm. Coral Reefs 10:199–209.

Litvak, M. K., and N. E. Mandrak. 1993. Ecology of freshwater baitfish use in Canada and the United States. Fisheries 18:6–13.

Livoreil, B., and C. Baudoin. 1996. Differences in food hoarding behaviour in two species of ground squirrels *Spermophilus tridecemlineatus* and *S. spilosoma*. Ethology, Ecology and Evolution 8:199–205.

Lloyd, P. H. 1975. A study of the Himalayan Tahr (*Hemitragus jemlahicus*) and its potential effects on the ecology of the Table Mountain Range. Report, Cape Department of Nature and Environmental Conservation. Cape Town, South Africa. 80 pp.

Loaiciga, H. A., J. B. Valdes, R. Vogel, J. Garvey, and H. Schwarz. 1995. Global warming and the hydrologic cycle. Journal of Hydrology 174:83–127.

Logan, T. J. 1990. Sustainable agriculture and water quality. Pp. 582–613 in C. A. Edwards, R. Lal, P. Madden, R. H. Miller, and G. House, eds., *Sustainable agricultural systems.* Soil and Water Conservation Society, Ankeny, IA, USA.

Longhurst, A. R. 1981. *Analysis of marine ecosystems.* Academic Press, London, UK.

Lonzarich, D. G., and T. P. Quinn. 1995. Experimental evidence for the effect of depth and structure on the distribution, growth, and survival of stream fishes. Canadian Journal of Zoology 73:2223–2230.

López de Casenave, J. L., J. P. Pelotto, and J. Protomastro. 1995. Edge-interior differences in vegetation structure and composition in a Chaco semi-arid forest, Argentina. Forest Ecology and Management 72:61–69.

Loreau, M. 1998. Ecosystem development explained by competition within and between material cycles. Proceedings of the Royal Society of London, B 265:33–38.

Loreau, M., and N. Mouquet. 1999. Immigration and the maintenance of local species diversity. American Naturalist 154:427–440.

Louw, G., and M. K. Seely. 1982. *Ecology of desert organisms.* Longman, London, UK.

Lovejoy, T. E. 1980. A projection of species extinctions. Pp. 328–331 in *The Global 2000 Report to the President: Entering the Twenty-First Century.* Council on Environmental Quality and the Department of State. Government Printing Office, Washington, DC, USA.

Lovett, G. M. 1994. Atmospheric deposition of nutrients and pollutants in North America: An ecological perspective. Ecological Applications 4:629–650.

Lowe-McConnell, R. H. 1987. *Ecological studies in tropical fish communities.* Cambridge University Press, Cambridge, UK.

Lowrance, R. R., R. A. Leonard, L. E. Asmussen, and R. L. Todd. 1985. Nutrient budgets for agricultural watersheds in the southeastern coastal plain. Ecology 66:287–296.

Loye, J. E. 1985a. Host-effects on feeding and survival of the polyphagous cliff swallow bug, *Oeciacus vicarious* (Hemiptera: Cimicidae). Bulletin of the Society of Vector Ecology 10: 7–13.

———. 1985b. The life history and ecology of the cliff swallow bug, *Oeciacus vicarious* (Hemiptera: Cimicidae). Cahiers O.R.S.T.O.M., Serie Entomologie Medicale et Parasitologie 23:133–139.

Lubchenco, J. L. 1978. Plant species diversity in a marine intertidal community: Importance of herbivore food preference and algal competitive abilities. American Naturalist 112:23–39.

———. 1986. Relative importance of competition and predation: Early colonization by seaweeds in New England. Pp. 537–555 in J. M. Diamond and T. Case, eds., *Community ecology.* Harper and Row, New York, USA.

Lubetkin, S. C. 1997. Multi-source mixing models: Food web determination using stable isotope tracers. M.S. thesis, University of Washington, Seattle, USA.

Ludwig, J. 1986. Primary production variability in desert ecosystems. Pp. 5–17 in W. Whitford, ed., *Pattern and process in desert ecosystems.* University of New Mexico Press, Albuquerque, USA.

———. 1987. Primary productivity in arid lands: Myths and realities. Journal of Arid Environments 13:1–7.

Ludwig, W., J.-L. Probst, and S. Kempe. 1997. Predicting the oceanic input of organic carbon by continental erosion. Global Biogeochemical Cycles 10:23–41.

Lükewille, A., and R. F. Wright. 1997. Experimentally increased soil temperature causes release of nitrogen at a boreal forest catchment in southern Norway. Global Change Biology 3:13–21.

Luque, M. H., and A. W. Stokes. 1976. Fishing behaviour of Alaska brown bear. Pp. 71–78 in M. R. Pelton, J. W. Lentfer, and G. E. Folk Jr., eds., *Bears: Their biology and management.* IUCN Publications, New Series 40. International Union for Conservation of Nature and Natural Resources, Morges, Switzerland.

Luther, G. C., H. R. Valenzuela, and J. Defrank. 1996. Impact of cruciferous trap crops on lepidopteran pests of cabbage in Hawaii. Environmental Entomology 25:39–47.

Luxmoore, R. J., S. D. Wullschleger, and P. J. Hanson. 1993. Forest responses to CO_2 enrichment and climate warming. Water, Air, and Soil Pollution 70:309–323.

L'vovich, M. I., and G. F. White. 1990. Use and transformation of terrestrial water systems. Pp. 235–252 in B. L. Turner II, W. C. Clark, R. W. Kates, J. F. Richards, J. T. Mathews, and W. B. Meyer, eds., *The earth as transformed by human action*. Cambridge University Press, Cambridge, UK.

Lyle, A. A., and J. M. Elliot. 1998. Migratory salmonids as vectors of carbon, nitrogen and phosphorus between marine and freshwater environments in north-east England. Science of the Total Environment 210/211:457–468.

Lynch, J. J. 1975. Winter ecology of snow geese on the Gulf Coast, 1925–1975. 37th Midwest Fish and Wildlife Conference, Toronto, Canada.

Lynch, J. J., T. O'Neil, and D. W. Lay. 1947. Management significance of damage by geese and muskrats to Gulf Coast marshes. Journal of Wildlife Management 11:50–77.

Lynch, M., and J. Shapiro. 1981. Predation, enrichment, and phytoplankton community structure. Limnology and Oceanography 26:86–102.

MacArthur, R. H., J. M. Diamond, and J. R. Karr. 1972. Density compensation in island faunas. Ecology 53:330–342.

MacArthur, R. H., and E. O. Wilson. 1967. *The theory of island biogeography*. Princeton University Press, Princeton, NJ, USA.

MacAvoy, S. E., S. A. Macko, S. P. McIninch, and G. C. Garman. 2000. Marine nutrient contributions to freshwater apex predators. Oecologia 122:568–573.

Macdonald, I. A. W., and G. W. Frame. 1988. The invasion of introduced species into nature reserves in tropical savannas and dry woodlands. Biological Conservation 44:67–93.

MacInnes, C. D., E. H. Dunn, D. H. Rusch, F. Cooke, and F. G. Cooch. 1990. Advancement of goose-nesting dates in the Hudson Bay region, 1951–1986. Canadian Field Naturalist 104:295–297.

Mackey, R. L., and R. M. R. Barclay. 1989. The influence of physical clutter and noise on the activity of bats over water. Canadian Journal of Zoology 67:1167–1170.

Macko, S. A., and M. L. F. Estep. 1994. Microbial alteration of stable nitrogen and carbon isotopic compositions of organic matter. Organic Geochemistry 6:787–790.

Macko, S. A., and N. E. Ostrom. 1994. Pollution studies using stable isotopes. Pp. 45–62 in K. Lajtha and R. H. Michener, eds., *Stable isotopes in ecology and environmental science*. Blackwell Scientific Publications, London, UK.

Madsen, T. V., and E. Warnke. 1983. Velocities of currents around and within submerged aquatic vegetation. Archiv für Hydrobiologie 97:389–394.

Magette, W. L., R. B. Brinsfield, R. E. Palmer, and J. D. Wood. 1989. Nutrient and sediment removal by vegetated filter strips. Transactions of the American Society of Agricultural Engineers 32:663–667.

Malanson, G. P. 1993. *Riparian landscapes*. Cambridge University Press, Cambridge, UK.

Mallin, M. A., M. H. Posey, G. C. Shank, M. R. McIver, S. H. Ensign, and T. D. Alphin. 1999. Hurricane effects on water quality and benthos in the Cape Fear watershed: Natural and anthropogenic impacts. Ecological Applications 9:350–362.

Malt, S. 1995. Epigeic spiders as an indicator system to evaluate biotope quality of riversides and floodplain grasslands on the river Ilm (Thuringia). Pp. 136–146 in V. Ruzicka, ed., *Proceedings of the 15th European Colloquium of Arachnology*. Ceske Budejovice, Czech Republic.

Manabe, S., and R. T. Wetherald. 1986. Reduction in summer soil wetness induced by an increase in atmospheric carbon dioxide. Science 232:626–628.

Mandrak, N. E. 1995. Biogeographic patterns of fish species richness in Ontario lakes in relation to historical and environmental factors. Canadian Journal of Fisheries and Aquatic Sciences 52:1462–1474.

Mandrak, N. E., and E. J. Crossman. 1992. Postglacial dispersal of freshwater fishes into Ontario. Canadian Journal of Zoology 70:2247–2259.

Mann, C. J., and R. G. Wetzel. 1995. Dissolved organic carbon and its utilization in a riverine wetland ecosystem. Biogeochemistry 31:99–120.

Mann, K. H., and J. R. N. Lazier. 1991. *Dynamics of marine ecosystems*. Blackwell, Boston, MA, USA.

Manny, B. A., W. C. Johnson, and R. G. Wetzel. 1994. Nutrient additions by waterfowl to lakes and reservoirs: Predicting their effects on productivity and water quality. Hydrobiologia 279/280:121–132.

Margat, J. 1996. Comprehensive assessment of the freshwater resources of the world: Groundwater component. Contribution to Chapter 2 of the Comprehensive Global Freshwater Assessment. United Nations, New York, USA.

Marinelli, L., and J. S. Millar. 1989. The ecology of beach-dwelling *Peromyscus maniculatus* on the Pacific Coast. Canadian Journal of Zoology 67:412–417.

Marino, R., R. Howarth, and R. Garritt. 1988. Determination of atmospheric gas exchange rates in the Hudson River Estuary. EOS 69:1108–1108.

Marko, P. B. 1998. Historical allopatry and the biogeography of speciation in the prosobranch snail genus *Nucella*. Evolution 52:757–774.

Maron, J. L., and P. G. Connors. 1996. A native nitrogen-fixing shrub facilitates weed invasion. Oecologia 105:302–312.

Maron, J. L., and R. L. Jefferies. 1999. Bush lupine mortality, altered resource availability. Ecology 80:443–454.

Marshall, E. J. P., and G. M. Arnold. 1995. Factors affecting field weed and field margins flora on a farm in Essex, UK. Landscape and Urban Planning 31:205–216.

Marshall, J. K. 1973. Drought, land-use and soil erosion. Pp. 55–80 in J. Lovett, ed., *Drought*. Angus and Robertson, Sydney, Australia.

Marston, B. H., M. F. Willson, and S. M. Gende. 2002. Predator aggregations at eulachon (*Thaleichthys pacificus*) spawning runs in southeast Alaska. Marine Ecology Progress Series 231:229–236.

Martin, N. V. 1952. A study of the lake trout, *Salvelinus namaycush*, in two Algonquin Park lakes. Transactions of the American Fisheries Society 81:111–137.

———. 1954. Catch and winter food of lake trout in certain Algonquin Park lakes. Journal of the Fisheries Research Board of Canada 11:5–10.

———. 1966. The significance of food habits in the biology, exploitation, and management of Algonquin Park, Ontario, lake trout. Transactions of the American Fisheries Society 96:415–422.

Martin, N. V., and N. S. Baldwin. 1960. Observations on the life history of the hybrid between eastern brook trout and Lake Trout in Algonquin Park, Ontario. Journal of the Fisheries Research Board of Canada 17:541–551.

Martin, N. V., and L. J. Chapman. 1965. Distribution of certain crustaceans and fishes in the region of Algonquin Park, Ontario. Journal of the Fisheries Research Board of Canada 22:969–976.

Martinelli, L. A., A. H. Devol, R. L. Victoria, and J. E. Richey. 1991. Stable carbon isotope variation in C_3 and C_4 plants along the Amazon River. Nature 353:57–59.

Mason, C. J., and M. L. McManus. 1981. Larval dispersal of the gypsy moth. Pp. 161–214 in C. C. Doane and M. L. McManus, eds., *The gypsy moth: Research toward integrated pest management*. USDA, Washington, DC, USA.

Mathews, W. A., and D. J. Keep. 1993. Ozone trends and variability, globally and over New Zealand. Paper read at National Science Strategy Committee for Climate Change Workshop, 20–21 May 1993, at International Antarctic Centre, Christchurch.

Mathisen, O. A. 1971. Escapement levels and productivity of the Nushagak sockeye salmon run from 1908-1966. Fisheries Bulletin 69:747–763.

———. 1972. Biogenic enrichment of sockeye salmon lakes and stock productivity. Verhandlungen Internationale Vereinigung für Theoretische und Angewandte Limnologie 18:1089–1095.

Mathisen, O. A., P. L. Parker, J. J. Goering, T. C. Kline, P. H. Poe, and R. S. Scalan. 1988. Recycling of marine elements transported into freshwater systems by anadromous salmon. Verhandlungen Internationale Vereinigung für Theoretische und Angewandte Limnologie 23:2249–2258.

Matlack, G. R. 1993. Microenvironment variation within and among forest edge sites in the eastern United States. Biological Conservation 66:185–194.

Matson, P. A., C. Billow, S. Hall, and J. Zachariassen. 1996. Fertilization practices and soil variations control nitrogen oxide emissions from tropical sugar cane. Journal of Geophysical Research 101 (D13):18533–18545.

Matson, P. A., R. Naylor, and I. Ortiz-Monasterio. 1998. Integration of environmental, agronomic, and economic aspects of fertilizer management. Science 280:112–115.

Matson, P. A., W. J. Parton, A. G. Power, and M. J. Swift. 1997. Agricultural intensification and ecosystem properties. Science 277:504–509.

Matthews, J. A. 1992. *The ecology of recently deglaciated terrain*. Cambridge University Press, Cambridge, UK.

Mattocks, J. G. 1971. Goose feeding and cellulose digestion. Wildfowl 22:107–113.

Mauremooto, J. R., S. D. Wratten, S. P. Worner, and G. L. A. Fry. 1994. Permeability of hedgerows to predatory beetles. Agriculture, Ecosystems & Environment 52:141–148.

May, R. M. 1973. *Stability and complexity in model ecosystems*. Princeton University Press, Princeton, NJ, USA.

———. 1974. *Stability and complexity in model ecosystems*. 2d ed. Princeton University Press, Princeton, NJ, USA.

Mayer, M. S., and G. E. Likens. 1987. The importance of algae in a shaded headwater stream as food for an abundant caddisfly (Trichoptera). Journal of the North American Benthological Society 6:262–269.

Mayewski, P. A., W. B. Lyons, M. J. Spencer, M. S. Twickler, C. F. Buck, and S. Whitlow. 1990. An ice-core record of atmospheric response to anthropogenic sulphate and nitrate. Nature 346:554–556.

Mayewski, P. A., W. B. Lyons, M. J. Spencer, M. S. Twickler, B. Koci, C. Dansgaard, C. Davidson, and R. Honrath. 1986. A detailed (1869–1984) record of sulfate and nitrate concentrations from south Greenland. Science 232:975–977.

Mazourek, J. C. 1998. Dynamics of seed banks and vegetation communities in existing and potential emergent marshes in the Missouri River flood plain. M.S. thesis, University of Missouri, Columbia, USA.

Mazumder, A. 1994. Patterns of algal biomass in dominant odd-link vs. even-link lake ecosystems. Ecology 75:1141–1149.

McArthur, J. V., and K. K. Moorhead. 1996. Characterization of riparian species and stream detritus using multiple stable isotopes. Oecologia 107:232–238.

McCabe, T. T., and I. M. Cowan. 1945. *Peromyscus maniculatus macrorhinus* and the problem of insularity. Transactions of the Royal Canadian Institute 25:117–216.

McCann, J. M., S. E. Mabey, L. J. Niles, C. Bartlett, and P. Kerlinger. 1993. A regional study of coastal migratory stopover habitat for Neotropical migrant songbirds: Land management implications. Transactions of North American Wildlife and Natural Resources Conference 58:398–407.

McCann, K. S., and A. Hastings. 1997. Re-evaluating the omnivory-stability relationship. Bulletin of the Ecological Society of America 78:142.

McCann, K. S, A. Hastings, and G. R. Huxel. 1998. Weak trophic interactions and the balance of nature. Nature 395:794–798.

McCann, K. S., A. Hastings, and D. R. Strong. 1997. Trophic cascades and trophic trickles in pelagic food webs. Proceedings of the Royal Society of London, Series B: Biological Sciences 265:205–209.

McCauley, E., B. E. Kendall, A. Janssen, S. Wood, W. W. Murdoch, P. Hosseini, C. J. Briggs, S. P. Ellner, R. M. Nisbet, M. W. Sabelis, and P. Turchin. 2000. Inferring colonization processes from population dynamics in spatially structured predator-prey systems. Ecology 81:3350–3361.

McCay, C. M., A. V. Tunison, M. Crowell, and H. Paul. 1936. The calcium and phosphorus content of the body of the brook trout in relation to age, growth, and food. Journal of Biological Chemistry 114:259–263.

McChesney, G. J., and B. R. Tershy. 1998. History and status of introduced mammals and impacts to breeding seabirds on the California channel and northwestern Baja California islands. Colonial Waterbirds 21:335–347.

McClelland, B. R., L. S. Young, P. T. McClelland, J. G. Crenshaw, H. L. Allen, and D. S. Shea. 1994. Migration ecology of Bald Eagles from autumn concentrations in Glacier National Park, Montana. Wildlife Monographs, no. 125. The Wildlife Society, Blacksburg, VA, USA.

McClelland, J. W., and I. Valiela. 1998. Linking nitrogen in estuarine producers to land-derived sources. Limnology and Oceanography 43:577–585.

McConnaughey, T., and C. P. McRoy. 1979. Food-web structure and the fractionation of carbon isotopes in the Bering Sea. Marine Biology 53:257–262.

McCracken, G. F., Y.-F. Lee, J. Westbrook, and W. W. Wolf. 1996. High altitude predation by Mexican free-tailed bats on migratory insect pests. Bat Research News 37:140–141.

McDowall, R. M. 1987. Evolution and importance of diadromy. American Fisheries Society Symposium 1:1–13.

———. 1988. *Diadromy in fishes: Migrations between freshwater and marine environments.* Croom Helm, London, UK.

McIlhenny, E. A. 1932. The blue goose in its winter home. Auk 49:279–306.

McIntire, C. D. 1973. Periphyton dynamics in laboratory streams: A simulation model and its implications. Ecological Monographs 43:399–420.

McIntosh, A. R., and C. R. Townsend. 1996. Interactions between fish, grazing invertebrates and algae in a New Zealand stream: A trophic cascade mediated by fish-induced changes to grazer behaviour. Oecologia 108:174–181.

McIntyre, A. D. 1992. The current state of the oceans. Marine Pollution Bulletin 25:28–31.

McKenney, R., R. B. Jacobson, and R. C. Wertheimer. 1995. Woody vegetation and channel morphogenesis in low gradient gravel-bed streams in Ozark Plateaus, Missouri and Arkansas. Geomorphology 13:175–198.

McNaughton, S. J. 1985. Ecology of a grazing ecosystem: The Serengeti. Ecological Monographs 55:259–294.

———. 1990. Mineral nutrition and seasonal movements of African migratory ungulates. Nature 345:613–615.

McNulty, S. G., J. M. Vose, and W. T. Swank. 1996. Potential climate change effects on Loblolly Pine forest productivity and drainage across the southern United States. Ambio 25: 449–453.

McQueen, D. J., J. R. Post, and E. L. Mills. 1986. Trophic relationships in freshwater pelagic ecosystems. Canadian Journal of Fisheries and Aquatic Sciences 43:1571–1581.

Medley, K. E., W. O. Okey, G. W. Barrett, M. F. Lucas, and W. H. Renwick. 1995. Landscape change with agricultural intensification in a rural watershed, southwestern Ohio, U.S.A. Landscape Ecology 10:161–176.

Meeus, J. H. A. 1993. The transformation of agricultural landscapes in western Europe. The Science of the Total Environment 129:171–190.

Meili, M., G. W. Kling, B. Fry, R. T. Bell, and I. Ahgren. 1996. Sources and partitioning of organic matter in a pelagic microbial food web inferred from the isotopic composition of (δ^{13}C and δ^{15}N) zooplankton species. Pp. 53-61 in M. Simmno, H. Gude, and T. Weisse, eds., *Aquatic microbial ecology.* E. Schweizerbart'she Verlagsbuchandlung, Konstanz.

Menge, B. A. 1976. Organization of the New England rocky intertidal community: Role of predation, competition and environmental heterogeneity. Ecological Monographs 46:355–393.

———. 1978a. Predation intensity in a rocky intertidal community: Effect of an algal canopy, wave action and desiccation on predator feeding rates. Oecologia 34:17–35.

———. 1978b. Predation intensity in a rocky intertidal community: Relation between predator foraging activity and environmental harshness. Oecologia 34:1–16.

———. 1992. Community regulation: Under what conditions are bottom-up factors important on rocky shores? Ecology 73:755–765.

———. 2000a. Recruitment vs. post-recruitment processes as determinants of barnacle population abundance on Oregon rocky shores. Ecological Monographs 70:265–288.

———. 2000b. Top-down and bottom-up community regulation in marine rocky intertidal habitats. Journal of Experimental Marine Biology and Ecology 250:257–289.

———. 2003. The overriding importance of environmental context in determining the outcome of species deletion experiments. Pp. 17–43 in S. A. Levin and P. Kareiva, eds., *The importance of species: Perspectives on expendability and triage*. Princeton University Press, Princeton, NJ, USA.

Menge, B. A., E. L. Berlow, C. A. Blanchette, S. A. Navarrete, and S. B. Yamada. 1994. The keystone species concept: Variation in interaction strength in a rocky intertidal habitat. Ecological Monographs 64:249–286.

Menge, B. A., B. A. Daley, J. Lubchenco, E. Sanford, E. Dahlhoff, P. M. Halpin, G. Hudson, and J. L. Burnaford. 1999. Top-down and bottom-up regulation of New Zealand rocky intertidal communities. Ecological Monographs 69:297–330.

Menge, B. A., B. A. Daley, and P. A. Wheeler. 1996. Control of interaction strength in marine benthic communities. Pp. 258–274 in G. A. Polis and K. O. Winemiller, eds., *Food webs: Integration of patterns and dynamics*. Chapman and Hall, New York, USA.

Menge, B. A., B. A. Daley, P. A. Wheeler, E. Dahlhoff, E. Sanford, and P. T. Strub. 1997a. Benthic-pelagic links and rocky intertidal communities: Bottom-up effects on top-down control? Proceedings of the National Academy of Sciences, USA 94:14530–14535.

Menge, B. A., B. A. Daley, P. A. Wheeler, and P. T. Strub. 1997b. Rocky intertidal oceanography: An association between community structure and nearshore phytoplankton concentration. Limnology and Oceanography 42:57–66.

Menge, B. A., and A. M. Olson. 1990. Role of scale and environmental factors in regulation of community structure. Trends in Ecology and Evolution 5:52–57.

Menge, B. A., E. Sanford, B. A. Daley, T. L. Freidenburg, G. Hudson, and J. Lubchenco. 2002. An inter-hemispheric comparison of bottom-up effects on community structure: Insights revealed using the comparative-experimental approach. Ecological Research 17:1–16.

Menge, B. A., and J. P. Sutherland. 1976. Species diversity gradients: Synthesis of the roles of predation, competition and temporal heterogeneity. American Naturalist 110:351–369.

———. 1987. Community regulation: Variation in disturbance, competition, and predation in relation to environmental stress and recruitment. American Naturalist 130:730–757.

Merna, J. W. 1986. Contamination of stream fishes with chlorinated hydrocarbons from eggs of Great Lakes salmon. Transactions of the American Fisheries Society 129:158–173.

Mérot, P. 1999. The influence of hedgerow systems on the hydrology of agricultural catchments in a temperate climate. Agronomie 19:655–669.

Merriam, H. G., and A. Lanoue. 1990. Corridor use by small mammals: Field measurements for three experimental types of *Peromyscus leucopus*. Landscape Ecology 4:123–131.

Merritt, R. W., and K. Cummins. 1996. Trophic relations of macroinvertebrates. Pp. 453–474 in F. R. Hauer and G. A. Lambert, eds., *Methods in stream ecology*. Academic Press, New York, USA.

Metais, O., and M. Lesieur, eds. 1991. *Turbulence and coherent structures*. Kluwer Press, Dordrecht, The Netherlands.

Meybeck, M. 1982. Carbon, nitrogen, and phosphorus transport by world rivers. American Journal of Science 282:401–450.

———. 1993. Riverine transport of atmospheric carbon: Sources, global typology and budget. Water, Air, and Soil Pollution 70:443–463.

Meybeck, M., D. V. Chapman, and R. Helmer. 1990. *Global freshwater quality: A first assessment*. Blackwell, Cambridge.

Meyer, J. L., A. Benke, R. T. Edwards, and J. B. Wallace. 1997. Organic matter dynamics in the Ogeechee River, a blackwater river in Georgia, USA. Journal of the North American Benthological Society 16:82–87.

Meyer, J. L., and C. Johnson. 1983. The influence of elevated nitrate concentration on rate of leaf decomposition in a stream. Freshwater Biology 13:177–183.

Meyer, J. L., E. T. Schultz, and G. S. Helfman. 1983. Fish schools: An asset to corals. Science 220:1047–1049.

Meyer, W. B., and B. L. Turner. 1992. Human population growth and global land-use/cover change. Annual Reviews of Ecology and Systematics 23:39–61.

Michael, J. H. 1995. Enhancement effects of spawning pink salmon on stream rearing juvenile coho salmon: Managing one resource to benefit another. Northwest Science 69:228–233.

———. 1998. Pacific salmon spawner escapement goals for the Skagit River watershed as determined by nutrient cycling concentrations. Northwest Science 72:239–248.

Michaletz, P. H. 1997. Factors affecting abundance, growth, and survival of age-0 gizzard shad. Transactions of the American Fisheries Society 126:84–100.

Michener, R. H., and D. M. Schell. 1994. Stable isotope ratios as tracers in marine aquatic food webs. Pp. 138–157 in K. Lajtha and R. H. Michener, eds., *Stable isotopes in ecology and environmental science*. Blackwell Scientific Publications, London, UK.

Michener, W. K., and R. A. Haeuber. 1998. Flooding: Natural and managed disturbances. BioScience 48:677–680.

Milakovic, B. 1999. Changes in aquatic and terrestrial invertebrate populations in response to habitat degradation by lesser snow geese. M.S. thesis, University of Toronto, Canada.

Milakovic, B., T. J. Carleton, and R. L. Jefferies. 2001. Changes in midge (Diptera: Chironomidae) populations of sub-arctic supratidal vernal ponds in response to goose foraging. Écoscience 8:58–67.

Miller, S. D. 1989. Population management of bears in North America. International Conference on Bear Research and Management 8:357–373.

Miller, S. D., G. C. White, R. A. Sellers, H. V. Reynolds, J. W. Schoen, K. Titus, V. G. Barnes, Jr., R. B. Smith, R. R. Nelson, W. B. Ballard, and C. C. Schwartz. 1997. Brown and black bear density estimation in Alaska using radiotelemetry and replicated mark-resight techniques. Wildlife Monographs, 133. The Wildlife Society, Bethesda, MD, USA.

Mills, E. L., J. L. Forney, and K. J. Wagner. 1987. Fish predation and its cascading effects on the Oneida Lake food chain. Pp. 118–131 in W. C. Kerfoot and A. Sih, eds., *Predation: Direct and indirect impacts on aquatic communities*. University Press of New England, Hanover, NH, USA.

Mills, E. L., J. H. Leach, J. T. Carlton, and C. L. Secor. 1993. Exotic species in the Great Lakes: A history of biotic crises and anthropogenic introductions. Journal of Great Lakes Research 19:1–54.

———. 1994. Exotic species and the integrity of the Great Lakes. BioScience 44:666–676.

Milner, A. M., E. E. Knudsen, C. Soiseth, A. L. Robertson, D. Schell, I. T. Phillips, and K. Magnusson. 2000. Colonization and development of stream communities across a 200-year gradient in Glacier Bay National Park, Alaska, USA. Canadian Journal of Fisheries and Aquatic Sciences 57:2319–2335.

Minagawa, M., and E. Wada. 1984. Stepwise enrichment of ^{15}N along food chains: Further evidence and the relation between $\delta^{15}N$ and animal age. Geochimica et Cosmochimica Acta 48:1135–1140.

Minshall, G. W. 1978. Autotrophy in stream ecosystems. BioScience 28:767–771.

Minshall, G. W., E. Hitchcock, and J. R. Barnes. 1991. Decomposition of rainbow trout (*Oncorhynchus mykiss*) carcasses in a forest stream ecosystem inhabited only by nonanadromous fish populations. Canadian Journal of Fisheries and Aquatic Sciences 48:191–195.

Miranda, L. E. 1983. Average ichthyomass in Texas large impoundments. Proceedings of the Texas Chapter of the American Fisheries Society 6:58–67.

Mishra, B. K., and P. S. Ramakrishnan. 1984. Nitrogen budget under rotational bush fallow agriculture (jhum) at higher elevations of Meghalaya in north-eastern India. Plant and Soil 81:37–46.

Mittlebach, G. G., C. F. Steiner, S. M. Scheiner, K. L. Gross, H. L. Reynolds, R. B. Waide, M. R. Willig, S. L. Dodson, and L. Gough. 2001. What is the observed relationship between species richness and productivity? Ecology 82:2381–2396.

Mizutani, H., and E. Wada. 1988. Nitrogen and carbon isotope ratios in seabird rookeries and their ecological implications. Ecology 69:340–349.

Moffat, A. S. 1998. Global nitrogen overload problem grows critical. Science 279:988–989.

Moline, M. A., and B. B. Prezelin. 1996. Long-term monitoring and analyses of physical factors regulating variability in coastal Antarctic phytoplankton biomass, in situ productivity and taxonomic composition over subseasonal, seasonal and interannual time scales. Marine Ecology Progress Series 145:143–160.

Moloney, C. L. 1992. Simulation studies of trophic flow and nutrient cycles in Benguela upwelling foodwebs. In A. I. L. Payne, K. H. Brink, K. H. Mann, and R. Hilborn, eds., *Benguela trophic functioning*. South African Journal of Marine Science 12:457–476.

Monger, B. C., J. M. Fischer, B. A. Grantham, V. Medland, B. Cai, and K. Higgins. 1997. Frequency response of a simple food-chain model with time-delayed recruitment: Implications for abiotic-biotic coupling. Pp. 435–450 in S. Tuljapurkar and H. Caswell, eds., *Structured-population models in marine, terrestrial, and freshwater systems*. Chapman and Hall, London, UK.

Moore, D. M. 1983. Human impact on island vegetation. Pp. 237–246 in W. Holzner, M. J. Werger, and I. Kusind, eds., *Man's impact on vegetation*. Junk, The Hague, The Netherlands.

Moore, T. R. 1981. Controls on the decomposition of organic matter in subarctic spruce-lichen soil. Soil Science 131:107–113.

Morin, P. J. 1999. *Community ecology*. Blackwell Science, Oxford, UK.

Morris, H. M. 1955. A new concept of flow in rough conduits. Transactions of the American Society of Civil Engineers 120:373–398.

Morse, D. H., and R. S. Fritz. 1982. Experimental and observational studies of patch choice at different scales by the crab spider *Misumena vatia*. Ecology 63:172–182.

Moss, B. 1990. Engineering and biological approaches to the restoration from eutrophication of shallow lakes in which aquatic plant communities are important components. Hydrobiologia 200/201:367–377.

Mulholland, P. J., and D. L. DeAngelis. 1999. Effect of surface/subsurface exchange on nutrient spiralling in streams. Pp. 149–166 in J. B. Jones, Jr., and P. J. Mulholland, eds., *Surface-subsurface interactions in streams*. Landes Bioscience, Austin, TX, USA.

Mundahl, N. D. 1988. Nutritional quality of foods consumed by gizzard shad in western Lake Erie. Ohio Journal of Science 88:110–113.

———. 1991. Sediment processing by gizzard shad, *Dorosoma cepedianum* (LeSueur), in Acton Lake, Ohio, U.S.A. Journal of Fish Biology 38:565–572.

Mundahl, N. D., and T. E. Wissing. 1987. Nutritional importance of detritivory in the growth and condition of gizzard shad in an Ohio reservoir. Environmental Biology of Fishes 20:129–142.

———. 1988. Selection and digestive efficiencies of gizzard shad feeding on natural detritus and two laboratory diets. Transactions of the American Fisheries Society 117:480–487.

Murcia, C. 1995. Edge effects in fragmented forests: Implications for conservation. Trends in Ecology and Evolution 10:58–62.

Murdoch, W. W., and C. J. Briggs. 1996. Theory for biological control: Recent developments. Ecology 77:2001–2013.

Murphy, G. I. 1981. Guano and the anchovetta fishery. Research and Management of Environmental Uncertainty 11:81–106.

Muscatine, L., and C. D'Elia. 1978. The uptake, retention and release of ammonium by reef corals. Limnology and Oceanography 22:725–734.

Mutikainen, P., M. Walls, and A. Ojala. 1994. Sexual differences in responses to simulated herbivory in *Urtica dioica*. Oikos 69:397–404.

Myers, J. H. 1993. Population outbreaks in forest Lepidoptera. American Scientist 81:240–251.

Myrcha, A., S. J. Pietr, and A. Tatur. 1985. The role of pygoscelid penguin rookeries in nutrient cycles at Admiralty Bay, King George Island. Pp. 156–162 in W. R. Siegfried, P. R. Condy, and R. M. Laws, eds., *Antarctic nutrient cycles and food webs*. Springer-Verlag, Berlin, Germany.

Nadelhoffer, K. J., J. D. Aber, and J. M. Melillo. 1983. Leaf-litter production and soil organic matter dynamics along a nitrogen-availability gradient in southern Wisconsin (U.S.A.). Canadian Journal of Forest Research 13:12–21.

Naiman, R. J., R. E. Bilby, D. E. Schindler, and J. M. Helfield. 2002. Pacific salmon, nutrients and the dynamics of freshwater ecosystems. Ecosystems 5:399–417.

Naiman, R. J., and H. Décamps, eds. 1990. *The ecology and management of aquatic-terrestrial ecotones*. Man and the Biosphere Series, vol. 4. Parthenon, New York, USA.

———. 1997. The ecology of interfaces: Riparian zones. Annual Review of Ecology and Systematics 28:621–658.

Naiman, R. J., J. J. Magnuson, D. M. McKnight, and J. A. Stanford. 1995a. *The freshwater imperative: A research agenda*. Island Press, Washington, DC, USA.

Naiman, R. J., J. J. Magnuson, D. M. McKnight, J. A. Stanford, and J. R. Karr. 1995b. Freshwater ecosystems and their management: A national initiative. Science 270:584–585.

Naiman, R. J., J. M. Melillo, and J. E. Hobbie. 1986. Ecosystem alteration of boreal forest streams by beaver (*Castor Canadensis*). Ecology 67:1254–1269.

Naiman, R. J., J. M. Melillo, M. A. Lock, T. E. Ford, and S. R. Reice. 1987. Longitudinal patterns of ecosystem processes and community structure in a subarctic river continuum. Ecology 68:1139–1156.

Naiman, R. J., and K. H. Rogers. 1997. Large animals and system-level characteristics in river corridors. BioScience 47:521–529.

Nakano, S., H. Miyasaka, and N. Kuhara. 1999. Terrestrial-aquatic linkages: Riparian arthropod inputs alter trophic cascades in a stream food web. Ecology 80:2435–2441.

Nakano, S., and M. Murakami. 2001. Reciprocal subsidies: Dynamic interdependence between terrestrial and aquatic food webs. Proceedings of the National Academy of Sciences, USA 98:166–170.

Nakashima, B. S., and W. C. Leggett. 1980. Natural sources and requirements of phosphorus for fishes. Canadian Journal of Fisheries and Aquatic Sciences 37:679–686.

National Research Council (NRC). 1992. *Restoration of aquatic ecosystems*. National Academy Press, Washington, DC, USA.

———. 1996. *Upstream: Salmon and society in the Pacific Northwest*. National Academy of Sciences, Washington, DC, USA.

Navarrete, S. A., and J. C. Castilla. 1993. Predation by Norway rats in the intertidal zone of central Chile. Marine Ecology Progress Series 92:187–199.

Navarrete, S. A., and B. A. Menge. 1996. Keystone predation and interaction strength: Interactive effects of predators on their main prey. Ecological Monographs 66:409–429.

Navarrete, S. A., B. A. Menge, and B. A. Daley. 2000. Species interactions in a rocky intertidal food web: Prey or predation regulation of intermediate predators? Ecology 80:2206–2224.

Navarrete, S. A., and E. A. Wieters. 2000. Variation in barnacle recruitment over small scales: Larval predation by adults and maintenance of community pattern. Journal of Experimental Marine Biology and Ecology 253:131–148.

Naylor, R. L., R. J. Goldburg, J. H. Primavera, N. Kautsky, M. C. M. Beveridge, J. Clay, C. Folke, J. Lubchenco, H. Mooney, and M. Troell. 2000. Effect of aquaculture on world fish supplies. Nature 405:1017–1024.

Ndiaye, P., D. Mailly, M. Pineau, and H. A. Margolis. 1993. Growth and yield of *Casuarina equisetifolia* plantations on the coastal sand dunes of Senegal as a function of microtopography. Forest Ecology and Management 56:13–28.

Nehlsen, W. 1997. Pacific salmon status and trends—a coastwide perspective. Pp. 41–50 in D. J. Stouder, P. A. Bisson, and R. J. Naiman, eds., *Pacific salmon and their ecosystems: Status and future options*. Chapman and Hall, New York, USA.

Nelson, J. S. 1994. *Fishes of the world*. 3d ed. John Wiley and Sons, New York, USA.

Nelson, P. R., and W. T. Edmondson. 1955. Limnological effects of fertilizing Bare Lake, Alaska. Fishery Bulletin 102 (from Fishery Bulletin of the Fish and Wildlife Service 56:415–436). U.S. Government Printing Office, Washington, DC, USA.

Neubert, M. G., and H. Caswell. 1997. Alternatives to resilience for measuring the responses of ecological systems to perturbations. Ecology 78:653–665.

Newbold, J. D., P. J. Mulholland, J. W. Elwood, and R. V. O'Neill. 1982. Organic carbon spiralling in stream ecosystems. Oikos 38:266–272.

Nickelson, T. E., J. W. Nicholas, A. M. McGie, R. B. Lindsay, D. L. Bottom, R. J. Kaiser, and S. E. Jacobs. 1992. Status of anadromous salmonids in Oregon coastal basins. Research and Development Section, Oregon Department of Fisheries and Wildlife, Corvallis, USA.

Nielsen, K. J. 2000. Bottom-up and top-down forces in tide pools: Test of a food chain model in an intertidal community. Ecological Monographs 71:187–217.

Nisbet, R. M., S. Diehl, W. G. Wilson, S. D. Cooper, D. D. Donalson, and K. Kratz. 1997. Primary productivity gradients and short-term population dynamics in open systems. Ecological Monographs 67:535–553.

Nisbet, R. M., and W. S. C. Gurney. 1982. *Modelling ecological fluctuations*. John Wiley and Sons, Chichester, UK.

Northcote, T. G. 1988. Fish in the structure and function of freshwater ecosystems: A "top-down" view. Canadian Journal of Fisheries and Aquatic Sciences 45:361–379.

Northcote, T. G., and P. A. Larkin. 1989. The Fraser River: A major salmonine production system. Pp. 172–204 in D. P. Dodge, ed., *Proceedings of the International large river symposium*. Canadian Special Publications in Fisheries and Aquatic Sciences 106. Department of Fisheries and Oceans, Ottawa, Ontario, Canada.

Nowak, R. M., and J. L. Paradiso. 1983. *Walker's mammals of the world*, 4th ed. Vol. 2. Johns Hopkins University Press, Baltimore, MD, USA.

Nowell, A. R. M., and P. A. Jumars. 1984. Flow environments of aquatic benthos. Annual Review of Ecology and Systematics 15:303–328.

Noy-Meir, I. 1973. Desert ecosystems: Environment and producers. Annual Review of Ecology and Systematics 4:25–51.

———. 1974. Desert ecosystems: Higher trophic levels. Annual Review of Ecology and Systematics 5:195–214.

Nürnberg, G. K. 1988. Prediction of phosphorus release rates from total and reductant soluble phosphorus in anoxic lake sediments. Canadian Journal of Fisheries and Aquatic Sciences 45:453–462.

Nürnberg, G. K., M. Shaw, P. J. Dillon, and D. J. McQueen. 1986. Internal phosphorus load in an oligotrophic Precambrian Shield lake with an anoxic hypolimnion. Canadian Journal of Fisheries and Aquatic Sciences 43:574–580.

Nyffeler, M., and G. Benz. 1987. Spiders in natural pest control: A review. Journal of Applied Entomology 103:321–339.

O'Brien, W. J. 1974. The dynamics of nutrient limitation of phytoplankton algae: A model reconsidered. Ecology 55:135–141.

Odum, E. P. 1971. *Fundamentals of ecology*. 3d ed. Saunders, Philadelphia, USA.

Odum, W. E., E. P. Odum, and H. T. Odum. 1995. Nature's pulsing paradigm. Estuaries 18:547–555.

Oechel, W. C., G. L. Vourlitis, S. J. Hastings, R. P. Ault, and P. Bryant. 1998. The effects of water table manipulation and elevated temperature on the net CO_2 flux of wet sedge tundra ecosystems. Global Change Biology 4:77–90.

Oelschaeger, M. 1991. *The idea of wilderness: From prehistory to the age of ecology*. Yale University Press, New Haven, CT, USA.

O'Farrell, M. J., and B. W. Miller. 1972. Pipistrelle bats attracted to vocalizing females and to a blacklight insect trap. American Midland Naturalist 88:462–463.

Ojima, D. S., K. A. Galvin, and B. L. Turner II. 1994. The global impact of land-use change. BioScience 44:300–304.

Ojima, D. S., D. S. Schimel, W. J. Parton, and C. E. Owensby. 1994. Long-term and short-term effects of fire on nitrogen cycling in tallgrass prairie. Biogeochemistry 24:67–84.

Oksanen, L., S. D. Fretwell, J. Arruda, and P. Niemalä. 1981. Exploitation ecosystems in gradients of productivity. American Naturalist 118:240–261.

Oksanen, L., and T. Oksanen. 2000. The logic and realism of the hypothesis of exploitation ecosystems. American Naturalist 155:703–723.

Oksanen, T., M. E. Power, and L. Oksanen. 1995. Ideal free habitat selection and consumer-resource dynamics. American Naturalist 146:565–585.

Oliver, B. G., and A. J. Niimi. 1988. Trophodynamic analysis of polychlorinated biphenyl congeners and other chlorinated hydrocarbons in the Lake Ontario ecosystem. Environmental Science and Technology 22:388–397.

Onuf, C. P., J. M. Teal, and I. Valiela. 1977. Interactions of nutrients, plant growth and herbivory in a mangrove ecosystem. Ecology 58:514–526.

Osborne, T. O., and W. A. Sheppe. 1971. Food habits of *Peromyscus maniculatus* on a California beach. Journal of Mammalogy 52:844–845.

Ostfeld, R. S., and F. Keesing. 2000. Pulsed resources and community dynamics of consumers in terrestrial ecosystems. Trends in Ecology and Evolution 15:232–237.

Ostrom, P. H., M. Colunga-Garcia, and S. H. Gage. 1997. Establishing pathways of energy flow for insect predators using stable isotope ratios: Field and laboratory evidence. Oecologia 109:108–113.

Oswood, M. W. 1997. Streams and rivers of Alaska: A high latitude perspective on running waters. Pp. 331–356 in A. M. Milner and M. W. Oswood, eds., *Freshwaters of Alaska: Ecological syntheses*. Springer-Verlag, New York, USA.

Ouin, A., G. Paillat, A. Butet, and F. Burel. 2000. Spatial dynamics of wood mouse (*Apodemus sylvaticus*) in an agricultural landscape under intensive use in the Mont-Saint-Michel Bay (France). Agriculture, Ecosystems and Environment 78:159–165.

Owen, M. 1971. The selection of feeding site by white-fronted geese in winter. Journal of Applied Ecology 8:905–917.

Owen, M. 1980. *Wild geese of the world*. Batsford, London.

Owens, N. J. P. 1987. Natural variations in ^{15}N in the marine environment. Advances in Marine Biology 24:389–451.

Paerl, H. W. 1985. Enhancement of marine primary production by nitrogen-enriched acid rain. Nature 315:747–749.

Paine, R. T. 1966. Food web complexity and species diversity. American Naturalist 100:65–75.

———. 1974. Intertidal community structure: Experimental studies on the relationship between a dominant competitor and its principal predator. Oecologia 15:93–120.

———. 1988. Food webs: Road maps of interactions or grist for theoretical development? Ecology 69:1648–1654.

———. 1992. Food-web analysis through field measurement of per capita interaction strength. Nature 355:73–75.

———. 1994. *Marine rocky shores and community ecology: An experimentalist's perspective*. Ecology Institute, Oldendorf/Luhe, Germany.

Paine, R. T., and S. A. Levin. 1981. Intertidal landscapes: Disturbance and the dynamics of pattern. Ecological Monographs 51:145–178.

Pair, S. D., J. R. Raulston, J. K. Westbrook, W. W. Wolf, and S. D. Adams. 1991. Fall armyworm (Lepidoptera: Noctuidae) outbreak originating in the lower Rio Grande Valley (Texas [USA] and Mexico). Florida Entomologist 74:200–213.

Palik, B. J., and P. G. Murphy. 1990. Disturbance versus edge effects in sugar-maple/beech forest fragments. Forest Ecology and Management 32:187–202.

Palmer, M., C. C. Hakenkamp, and K. Nelson-Baker. 1997. Ecological heterogeneity in streams: Why variance matters. Journal of the North American Benthological Society 16:189–202.

Palmer, M., and G. X. Pons. 1996. Diversity in western Mediterranean islets: Effects of rat presence on a beetle guild. Acta Oecologia 17:297–305.

Palmisano, J. F., R. H. Ellis, and V. W. Kaczynski. 1993. The impact of environmental and management factors on Washington's wild anadromous salmon and trout. Washington Forest Protection Association and Washington Department of Natural Resources, Olympia, USA.

Panagis, K. 1985. The influence of elephant seals on the terrestrial ecosystem at Marion Island. Pp. 173–179 in W. R. Siegfried, P. R. Condy, and R. M. Laws, eds., *Antarctic nutrient cycles and food webs*. Springer-Verlag, Berlin, Germany.

Pandey, C. B., and J. S. Singh. 1992. Rainfall and grazing effects on net primary production in a tropical savanna, India. Ecology 73:2007–2021.

Paoletti, M. G., and D. Pimentel, eds. 1992. *Biotic diversity in agroecosystems*. Elsevier, Amsterdam, The Hague, The Netherlands.

Papy, F., and V. Souchere. 1993. Control of overland runoff and talweg erosion: A land management approach. Pp. 87–98 in J. Brossier, L. De Bonneval, and E. Landais, eds., *Systems studies in agriculture and rural development*. INRA, Paris, France.

Parker, M. S., and M. E. Power. 1993. Algal-mediated differences in aquatic insect emergence and the effect on a terrestrial predator. Bulletin of the North American Benthological Society 10:171.

Parsons, J. 1972. Spread of African pasture grasses to the American Tropics. Journal of Range Management 25:7–12.

Paterson, I. W., and C. D. Coleman. 1982. Activity patterns of seaweed-eating sheep of North Ronaldsay, Orkney. Applied Animal Ethology 8:137–146.

Patterson, K. R., J. Zuzunaga, and G. Cardenas. 1992. Size of the South American sardine (*Sardinops agax*) population in the northern part of the Peru upwelling ecosystem after collapse of anchoveta (*Engraulis ingens*) stocks. Canadian Journal of Fisheries and Aquatic Sciences 49:1762–1769.

Patterson, M. R. 1980. Hydromechanical adaptation in *Alcyonium siderium* (Octocorallia). Pp. 183–201 in D. J. Schneck, ed., *Biofluid Mechanics* 2. Plenum Press, New York, USA.

Patterson, M. R. 1984. Patterns of whole coral prey capture in an octocoral *Alcyonium siderium*. Biological Bulletin 180:93–102.

Patterson, M. R., and J. D. Witman. In press. Internal waves and scope for growth of suspension feeding benthos at an offshore pinnacle. Limnology and Oceanography.

Patton, G. W., D. A. Hinckley, M. D. Walla, T. F. Bidleman, and B. T. Hargrave. 1989. Airborne organochlorines in the Canadian high arctic. Tellus 41B:243–245.

Paw, J. N., and T. E. Chua. 1991. Climate changes and sea level rise: Implications on coastal area utilization and management in south-east Asia. Ocean Shoreline Management 15:205–232.

Pazzia, I., M. Trudel, M. Ridgway, and J. B. Rasmussen. 2002. Influence of food web structure on the growth and bioenergetics of lake trout (*Salvelinus namaycush*). Canadian Journal of Fisheries and Aquatic Sciences 59:1593–1605.

Pedgley, D. E. 1982. Windborne pests and diseases: Meteorology of airborne organisms. Horwood, Chichester, UK.

Peierls, B. L., N. F. Caraco, M. L. Pace, and J. J. Cole. 1991. Human influence on river nitrogen. Nature 350:386–387.

Perlin, J. 1991. *A forest journey*. Harvard University Press, Cambridge, MA, USA.

Perrin, C. J., and J. S. Richardson. 1997. N and P limitation of benthos abundance in the Nechako River, British Columbia. Canadian Journal of Fisheries and Aquatic Sciences 54:2574–2583.

Persson, A. 1997. Phosphorus release by fish in relation to external and internal load in a eutrophic lake. Limnology and Oceanography 42:577–583.

Persson, L., G. Andersson, S. F. Hamrin, and L. Johansson. 1988. Predator regulation and primary production along the productivity gradient of temperate lake ecosystems. Pp. 45–68 in S. R. Carpenter, ed., *Complex interactions in lake communities*. Springer-Verlag, New York, USA.

Persson, L., J. Bengtsson, B. A. Menge, and M. E. Power. 1996. Productivity and consumer regulation-concepts, patterns, and mechanisms. Pp. 396–434 in G. A. Polis and K. O. Winemiller, eds., *Food webs: Integration of patterns and dynamics*. Chapman and Hall, New York, USA.

Peterjohn, W. T., and D. L. Correll. 1984. Nutrient dynamics in an agricultural watershed: Observations on the role of a riparian forest. Ecology 65:1466–1475.

Peterjohn, W. T., and W. H. Schlesinger. 1990. Nitrogen loss from deserts in the southwestern United States. Biogeochemistry 10:67–79.

———. 1991. Factors controlling denitrification in a Chihuahuan desert ecosystem. Soil Science Society of America Journal 55:1694–1701.

Peterson, B. J., L. Deegan, J. Helfrich, J. E. Hobbie, M. Hullar, B. Moller, T. E. Ford, A. Hershey, A. Hiltner, G. Kipphut, M. A. Lock, D. M. Fiebig, V. McKinley, M. C. Miller, J. R. Vestal, R. Ventullo, and G. Volk. 1993. Biological responses of a tundra river to fertilization. Ecology 74:653–672.

Peterson, B. J., and B. Fry. 1987. Stable isotopes in ecosystem studies. Annual Review of Ecology and Systematics 18:293–320.

Peterson, B. J., J. E. Hobbie, and T. L. Corliss. 1986. Carbon flow in a tundra stream ecosystem. Canadian Journal of Fisheries and Aquatic Sciences 43:1259–1270.

Peterson, B. J., and R. W. Howarth. 1987. Sulfur, carbon, and nitrogen isotopes used to trace organic matter flow in the salt-marsh estuaries of Sapelo Island, Georgia. Limnology and Oceanography 32:1195–1213.

Peterson, B. J., R. W. Howarth, and R. H. Garritt. 1985. Multiple stable isotopes used to trace the flow of organic matter in estuarine food webs. Science 227:1361–1363.

———. 1986. Sulfur and carbon isotopes as tracers of salt-marsh organic matter flow. Ecology 67:865–874.

Peterson, C. H. 1982. The importance of predation and intra- and interspecific competition in the population biology of two infaunal suspension-feeding bivalves, *Protothaca staminea* and *Chione undatella*. Ecological Monographs 52:437–475.

Peterson, M. S. 1986. *River engineering.* Prentice Hall, Englewood Cliffs, NJ, USA.

Petit, S., and F. Burel. 1993. Movement of *Abax ater* (Col. Carabidae): Do forest species survive in hedgerow networks? Vie et Milieu 1:119–124.

———. 1998a. Connectivity in fragmented populations: *Abax parallelepipedus* in a hedgerow network landscape. Compte rendu Académie des Sciences Paris, Sciences de la vie 321:55–61.

———. 1998b. Effects of landscape dynamics on the metapopulation of a ground beetle (Coleoptera, Carabidae) in a hedgerow network. Agriculture, Ecosystems and Environment 69:243–252.

Petts, G. E. 1984. *Impounded rivers: Perspectives for ecological management.* John Wiley and Sons, New York, USA.

Phillips, N. E. In press. Effects of nutrition-mediated larval condition on juvenile performance in a marine mussel. Ecology 83.

Phillips, N. E., and S. D. Gaines. In press. Spatial and temporal variability in size at settlement of intertidal mytilid mussels from around Pt. Conception, California. Invertebrate Reproduction and Development.

Pickett, S. T. A., and M. L. Cadenasso. 1995. Landscape ecology: Spatial heterogeneity in ecological systems. Science 269:331–334.

Pickett, S. T. A., and P. S. White, eds. 1985. *The ecology of natural disturbance and patch dynamics.* Academic Press, Orlando, FL, USA.

Pierce, R. J., T. E. Wissing, and B. A. Megrey. 1981. Aspects of the feeding ecology of gizzard shad in Acton Lake, Ohio. Transactions of the American Fisheries Society 110:391–395.

Pierson, E. D. 1998. Tall trees, deep holes, and scarred landscapes: Conservation biology of North American bats. Pp. 309–324 in T. H. Kunz and P. A. Racey, eds., *Bats: Phylogeny, morphology, echolocation, and conservation biology.* Smithsonian Institute Press, Washington, DC, USA.

Pierson, E. D., W. E. Rainey, and R. M. Miller. 1996. Night roost sampling: A window on the forest bat community in northern California. Pp. 151–163 in R. M. R. Barclay and R. M. Brigham, eds., *Bats and forests symposium, October 19–21, 1995, Victoria, British Columbia, Canada.* Working Paper 23/1996. British Columbia, Ministry of Forests Research Program, Victoria, BC, Canada.

Pile, A. J. 1996. The role of microbial food webs in benthic-pelagic coupling in freshwater and marine ecosystems. Ph.D. dissertation, College of William and Mary, Williamsburg, VA, USA.

Pile, A. J., M. R. Patterson, and J. D. Witman. 1996. In situ grazing on plankton < 10 μm by the boreal sponge *Mycale lingua*. Marine Ecology Progress Series 141:95–102.

Pimentel, D. 1976. Land degradation: Effects of food and energy resources. Science 194: 149–155

Pimentel, D., C. Harvey, P. Resosudarmo, K. Sinclair, D. Kurz, M. McNair, S. Crist, L. Shoritz, L. Fitton, R. Saffouri, and R. Blair. 1995. Environmental and economic costs of soil erosion and conservation benefits. Science 267:1117–1123.

Pimm, S. L. 1982. *Food webs.* Chapman and Hall, London, UK.

———. 1984. The complexity and stability of ecosystems. Nature 307:321–326.

———. 1992. *The balance of nature?* University of Chicago Press, Chicago, IL, USA.

Pimm, S. L., and J. H. Lawton. 1977. The numbers of trophic levels in ecological communities. Nature 268:329–331.

Pimm, S. L., and A. Redfearn. 1988. The variability of population densities. Nature 334:613–614.

Pineda, J. 1991. Predictable upwelling and the shoreward transport of planktonic larvae by internal tidal bores. Science 253:548–551.

———. 1995. An internal tidal bore regime at nearshore stations along western USA: Predictable upwelling within the lunar cycle. Continental Shelf Research 15:1023–1041.

———. 1999. Circulation and larval distribution in internal tidal bore warm fronts. Limnology and Oceanography 44:1400–1414.

Piorkowski, R. J. 1995. Ecological effects of spawning salmon on several south central Alaskan streams. Ph.D. dissertation, University of Alaska, Fairbanks, USA.

Platt, A. E. 1996. Infecting ourselves: How environmental and social disruptions trigger disease. Worldwatch paper 129. Worldwatch Institute, Washington, DC, USA.

Poiani, K. A., B. L. Bedford, and M. D. Merrill. 1996. A GIS-based index for relating landscape characteristics to potential nitrogen leaching to wetlands. Landscape Ecology 11:237–255.

Policansky, D. 1998. Science and decision making for water resources. Ecological Applications 8:610–618.

Polis, G. A. 1991. Complex trophic interactions in deserts: An empirical critique of food web theory. American Naturalist 138:123–155.

———. 1994. Food webs, trophic cascades and community structure. Australian Journal of Ecology 19:121–136.

———. 1998. Stability is woven by complex webs. Nature 395:744–745.

———. 1999. Why are parts of the world green? Multiple factors control productivity and the distribution of biomass. Oikos 86:3–15.

Polis, G. A., W. B. Anderson, and R. D. Holt. 1997. Toward an integration of landscape and food web ecology: The dynamics of spatially subsidized food webs. Annual Review of Ecology and Systematics 28:289–316.

Polis, G. A., and R. D. Holt. 1992. Intraguild predation: The dynamics of complex trophic interactions. Trends in Ecology and Evolution 7:151–154.

Polis, G. A., R. D. Holt, B. A. Menge, and K. O. Winemiller. 1996. Time, space, and life history: Influences on food webs. Pp. 435–460 in G. A. Polis and K. O. Winemiller, eds., *Food webs: Integration of patterns and dynamics.* Chapman and Hall, New York, USA.

Polis, G. A., and S. D. Hurd. 1994. High densities of spiders, scorpions and lizards on islands in the Gulf of California: Flow of energy from the marine to terrestrial food webs and the absence of predation. Bulletin of the Ecological Society of America 75:183.

———. 1995. Extraordinarily high spider densities on islands: Flow of energy from the marine to terrestrial food webs and the absence of predation. Proceedings of the National Academy of Sciences, USA 92:4382–4386.

———. 1996a. Allochthonous inputs across habitats, subsidized consumers, and apparent trophic cascades: Examples from the ocean-land interface. Pp. 435–460 in G. A. Polis and K. O. Winemiller, eds., *Food webs: Integration of patterns and dynamics*. Chapman and Hall, New York, USA.

———. 1996b. Linking marine and terrestrial food webs: Allochthonous input from the ocean supports high secondary productivity on small islands and coastal land communities. American Naturalist 147:396–423.

Polis, G. A., S. D. Hurd, C. T. Jackson, and F. Sánchez-Piñero. 1997. El Niño effects on the dynamics and control of an island ecosystem in the Gulf of California. Ecology 78:1884–1897.

———. 1998. Multifactor population limitation: Variable spatial and temporal control of spiders on Gulf of California Island. Ecology 79:490–502.

Polis, G. A., C. A. Myers, and R. Holt. 1989. The evolution and ecology of intraguild predation: Competitors that eat each other. Annual Review of Ecology and Systematics 20.297–330

Polis, G. A., A. L. W. Sears, G. R. Huxel, D. R. Strong, and J. Maron. 2000. When is a trophic cascade a trophic cascade? Trends in Ecology and Evolution 15:473–475.

Polis, G. A., and D. R. Strong. 1996. Food web complexity and community dynamics. American Naturalist 147:813–846.

Polis, G. A., and K. O. Winemiller, eds. 1996. *Food webs: Integration of patterns and dynamics*. Chapman and Hall, New York, USA.

Pollard, A. I., M. J. González, M. J. Vanni, and J. L. Headworth. 1998. Effects of turbidity and biotic factors on the rotifer community in an Ohio reservoir. Hydrobiologia 387:215–223.

Pond, C. M., C. A. Mattacks, I. Gilmour, M. A. Johnston, C. T. Pillinger, and P. Prestrud. 1995. Chemical and carbon isotopic composition of fatty acids in the adipose tissue as indicators of dietary history in wild arctic foxes (*Alopex lagopus*) on Svalbard. Journal of Zoology 236:611–623.

Pond, D. W., D. R. Dixon, M. V. Bell, A. E. Fallick, and J. R. Sargent. 1997. Occurrence of 16:2(n-4) and 18:2(n-4) fatty acids in the lipids of the hydrothermal vent shrimps *Rimicaris exoculata* and *Alvinocaris markensis:* Nutritional and trophic implications. Marine Ecology Progress Series 156:167–174.

Pond, D. W., M. Segonzac, M. V. Bell, D. R. Dixon, A. E. Fallick, and J. R. Sargent. 1997. Lipid and lipid carbon stable isotope composition of the hydrothermal vent shrimp *Mirocaris fortunata:* Evidence for nutritional dependence on photosynthetically fixed carbon. Marine Ecology Progress Series 157:221–231.

Pons, A., J. P. Suc, M. Reille, and N. Combourieu-Nebout. 1995. The history of dryness in regions with a mediterranean climate. Pp. 169–188 in J. Roy, J. Aronson, and F. di Castri, eds., *Time scales of biological responses to water constraints*. SPB Academic Publishing, Amsterdam, The Netherlands.

Poorter, E. P. R. 1981. *Cygnus columbianus bewickii* in the border lakes of the Ilsselmeerpolders. Pp. 49–57 in *Proceedings of the Second International Swan Symposium, Sapporo, Japan*. International Waterfowl Research Bureau, Slimbridge, UK.

Portt, C. B., E. K. Balon, and D. L. Noakes. 1986. Biomass and production of fishes in natural and channelized streams. Canadian Journal of Fisheries and Aquatic Sciences 43:1926–1934.

Post, D. M., J. P. Taylor, J. F. Kitchell, M. H. Olson, D. E. Schindler, and B. R. Herwig. 1998. The role of migratory waterfowl as nutrient vectors in a managed wetland. Conservation Biology 12:910–920.

Post, J. R., and D. J. McQueen. 1987. The impact of planktivorous fish on the structure of a plankton community. Freshwater Biology 17:79–89.

Potts, M. J. 1978. The pattern of deposition of air-borne salt of marine origin under a forest canopy. Plant and Soil 50:233–236.

Powell, E. O. 1958. Criteria for the growth of contaminants and mutants in continuous culture. Journal of General Microbiology 18:259–268.

Power, M. E. 1990a. Benthic turfs versus floating mats of algae in river food webs. Oikos 58:67–79.

———. 1990b. Effects of fish in river food webs. Science 250:411–415.

———. 1992a. Hydrologic and trophic controls of seasonal algal blooms in northern California rivers. Archiv für Hydrobiologie 125:385–410.

———. 1992b. Top-down and bottom-up forces in food webs: Do plants have primacy? Ecology 73:733–746.

———. 1995. Floods, food chains, and ecosystem processes in rivers. Pp. 52–60 in C. G. Jones and J. H. Lawton, eds., *Linking species and ecosystems*. Chapman and Hall, New York, USA.

———. 2001. Prey exchange between a stream and its forested watershed elevate predator densities in both habitats. Proceedings of the National Academy of Sciences, USA 98:14–15.

Power, M. E., W. E. Dietrich, and K. O. Sullivan. 1998. Experiment, observation, and inference in river and watershed investigations. Pp. 113–132 in W. J. Resetarits and J. Bernardo, eds., *Experimental ecology: Issues and perspectives*. Oxford University Press, Oxford, UK.

Power, M. E., S. J. Kupferberg, G. W. Minshall, M. C. Molles, and M. S. Parker. 1997. Sustaining western aquatic food webs. Pp. 45–61 in W. C. Minckley, ed., *Aquatic Ecosystems Symposium, Tempe, AZ. Report to the Western Water Policy Review Advisory Commission*. U.S. Bureau of Reclamation, Denver, CO, USA.

Power, M. E., and W. E. Rainey. 2000. Food webs and resource sheds: Towards spatially delimiting trophic interactions. Pp. 291–314 in M. J. Hutchings, E. A. John, and A. J. A. Stewart, eds., *Ecological consequences of habitat heterogeneity*. Blackwell Scientific, Oxford, UK.

Power, M. E., J. L. Sabo, M. S. Parker, W. E. Rainey, A. Smyth, P. Bernazzini, J. C. Finlay, G. Cabana, E. D. Pierson, and W. E. Dietrich. 1998. Consequences of trophic exchange from a river to its watershed. Page 341 in Proceedings VII International Congress of Ecology, 19–25 July 1990, Florence, Italy.

Power, M. E., A. Sun, G. Parker, W. E. Dietrich, and J. T. Wootton. 1995. Hydraulic food chain models. BioScience 45:159–167.

Power, M. E., D. Tilman, J. A. Estes, B. A. Menge, W. J. Bond, L. S. Mills, G. Daily, J. C. Castilla, J. Lubchenco, and R. T. Paine. 1996. Challenges in the quest for keystones. BioScience 46:609–620.

Prescott, C. E., B. R. Taylor, W. F. J. Parsons, D. M. Durall, and D. Parkinson. 1993. Nutrient release from decomposing litter in Rocky Mountain coniferous forests: Influence of nutrient availability. Canadian Journal of Forest Research 23:1576–1586.

Prop, J., and J. T. Vulink. 1992. Digestion by barnacle geese in the annual cycle: The interplay between retention time and food quality. Functional Ecology 6:180–189.

Puckridge, J. T., F. Sheldon, K. F. Walker, and A. J. Boulton. 1998. Flow variability and the ecology of large rivers. Marine and Freshwater Research 49:55–72.

Pullin, A. S., and J. E. Gilbert. 1989. The stinging nettle, *Urtica dioica*, increases trichome density after herbivore and mechanical damage. Oikos 54:275–280.

Rabalais, N. N., R. E. Turner, D. Justic, Q. Dortch, W. J. Wiseman, and B. D. Sen Gupta. 1996. Nutrient changes in the Mississippi River and system responses on the adjacent continental shelf. Estuaries 19:386–407.

Raffaelli, D. G. 1998. Impact of catchment land use on an estuarine benthic food web. Pp. 161–172 in J. Gray, W. Ambrose, and A. Szaniawska, eds., *Biogeochemical cycling and sediment ecology*. Kluwer, Dordrecht, The Netherlands.

———. 1999. Nutrient enrichment and trophic organization in an estuarine food web. Acta Oecologia 20:449–461.

Raffaelli, D. G., P. Balls, S. Way, I. J. Patterson, S. A. Hohmann, and N. Corp. 1999. Eutrophication-related trends in the ecology of the Ythan estuary, Aberdeenshire, Scotland. Aquatic Conservation: Marine and Freshwater Ecosystems 9:219–236.

Raffaelli, D. G., V. Falcy, and C. Galbraith. 1990. Eider predation and the dynamics of mussel bed communities. Pp. 157–169 in M. Barnes and R. N. Gibson, eds., *Proceedings of the 24th European Marine Biological Symposium*. Aberdeen University Press, Aberdeen, UK.

Raffaelli, D. G., S. Hull, and H. Milne. 1989. Long-term changes in nutrients, weed mats and shorebirds in an estuarine system. Cahiers Biologie Marine 30:259–270.

Raffaelli, D. G., J. Limia, S. Hull, and S. Pont. 1991. Interactions between invertebrates and macro-algal mats on estuarine mudflats. Journal of the Marine Biological Association of the United Kingdom 71:899–908.

Raffaelli, D. G., J. R. Raven, and L. Poole. 1998. Ecological impact of green macro-algal blooms. Annual Review of Marine Biology and Oceanography 36:97–125.

Ragnarsson, S. A. 1996. Successional patterns and biotic interactions in intertidal sediments. Ph.D. thesis, University of Aberdeen, Scotland, UK.

Rahel, F. J. 2000. Homogenization of fish faunas across the United States. Science 288: 854–856.

Rainey, R. C. 1963. Meteorology and the migration of desert locusts: Applications of synoptic meteorology in locust control. WMO Technical Note no. 54. Secretariat of the World Meteorological Organization, Geneva, Switzerland.

Rainey, W. E., E. D. Pierson, M. Colberg, and J. H. Barclay. 1992. Bats in hollow redwoods: Seasonal use and role in nutrient transfer into old growth communities. Bat Research News 33:71.

Ramakrishnan, P. S., and P. M. Vitousek. 1989. Ecosystem-level processes and the consequences of biological invasions. Pp. 281–300 in J. A. Drake, H. A. Mooney, F. di Castri, R. H. Groves, F. J. Kruger, M. Rejmánek and M. Williamson, eds., *Biological invasions, a global perspective*. John Wiley and Sons, New York, USA.

Ramsay, M. A., and K. A. Hobson. 1991. Polar bears make little use of terrestrial food webs: Evidence from stable-carbon isotope analysis. Oecologia 86:598–600.

Rand, P. S., C. A. S. Hall, W. H. McDowell, N. H. Ringler, and J. G. Kennen. 1992. Factors limiting primary productivity in Lake Ontario tributaries receiving salmon migrations. Canadian Journal of Fisheries and Aquatic Sciences 49:2377–2385.

Ranney, J. W., M. C. Bruner, and J. B. Levenson. 1981. The importance of edge in the structure and dynamics of forest islands. Pp. 67–96 in R. L. Burgess and D. M. Sharpe, eds., *Forest island dynamics in man-dominated landscapes*. Springer-Verlag, New York, USA.

Rasmussen, J. B. 1988. Littoral zoobenthic biomass in lakes, and its relationship to physical, chemical, and trophic factors. Canadian Journal of Fisheries and Aquatic Sciences 45:1436–1447.

Rasmussen, J. B., D. J. Rowan, D. R. S. Lean, and J. H. Carey. 1990. Food chain structure affects PCB levels in pelagic fish from Ontario lakes. Canadian Journal of Fisheries and Aquatic Sciences 47:2030–2038.

Rau, G. H. 1976. Dispersal of terrestrial plant litter into a subalpine lake. Oikos 27:153–160.

———. 1981. Low ^{15}N/^{14}N of hydrothermal vent animals: Ecological implications. Nature 289:484–485.

Rau, G. H., D. G. Ainley, J. L. Bengtsson, J. J. Torres, and T. L. Hopkins. 1992. ^{15}N/^{14}N and ^{13}C/^{12}C in Weddell Sea birds, seals, and fish: Implications for diet and trophic structure. Marine Ecology Progress Series 84:1–8.

Rau, G. H., R. E. Sweeney, I. R. Kaplan, A. J. Mearns, and D. R. Young. 1981. Differences in animal ^{13}C, ^{15}N, and D abundance between a polluted and an unpolluted coastal site: Likely indicators of sewage uptake by a marine food web. Estuarine Coastal Shelf Science 13:701–707.

Raymond, P. A., N. F. Caraco, and J. J. Cole. 1997. CO_2 concentration and atmospheric flux in the Hudson River. Estuaries 20:381–390.

Readshaw, J. L. 1973. The numerical response of predators to prey density. Journal of Applied Ecology 10:342–351.

Redak, R. A., J. T. Trumble, and T. D. Paine. 1997. Interactions between the *Encelia* leaf beetle and its host plant, *Encelia farinosa:* The influence of acidic fog on insect growth and plant chemistry. Environmental Pollution 95:241–248.

Redfield, A. C., B. H. Ketchum, and F. A. Richards. 1963. The influence of organisms on the composition of seawater. Pp. 26–77 in M. N. Hill, ed., *The sea*, vol. 2. Interscience Publishers, New York, USA.

Regier, H. 1997. Old traditions that lead to abuses of salmon and their ecosystems. Pp. 17–28 in D. J. Stouder, P. A. Bisson, and R. J. Naiman, eds., *Pacific salmon and their ecosystems: Status and future options*. Chapman and Hall, New York, USA.

Reimchen, T. E. 1994. Further studies of predator and scavenger use of chum salmon in stream and estuarine habitats at Bag Harbour, Gwaii Haanas. Unpublished report. Canadian Parks Service, Queen Charlotte City, B.C., Canada.

———. 1998. Nocturnal foraging behaviour of black bear, *Ursus americanus*, on Moreby Island, Canada. Canadian Field-Naturalist 112:446–450.

———. 2000. Some ecological and evolutionary aspects of bear-salmon interactions in coastal British Columbia. Canadian Journal of Zoology 78:448–457.

Reimchen, T. E., D. Mathewson, M. D. Hocking, and J. Moran. In press. Isotopic evidence for enrichment of salmon-derived nutrients in vegetation, soil, and insects in riparian zones in coastal British Columbia. American Fisheries Society Symposium.

Reisner, M. 1986. *Cadillac desert*. Viking-Penguin, New York, USA.

———. 1990. *Overtapped oasis*. Island Press, Washington, DC, USA.

Rennenberg, H., and A. Gessler. 1999. Consequences of N deposition to forest ecosystems: Recent results and future research needs. Water, Air and Soil Pollution 116:47–64.

Renshaw, E. 1991. *Modelling biological populations in space and time*. Cambridge University Press, Cambridge, UK.

Renwick, W. H. 1996. Continent-scale reservoir sedimentation patterns in the United States. Pp. 513–522 in D. E. Walling and B. W. Webb, eds., *Erosion and sediment yield: Global and regional perspectives*. IAHS Publication 236. International Association of Hydrological Sciences, Wallingford, UK.

Reynolds, D. R., and J. R. Riley. 1979. Radar observations of concentrations of insects above a river in Mali, West Africa. Ecological Entomology 4:161–174.

Ribeiro, M. C. L. de B. 1983. As migrações dos jaraquis (Pisces, Prochilodontidae) no rio Negro, Amazonas, Brasil. M.S. thesis, Univesidade do Amazonas, Manaus, Brazil.

Ribeiro, M. C. L. de B., and M. Petrere, Jr. 1990. Fisheries ecology and management of the jaraqui (*Semaprochilodus taeniurus*, *S. insignis*) in central Amazonia. Regulated Rivers: Research and Management 5:195–215.

Richards, F. A. 1981. *Coastal upwelling*. American Geophysical Union, Washington, DC, USA.

Richards, O. W. 1948. Insects and fungi associated with *Urtica*: Insects. Journal of Ecology 36:340–343.

Richey, J. E., J. I. Hedges, A. H. Devol, P. D. Quay, L. A. Martinelli, and B. R. Forsberg. 1990. Biogeochemistry of carbon in the Amazon River. Limnology and Oceanography 35:352–371.

Richey, J. E., M. A. Perkins, and C. R. Goldman. 1975. Effects of kokanee salmon (*Oncorhynchus nerka*) decomposition on the ecology of a subalpine stream. Journal of the Fisheries Research Board of Canada 32:817–820.

Ricker, W. E. 1987. Effects of the fishery and of obstacles to migration on the abundance of Fraser River sockeye salmon (*Oncorhynchus nerka*). Canadian Technical Report in Fisheries and Aquatic Sciences, no. 1522. Department of Fisheries and Oceans, Nanaimo, B.C., Canada.

Ricklefs, R. E., and D. Schluter. 1993. Species diversity: Regional and historical influences. Pp. 350–362 in R. Ricklefs and D. Schluter, eds., *Species diversity in ecological communities*. University of Chicago Press, Chicago, IL, USA.

Riechert, S. E., and L. Bishop. 1990. Prey control by an assemblage of generalist predators: Spiders in garden test systems. Ecology 71:1441–1450.

Riechert, S. E., and T. Lockley. 1984. Spiders as biological control agents. Annual Review of Entomology 29:299–320.

Rieux, R., H. Simon, and H. Defrance. 1999. Role of hedgerows and ground cover management on arthropod populations in pear orchards. Agriculture, Ecosystems & Environment 73:119–127.

Riley, R. H., C. Mitchell, and J. Thorp. 1997. UV and stream ecology. Pp. 77–78 in *UV radiation and its effects: An update*. The Royal Society of New Zealand, Christchurch, New Zealand.

Rind, D., R. Goldberg, J. Hansen, C. Rosenzweig, and R. Ruedy. 1990. Potential evapotranspiration and the likelihood of future drought. Journal of Geophysical Research 95:9983–10004.

Ripa, J., and M. Heino. 1999. Linear analysis solves two puzzles in population dynamics: The route to extinction, and extinction in coloured environments. Ecology Letters 2:219–222.

Ripa, J., P. Lundberg, and V. Kaitala. 1998. A general theory of environmental noise in ecological food webs. American Naturalist 151:256–263.

Risley, L. S., and D. A. Crossley. 1993. Contribution of herbivore-caused greenfall to litterfall nitrogen flux in several southern Appalachian forested watersheds. American Midland Naturalist 129:67–74.

Risser, P. G. 1987. Landscape ecology: State of the art. Pp. 3–14 in M. G. Turner, ed., *Landscape heterogeneity and disturbance*. Springer-Verlag, New York, USA.

Roberts, D. A., and C. S. Boothroyd. 1972. *Fundamentals of plant pathology*. W. H. Freeman, San Francisco, CA, USA.

Roberts, N. 1994. *The changing global environment*. Blackwell Publishers, Oxford, UK.

Robertson, D. G., and R. D. Slack. 1995. Landscape change and its effects on the wintering range of a lesser snow goose *Chen caerulescens caerulescens* population: A review. Biological Conservation 71:179–185.

Robinson, D. 2001. (δ^{15}N as an integrator of the nitrogen cycle. Trends in Ecology and Evolution 16:153–162.

Robinson, D. G., and W. E. Barraclough. 1978. Population estimates of sockeye salmon (*Oncorhynchus nerka*) in a fertilized oligotrophic lake. Journal of the Fisheries Research Board of Canada 35:851–860.

Robinson, S. M. 1962. Computing wind profile parameters. Journal of Atmospheric Science 19:189–190.

Robles, C. D. 1987. Predator foraging characteristics and prey population structure on a sheltered shore. Ecology 68:1502–1514.

———. 1997. Changing recruitment in constant species assemblages: Implications for predation theory in intertidal communities. Ecology 78:1400–1414.

Rockwell, R. F., C. R. Witte, R. L. Jefferies, and P. J. Weatherhead. 2003. Response of nesting Savannah sparrows to 25 years of habitat change in a Snow Goose colony. Écoscience 10.

Rodriguez, S. R. 1999. Trophic subsidies on marine environments: The importance of brown drift algae as an exogenous source of food for the urchin *Tetrapygus niger* (Echinodermata: Echinoidea) in the rocky intertidal zone of central Chile. Ph.D. dissertation, Pontificia Universidad Catolica de Chile, Santiago.

Rodriguez, S. R., and J. M. Fariña. 2001. Effect of drift kelp and substratum irregularities on the spatial distribution pattern of the sea urchin *Tetrapygus niger* (Molina): A geostatistical approach. Journal of the Marine Biological Association of the United Kingdom 81:179–180.

Roff, J. C., W. G. Sprules, J. C. H. Carter, and M. J. Dadswell. 1981. The structure of crustacean zooplankton communities in glaciated eastern North America. Canadian Journal of Fisheries and Aquatic Sciences 38:1428–1437.

Rogers, C. M., and J. N. M. Smith. 1993. Life-history theory in the nonbreeding period: Trade-offs in avian fat reserves? Ecology 74:419–426.

Rogers, L. E., W. T. Hinds, and R. L. Buschbom. 1976. A general weight vs. length relationship for insects. Annals of the Entomological Society of America 69:387–389.

Roos, J. F. 1991. Restoring Fraser River salmon. Pacific Salmon Commission, Vancouver, BC, Canada.

Rose, M., and G. A. Polis. 1998. The distribution and abundance of coyotes: The effects of allochthonous food subsidies from the sea. Ecology 79:998–1007.

Roseman, E. F., E. L. Mills, J. L. Forney, and L. G. Rudstam. 1996. Evaluation of competition between age-0 yellow perch (*Perca flavescens*) and gizzard shad (*Dorosoma cepedianum*) in Oneida Lake, New York. Canadian Journal of Fisheries and Aquatic Sciences 53:865–874.

Rosemond, A. D., P. J. Mulholland, and J. W. Elwood. 1993. Top-down and bottom-up control of stream periphyton: Effects of nutrients and herbivores. Ecology 74:1264–1280.

Rosenberg, N. J. 1974. *Microclimate: The biological environment*. John Wiley and Sons, New York, USA.

Rosenberg, N. J., B. L. Blad, and S. B. Verma. 1983. *Microclimate: The biological environment*. 2d ed. John Wiley and Sons, New York, USA.

Rosenzweig, M. L. 1968. Net primary productivity of terrestrial communities: Prediction from climatological data. American Naturalist 102:67–74.

———. 1971. Paradox of enrichment: Destabilization of exploitation ecosystems in ecological time. Science 171:385–387.

———. 1995. *Species diversity in space and time*. Cambridge University Press, Cambridge, UK.

Roughgarden, J. 1975. A simple model for population dynamics in stochastic environments. American Naturalist 109:713–736.

Roughgarden, J., S. D. Gaines, and H. Possingham. 1988. Recruitment dynamics in complex life cycles. Science 241:1460–1466.

Roughgarden, J., J. T. Pennington, D. Stoner, S. Alexander, and K. Miller. 1991. Collisions of upwelling fronts with the intertidal zone: The cause of recruitment pulses in barnacle populations of central California. Acta Oecologia 12:35–51.

Rounick, J. S., and M. J. Winterbourne. 1986. Stable carbon isotopes and carbon flow in ecosystems. Bioscience 36:171–177.

Rowan, D. J., and J. B. Rasmussen. 1992. Why don't Great Lakes fish reflect environmental concentrations of organic contaminants? An analysis of between-lake variability in the ecological partitioning of PCBs and DDT. Journal of Great Lakes Research 18:724–741.

———. 1994. The bioaccumulation of radiocesium by fish: The influence of physico-chemical factors and trophic structure. Canadian Journal of Fisheries and Aquatic Sciences 51:2388–2410.

———. 1995. The elimination of radiocesium from fish. Journal of Applied Ecology 32: 739–744.

———. 1996. Measuring the bioenergetic cost of fish activity in situ using a globally dispersed radiotracer [137]Cs. Canadian Journal of Fisheries and Aquatic Sciences 53: 734–745.

Rowe-Rowe, D. T., and J. E. Crafford. 1992. Density, body size, and reproduction of feral house mice on Gough Island. South African Journal of Zoology 27:1–5.

Rowe-Rowe, D. T., B. Green, and J. E. Crafford. 1989. Estimated impact of feral house mice on Sub-Antarctic invertebrates at Marion Island. Polar Biology 9:457–460.

Roy, K., J. W. Valentine, D. Jablonski, and S. M. Kidwell. 1996. Scales of climatic variability and time averaging in Pleistocene biotas: Implications for ecology and evolution. Trends in Ecology and Evolution 11:458–463.

Rozema, J., J. van de Staaij, M. M. Caldwell, and L. O. Björn. 1997. UV-B as an environmental factor in plant life: Stress and regulation. Trends in Ecology and Evolution 12:22–28.

Rudemann, R., and W. J. Schoonmaker. 1938. Beaver dams as geologic agents. Science 88:523–525.

Ruel, J. J., and M. P. Ayres. 1999. Jensen's inequality predicts effects of environmental variation. Trends in Ecology and Evolution 14:361–366.

Ruesink, J. L. 1998. Variation in per capita interaction strength: Thresholds due to nonlinear dynamics and nonequilibrium conditions. Proceedings of the National Academy of Sciences, USA 95:6843–6847.

Runkel, R. L., and S. C. Chapra. 1993. An efficient numerical solution of the transient storage equations for solute transport in small streams. Water Resources Research 29(1):211–215.

Russell, E. W. 1966. *Soil conditions and plant growth*. John Wiley and Sons, New York, USA.

Ryan, P. G., C. L. Moloney, and B. P. Watkins. 1989. Concern about the adverse effect of introduced mice on island tree *Phylica arborea* regeneration. South African Journal of Science 85:626–627.

Ryan, P. G., and B. P. Watkins. 1989. The influence of physical factors and ornithogenic products on plant and arthropod abundance at an inland nunatak group in Antarctica. Polar Biology 10:151–160.

Ryszkowski, L. 1992. Energy and material flows across boundaries in agricultural landscapes. Pp. 270–284 in A. J. Hansen and F. di Castri, eds., *Landscape boundaries: Consequences for biotic diversity and ecological flows*. Ecological studies, vol. 92. Springer-Verlag, New York, USA.

Ryszkowski, L., A. Bartoszewicz, and A. Kedzoria. 1999. Management of matter fluxes by biogeochemical barriers at the agricultural landscape level. Landscape Ecology 14:479–492.

Ryther, J. H. 1962. Geographic variations in productivity. Pp. 347–380 in M. Hill, ed., *The sea*, vol. 2. Interscience Publishers, New York, USA.

Sabo, J. L., and M. E. Power. 2002a. River-watershed exchange: Effects of riverine subsidies on riparian lizards and their terrestrial prey. Ecology 83:1860–1869.

———. 2002b. Numerical response of riparian lizards to aquatic insects and the short-term consequences for alternate terrestrial prey. Ecology 83:3023–3236.

Sackett, W. M. 1989. Stable carbon isotope studies on organic matter in the marine environment. Pp. 139–170 in P. Fritz and J. Ch. Fontes, eds., *Handbook of environmental isotope geochemistry*, vol. 3, *The marine environment*. Elsevier, Amsterdam, The Netherlands.

Sæther, B.-E. 1997. Environmental stochasticity and population dynamics of large herbivores: A search for mechanisms. Trends in Evolution and Ecology 12:143–149.

Sala, O. E., W. J. Parton, L. A. Joyce, and W. K. Lauenroth. 1988. Primary production of the central grassland region of the United States. Ecology 69:40–45.

Salvatore, S. R., N. D. Mundahl, and T. E. Wissing. 1987. Effect of water temperature on food evacuation rate and feeding activity of age-0 gizzard shad. Transactions of the American Fisheries Society 116:67–70.

Sánchez-Piñero, F., and G. A. Polis. 2000. Donor controlled dynamics on islands: Direct and indirect effects of seabirds on tenebrionids. Ecology 81:3117–3132.

Sanford, E., D. Bermudez, M. D. Bertness, and S. D. Gaines. 1994. Flow, food supply and acorn barnacle population dynamics. Marine Ecology Progress Series 104:49–62.

Sanford, E., and B. A. Menge. 2001. Spatial and temporal variation in barnacle growth in a coastal upwelling system. Marine Ecology Progress Series 209:143–157.

Sanzone, D. M., J. L. Meyer, J. L. Tank, P. J. Mulholland, N. B. Grimm, S. V. Gregory, W. H. McDowell, W. B. Bowden, and W. K. Dodds. 2000. Nitrogen transfer from stream to riparian foodwebs: Results from eight ^{15}N tracer experiments. Abstract, Ecological Society of America Annual Meeting.

Sarmiento, J. L., T. Herbert, and J. R. Toggweiler. 1988. Mediterranean nutrient balance and episodes of anoxia. Global Biogeochemical Cycles 2:427–444.

Sauer, C. 1967. *Land and life: A selections from the writings of Carl Ortwin Sauer*. University of California Press, Berkeley, USA.

Savidge, J. A. 1987. Extinction of an island forest avifauna by an introduced snake. Ecology 68:660–668.

Saxena, K. G., and P. S. Ramakrishnan. 1986. Nitrification during slash and burn agriculture (jhum) in north-eastern India. Oecologia 7:307–319.

Schaller, G. B. 1972. *The Serengeti lion: A study of predator-prey relations*. University of Chicago Press, Chicago, IL, USA.

Schaus, M. H. 1998. Effects of gizzard shad on nutrient cycles and phytoplankton in a reservoir ecosystem: Roles of diet, biomass and population size-structure. Ph.D. dissertation, Miami University, Oxford, OH, USA.

Schaus, M. H., and M. J. Vanni. 2000. Effects of omnivorous gizzard shad on phytoplankton and nutrient dynamics: Role of sediment feeding and fish size. Ecology 81:1701–1719.

Schaus, M. H., M. J. Vanni, and T. E. Wissing. 2002. Biomass-dependent diet shifts in omnivorous gizzard shad: Implications for growth, food web, and ecosystem effects. Transactions of the American Fisheries Society 131:40–54.

Schaus, M. H., M. J. Vanni, T. E. Wissing, M. T. Bremigan, J. A. Garvey, and R. A. Stein. 1997. Nitrogen and phosphorus excretion by detritivorous gizzard shad in a reservoir ecosystem. Limnology and Oceanography 42:1386–1397.

Schell, D. M. 1983. Carbon-13 and carbon-14 abundances in Alaskan aquatic organisms: Delayed production from peat in Arctic food webs. Science 219:1068–1071.

Schimel, D. S. 1995. Terrestrial ecosystems and the carbon cycle. Global Change Biology 1:77–91.

Schimel, D. S., B. H. Braswell, and W. J. Parton. 1997. Equilibration of the terrestrial water, nitrogen, and carbon cycles. Proceedings of the National Academy of Sciences, USA 94:8280–8283.

Schindler, D. E., S. R. Carpenter, J. J. Cole, J. F. Kitchell, and M. L. Pace. 1997. Influence of food web structure on carbon exchange between lakes and the atmosphere. Science 277:248–251.

Schindler, D. E., S. R. Carpenter, K. L. Cottingham, X. He, J. R. Hodgson, J. F. Kitchell, and P. A. Soranno. 1996. Food web structure and littoral zone coupling to pelagic trophic cascades. Pp. 96–105 in G. A. Polis and K. O. Winemiller, eds., *Food webs: Integration of patterns and dynamics*. Chapman and Hall, New York, USA.

Schindler, D. E., and L. A. Eby. 1997. Stoichiometry of fishes and their prey: Implications for nutrient recycling. Ecology 78:1816–1831.

Schindler, D. E., J. F. Kitchell, X. He, S. R. Carpenter, J. R. Hodgson, and K. L. Cottingham. 1993. Food web structure and phosphorus cycling in lakes. Transactions of the American Fisheries Society 122:756–772.

Schindler, D. W., and S. E. Bayley. 1993. The biosphere as an increasing sink of atmospheric carbon: Estimates from increased nitrogen deposition. Global Biogeochemical Cycling 7:717–733.

Schindler, D. W., K. G. Beaty, E. J. Fee, D. R. Cruikshank, E. R. DeBruyn, D. L Findlay, G. A. Linsey, J. A. Shearer, M. P. Stainton, and M. A. Turner. 1990. Effects of climate warming on lakes of the central boreal forest. Science 250:967–970.

Schindler, D. W., P. J. Curtis, B. R. Parker, and M. P. Stainton. 1996. Consequences of climate warming and lake acidification for UV-B penetration in North American lakes. Nature 379:705–708.

Schlacher, T. A., and T. H. Wooldridge. 1996. Origin and trophic importance of detritus— evidence from stable isotopes in the benthos of a small, temperate estuary. Oecologia 106:382–388.

Schlesinger, W. H. 1997. *Biogeochemistry: An analysis of global change*. 2d ed. Academic Press, San Diego, CA, USA.

Schlesinger, W. H., and J. M. Melack. 1981. Transport of organic carbon in the world's rivers. Tellus 33:172–187.

Schlesinger, W. H., and W. T. Peterjohn. 1991. Processes controlling ammonia volatilization from Chihuahuan desert soils. Soil Biology and Biogeochemistry 23:637–642.

Schmidt, G. W. 1973. Primary production of phytoplankton in the three types of Amazonian waters. III. Primary production of phytoplankton in a tropical flood-plain lake of Central Amazonia, Lago do Castanho, Amazon, Brazil. Amazoniana 4:379–404.

Schneider, S. H., and T. L. Root. 1996. Ecological implications of climate change will include surprises. Biodiversity and Conservation 5:1109–1119.

Schoener, T. W. 1971. Theory of feeding strategies. Annual Review of Ecology and Systematics 11:369–404.

———. 1973. Population growth regulated by intraspecific competition for energy or time: Some simple representations. Theoretical Population Biology 4:56–84.

Schoener, T. W., and D. A. Spiller. 1995. Effect of predators and area on invasion: An experiment with island spiders. Science 267:1811–1813.

Schoener, T. W., and C. A. Toft. 1983. Spider populations: Extraordinary high densities on islands without top predators. Science 219:1353–1355.

Schoeninger, M. J., M. J. DeNiro, and H. Tauber. 1983. Stable nitrogen isotope ratios of bone collagen reflect marine and terrestrial components of prehistoric human diet. Science 220:1381–1383.

Scholte, K. 2000. Effect of potato used as a trap crop on potato cyst nematodes and other soil pathogens and on the growth of a subsequent main potato crop. Annals of Applied Biology 136:229–238.

Schuldt, J. A., and A. E. Hershey. 1995. Effect of salmon carcass decomposition on Lake Superior tributary streams. Journal of the North American Benthological Society 14:259–268.

Schulz-Bull, D. E., G. Petrick, N. Kamann and J. C. Duinker. 1995. Distribution of individual chlorobiphenyls (PCB) in solution and suspension in the Baltic Sea. Marine Chemistry 48:245–270.

Scientific Assessment and Strategy Team (S.A.S.T.). 1994. Science for floodplain management into the 21st Century. U.S. Government Printing Office, Washington, DC, USA.

Scott, W. B., and E. J. Crossman. 1973. Freshwater fishes of Canada. Bulletin no. 184. Fisheries Research Board of Canada, Ottawa, Ontario, Canada.

Scrudato, R. J., and W. H. McDowell. 1989. Upstream transport of Mirex by migrating salmonids. Canadian Journal of Fisheries and Aquatic Sciences 46:1484–1488.

Sealy, S. G. 1974. Ecological segregation of Swainson's and Hermit Thrushes on Langara Island, British Columbia. Condor 76:350–351.

Sebens, K. P., and M. A. R. Koehl. 1984. Predation on zooplankton by the benthic anthozoans *Alcyonium siderium and Metridium senile* in the New England subtidal. Marine Biology 81:255–271.

Seed, R. 1976. Ecology. Pp. 13–65 in B. L. Bayne (ed.)., *Marine mussels: Their ecology and physiology.* Cambridge University Press, Cambridge, UK.

Seehausen, O., J. J. M. van Alphen, and F. White. 1997. Cichlid fish diversity threatened by eutrophication that curbs sexual selection. Science 277:1808–1811.

Seginer, I. 1975. Flow around a windbreak in oblique wind. Boundary Layer Meteorology 9:133–141.

Seidl, M. A., and W. E. Dietrich. 1992. The problem of channel erosion into bedrock. Catena Supplement 23:101–124.

Semlitsch, R. D., and J. R. Bodie. 1998. Are small, isolated wetlands expendable? Conservation Biology 12:1129–1133.

Senft, R. L., M. B. Coughenour, D. W. Bailey, L. R. Rittenhouse, O. E. Sala, and D. M. Swift. 1987. Large herbivore foraging and ecological hierarchies. BioScience 37:789–799.

Shanks, A. L. 1983. Surface slicks associated with tidally forced internal waves may transport pelagic larvae of invertebrates and fishes onshore. Marine Ecology Progress Series 13:311–315.

———. 1986. Tidal periodicity in the daily settlement of intertidal barnacle larvae and an hypothesized mechanism for the cross shelf transport of cyprids. Biological Bulletin 170:429–440.

———. 1995. Mechanisms of cross shelf dispersal of larval invertebrates and fish. Pp. 323–367 in L. McEdward, ed., *Ecology of marine invertebrate larvae.* CRC Press, Boca Raton, FL, USA.

Shea, S. R., and R. Dell. 1981. Structure of the surface root system of *Eucalyptus marginata* Sm. and its infection by *Phytophthora cinnamomi* Rands. Australian Journal of Botany 29:49–58.

Shearer, G. 1997. Interactions between macroalgae and the mudsnail *Hydrobia ulvae*. M.S. thesis, University of Aberdeen, UK.

Shearer, K. D., T. Åsgård, G. Andorsdottir, and G. A. Aas. 1994. Whole body elemental and proximate composition of Atlantic salmon (*Salmo salar*) during the life cycle. Journal of Fish Biology 44:785–797.

Shellenbarger, G. G. 1994. Defenses of Gulf of Maine sponges: Implications for predation and population structure. M.S. thesis, Northeastern University, Boston, MA, USA.

Shepherd, W. C., and E. M. Mills. 1996. Diel feeding, daily food intake, and *Daphnia* consumption by age-0 gizzard shad in Oneida Lake, New York. Transactions of the American Fisheries Society 125:411–421.

Sherr, B. F., E. B. Sherr, and C. S. Hopkinson. 1988. Trophic interactions within pelagic microbial communities: Indications of feedback regulation of carbon flow. Hydrobiologia 159:19–26.

Sherwood, G., J. C. Brodeur, A. Hontela, and J. B. Rasmussen. 2000. The bioenergetic cost of pollution by heavy metals in yellow perch (*Perca flavescens*) as measured with a radionuclide mass-balance. Canadian Journal of Fisheries and Aquatic Sciences 57: 441–450.

Shkedy, Y., and J. Roughgarden. 1997. Barnacle recruitment and population dynamics predicted from coastal upwelling. Oikos 80:487–498.

Shuman, J. R. 1995. Environmental considerations for assessing dam removal alternatives for river restoration. Regulated Rivers: Research and Management 11:249–261.

Shuman, R. F. 1950. Bear depredations on red salmon spawning populations in the Karluk River system, 1947. Journal of Wildlife Management 14:1–9.

Shuter, B. J., M. L. Jones, R. M. Korver, and N. P. Lester. 1998. A general life history based model for regional management of fish stocks: The inland lake trout (*Salvelinus namaycush*) fisheries of Ontario. Canadian Journal of Fisheries and Aquatic Sciences 55:2161–2177.

Sidle, R. C. 1986. Seasonal patterns of allochthonous debris in three riparian zones of a coastal Alaska drainage. Pp. 283–304 in D. L. Correll, ed., *Watershed research perspectives*. Smithsonian Institution Press, Washington, DC, USA.

Siegfried, W. R. 1981. The role of birds in ecological processes affecting the functioning of the terrestrial ecosystem at sub-antarctic Marion Island. Comité National Français de Recherches Antarctiques 51:493–499.

Simberloff, D. 1986. Are we on the verge of a mass extinction in tropical rainforests? Pp. 165–180 in D. K. Elliott, ed., *Dynamics of extinction*. John Wiley and Sons, New York, USA.

Simberloff, D., and J. Cox. 1987. Consequences and costs of conservation corridors. Conservation Biology 1:63–71.

Simberloff, D., J. A. Farr, J. Cox, and D. W. Mehlman. 1992. Movement corridors: Conservation bargains or poor investments? Conservation Biology 6:493–504.

Sims, P. L., and J. S. Singh. 1978. The structure and function of ten western North American grasslands. III. Net primary production, turnover and efficiencies of energy capture and water use. Journal of Ecology 66:573–597.

Sinclair, A. R. E., and P. Arcese. 1995. Population consequences of predation-sensitive foraging: The Serengeti wildebeest. Ecology 76:882–891.

Sinclair, A. R. E., and J. M. Fryxell. 1985. The Sahel of Africa: Ecology of a disaster. Canadian Journal of Zoology 63:987–994.

Sinclair, A. R. E., and M. Norton-Griffiths. 1979. Serengeti: Dynamics of an ecosystem. University of Chicago Press, Chicago, IL, USA.

Sioli, H. 1975. Tropical rivers as an expression of their terrestrial environment. Pp. 275–288 in F. B. Golly and E. Medina, eds., *Tropical ecological systems*. Springer-Verlag, Berlin, Germany.

Skinner, W. R., R. L. Jefferies, T. J. Carleton, R. F. Rockwell, and K. F. Abraham. 1998. Prediction of reproductive success and failure in lesser snow geese based on early season climatic variables. Global Change Biology 4:3–16.

Sklar, F. H., and J. A. Browder. 1998. Coastal environmental impacts brought about by alterations to freshwater flow in the Gulf of Mexico. Environmental Management 22: 547–562.

Slaney, T. L., K. D. Hyatt, T. G. Northcote, and R. J. Fielden. 1996. Status of anadromous salmon and trout in British Columbia and Yukon. Fisheries 21:2035.

Small, L. F., and D. W. Menzies. 1981. Patterns of primary productivity and biomass in a coastal upwelling region. Deep Sea Research 28A:123–149.

Smetacek, V. 1984. The supply of food to the benthos. Pp. 517–548 in M. J. Fasham, ed., *Flows of energy and materials in marine ecosystems: Theory and practice.* Plenum Press, New York, USA.

Smil, V. 1997. Global population and the nitrogen cycle. Scientific American 277:58–63.

Smith, C. R. 1985. Food for the deepsea: Utilization, dispersal and flux of nekton falls in the Santa Catalina Basin Floor. Deep Sea Research 32:417–442.

Smith, C. W. 1985. Impact of alien plants on Hawaii's native biota. Pp. 180–250 in C. P. Stone and J. M. Scott, eds., *Hawaii's terrestrial ecosystems: Preservation and management.* Cooperative National Park Resources Study Unit, University of Hawaii, Honolulu, USA.

Smith, H. L., and P. Waltman. 1995. *The theory of the chemostat: Dynamics of microbial competition.* Cambridge University Press, Cambridge, UK.

Smith, J. S., and C. R. Johnson. 1995. Nutrient inputs from seabirds and humans on a populated coral cay. Marine Ecology Progress Series 124:189–200.

Smith, R. 1981. A comparison of the structure and variability of the flow fields in three coastal upwelling regions: Oregon, northwest Africa, and Peru. Pp. 107–118 in F. A. Richards, ed., *Coastal upwelling.* American Geophysical Union, Washington, DC, USA.

Smith, S. V., W. J. Kimmer, E. A. Laws, R. E. Brock, and T. W. Walsh. 1981. Kanohoe Bay sewage diversion experiment: Perspectives on ecosystem response to nutritional perturbation. Pacific Science 35:279–397.

Smith, V. H. 1979. Nutrient dependence of primary productivity in lakes. Limnology and Oceanography 24:1051–1064.

———. 1983. Low nitrogen to phosphorus ratios favor dominance by blue-green algae in lake phytoplankton. Science 221:669–671.

———. 1998. Cultural eutrophication of inland, estuarine, and coastal waters. Pp. 7–49 in M. L. Pace and P. M. Groffman, eds., *Successes, limitations, and frontiers in ecosystem science.* Springer-Verlag, New York, USA.

Smith, V. R. 1979. The influence of seabird manuring on the phosphorus status of Marion Island (sub-antarctic) soils. Oecologia 41:123–126.

Smoot, J. C. 1999. A field study of sedimentary microbiota as food for detritivorous gizzard shad, *Dorosoma cepedianum,* in Acton Lake: A biomarker approach. Ph.D. dissertation, Miami University, Oxford, OH, USA.

Sommaggio, D., M. G. Paoletti, and S. Ragusa. 1995. The effects of microhabitat conditions, nutrients and predators on the abundance of herbivores on stinging nettles (*Urtica dioica* L.). Acta Oecologica 16:671–686.

Sommer, U. 1983. Nutrient competition between phytoplankton species in multispecies chemostat experiments. Archiv für Hydrobiologie 96:399–416.

———. 1984. The paradox of the plankton: Fluctuations of phosphorus availability maintain diversity of phytoplankton in flow-through cultures. Limnology and Oceanography 29:633–636.

———. 1985. Comparison between steady-state and non-steady state competition: Experiments with natural phytoplankton. Limnology and Oceanography 30:335–346.

———. 1988. Phytoplankton succession in microcosm experiments under simultaneous grazing pressure and resource limitation. Limnology and Oceanography 33:1037–1054.

Sorokin, Y. I., and E. B. Paveljeva. 1978. On structure and functioning of ecosystem in a salmon lake. Hydrobiologia 57:25–48.

Sousa, W. P. 1984. Intertidal mosaics: Patch size, propagule availability, and spatially variable patterns of succession. Ecology 65:1918–1935.

Southwick, C. H. 1996. *Global ecology in human perspective.* Oxford University Press, New York, USA.

Southwood, T. R. E. 1977. Habitat, the templet for ecological strategies? Journal of Animal Ecology 46:337–365.

Sparks, R. E. 1995. Need for ecosystems management of large rivers and their floodplains. Bioscience 45:168–181.

Spencer, C. N., B. R. McClelland, and J. A. Stanford. 1991. Shrimp stocking, salmon collapse, and eagle displacement. BioScience 41:14–21.

Spencer-Booth, Y. 1963. A coastal population of shrews (*Crocidura suaveolens cassiteridum*). Proceedings of the Zoological Society of London 140:322–326.

Spiller, D. A. 1986. Interspecific competition between spiders and its relevance to biological control by general predators. Environmental Entomology 15:177–181.

———. 1992. Numerical response to prey abundance by *Zygiella x-notata* (Araneae, Araneidae). Journal of Arachnology 20:179–188.

Spiller, D. A., and T. W. Schoener. 1988. An experimental study of the effect of lizards on web-spider communities. Ecological Monographs 58:57–77.

———. 1990. A terrestrial field experiment showing the impact of eliminating top predators on foliage damage. Nature 347:469–472.

———. 1994. Effects of top and intermediate predators in a terrestrial food web. Ecology 75:182–196.

Spraker, T. H., W. B. Ballard, and S. D. Miller. 1981. Game management unit 13 brown bear studies. Alaska Department of Fish and Game, Juneau, USA.

Sprules, W. G., and J. E. Bowerman. 1988. Omnivory and food chain length in zooplankton food webs. Ecology 69:418–425.

Srivastava, D. S., and R. L. Jefferies. 1995a. The effects of salinity on the leaf and shoot demography of two arctic forage species. Journal of Ecology 83:421–430.

———. 1995b. Mosaics of vegetation and soil salinity: A consequence of goose foraging in an arctic salt marsh. Canadian Journal of Botany 73:75–85.

———. 1996. A positive feedback: Herbivory, plant growth, salinity and the desertification of an arctic salt marsh. Journal of Ecology 84:31–42.

Stamps, J. A., M. Buechner, and V. V. Krishnan. 1987. The effects of edge permeability and habitat geometry on emigration from patches of habitat. American Naturalist 129:533–552.

Stanford, J. A., and J. V. Ward. 1986. Fish of the Colorado System. Pp. 385–402 in B. R. Davies and K. F. Walker, eds., *Ecology of river systems*. Junk, Dordrecht, The Netherlands.

Stapp, P. T. 2002. Stable isotopes reveal evidence of predation by ship rats on seabirds on the Shiant Islands, Scotland. Journal of Applied Ecology 39:831–840.

Stapp, P. T., and G. A. Polis. In press. Effects of marine subsidies on insular rodent populations in the Gulf of California, México. Ecology.

Stapp, P. T., G. A. Polis, and F. Sánchez-Piñero. 1999. Stable isotopes reveal strong marine and El Nino effects on island food webs. Nature 401:467–469.

Starfield, A. M., and A. L. Bleloch. 1986. *Building models for conservation and wildlife management*. Macmillan, New York, USA.

Starfield, A. M., N. Owen-Smith, and A. L. Bleloch. 1985. A rule-based population model for adaptive management. South African Journal of Wildlife Research 15:59–62.

Stearns, S. C. 1992. *The evolution of life histories*. Oxford University Press, Oxford, UK.

Stedman, R. M., and R. L. Argyle. 1985. Rainbow smelt (*Osmerus mordax*) as predators on young bloaters (*Coregonus hoyi*) in Lake Michigan. Journal of Great Lakes Research 11:40–42.

Stein, R. A., M. T. Bremigan, and J. M. Dettmers. 1996. Understanding reservoir systems with experimental tests of ecological theory: A prescription for management. Pp. 12–22 in L. E. Miranda and D. R. DeVries, eds., *Multidimensional approaches to reservoir fisheries management*. American Fisheries Society, Bethesda, MD, USA.

Stein, R. A., D. R. DeVries, and J. M. Dettmers. 1995. Food-web regulation by a planktivore: Exploring the generality of the trophic cascade hypothesis. Canadian Journal of Fisheries and Aquatic Sciences 52:2518–2526.

Steneck, R. S., and J. T. Carlton. 2001. Human alterations of marine communities. Pp. 445–468 in M. D. Bertness, S. D. Gaines, and M. E. Hay, eds., *Marine community ecology*. Sinauer Associates, Sunderland, MA, USA.

Stenton-Dozey, J., and C. L. Griffiths. 1983. The fauna associated with kelp stranded on a sandy beach. Pp. 557–568 in A. McLachlan and T. Erasmus, eds., *Sandy beaches as ecosystems.* Developments in Hydrobiology 19. Junk, The Hague, The Netherlands.

Stephenson, T. A., and A. Stephenson. 1972. *Life between tidemarks on rocky shores.* Freeman, San Francisco, CA, USA.

Stewart, D. T., T. B. Herman, and T. Teferi. 1989. Littoral feeding in a high-density insular population of *Sorex cinereus.* Canadian Journal of Zoology 67:2074–2077.

Stewart, J., and H. Tiessen. 1990. Grasslands into deserts? Pp. 188–206 in C. Mungall and D. McLaren, eds., *Planet under stress.* Oxford University Press, New York, USA.

Stockner, J. G. 1987. Lake fertilization: The enrichment cycle and lake sockeye salmon (*Oncorhynchus nerka*) production. Pp. 198–215 in H. D. Smith, L. Margolis, and C. C. Wood, eds., *Sockeye salmon* (Oncorhynchus nerka) *population biology and future management.* Canadian Special Publication in Fisheries and Aquatic Sciences 96. Canadian Special Publication in Fisheries and Aquatic Sciences 96. Department of Fisheries and Oceans, Ottawa, Ontario, Canada.

Stockner, J. G., and E. A. MacIsaac. 1996. British Columbia lake enrichment programme: Two decades of habitat enhancement for sockeye salmon. Regulated Rivers: Research & Management 12:547–561.

Stockton, W. L., and T. E. DeLaca. 1982. Food falls in the deep sea: Occurrence, quality and significance. Deep Sea Research 29:157–169.

Stoddard, J. L., D. S. Jeffries, A. Lukewille, T. A. Clair, P. J. Dillon, C. T. Driscoll, M. Forsius, M. Johannessen, J. S. Kahl, J. H. Kellogg, A. Kemp, J. Mannio, D. T. Monteith, P. S. Murdoch, S. Patrick, A. Rebsdorf, B. L. Skjelkvale, M. P. Stainton, T. Traaen, H. van Dam, K. E. Webster, J. Wieting, and A. Wilander. 1999. Regional trends in aquatic recovery from acidification in North America and Europe. Nature 401:575–578.

Storer, T. I., and L. P. Tevis, Jr. 1955. *California grizzly.* University of Nebraska Press, Lincoln, USA.

Strayer, D. L., N. F. Caraco, J. J. Cole, S. Findlay, and M. L. Pace. 1999. Transformation of freshwater ecosystems by bivalves. BioScience 49:19–28.

Strayer, D. L., J. Powell, P. Ambrose, M. L. Pace, and D. T. Fischer. 1996. Arrival, spread, and early dynamics of a zebra mussel (*Dreissena polymorpha*) population in the Hudson River estuary. Canadian Journal of Fisheries and Aquatic Sciences 53:1143–1149.

Strong, D. R. 1992. Are trophic cascades all wet? Differentiation and donor control in species ecosystems. Ecology 73:747–754.

Strong, D. R., J. L. Maron, P. G. Connors, A. V. Whipple, S. Harrison, and R. L. Jefferies. 1995. High mortality, fluctuations in numbers and heavy subterranean insect herbivory in bush lupine *Lupinus arboreus.* Oecologia 104:85–92.

Strub, P. T., J. S. Allen, A. Huyer, and R. L. Smith. 1987a. Large scale structure of the spring transition in the coastal ocean off western North America. Journal of Geophysical Research 92:1527–1544.

Strub, P. T., J. S. Allen, A. Huyer, R. L. Smith, and R. Beardsley. 1987b. Seasonal cycles of currents, temperatures, winds, and sea level over the Northeast Pacific continental shelf: 35 N to 48 N. Journal of Geophysical Research 92:1507–1526.

Strub, P. T., P. M. Kosro, A. Huyer, and CTZ Collaborators. 1991. The nature of the cold filaments in the California Current System. Journal of Geophysical Research 96:14743–14768.

Stuart, V., J. G. Field, and R. C. Newell. 1982. Evidence for the absorption of kelp detritus by the ribbed mussel *Aulacomya* after using new ^{15}C-labelled microsphere technique. Marine Ecology Progress Series 49:57–64.

Stull, R. B. 1988. *Introduction to boundary layer meteorology.* Kluwer Academic Publishers, Dordrecht, The Netherlands.

Stutzenbaker, C. D., and R. J. Buller. 1974. Goose depredation on ryegrass pastures along the Texas Gulf coast. Special Report, Federal Aid Project 106R, Texas Parks and Wildlife Department, Austin, USA.

Suberkropp, K., and E. Chauvet. 1995. Regulation of leaf breakdown by fungi in streams: Influences of water chemistry. Ecology 76:1433–1445.

Suchanek, T. H., S. L. Williams, J. C. Ogden, D. K. Hubbard, and I. P. Gill. 1985. Utilization of shallow water sea grass detritus by Caribbean deep sea macrofauna: ^{13}C evidence. Deep Sea Research 32:201–214.

Sugai, S. F., and D. C. Burrell. 1984. Transport of dissolved organic carbon, nutrients, and trace metals from the Wilson and Blossom Rivers to Smeaton Bay, southeast Alaska. Canadian Journal of Fisheries and Aquatic Sciences 41:180–190.

Sugihara, G. 1983. Holes in niche space: A derived assembly rule and its relation to intervality. Pp. 25–33 in D. DeAngelis, W. M. Post, and G. Sugihara, eds., *Current trends in food web theory*. Oak Ridge National Laboratory, Oak Ridge, TN, USA.

Sullivan, J. H. 1997. Effects of increasing UV-B radiation and atmospheric CO_2 on photosynthesis and growth: Implications for terrestrial ecosystems. Plant Ecology 128: 194–206.

Sullivan, M. J., and C. A. Moncreiff. 1990. Edaphic algae are an important component of salt marsh food-webs: Evidence from multiple stable isotope analyses. Marine Ecology Progress Series 62:149–159.

Summerhayes, V. S., and C. S. Elton. 1923. Contribution to the ecology of Spitsbergen and Bear Island. Journal of Ecology 11:214–286.

Sverdrup, H. U., M. W. Johnson, and R. H. Fleming. 1970. *The oceans: Their physics, chemistry, and general biology*. Prentice-Hall, Englewood Cliffs, NJ, USA.

Swap, R., M. Garstang, and P. Greco-Sears. 1935. *Deserts on the march*. University Oklahoma Press, Norman, USA.

Sweeney, B. W. 1984. Factors influencing life-history patterns of aquatic insects. Pp. 56–100 in V. H. Resh and D. M. Rosenberg, eds., *The ecology of aquatic insects*. Plenum, New York, USA.

Talbot, C., T. Preston, and B. W. East. 1986. Body composition of Atlantic salmon (*Salmo salar* L.) studied by neutron activation analysis. Comparative Biochemistry and Physiology 85A:445–450.

Tan, F. C., and J. M. Edmond. 1993. Carbon isotope geochemistry of the Orinoco basin. Estuarine, Coastal and Shelf Science 36:541–547.

Tanner, E. V. J. 1980. Litterfall in montane rain forests of Jamaica and its relation to climate. Journal of Ecology 68:833–848.

Taphorn, D. C., and A. Barbarino. 1993. Evaluacion de la situación actual de los pavones, *Cichla* spp., en el Parque Nacional Capanaparo-Cinaruco, Estado Apure, Venezuela. Natura 96:10–25.

Tarrant, R. F., and R. E. Miller. 1963. Accumulation of organic matter and soil nitrogen beneath a plantation of red alder and Douglas-fir. Proceedings of the Soil Science Society of America 27:231–234.

Teferi, T., and T. B. Herman. 1995. Epigeal movements by *Sorex cinereus* on Bon Portage Island, Nova Scotia. Journal of Mammalogy 76:137–140.

Thimonier, A., J. L. Dupouey, and J. Timbal. 1992. Floristic changes in the herb-layer vegetation of a deciduous forest in the Lorraine Plain under the influence of atmospheric deposition. Forest Ecology and Management 55:149–167.

Thomas, B. 1971. Evolutionary relationships among *Peromyscus* from the Georgia Strait, Gordon, Goletas, and Scott Islands of British Columbia, Canada. Ph.D. dissertation, University of British Columbia, Vancouver, Canada.

Thomas, D. W. 1988. The distribution of bats in different ages of Douglas-fir forests. Journal of Wildlife Management 52:619–626.

Thomas, F. I. M., and M. J. Atkinson. 1997. Ammonium uptake by coral reefs: Effects of water velocity and roughness on mass transfer. Limnology and Oceanography 42: 81–88.

Thomas, F. I. M., C. D. Cornelisen, and J. M. Zande. 2000. Effects of water velocity and canopy morphology on ammonium uptake by seagrass communities. Ecology 81: 2704–2713.

Thornton, I. W., and T. R. New. 1988. Krakatau invertebrates: The 1980s fauna in the context of a century of recolonization. Philosophical Transactions of the Royal Society of London B 328:131–165.

Thornton, K. W. 1990. Perspective on reservoir limnology. Pp. 1–13 in K. W. Thornton, B. L. Kimmel, and F. E. Payne, eds., *Reservoir limnology: Ecological perspectives*. John Wiley and Sons, New York, USA.

Tieszen, L. L., ed. 1978. *Vegetation and production ecology of an Alaskan Arctic tundra*. Ecological studies, 29. Springer-Verlag, New York, USA.

Tietema, A., L. Riemer, J. M. Verstraten, M. P. van der Maas, A. J. van Wijk, and J. van Voorthuyzen. 1993. Nitrogen cycling in acid forest soils subject to increased atmospheric nitrogen input. Forest Ecology and Management 57:29–44.

Tilman, D. 1976. Ecological competition between algae: Experimental confirmation of resource-based competition theory. Science 192:463–466.

———. 1977. Resource competition between planktonic algae: An experimental and theoretical approach. Ecology 58:338–348.

———. 1982. *Resource competition and community structure*. Princeton University Press, Princeton, NJ, USA.

———. 1990. Mechanisms of plant competition for nutrients: The elements of a predictive theory of competition. Pp. 117–141 in J. B. Grace and D. Tilman, eds., *Perspectives on plant competition*. Academic Press, San Diego, CA, USA.

Tilman, D., and J. A. Downing. 1994. Biodiversity and stability in grasslands. Nature 367: 363–365.

Tilman, D., S. S. Kilham, and P. Kilham. 1982. Phytoplankton community ecology: The role of limiting nutrients. Annual Review of Ecology and Systematics 13:349–372.

Timmermann, A., J. Oberhuber, A. Bacher, M. Esch, M. Latif, and E. Roeckner. 1999. Increased El Niño frequency in a climate model forced by future greenhouse warming. Nature 398:694–697.

Tonn, W. M., and J. J. Magnuson. 1982. Patterns in the species composition and richness of fish assemblages in northern Wisconsin lakes. Ecology 63:1149–66.

Towns, D. R. 1975. Ecology of the black shore skink, *Leiolopisma suteri* (Lacertilia: Scincidae), in boulder beach habitats. New Zealand Journal of Zoology 2:389–408.

———. 1991. Response of lizard assemblages in the Mercury Islands, New Zealand, to removal of an introduced rodent: The kiore (*Rattus exulans*). Journal of the Royal Society of New Zealand 21:119–136.

Townsend, C. R., C. J. Arbuckle, T. A. Crowl, and M. R. Scarsbrook. 1997. The relationship between land use and physicochemistry, food resources and macroinvertebrate communities in tributaries of the Taieri River, New Zealand: A hierarchically scaled approach. Freshwater Biology 37:177–191.

Townsend, C. R., and T. A. Crowl. 1991. Fragmented population structure in a native New Zealand fish: An effect of introduced brown trout? Oikos 61:347–354.

Townsend, C. R., and R. H. Riley. 1999. Assessment of river integrity—accounting for perturbation pathways in physical and ecological space. Freshwater Biology 41: 393–405.

Townsend, C. R., M. R. Scarsbrook, and S. Dolédec. 1997. Quantifying disturbance in streams: Alternative measures of disturbance in relation to macroinvertebrate species traits and species richness. Journal of the North American Benthological Society 16:531–544.

Townsend, C. R., R. Thompson, A. R. McIntosh, C. Kilroy, E. Edwards, and M. R. Scarsbrook. 1998. Disturbance, resource supply and food-web architecture in streams. Ecology Letters 1:200–209.

Townsend, D. W., T. L. Cucci, and T. Berman. 1984. Subsurface chlorophyll maxima and vertical distribution of zooplankton in the Gulf of Maine. Journal of Plankton Research 6:793–802.

Treseder, K. K., D. W. Davidson, and J. R. Ehleringer. 1995. Absorption of ant-provided carbon dioxide and nitrogen by a tropical epiphyte. Nature 375:137–139.

Treseder, K. K., and P. M. Vitousek. 2001. Effects of soil nutrient availability on investment in acquisition of N and P in Hawaiian rain forests. Ecology 82:946–954.

Trippel, E. A., and F. W. H. Beamish. 1993. Multiple trophic level structuring in *Salvelinus-Coregonus* assemblages in boreal forest lakes. Canadian Journal of Fisheries and Aquatic Sciences 50:1442–1455.

Triska, F. J., J. H. Duff, and R. J. Avazino. 1990. Influence of exchange flow between channel and hyporrheic zone in nitrate production in a small mountain stream. Canadian Journal of Fisheries and Aquatic Sciences 47:2099–2111.

Trudel, M., and J. B. Rasmussen. 1997. Modeling the elimination of mercury by fish. Environmental Science and Technology 31:1716–1722.

———. 2001. Predicting mercury concentrations in fish using mass-balance models: The importance of food consumption rates and activity costs. Ecological Applications 11: 517–529.

Trudel, M., A. Tremblay, R. Schetagne, and J. B. Rasmussen. 2000. Estimating food consumption rates of fish using a mercury mass balance model. Canadian Journal of Fisheries and Aquatic Sciences 57:414–428.

———. 2001. Why are dwarf fish so small? An energetic analysis of dwarfism in lake whitefish (*Coregonus clupeaformis*). Canadian Journal of Fisheries and Aquatic Sciences 58: 394–405.

Trumbore, S. E. 1997. Potential responses of soil organic carbon to global environmental change. Proceedings of the National Academy of Sciences, USA 94:8284–8291.

Tsuchiya, H. M., J. F. Drake, J. L. Jost, and A. G. Fredrickson. 1972. Predator-prey interactions of *Dictyostelium discoideum* and *Escherichia coli* in a continuous culture. Journal of Bacteriology 110:1147–1153.

Tucker, C., H. Dregne, and W. Newcomb. 1991. Expansion and contraction of the Sahara Desert from 1980 to 1990. Science 253:299–301.

Turner, B. L. II, W. C. Clark, R. W. Kates, J. F. Richards, J. T. Mathews, and W. B. Meyer, eds. 1990. *The earth as transformed by human actions..* Cambridge University Press, Cambridge, UK.

Turner, F. B., D. C. Weaver, and J. C. Rorabaugh. 1984. Effects of reduction in windblown sand on the abundance of the fringe-toed lizard (*Uma inornata*) in the Coachella Valley, California. Copeia 1984(2):370-378.

Turner, M. G. 1989. Landscape ecology—the effect of pattern on process. Annual Review of Ecology and Systematics 20:171–197.

Turner, R. D. 1973. Wood-boring bivalves, opportunistic species in the deep sea. Science 180: 1377–1379.

Urabe, J., and R. W. Sterner. 1996. Regulation of herbivore growth by the balance of light and nutrients. Proceedings of the National Academy of Sciences, USA 93:8465–8469.

Ulrich, A., and P. L. Gersper. 1978. Plant nutrient limitations of tundra plant growth. Pp. 456–481 in L. L. Tieszen, ed., *Vegetation and production ecology of an Alaskan Arctic tundra.* Ecological studies, 29. Springer-Verlag, New York, USA.

Underwood, A. J., E. J. Denley, and M. J. Moran. 1983. Experimental analyses of the structure and dynamics of mid-shore rocky intertidal communities in New South Wales. Oecologia 56:202–219.

United Nations Environment Programme Ozone Secretariat. 1998. *Environmental effects of ozone depletion.* United Nations Environment Programme, Germany.

Vance, R. R. 1978. A mutualistic interaction between a sessile marine clam and its epibionts. Ecology 59:679–685.

Van Cleve, K., R. Barney, and R. Schlentner. 1981. Evidence of temperature control of production and nutrient cycling in two interior Alaska black spruce ecosystems. Canadian Journal of Forest Research 11:258–273.

van den Ende, P. 1973. Predator-prey interactions in continuous culture. Science 181: 562–564.

van der Leeden, F., F. L. Troise, and D. K. Todd. 1990. *The water encyclopedia.* Lewis, Chelsea, MI, USA.

Vandermeer, J., and A. Power. 1990. An epidemiologic model of the corn stunt system in Central America. Ecological Modelling 52:235–248.

Van der Meijden, E., M. Wijn, and H. J. Verkaar. 1988. Defence and regrowth, alternative plant strategies in the struggle against herbivores. Oikos 51:355–363.

Vander Zanden, M. J., G. Cabana, and J. B. Rasmussen. 1997. Comparing trophic position of freshwater fish calculated using stable nitrogen isotope ratios (δ^{15}N) and literature dietary data. Canadian Journal of Fisheries and Aquatic Sciences 54:1142–1158.

Vander Zanden, M. J., J. M. Casselman, and J. B. Rasmussen. 1999. Stable isotope evidence for the food web consequences of species invasions in lakes. Nature 401:464–467.

Vander Zanden, M. J., and J. B. Rasmussen. 1996. A trophic position model of pelagic food webs: Impact on contaminant bioaccumulation in lake trout. Ecological Monographs 66:451–477.

———. 1999a. Patterns of food chain length in lakes. American Naturalist 154:406–416.

———. 1999b. Primary consumer δ^{15}N and δ^{13}C and the trophic position of aquatic consumers. Ecology 80:1395–1404.

———. In press. Accounting for the littoral-pelagic-profundal gradient in δ^{15}N and δ^{13}C to estimate the trophic position of aquatic consumers. Ecology.

Vander Zanden, M. J., B. J. Shuter, N. P. Lester, and J. B. Rasmussen. 1999. Patterns of food chain length in lakes: A stable isotope study. American Naturalist 154:406–416.

———. 2000. Within- and among-population variation in the trophic position of the aquatic top predator, lake trout. Canadian Journal of Fisheries and Aquatic Sciences 57: 725–731.

Van Dover, C. L. 2000. *Ecology of deep sea hydrothermal vents.* Princeton University Press, Princeton, NJ, USA.

Van Dover, C. L., J. F. Grassle, B. Fry, R. H. Garritt, and V. R. Starczak. 1992. Stable isotope evidence for entry of sewage-derived organic material into a deep-sea food web. Nature 360:153–155.

Van Eerden, M. R. 1984. Waterfowl movement in relation to food stocks. Pp. 84–100 in P. R. Evans, J. D. Goss-Custard, and W. G. Hale, eds., *Coastal waders and wildfowl in winter.* Cambridge University Press, Cambridge, UK.

———. 1990. The solution of goose damage problems in the Netherlands, with special reference to compensation schemes. Ibis 132:253–261.

Van Es, J., J.-M. Paillisson, and F. Burel. 1999. Impacts de l'eutrophisation de la végétation des zones humides de fonds de vallées sur la biodiversité des rhopalocères (Lepidoptera). Vie et Milieu 49:107–116.

Vanni, M. J. 1987. Effects of nutrients and zooplankton size on the structure of a phytoplankton community. Ecology 68:624–635.

———. 1996. Nutrient transport and recycling by consumers in lake food webs: Implications for algal communities. Pp. 81–95 in G. A. Polis and K. O. Winemiller, eds., *Food webs: Integration of patterns and dynamics.* Chapman and Hall, New York, USA.

———. 2002. Nutrient cycling by animals in freshwater ecosystems. Annual Review of Ecology and Systematics 33:341–370.

Vanni, M. J., and P. C. DeRuiter. 1996. Detritus and nutrients in food webs. Pp. 25–29 in G. A. Polis and K. O Winemiller, eds., *Food webs: Integration of patterns and dynamics.* Chapman and Hall, New York, USA.

Vanni, M. J., and C. D. Layne. 1997. Nutrient recycling and herbivory as mechanisms in the "top-down" effect of fish on algae in lakes. Ecology 78:21–40.

Vanni, M. J., C. D. Layne, and S. E. Arnott. 1997. "Top-down" trophic interactions in lakes: Effects of fish on nutrient dynamics. Ecology 78:1–20.

Vanni, M. J., W. H. Renwick, J. L. Headworth, J. D. Auch, and M. H. Schaus. 2001. Dissolved and particulate nutrient flux from three adjacent agricultural watersheds: A five-year study. Biogeochemistry 54:85–114.

Vannote, R. L., G. W. Minshall, K. W. Cummings, J. R. Sedell, and C. E. Cushing. 1980. The river continuum concept. Canadian Journal of Fisheries and Aquatic Sciences 37: 130–137.

Van Steenbergen, T. 1977. Influence of type of soil and year on the effect of nitrogen fertilization on the yield of grassland. Stikstof 20:29–35.

Van Wilgen, B. W., and D. M. Richardson. 1985. The effects of alien shrub invasions on vegetation structure and fire behaviour in South African fynbos shrub lands: A simulation study. Journal of Applied Ecology 22:955–966.

Vaughan, T. A. 1980. Opportunistic feeding by two species of *Myotis*. Journal of Mammalogy 61:118–119.

Vazzoler, A. E. de M., S. A. Amadio, and M. C. Daraciolo-Malta. 1989. Aspectos biológicos de peixes Amazônicos. XI. Reprodução das espécies do gênero *Semaprochilodus* (Characiformes, Prochilodontidae) no baixo Rio Negro, Amazonas, Brasil. Revista Brasiliera de Biologia 49:165–173.

Ventura, L. 1997. The effects of the feeding and burrowing activities of *Corophium volutator* (Pallas) and *Neries diversicolor* (O. F. Muller) on the establishment of the green algae *Enteromorpha* spp. in an estuarine mudflat. M.S. thesis, University of Aberdeen, UK.

Verbeek, N. A. M., and R. W. Butler. 1989. Feeding ecology of shoreline birds in the Strait of Georgia. Pp. 74–78 in K. Vermeer and R. W. Butler, eds., *The ecology and status of marine and shoreline birds in the Strait of Georgia, British Columbia*. Special publication. Canadian Wildlife Service, Ottawa, Canada.

Vetter, E. 1994. Hotspots of benthic production. Nature 372:47.

Vink, A. P. A. 1983. Landscape ecology and land use. Longman, London, UK.

Vitousek, P. M. 1986. Biological invasions and ecosystem properties: Can species make a difference? Pp. 163–176 in J. A. Drake and H. A. Mooney, eds., *Ecology of biological invasions of North America and Hawaii*. Springer-Verlag, New York, USA.

———. 1987. Biological invasion by *Myrica faya* alters ecosystem development in Hawaii. Science 238:802–804.

———. 1994. Beyond global warming: Ecology and global change. Ecology 75:1861–1876.

Vitousek, P. M., D. Aber, R. W. Howarth, G. E. Likens, P. A. Matson, D. W. Schindler, W. H. Schlesinger, and D. G. Tilman. 1997. Human alteration of the global nitrogen cycle: Sources and consequences. Ecological Applications 7:737–750.

Vitousek, P. M., H. A. Mooney, J. Lubchenco, and J. M. Melillo. 1997. Human domination of earth's ecosystems. Science 227:494–499.

Vogel, S. 1981. *Life in moving fluids*. Willard Grant Press, Boston, MA, USA.

Vollenweider, R. A. 1968. Scientific fundamentals of the eutrophication of lakes and flowing waters with particular reference to nitrogen and phosphorus as factors in eutrophication. DAS/CSI/ 68:27. Organization for Economic Cooperation and Development, Paris, France.

Vollenweider, R. A., and J. Kerekes. 1980. The loading concept as basis for controlling eutrophication: Philosophy and preliminary results of the OECD programme on eutrophication. Progress in Water Technology 12:5–38.

von Frenckell, B., and R. M. R. Barclay. 1987. Bat activity over calm and turbulent water. Canadian Journal of Zoology 65:219–222.

von Gunten, H. R., and C. Lienert. 1993. Decreased metal concentrations in ground water caused by controls of phosphate emissions. Nature 364:220–222.

Vorosmarty, C. J., K. P. Sharma, B. M. Fekete, A. H. Copeland, J. Marble, and J. A. Lough. 1997. The storage and aging of continental runoff in large reservoir systems of the world. Ambio 26:210–219.

Wade, N. 1974. Sahelian drought: No victory for Western aid. Science 185:234–237.

Wales, B. A. 1972. Vegetation analysis of north and south edges in a mature oak-hickory forest. Ecological Monographs 42:451–471.

Walker, J. C. 1969. *Plant pathology*. McGraw-Hill, New York, USA.

Wallace, J. B., S. L. Eggert, J. L. Meyer, and J. R. Webster. 1997. Multiple trophic levels of a forest stream linked to terrestrial litter inputs. Science 277:102–104.

Walters, C., V. Christensen, and D. Pauly. 1997. Structuring dynamic models of exploited ecosystems from trophic mass-balance assessments. Reviews in Fish Biology and Fisheries 7:139–172.

Waltman, P., S. P. Hubbell, and S. B. Hsu. 1980. Theoretical and experimental investigations of microbial competition in continuous culture. Pp. 107–152 in T. Burton, ed., *Modeling and differential equations in biology.* Marcel Dekker, New York, USA.

Ward, J. B., I. M. Henderson, B. H. Patrick, and P. H. Norrie. 1996. Seasonality, sex ratios and arrival pattern of some New Zealand Caddis (Trichoptera) to light-traps. Aquatic Insects 18:157–174.

Ward, J. V. 1989. The four dimensional nature of lotic ecosystems. Journal of the North American Benthological Society 8:2–8.

Ward, J. V., and J. A. Stanford. 1995. Ecological connectivity in alluvial river ecosystems and its disruption by flow regulation. Regulated Rivers: Research and Management 11: 105–119.

Ward, J. V., K. Tockner, and F. Schiemer. 1999. Biodiversity of floodplain river ecosystems: Ecotones and connectivity. Regulated Rivers: Research and Management 15:125–139.

Ward, R. C. 1978. *Floods: A geographical perspective.* Macmillan, London, UK.

Waringer, J. A. 1991. Phenology and the influence of meteorological parameters on the catching success of light-trapping for Trichoptera. Freshwater Biology 25:307–320.

Washington Department of Fisheries, Washington Department of Wildlife, and Western Washington Treaty Indian Tribes. 1993. 1992 Washington State salmon and steelhead stock inventory. Washington Department of Fisheries and Wildlife, Olympia, USA.

Watson, A. J., D. C. E. Bakker, A. J. Ridgwell, P. W. Boyd, and C. S. Law. 2000. Effect of iron supply on Southern Ocean CO_2 uptake and implications for glacial atmospheric CO_2. Nature 407:730–733.

Watts, J. F. D., and G. D. Watts. 1990. Seasonal change in aquatic vegetation and its effect on river channel flow. Pp. 257–267 in J. B. Thornes, ed., *Vegetation and erosion.* John Wiley and Sons, New York, USA.

Weathers, K. C., and M. L. Cadenasso. 1996. The function of forest edges: Nutrient inputs. Bulletin of the Ecological Society of America 77:471.

Weathers, K. C., M. L. Cadenasso, and S. T. A. Pickett. 2001. Forest edges as nutrient and pollutant concentrators: Potential synergisms between fragmentation, forest canopies, and the atmosphere. Conservation Biology 15:1506–1514.

Weathers, K. C., G. M. Lovett, and G. E. Likens. 1992. The influence of a forest edge on cloud deposition. Pp. 1415–1423 in S. E. Schwartz and W. G. N. Slinn, eds., *Precipitation scavenging and atmosphere-surface exchange.* Volume 3. *The Summers volume: Applications and appraisals.* Hemisphere, Washington, DC, USA.

———. 1995. Cloud deposition to a spruce forest edge. Atmospheric Environment 29: 665–672.

Weathers, K. C., G. M. Lovett, G. E. Likens, and N. F. M. Caraco. 2000a. Cloudwater inputs of nitrogen to forest ecosystems in southern Chile: Forms, fluxes and sources. Ecosystems 3:590–595.

Weathers, K. C., G. M. Lovett, G. E. Likens, and R. Lathrop. 2000b. The effect of landscape features on deposition to Hunter Mountain, Catskill Mountains, New York. Ecological Applications 10:528–540.

Webster, J. R., and E. F. Benfield. 1986. Vascular plant breakdown in freshwater ecosystems. Annual Review of Ecology and Systematics 17:567–594.

Webster, J. R., and J. L. Meyer. 1997a. Organic matter budgets for streams: A synthesis. Journal of the North American Benthological Society 16:141–161.

———. 1997b. Stream organic matter budgets: Introduction. Journal of the North American Benthological Society 16:5–13.

Wedin, D. A. 1995. Species, nitrogen and grassland dynamics: The constraints of stuff. Pp. 253–262 in C. G. Jones and J. H. Lawton, eds., *Linking species and ecosystems*. Chapman and Hall, London, UK.

Wedin, D. A., and J. Pastor. 1993. Nitrogen mineralization dynamics in grass monocultures. Oecologia 96:186–192.

Wedin, D. A., and D. Tilman. 1990. Species effects on nitrogen cycling: A test with perennial grasses. Oecologia 84:443–441.

Weiher, E., and P. A. Keddy. 1995. Assembly rules, null models, and trait dispersion: New questions from old patterns. Oikos 74:159–164.

Welch, C. A., J. Keay, K. C. Kendall, and C. T. Robbins. 1997. Constraints on frugivory by bears. Ecology 78:1105–1119.

Welcomme, R. L. 1979. *Fisheries ecology of floodplain rivers*. Longman, New York, USA.

———. 1989. Floodplain fisheries management. Pp. 209–233 in J. A. Gore and G. E. Petts, eds., *Alternatives to regulated river management*. CRC Press, Boca Raton, FL, USA.

Werner, E. E., and J. F. Gilliam. 1984. The ontogenetic niche and species interactions in size-structured populations. Annual Review of Ecology and Systematics 15:393–425.

West, N. E., and J. J. Skujins. 1978. *Nitrogen in desert ecosystems*. Dowden, Hutchinson and Ross, Stroudsburg, PA, USA.

Westbrook, J. K., and S. A. Isard. 1999. Atmospheric scales of biotic dispersal. Agricultural and Forest Meteorology 97:263–274.

Wetzel, R. G. 1975. *Limnology*. Saunders, London, UK.

———. 1983. *Limnology*. 2d ed. W. B. Saunders, Philadelphia, PA, USA.

Wetzel, R. G., P. G. Hatcher, and T. S. Bianchi. 1995. Natural photolysis by ultraviolet irradiance of recalcitrant dissolved organic matter to simple substrates for rapid bacterial metabolism. Limnology and Oceanography 40:1369–1380.

Whelan, R. J. 1995. *The ecology of fire*. Cambridge University Press, Cambridge, UK.

White, T. C. R. 1978. The importance of a relative shortage of food in animal ecology. Oecologia 3:71–86.

———. 1993. *The inadequate environment: Nitrogen and the abundance of animals*. Springer-Verlag, Barcelona, Spain.

Whitney, G. G., and J. R. Runkle. 1981. Edge versus age effects in the development of beech-maple forest. Oikos 37:377–381.

Whittaker, R. H. 1975. *Communities and ecosystems*. Macmillan, New York, USA.

Wickens, P. A., and J. G. Field. 1986. The effect of water transport on nitrogen flow through a kelp-bed community. South African Journal of Marine Science 4:79–92.

Widner, M. R., and S. C. Yaich. 1990. Distribution and habitat use of snow and white-fronted geese in Arkansas. Proceedings of the Annual Conference of Southeastern Fish and Wildlife Agencies 44:328–336.

Wiens, J. A. 1977. On competition and variable environments. American Scientist 65: 590–597.

———. 1986. Spatial scale and temporal variation in studies of shrubsteppe birds. Pp. 154–172 in J. Diamond and T. J. Case, eds., *Community ecology*. Harper and Row, New York, USA.

———. 1992. Ecological flows across landscape boundaries: A conceptual overview. Pp. 216–235 in F. di Castri and A. J. Hansen, eds., *Landscape boundaries*. Ecological studies, 92. Springer-Verlag, New York, USA.

———. Landscape mosaics and ecological theory. Pp. 1–26 in L. Hansson, L. Fahrig, and G. Merriam, eds., *Mosaic landscapes and ecological processes*. Chapman and Hall, New York, USA.

Wiens, J. A., J. F. Addicott, T. J. Case, and J. Diamond. 1986. Overview: The importance of spatial and temporal scale in ecological investigations. Pp. 145–153 in J. Diamond and T. J. Case, eds., *Community ecology*. Harper and Row, New York, USA.

Wiens, J. A., C. S. Crawford, and J. R. Gosz. 1985. Boundary dynamics: A conceptual framework for studying landscape ecosystems. Oikos 45:421–427.

Wiens, J. A., N. C. Stenseth, B. Van Horne, and R. A. Ims. 1993. Ecological mechanisms and landscape ecology. Oikos 66:369–380.

Wilcove, D. S., C. H. McLellan, and A. P. Dodson. 1986. Habitat fragmentation in the temperate zone. Pp. 237–256 in M. E. Soulé, ed., *Conservation biology: The science of scarcity and diversity*. Sinauer Associates, Sunderland, MA, USA.

Wilcox, B. A. 1980. Species number, stability, and equilibrium status of reptile faunas on the California Islands. Pp. 551–564 in D. M. Power, ed., *The California Islands: Proceedings of a multidisciplinary symposium*. Santa Barbara Museum of Natural History, Santa Barbara, CA, USA.

Wildish, D., and D. Kristmanson. 1997. *Benthic suspension feeders and flow*. Cambridge University Press, Cambridge, UK.

Williams, A, J. 1978. Mineral and energy contributions of petrels (Procellariiformes) killed by cats to the Marion Island terrestrial ecosystem. South African Journal of Antarctic Research 8:71–74.

Williams, A. J., A. E. Burger, and A. Berruti. 1978. Mineral and energy contributions of carcasses of selected species of seabirds to the Marion Island terrestrial ecosystems. South African Journal of Antarctic Research 8:53–59.

Williams, D. D., L. G. Ambrose, and L. N. Browning. 1995. Trophic dynamics of two sympatric species of riparian spider (Araneae, Tetragnathidae). Canadian Journal of Zoology 73:1545–1553.

Williams, J. L. 1989. Measurements of turbulence in the benthic boundary layer over a gravel bed. Sedimentology 36:959–971.

Williams, K. S., and C. Simon. 1995. The ecology, behavior, and evolution of periodical cicadas. Annual Review of Entomology 40:269–295.

Williams-Linera, G. 1990. Vegetation structure and environmental conditions of forest edges in Panama. Journal of Ecology 78:356–373.

Williamson, C. E. 1996. Effects of UV radiation on freshwater ecosystems. International Journal of Environmental Studies 51:245–256.

Williamson, M. 1981. *Island populations*. Oxford University Press, Oxford, UK.

Willson, M. F. 1997. Variation in salmonid life histories: Patterns and perspectives. Pacific Northwest Research Station Research Paper PNW-RP-498:1–50.

Willson, M. F., and S. M. Gende. 2001. Passerine densities in riparian forests of southeast Alaska: Potential effects of anadromous spawning salmon. Condor 103:624–629.

Willson, M. F., S. M. Gende, and B. H. Marston. 1998. Fishes and the forest: Expanding perspectives on fish/wildlife interactions. BioScience 48:455–462.

Willson, M. F., and K. C. Halupka. 1995. Anadromous fish as keystone species in vertebrate communities. Conservation Biology 9:489–497.

Wilson, D. J., and R. L. Jefferies. 1996. Nitrogen mineralization, plant growth and goose herbivory in an Arctic coastal ecosystem. Journal of Ecology 84:841–851.

Wilson, J. D., A. J. Morris, B. E. Arroyo, S. C. Clark, and R. B. Bradbury. 1999. A review of the abundance and diversity of invertebrate and plant foods of granivorous birds in northern Europe in relation to agricultural change. Agriculture, Ecosystems and Environment 75:13–30.

Wiman, B. L. B., and G. I. Ågren. 1985. Aerosol depletion and deposition in forests—a model analysis. Atmospheric Environment 19:335–347.

Winemiller, K. O. 1989. Patterns of variation in life history among South American fishes in seasonal environments. Oecologia 81:225–241.

———. 1996. Factors driving spatial and temporal variation in aquatic floodplain food webs. Pp. 298–312 in G. A. Polis and K. O. Winemiller, eds., *Food webs: Integration of patterns and dynamics*. Chapman and Hall, New York, USA.

Winemiller, K. O., and D. B. Jepsen. 1998. Effects of seasonality and fish movement on tropical river food webs. Journal of Fish Biology 53:267–296.

Winemiller, K. O., and G. A. Polis. 1996. Food webs: What do they tell us about the world? Pp. 1–24 in G. A. Polis and K. O. Winemiller, eds., *Food webs: Integration of patterns and dynamics*. Chapman and Hall, New York, USA.

Winemiller, K. O., D. C. Taphorn, and A. Barbarino. 1997. Ecology of *Cichla* (Cichlidae) in two blackwater rivers of southern Venezuela. Copeia 1997:690–696.

Wipfli, M. S. 1997. Terrestrial invertebrates as salmonid prey and nitrogen sources in streams: Contrasting old-growth and young-growth riparian forests in southeastern Alaska, U.S.A. Canadian Journal of Fisheries and Aquatic Sciences 54:1259–1269.

Wipfli, M. S., J. P. Hudson, and J. P. Caouette. 1998. Influence of salmon carcasses on stream productivity: Response of biofilm and benthic macroinvertebrates in southeastern Alaska, U.S.A. Canadian Journal of Fisheries and Aquatic Sciences 55: 1503–1511.

Wipfli, M. S., J. P. Hudson, D. T. Chaloner, and J. P. Caouette. 1999. Influence of salmon spawner densities on stream productivity in southeast Alaska. Canadian Journal of Fisheries and Aquatic Sciences 56:1600–1611.

Wise, D. H. 1993. *Spiders in ecological webs*. Cambridge University Press, Cambridge, UK.

Witman, J. D. 1987. Subtidal coexistence: Storms, grazing, mutualism, and the zonation of kelps and mussels. Ecological Monographs 57:167–187.

———. 1998. Natural disturbance and colonization on subtidal hard substrates in the Gulf of Maine. Pp. 30–37 in E. M. Dorsey and J. Pederson, eds., *Effects of fishing gear on the sea floor of New England*. MIT Sea Grant Publication 98–4. Conservation Law Foundation, Boston, MA, USA.

Witman, J. D., and P. K. Dayton. 2001. Rocky subtidal communities. Page 339–366 in M. D. Bertness, S. D. Gaines, and M. E. Hay, eds., *Marine community ecology*. Sinauer Associates, Sunderland, MA, USA.

Witman, J. D., S. J. Genovese, J. F. Bruno, J. W. McLaughlin, and B. I. Pavlin. 2003. Massive prey recruitment and the control of rocky subtidal communities on large spatial scales. Ecological Monographs. In press.

Witman, J. D., and K. R. Grange. 1998. Links between rain, salinity, and predation in a rocky subtidal community. Ecology 79:2429–2447.

Witman, J. D., J. J. Leichter, S. J. Genovese, and D. A. Brooks. 1993. Pulsed phytoplankton supply to the rocky subtidal zone: Influence of internal waves. Proceedings of the National Academy of Sciences, USA 90:1686–1690.

Witman, J. D., and K. P. Sebens. 1988. Benthic community structure at a subtidal rock pinnacle in the central Gulf of Maine. Pp. 67–104 in I. Babb and M. De Luca, eds., *Benthic productivity and marine resources of the Gulf of Maine*. National Undersea Research Program Research Report 88–3.

———. 1992. Regional variation in fish predation intensity: A historical perspective in the Gulf of Maine. Oecologia 90:305–315.

Witman, J. D., and F. Smith. 2003. Rapid community change at a tropical upwelling site in the Galápagos Marine Reserve. Biodiversity and Conservation 12:25–45.

Wobeser, G. A. 1981. *Diseases of wild waterfowl*. Plenum Press, New York, USA.

Wolanski, E., and W. M. Hamner. 1988. Topographically controlled fronts in the ocean and their biological influence. Science 241:177–181.

Wolanski, E., and G. L Pickard. 1983. Upwelling by internal tides and Kelvin waves at the continental shelf break on the Great Barrier Reef. Australian Journal of Marine Research 34:65–80.

Wolinski, R. A. 1980. Rough-winged swallow feeding on fly larvae. Wilson Bulletin 92: 121–122.

Wollum, A. G. II, and C. B. Davey. 1975. Nitrogen accumulation, transformation, and transport in forest soils. Pp. 67–106 in B. Bernier and C. H. Winget, eds., *Forest soils and forest land management*. University of Laval Press, Laval, Quebec, Canada.

Woodin, S. A. 1978. Refuges, disturbance and community structure: A marine soft bottom example. Ecology 59:274–284.

Wootton, J. T., M. S. Parker, and M. E. Power. 1996. Effects of disturbance on river food webs. Science 273:1558–1561.

Wratten, S. D., and H. F. van Emden. 1995. Habitat management for enhanced activity of natural enemies of insect pests. Pp. 117–145 in D. M. Glen, M. P. Greaves, and H. M. Anderson, eds., *Ecology and integrated farming systems*. John Wiley and Sons, New York, USA.

Wright, S. J., C. Carrasco, O. Calderon, and S. Paton. 1999. The El Niño Southern Oscillation, variable fruit production, and famine in a tropical forest. Ecology 80:1632–1647.

Yako, L. A., J. M. Dettmers, and R. A. Stein. 1996. Feeding preferences of omnivorous gizzard shad as influenced by fish size and zooplankton density. Transactions of the American Fisheries Society 125:735–759.

Yoshioka, T., E. Wada, and H. Hayashi. 1994. A stable isotope study on seasonal food web dynamics in a eutrophic lake. Ecology 75:835–846.

Young, C. M. 1987. Novelty of supply-side ecology. Science 255:415–416.

Young, R. G., and A. D. Huryn. 1996. Inter-annual variation in discharge controls ecosystem metabolism along a grassland river continuum. Canadian Journal of Fisheries and Aquatic Sciences 53:2199–2211.

———. 1997. Longitudinal patterns of organic mater transport and turnover along a New Zealand grassland river. Freshwater Biology 38:93–107.

Young, R. G., A. D. Huryn, and C. R. Townsend. 1994. Effects of agricultural development on processing of tussock leaf litter in high country New Zealand streams. Freshwater Biology 32:413–427.

Zabel, C. J., and S. J. Taggart. 1989. Shift in red fox, *Vulpes vulpes*, mating system associated with El Niño in the Bering Sea. Animal Behavior 38:830–838.

Zann, R. A., M. V. Walker, A. S. Adhikerana, G. W. Davison, E. B. Male, and Darjono. 1990. The birds of the Krakatau Islands (Indonesia) 1984–86. Philosophical Transactions of the Royal Society of London 328:29–54.

Zaranko, D. T., R. W. Griffiths, and N. K. Kaushik. 1997. Biomagnification of polychorinated biphenyls through a riverine food web. Environmental Toxicology and Chemistry 16:1463–1471.

Zardus, J. D., L. F. Braithwaite, and B. A. Maurer. 1991. Diet of the barnacle *Balanus nubilus* in the presence and absence of *Metridium senile*. American Zoologist 31:103A.

Zeldis, J. R., and J. B. Jillett. 1982. Aggregation of pelagic *Munida gregaria* by coastal fronts and internal waves. Journal of Plankton Research 4:839–857.

Ziemer, R. R., and T. E. Lisle. 1998. Hydrology. Pp. 43–68 in R. J. Naiman and R. E. Bilby, eds., *River ecology and management: Lessons from the Pacific coastal ecoregion.* Springer-Verlag, New York, USA.

Zimmerman, R. C., and J. N. Kremer. 1984. Episodic nutrient supply to a kelp forest ecosystem in southern California. Journal of Marine Research 42:591–604.

Zobel, M. 1997. The relative role of species pools in determining plant species richness: An alternative explanation of species coexistence. Trends in Ecology and Evolution 12:266–269.

Contributors

Kenneth F. Abraham
Ontario Ministry of Natural Resources
300 Water Street
Peterborough, Ontario K9J 8M5
Canada

Clarissa Anderson
Department of Biological Sciences
University of California
Santa Barbara, CA 93106
U.S.A.

Wendy B. Anderson
Department of Biology
Drury University
Springfield, MO 65802
U.S.A.
wanderso@drury.edu

Jacques Baudry
Institut National de la Recherche
   Agronomique
SAD-Armorique
65 rue de Saint-Brieuc
35042 Rennes Cedex
France
jbaudry@roazhon.inra.fr

Peter A. Bisson
Forestry Sciences Laboratory
3635 93rd Avenue
Olympia, WA 98512
U.S.A.

Françoise Burel
Centre National de la Recherche
   Scientifique
Université de Rennes 1
UMR ECOBIO
Campus de Beaulieu
35042 Rennes Cedex
France
francoise.burel@univ-rennes1.fr

M. L. Cadenasso
Institute of Ecosystem Studies
Box AB
Millbrook, NY 12545
U.S.A.
cadenassom@ecostudies.org

Nina Caraco
Institute of Ecosystem Studies
Millbrook, NY 12545
U.S.A.
caracon@ecostudies.org

Jonathan Cole
Institute of Ecosystem Studies
Millbrook, NY 12545
U.S.A.

Donald L. DeAngelis
Biological Resources Division, USGS
Department of Biology
University of Miami
P.O. Box 249118
Coral Gables, FL 33124
U.S.A.
ddeangelis@umiami.ir.miami.edu

Julie C. Ellis
Department of Ecology and Evolutionary
   Biology
Brown University
Providence, RI 02912
U.S.A.

Jacques C. Finlay
Department of Ecology, Evolution and
   Behavior
University of Minnesota
1987 Upper Buford Circle
St. Paul, MN 55108
U.S.A.

Alex S. Flecker
Section of Ecology and Systematics
Corson Hall
Cornell University
Ithaca, NY 14853
U.S.A.

Scott M. Gende
Forestry Sciences Laboratory
2770 Sherwood Lane
Juneau, AK 99801
U.S.A.
and
School of Fisheries
University of Washington
Seattle, WA 98185
U.S.A.

Salvatore J. Genovese
Marine Science Center
Northeastern University
Nahant, MA 01908
U.S.A.
sgenoves@lynx.neu.edu

Jenifer L. Headworth
Department of Zoology
Miami University
Oxford, OH 45056
U.S.A.

Hugh A. L. Henry
Department of Botany
University of Toronto
25 Willcocks Street
Toronto, Ontario M5S 3B2
Canada

Joh R. Henschel
Desert Research Foundation of Namibia
Gobabeb Training and Research Centre
P.O. Box 20232
Windhoek, Namibia
jhenschel@drfn.org.na
and
Theodor-Boveri-Institut für
    Biowissenschaften der Universität
Lehrstuhl für Tierökologie und
    Tropenbiologie
Biozentrum: Am Hubland
Universität Würzburg
D-97074 Würzburg
Germany

Robert D. Holt
Department of Zoology
University of Florida
Gainesville, FL 32611
U.S.A.
rdholt@zoo.ufl.edu

Gary R. Huxel
Department of Biology, SCA110
4202 E. Fowler Avenue
University of South Florida
Tampa, FL 33620
U.S.A.
ghuxel@mail.cas.usf.edu

Robert L. Jefferies
Department of Botany
University of Toronto
25 Willcocks Street
Toronto, Ontario M5S 3B2
Canada
jefferie@botany.utoronto.ca

David B. Jepsen
Oregon Cooperative Fish and Wildlife
    Research Unit
Department of Fisheries and Wildlife
    Sciences
104 Nash Hall
Oregon State University
Corvallis, OR 97331
U.S.A.
david.jepsen@orst.edu

Sapna Khandwala
Stillwater Sciences
2532 Durant Ave.
Berkeley, CA 94704
U.S.A.

Susan C. Lubetkin
Program in Quantitative Ecology and
    Resource Management
University of Washington
Box 351720
Seattle, WA 98195
U.S.A.

Kevin Marsee
Department of Integrative Biology
University of California
Berkeley, CA 94720
U.S.A.

F. Camille McNeely
Department of Integrative Biology
University of California
Berkeley, CA 94720
U.S.A.

Bruce A. Menge
Department of Zoology
Oregon State University
Corvallis, OR 97331
U.S.A.
mengeb@science.oregonstate.edu

Patrick J. Mulholland
Environmental Sciences Division,
    Building 1505
Oak Ridge National Laboratory
P.O. Box 2008
Oak Ridge, TN 37831
U.S.A.

Michael S. Parker
Department of Biology
Southern Oregon University
Ashland, OR 97520
U.S.A.

Mark R. Patterson
School of Marine Science
Virginia Institute of Marine Science
Gloucester Point, VA 23062
U.S.A.
mrp@vims.edu

S. T. A. Pickett
Institute of Ecosystem Studies
Box AB
Millbrook, NY 12545
U.S.A.

Mary E. Power
Department of Integrative Biology
University of California
Berkeley, CA 94720
U.S.A.
mepower@socrates.berkeley.edu

Dave A. Raffaelli
Culterty Field Station
University of Aberdeen
Newburgh, Ellon
Aberdeen, Scotland AB41 0AA
United Kingdom

William E. Rainey
Department of Integrative Biology
University of California
Berkeley, CA 94720
U.S.A.

Joseph B. Rasmussen
Department of Biology
McGill University
1205 Docteur Penfield Ave.
Montreal, Quebec H3A 1B1
Canada
jrasmu@bio1.lan.mcgill.ca

Ralph H. Riley
407 Main Street, Suite 206
Mt. Vernon, WA 98273
U.S.A.
ewatch@nwlink.com

John L. Sabo
Department of Zoology
Arizona State University
Tempe, AZ 85287
U.S.A.

Francisco Sánchez-Piñero
Dpto. Biología Animal y Ecología
Facultad de Ciencias
Universidad de Granada
18071 Granada
Spain
fspinero@ugr.es

Diane M. Sanzone
Marine Biology Laboratory
Starr 119
Woods Hole, MA 02543
U.S.A.
dsanzone@mbl.edu

Daniel E. Schindler
Department of Biology
University of Washington
Box 351800
Seattle, WA 98195
U.S.A.
deschind@u.washington.edu

Anna L. W. Sears
Center for Population Biology
University of California
Davis, CA 95616
U.S.A.
alsears@ucdavis.edu

Adrianna Smyth
Department of Integrative Biology
University of California
Berkeley, CA 94720
U.S.A.

Paul T. Stapp
Department of Biological Science
California State University
Fullerton, CA 92834
U.S.A.

Colin R. Townsend
Department of Zoology
University of Otago
P.O. Box 56
Dunedin
New Zealand

M. Jake Vander Zanden
Center for Limnology
University of Wisconsin–Madison
680 North Park Street
Madison, WI 53706
U.S.A.
mjvanderzand@wisc.edu

Michael J. Vanni
Department of Zoology
Miami University
Oxford, OH 45056
U.S.A.
vannimj@muohio.edu

K. C. Weathers
Institute of Ecosystem Studies
Box AB
Millbrook, NY 12545
U.S.A.

Mary F. Willson
5230 Terrace Place
Juneau, AK 99801
U.S.A.
mwillson@gci.net

Kirk O. Winemiller
Department of Wildlife and Fisheries
   Sciences
Texas A&M University
College Station, TX 77843
U.S.A.
k-winemiller@tamu.edu

Jon D. Witman
Department of Ecology and Evolutionary
   Biology
Brown University
Providence, RI 02912
U.S.A.
jon_witman@brown.edu

Index

Note: Page numbers followed by f and t denote figures and tables, respectively.